T0213232

SURFACE AND
INTERFACIAL TENSION

SURFACTANT SCIENCE SERIES

ADDITIONAL VOLUMES IN PREPARATION

SURFACE AND INTERFACIAL TENSION
Measurement, Theory, and Applications

edited by
Stanley Hartland
ETH Swiss Federal Institute of Technology
Zurich, Switzerland

CRC Press
Taylor & Francis Group
Boca Raton London New York

CRC Press is an imprint of the
Taylor & Francis Group, an **informa** business

CRC Press
Taylor & Francis Group
6000 Broken Sound Parkway NW, Suite 300
Boca Raton, FL 33487-2742

First issued in paperback 2019

ISBN-13: 978-0-8247-5034-3 (hbk)
ISBN-13: 978-0-367-39449-3 (pbk)

Library of Congress Cataloging-in-Publication Data
A catalog record for this book is available from the Library of Congress.

Visit the Taylor & Francis Web site at
http://www.taylorandfrancis.com

and the CRC Press Web site at
http://www.crcpress.com

Preface

The effects of surface and interfacial tension are seen throughout nature and science in disciplines ranging from medicine to chemical engineering. The specialist contributions presented in this volume draw together current knowledge on a wide range of topics: the measurement of tensions and contact angles, shapes of interfaces, the motion and nucleation of bubbles, the effect of surfactants, the formation and stability of foams, and the wetting of fibers. The emphasis is on physical understanding backed by careful experimentation and theoretical interpretation.

The excess pressure inside a liquid surface is determined by its curvature and tension according to the Laplace equation. Together with gravity and the boundary conditions with other surfaces, this equation determines the shape of sessile and pendant interfaces and liquid bridges. The interfacial shape may also be used to infer the magnitude of the surface or interfacial tension. Sessile or pendant interfaces are generally used for direct measurements as they are easier to define than external menisci. However, it is the latter shape that determines the force required to pull a body from an interface. This is another commonly used method, as it allows the tension to be inferred from an accurately measured force rather than a less precise dimension. The pressure arising from the interfacial curvature may also be employed. In its simplest form capillary rise may be used, but the more complicated measurement of maximum bubble pressure may be more readily adapted to dynamic variations. Many careful measurements were made by classical physical chemists at the beginning of the twentieth century, but these have since been revolutionized by the advent of modern measuring techniques aided by powerful computers.

Surfactants invariably lower the value of the surface tension, creating variations with time as the surface concentration rises through molecular diffusion and with space when the surface concentration falls as the interface expands. Alternatively, the shear stresses associated with a liquid flow adjacent to the interface may concentrate the surfactant molecules, thereby progressively lowering the tension and creating a gradient opposing the flow. Such negative gradients form the basis of the Marangoni effect, which, together with the mechanical barrier provided by the surfactant molecules at the interface, is responsible for the stability of emulsions and foams. The behavior of gas–liquid foams depends on the surface tension gradients opposing the drainage in the thin liquid films between the bubbles. The driving force for drainage arises from the negative excess pressure in the Plateau borders linking the films, which provide a network of interconnected channels through which the continuous phase can return to the bulk liquid.

Foams have many industrial applications. The large gas–liquid interfacial area available is exploited in foam fractionation to efficiently separate surface active substances, such as proteins. However, when foam formation interferes with physical and chemical processes, stabilizing gradients must be avoided. Bubbles play an important role in chemical reactors, distillation, froth flotation, and aeration. In low-gravity environments surface tension gradients may provide the primary mechanism for bubble motion itself. The advent of modern computers has enabled simulations of complex interfacial motions to be performed, yielding information that is difficult to obtain by experiment or theoretical analysis.

Drainage of the films between liquid drops is also affected by surface-tension gradients. The negative gradients resulting from the presence of surfactants oppose the drainage and reduce coalescence. Conversely, coalescence is promoted if gradients enhancing the drainage can be created within the film. In liquid–liquid extraction one phase is dispersed in the other to provide and maintain a large suface area for mass transfer, so drop coalescence must be prevented. Since most solutes are surface active, this is usually achieved by transferring the solute from the continuous to the dispersed phase to create negative gradients of surface tension in the films between the drops.

The energies between the liquid, gas, and solid surfaces determine the contact angle and hence the wettability; this is important in many technological processes. In some processes, liquids must spread over solid surfaces, such as paint on paper and lubricating oil on metal. In others, hydrophobic coatings, such as Teflon film on frying pans, must spread cooking oil but repel water. The behavior of bubbles and drops on solid surfaces affects the performance of industrial apparatus during boiling and condensation. Surfactants are widely used to control wetting, capillary penetration, and evaporation. Adsorption of surfactants accelerates transfer processes such as the impregnation of hydrophobic porous bodies by aqueous solutions, cleaning of greasy oiled surfaces, and crude oil recovery. The surface properties of adsorbents, membranes, and catalysts are modified by surfactants.

The study of fiber wettability has important implications in detergency, water and oil repellancy of textiles, spinning, optical fiber processing, and the chemical design of fiber-reinforced composite systems. The degree to which a liquid wets the fiber determines how easily the liquid can penetrate fiber assemblies; this is important both during processing and for the performance of textiles and polymeric composite systems.

Since surface and interfacial tensions and their gradients control the shape and movement of interfaces, they have a profound effect on liquid transfer and the motion of drops and bubbles. They have application in crude oil extraction and its subsequent separation from associated water and gas. Exceedingly low tensions are used in tertiary oil extraction to free the remaining globules of oil from the porous rock. Wetting phenomena are important in detergency, adhesion, lubrication, and the coating of surfaces. Positive gradients are also used to promote spreading and coalescence in the pharmaceutical and oil industries, and negative gradients to stabilize emulsions and foams in the food and chemical industries. Capillary action also plays a key role in dyeing, paper and textile manufacturing, and soil science.

This volume provides essential reading for scientists and engineers concerned with problems involving surface effects associated with surface tension, contact angles, and wetting as outlined above. It is hoped that the information presented will clarify existing uncertainties and point the way to further advances and applications.

Stanley Hartland

Contents

Contributors

Suresh G. Advani   Department of Mechanical Engineering, University of Delaware, Newark, Delaware, U.S.A.

Ashok Bhakta   Aspen Technology Inc., Cambridge, Massachusetts, U.S.A.

Alexander Bismarck   Department of Chemical Engineering and Chemical Technology, Imperial College London, London, United Kingdom

Edgar M. Blokhuis   Colloid and Interface Science, Leiden Institute of Chemistry, Leiden University, Leiden, The Netherlands

N. V. Churaev   Institute of Physical Chemistry, Russian Academy of Sciences, Moscow, Russia

Zuzana Dimitrovova   Department of Mechanical Engineering, Technical University of Lisbon, Lisbon, Portugal

Jan Christer Eriksson   Department of Chemistry, Surface Chemistry, Royal Institute of Technology, Stockholm, Sweden

Yukihiro Kaneko   Material Science Research Center, Lion Corporation, Tokyo, Japan

Kenji Katoh   Department of Mechanical Engineering, Osaka City University, Osaka, Japan

Ying Liao   Department of Chemical Engineering, Clarkson University, Potsdam, New York, U.S.A.

Stig Ljunggren   Department of Chemistry, Surface Chemistry, Royal Institute of Technology, Stockholm, Sweden

Steven D. Lubetkin   Eli Lilly & Company, Greenfield, Indiana, U.S.A.

John B. McLaughlin   Department of Chemical Engineering, Clarkson University, Potsdam, New York, U.S.A.

Eli Ruckenstein   Department of Chemical Engineering, State University of New York at Buffalo, Buffalo, New York, U.S.A.

Bihai Song   Krüss GmbH, Hamburg, Germany

Jürgen Springer   Institute of Chemistry, Technical University of Berlin, Berlin, Germany

Takamitsu Tamura   Material Science Research Center, Lion Corporation, Tokyo, Japan

1
Drainage and Collapse in Standing Foams

ASHOK BHAKTA Aspen Technology Inc., Cambridge, Massachusetts, U.S.A.

ELI RUCKENSTEIN State University of New York at Buffalo, Buffalo, New York, U.S.A.

I. INTRODUCTION

Foams are highly concentrated dispersions of gas (dispersed phase) in a liquid (continuous phase). Concentrated emulsions (biliquid foams) are similar systems in which the dispersed phase is a liquid. The terms "foam" and "concentrated emulsion" are often used interchangeably in the literature because the phenomena responsible for their behavior are essentially the same.

Foams have several very interesting and unusual properties that make them useful in many industrial applications. The large gas/liquid interfacial area available is exploited in foam fractionation to efficiently separate surface-active substances, such as proteins, from their solutions [1–5]. Foams are also being considered for use in enhanced oil recovery [6,7], insulation, and reduction of the impact of explosions [6]. On the other hand, foams can often be a nuisance [8,9] in the chemical process industry. When they last long enough, they interfere with physical and chemical processes and adversely impact productivity and efficiency. Liquid–liquid concentrated emulsions have received relatively less attention in literature. However, interest in them is growing. Concentrated emulsions have been used to prepare high-molecular-weight polymers and composites as well as membranes for separation [10,11].

The structure of a foam depends on the relative proportions of the gas and liquid. Bubbles are spherical in foams containing a large amount of liquid (>25%). As the percentage of liquid decreases, the bubbles become less spherical. Foams with less than 2% liquid are almost completely polyhedral. For monodispersed foams (composed of identical bubbles), a gas volume fraction of 0.74 is considered to be the limit beyond which the bubbles cease to be spherical. This is because the highest volume fraction that identical

1

spheres can occupy is 0.74. It must be emphasized, however, that this criterion for determining the onset of polyhedricity is applicable only to monodisperse foams. In polydisperse foams, the bubbles can remain spherical at much higher volume fractions. Although foams are almost never monodisperse, it is possible to produce foams with a narrow-enough size distribution that the assumption of monodispersity in theoretical models is often reasonable.

Polyhedral foams have a complex structure composed of liquid films (formed between the faces of adjacent polyhedral bubbles) and Plateau border (PB) channels (formed at the edges). Although the dimensions of the films and Plateau border channels vary almost randomly in a typical foam, all foams are known to obey a set of geometrical rules formulated by the Belgian scientist Plateau:

1. Three and only three films meet at an edge at an angle of 120°.
2. Four and only four PB channels meet at a node at an angle of approximately 109°.

The PB channels, which contain most of the liquid, form a complex interconnected network within the framework of Plateau's laws. Fig. 1 shows a typical foam column, a polyhedral bubble, and the cross-section of a typical Plateau border channel formed where three films meet.

For purposes of modeling foam decay, the details of foam structure are relevant only to the extent that they are useful in determining the mean dimensions of the PB channels and films in terms of the bubble volume that can be measured experimentally. Kelvin's Tetrakaidecahedron (four quadrilateral, four pentagonal, and six hexagonal faces) is the only space-filling regular polyhedron that satisfies Plateau's laws. However, it has become common practice to assume a foam to consist of regular pentagonal dodecahedra for modeling purposes. Although it is not space-filling, the regular pentagonal dodecahedron comes remarkably close to satisfying Plateau's laws. The angle between the faces is about 116° and the angle between the edges is about 108°. In fact, in his experimental study of agglomerations of monodispersed bubbles, Matzke [13] found that a majority of the polyhedra had 12 faces, and pentagonal faces were the most common. In any case, because the structure assumed for the polyhedron is only likely to affect a few geometrical constants to a small extent, it is probably not worth the effort to refine these assumptions. Several interesting discussions of foam structure are available in the literature [12–20].

Foams are metastable. They show a spontaneous tendency to separate into two distinct phases. The time scale for the disintegration, however, varies widely. Foams can persist for a few minutes or several days, depending on the conditions. It is important to have some control over the stability of foams if

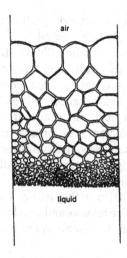

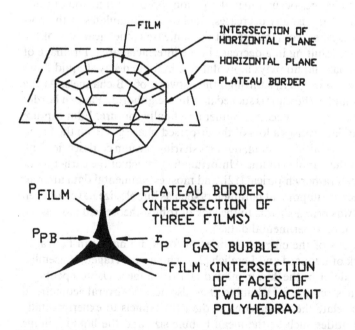

FIG. 1 A foam column, pentagonal dodecahedral bubble, and the cross-section of a PB channel.

they are to be effectively utilized. A detailed understanding of the mechanisms involved in foam persistence and decay is therefore desirable.

Experimental studies of foam stability can be broadly divided into two categories: steady state and unsteady state. In steady-state experiments, gas is bubbled through a surfactant solution at a constant flow rate. Foam is formed at the surface of the liquid and moves up at a rate that depends on the gas flow rate. Foam is continuously skimmed off at a certain distance from the liquid pool, which is replenished by adding a surfactant solution at a constant rate. This results in the formation of a steady-state moving foam column in which the properties at any position do not change with time. The goal is to study properties such as liquid holdup and bubble size along the length of the foam. Steady-state foams will not be discussed further in this article. The interested reader is encouraged to read several extensive reviews and articles available in the literature [21–29]. We shall restrict our discussion to the unsteady-state decay of standing foams.

In a typical experiment with standing foams, a certain volume of foam is prepared and the changes occurring in it are monitored as a function of time. The length of the foam usually decreases; the liquid accumulates at the bottom and the gas escapes from the top. At the same time, the mean size of the bubbles increases, resulting in a decrease in the interfacial area. Drainage of the continuous-phase liquid plays a pivotal role in foam decay. Liquid flows out under the action of gravity through the network of PB channels. At the same time, the liquid in the films is sucked into the neighboring Plateau border channels. As a result, films become thinner and finally rupture. Film rupture at the foam boundary causes a loss of the dispersed phase gas from the foam. When a film ruptures within a foam, bubbles sharing the film coalesce, leading to an increase in the bubble volume. Unfortunately, much of the earlier work on foams has been rather empirical [12], and most experimental data are rendered useless because important parameters such as bubble size have not been measured. This was primarily due to the lack of robust theoretical models for the interpretation of experimental data.

The foundations of the current theories of foam drainage can be traced back to the work of Leonard and Lemlich [21], who used a Hagen–Poiseuille-type equation to describe the flow in individual PB channels. Using a pentagonal dodecahedral structure for a bubble, they also derived several geometrical relationships to relate the dimensions of the PB channels to experimentally measurable quantities such as the mean bubble size and the liquid volume fraction. Their work was restricted to steady-state foam columns. Subsequently, several models [30–40] using the same approach for the drainage of standing foams have appeared in the literature. The most notable advances in this area were made by Krotov [31] and Narsimhan [32]. Krotov [31] was the first to recognize the effect of Plateau border suction (capillary pressure) on foam drainage and formulated the basic equations and boundary conditions. Krotov showed that drainage in a foam eventually ceases when the gradient of capillary pressure, which is set up as drainage proceeds, balances gravity. However, although Krotov formulated the basic equations, he did not attempt to solve them. Narsimhan [32] recognized the importance of the method of foam generation and numerically solved the drainage equations for foams produced by bubbling using a quasi-steady-state approximation to simulate the drainage during bubbling. He studied the effect of various parameters, including the bubble size and viscosity, on foam drainage. In a series of papers [35–39] over the past couple of years, we have theoretically studied the problem of standing foam decay in detail. For pneumatic foams produced by bubbling, we used an unsteady-state model [35] to simulate the drainage during bubbling. More importantly, the effect of film rupture was included in our models [36–39]. This has enabled us to directly evaluate the effect of drainage on the various phenomena associated with foam decay such as bubble coalescence and foam collapse.

In the sections that follow, our theoretical model for the drainage, coalescence, and collapse in standing foams is discussed in detail.

II. THEORETICAL MODEL

As mentioned in Section I, drainage of the continuous-phase liquid plays a crucial role in the decay of a standing foam. Two mechanisms are responsible for fluid flow in a foam. The capillary pressure drives flow in the films, whereas the flow in the Plateau border channels occurs due to gravity. We will discuss the flow in individual films and Plateau border channels and then formulate the bulk conservation equations.

A. Film Drainage

Fig. 2 shows a cross-section of a draining film. The surface is curved at the edges where neighboring films come together to form a Plateau border channel. Due to this curvature, the pressure is smaller at the edges than at the center of the film, and a radial flow is induced, leading to a reduction of the film thickness with time.

Film thinning is modeled using a modified Reynolds equation [41] for the flow between two circular parallel disks, which gives the rate of film thinning as:

$$-\frac{dx_F}{dt} = \frac{2c_f \Delta P x_F^3}{3\mu R_F^2} = V_f \tag{1}$$

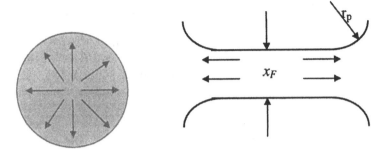

FIG. 2 A draining film.

In Eq. (1), t is the time, μ is the viscosity of the continuous phase, R_F is the radius of the disk, and ΔP is the pressure difference causing the flow. The driving force ΔP is a net result of the suction pressure in the adjacent Plateau border channels and the disjoining pressure (Π) in the films, and is given by $\Delta P = \sigma/r_p - \Pi$, where σ is the surface tension and r_p is the radius of curvature of the Plateau border channels. The coefficient c_f is a correction factor that accounts for the mobility of the film surfaces and is given by [42]:

$$\frac{1}{c_f} = 16 \sum_{n=1}^{\infty} \frac{\left(6\mu + \eta_s k_n^2 \alpha_s x_F\right)^2}{\left(6\mu + 6\mu\alpha_s + \eta_s k_n^2 \alpha_s x_F\right)\lambda_n^4} \tag{2}$$

In Eq. (2), λ_n is the nth root of the equation $J_0(\lambda_n) = 0$, $k_n = \lambda_n/R_F$:

$$\alpha_s = \frac{-3D\mu}{\Gamma\left(\dfrac{\partial\sigma}{\partial c_s}\right)}\left[1 + \frac{2D_s\left(\dfrac{\partial\Gamma}{\partial c_s}\right)}{Dx_F}\right] \approx -\frac{3D\mu}{\Gamma\left(\dfrac{\partial\sigma}{\partial c_s}\right)}$$

where J_0 is the zeroth-order Bessel function of the first kind, c_s is the surfactant concentration near the film surface, D is the bulk diffusivity and D_s is the surface diffusivity of the surfactant.

The disjoining pressure (Π) refers to the repulsive force that arises when the film surfaces are close enough ($x_F < 1000$ Å) to interact with each other. When Π is positive (repulsive), it opposes film thinning, whereas when it is negative (attractive), it increases the driving force (ΔP) and accelerates film thinning. In the most general case, Π is computed as a sum of an attractive van der Waals force (Π_{VDW}), a repulsive force (Π_{DL}) due to the interaction of the electrical double layers on the two surfaces and a short-range repulsive force (Π_{SR}), which could result from steric interaction when long-chained molecules are adsorbed on the surfaces, or due to hydration forces that are set up because of the ordering of the water molecules near charged surfaces. The former are not observed with ionic surfactants. They arise only when long-chained nonionic surfactants or polymeric molecules such as proteins are adsorbed on the film surfaces. Thus we have:

$$\Pi = \Pi_{VDW} + \Pi_{DL} + \Pi_{SR} \tag{3}$$

Unlike the other two forces that are reasonably well understood within the realm of the DLVO theory, the origin of the short-range forces is still a subject of discussion. It is generally agreed that this force decays roughly exponentially with distance.

The plot of Π versus x_F is referred to as the disjoining pressure isotherm and plays a crucial role in determining the rate of film thinning and its stability to rupture. Film rupture will be discussed next.

B. Film Rupture

The rupture of a film occurs when waves generated on the surface due to mechanical and thermal perturbations grow in an unbounded fashion. Whether a surface wave is damped or undergoes catastrophical growth is determined mainly by the shape of the disjoining pressure isotherm. If the repulsive disjoining force increases in response to the local thinning due to the disturbance, the wave is damped and no rupture occurs. On the other hand, if the disjoining pressure decreases, local thinning is accelerated and leads to rupture. Film rupture can therefore occur only if the disjoining pressure decreases when the film thickness decreases (i.e., when its derivative with respect to film thickness $d\Pi/dx_F$ is positive). Fig. 3 shows some typical plots of disjoining pressure (Π) versus film thickness (x_F). For the rupture of a thin film to occur via the growth of instabilities, $d\Pi/dx_F$ must be positive. Thus, a thin film with an isotherm such as the one in Fig. 3a will rupture only for $x_F < x_{Fm}$. A film with an initial thickness $x_F > x_{Fm}$ can arrive to thicknesses less than x_{Fm} only if the capillary pressure (σ/r_p) in the Plateau borders, which drives film thinning, exceeds the maximum disjoining pressure. It can therefore rupture only if the capillary pressure exceeds the maximum disjoining pressure (Π_{max}). On the other hand, a film with an isotherm such as that in Fig. 3b will definitely rupture because $d\Pi/dx_F$ is always positive. Fig. 3c shows an isotherm in which there are two maxima. The smaller one to the right is primarily due to the electrical double-layer forces, whereas the other is due to a short-range repulsive force. At high electrolyte concentrations, the electrical double layer is compressed and the double-layer forces are overwhelmed by the van der Waals forces at all thicknesses. If the short-range force is significant, we will either have an isotherm such as Fig. 3a, where the

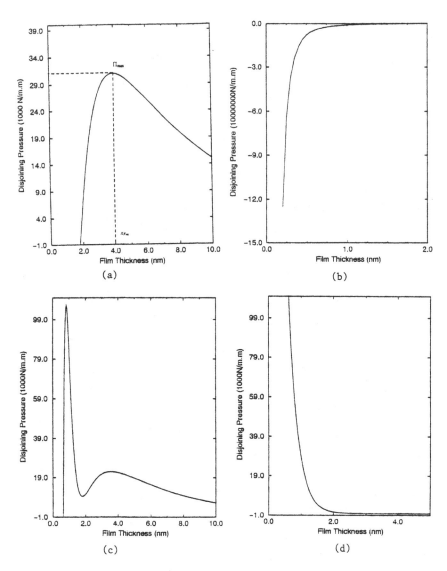

FIG. 3 Various types of disjoining pressure isotherms.

maximum is mainly due to the short-range repulsive forces, or one such as Fig. 3d, if the short-range forces are large enough to overcome the van der Waals forces at short distances (few angstroms).

The latter case corresponds to an extremely stable film because $d\Pi/dx_F$ is always negative. We will primarily consider isotherms of the type shown in Fig. 3a. For this kind of isotherm, the situation regarding the stability of the film to rupture can be summarized as follows:

$$x_F > x_{Fm} \Rightarrow \frac{d\Pi}{dx_F} < 0 \text{ \{Stable\}}$$

$$x_F < x_{Fm} \Rightarrow \frac{d\Pi}{dx_F} > 0 \text{ \{Unstable\}} \qquad (4)$$

$$x_F = x_{Fm} \Rightarrow \frac{d\Pi}{dx_F} = 0 \text{ \{Metastable\}}$$

A number of theories that attempt to predict the lifetime of a film using linear and nonlinear stability theories have appeared in the literature [42–47]. The emphasis in these theories is on computing the time elapsed between the onset of instability (the point at which $d\Pi/dx_F = 0$ and $x_F = x_{Fm}$) and the actual rupture of the film, which is deemed to occur when the waves on the surface become large enough for the two film surfaces to touch (Fig. 4). The mean film thickness at which film rupture occurs (x_{Fc}) is actually smaller than the mean thickness (x_{Fm}) when the instability begins (i.e., $x_{Fc} < x_{Fm}$). Needless to say, these theories are extremely complex and it is practically impossible to incorporate them into a global model for foam collapse. We therefore assume that a film ruptures at the moment its surfaces become unstable. In other words, we assume that $x_{Fc} \approx x_{Fm}$. This assumption is reasonable because the time scale for the growth of the instability is much shorter than the time for film drainage. For an isotherm of the type shown in Fig. 3a, x_{Fc} is therefore the film thickness corresponding to the maximum disjoining pressure (Π_{max}). The disjoining pressure isotherm and especially the value of the maximum disjoining pressure (Π_{max}) therefore play a critical role in our model for foam collapse.

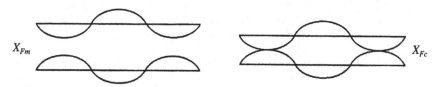

FIG. 4 Film rupture.

It must be emphasized here that a film can also rupture when it is very thick, much before the film surfaces are close enough to interact. The stability of the film in this situation will be independent of the disjoining pressure and will be determined primarily by the elasticity of the film [48] and the nature of the mechanical perturbations involved. This mechanism of film rupture is not considered in our model.

Details regarding the calculation of the disjoining pressure isotherm are discussed next.

C. Calculation of the Disjoining Pressure

1.   van der Waals Force

The attractive van der Waals component of the disjoining pressure (Π_{VDW}) can be obtained using the expression:

$$\Pi_{\mathrm{VDW}} = -\frac{A_{\mathrm{h}}}{6\pi x_{\mathrm{F}}^3} \tag{5}$$

The Hamaker constant (A_{h}) has a weak dependence on film thickness, but is often taken to be constant. Donners et al. [49] have obtained a simple expression for A_{h}:

$$A_{\mathrm{h}} = 6\pi \left[\frac{b + cx_{\mathrm{F}}}{1 + \mathrm{d}x_{\mathrm{F}} + ex_{\mathrm{F}}^2} + q \right] \tag{6}$$

where b, c, d, e, and q are constants, which depend on the specifics of the particular system such as the media involved and the thickness p of the surfactant film (see Fig. 5). In some of our calculations, sodium dodecyl sulfate was used as the model surfactant. For this system, the values available for air–

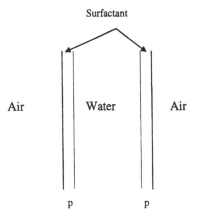

FIG. 5   The three layers involved in the calculation of the van der Waals forces.

water–air systems with dodecane monolayers ($p = 0.9$ nm) at the air–water interfaces were used. These values are: $b = -3.96 \times 10^{-23}$ J, $c = -2.05 \times 10^{-13}$ J/m, $d = 8.86 \times 10^{7}$ J/m, $e = 6.61 \times 10^{15}$ J/m², and $q = -1.8 \times 10^{-22}$ J. It must be emphasized that x_F refers to the thickness of the aqueous core of the film. Experimentally obtained disjoining pressure isotherms [50,51] are usually presented in terms of the total film thickness, which include the thicknesses of the surfactant layer. Thus, a correction (2.3 nm in case of sodium dodecyl sulfate) [50] must be applied to x_F to make a comparison with experimental results.

2. Electrical Double-Layer Force (Π_{DL})

When the film surfaces are charged, a repulsive force arises due to the confinement of ions in the electrical double layer as the film thickness decreases. The calculation of Π_{DL} involves the solution of the Poisson–Boltzmann equation to obtain the concentration of ions at the midplane. An approximate power series solution to the Poisson–Boltzmann equation at moderate potentials has been obtained by Oshima and Kondo [52] and Narsimhan [53]. The potential ψ at a distance x from the midplane is given by:

$$\tanh\left[\frac{e\psi}{k_B T}\right] = \gamma A_1(\kappa x) + \gamma^3 A_2(\kappa x) + \gamma^5 A_3(\kappa x) \tag{7}$$

where:

$$\gamma = \tanh\left(\frac{e\psi_s}{4k_B T}\right); \quad A_1(x) = \frac{\cosh(\kappa x)}{\cosh\left(\frac{\kappa x_F}{2}\right)};$$

$$A_2(x) = \frac{(\kappa x)\sinh(\kappa x) - \frac{\kappa x_F}{2}\tanh\left(\frac{\kappa x_F}{2}\right)\cosh(\kappa x)}{\cosh^3\left(\frac{\kappa x_F}{2}\right)} \tag{8a}$$

$$A_3(x) = \frac{A_1(x) - (A_1(x))^3}{4\cosh^2\left(\frac{\kappa x_F}{2}\right)} + \frac{3A_2(x)}{4\cosh^2\left(\frac{\kappa x_F}{2}\right)}\left[1 - 4\frac{\kappa x_F}{2}\tanh\left(\frac{\kappa x_F}{2}\right)\right]$$

$$- \frac{A_1(x)\left\{\left(\frac{\kappa x_F}{2}\right)^2 - (\kappa x)^2\right\}}{2\cosh^4\left(\frac{\kappa x_F}{2}\right)} \tag{8b}$$

The midplane potential (ψ_m) is therefore:

$$\tanh\left[\frac{e\psi_m}{k_B T}\right] = \gamma A_1(0) + \gamma^3 A_2(0) + \gamma^5 A_3(0) \tag{9}$$

The repulsive force per unit area is given by:

$$\Pi_{DL} = \frac{16c_{el}RTY_m^2}{\left(1 - Y_m^2\right)^2} \approx 16c_{el}RTY_m^2\left[1 + 2Y_m^2 + 3Y_m^4\right] \tag{10}$$

where $Y_m = \tanh\left[e\psi_m/k_BT\right]$ and c_{el} is the electrolyte concentration. Using Eq. (9), Eq. (10) can therefore be rewritten as:

$$\begin{aligned}\Pi_{DL} = 16c_{el}R_GT\Big[&\gamma^2A_1^2(0) + 2\gamma^4A_1(0)\left(A_2(0) + A_1^3(0)\right)\\ &+ \gamma^6\left(2A_1(0)A_3(0) + 8A_1^3(0)A_2(0) + 3A_1^6(0) + A_2^2(0)\right)\Big]\end{aligned} \tag{11}$$

If the surface potential (ψ_s) is fixed and known, Eqs. (7) (8a) (8b) (9)–(11) suffice for the calculation of Π_{DL}. However, the surface potential is often not fixed and depends on the surface charge. For nonionic surfactants, the surface charge arises due to the binding of ions present in the water and is probably independent of surfactant concentration. On the other hand, for ionic surfactants, the surface charge results from the dissociation of the adsorbed surfactant molecules and is likely to be strongly dependent on the surfactant concentration. The calculation of the surface charge and the surface potential for ionic surfactants is discussed below.

(a)   Calculation of the Surface Charge and Surface Potential for Ionic Surfactants.   In films stabilized by ionic surfactants, the charge on the surface of a film is a result of the adsorbed surfactant molecules that are dissociated. Let us consider a model system of sodium dodecyl sulfate and sodium chloride. If α_d is the degree of dissociation of the adsorbed surfactant molecules, the surface charge per unit area (σ_c) can be expressed in terms of the surfactant surface density (Γ) as:

$$\sigma_c = -\Gamma\alpha_d e \tag{12}$$

where e is the protonic charge and the degree of dissociation α_d is provided by the equilibrium of the following reaction: R-Na$|_{\text{surface}}$⇆R$^-|_{\text{surface}}$ + Na$^+|_{\text{aqueous phase}}$. Thus, denoting K_d as the equilibrium constant, we have:

$$K_d = \frac{(\alpha_d\Gamma)c_{Na^+}}{(1 - \alpha_d)\Gamma} = \frac{\alpha_d c_{Na^+}}{1 - \alpha_d} \tag{13}$$

In Eq. (13), the concentration c_{Na^+} of the sodium ions near the surface can be expressed in terms of the surfactant concentration (c_s) and the salt (NaCl) concentration (c_e) in the Plateau border channels as:

$$c_{Na^+} = c_{el}\exp\left(-\frac{e\psi_s}{k_BT}\right) \tag{14}$$

where $c_{el} = c_e + c_s$ and Γ can be computed using the Frumkin adsorption isotherm:

$$b_1 c_{R^-} = \frac{\dfrac{\Gamma}{\Gamma_\infty} \exp\left(-2a_1 \dfrac{\Gamma}{\Gamma_\infty}\right)}{1 - \dfrac{\Gamma}{\Gamma_\infty}} \tag{15}$$

where $c_{R^-} = c_s \exp(e\psi_s/k_B T)$ is the concentration of the surfactant anions near the interface, Γ_∞ is the surface excess at saturation, and the constants a_1 and b_1 are empirical parameters, which are available in the literature from experiments carried out for air–water interfaces in contact with a large amount of water. The experimental results, however, relate Γ to the bulk surfactant concentration c_s and not c_{R^-}. At large ionic strengths, the double layer is completely compressed and c_{R^-} is equal to c_s. The highest salt (NaCl) concentration for which data on sodium dodecyl sulfate are available [54] is 1 M, for which $\Gamma_\infty = 5 \times 10^{-6}$ mol/m^2, $a_1 = -1.53$, and $b_1 = 881$ m^3/mol. These values were used in our calculations with ionic surfactants. The Frumkin isotherm, when combined with the Gibbs adsorption equation, provides the following relation for the surface tension:

$$\sigma_0 - \sigma = -\Gamma_\infty R_G T \left[\log\left(1 - \frac{\Gamma}{\Gamma_\infty}\right) + a_1\left(\frac{\Gamma}{\Gamma_\infty}\right)^2\right]$$
$$+ \frac{4R_G T(2\varepsilon_{d0}\varepsilon_d R_G T(c_e + c_s))^{\frac{1}{2}}}{N_A e}\left[\cosh\left(\frac{e\psi_s}{2k_B T}\right) - 1\right] \tag{16}$$

The surface charge can also be expressed in terms of the surface potential and the midplane potential as:

$$\sigma_c = -\left[\sqrt{2\left\{\cosh\left(\frac{e\psi_s}{k_B T}\right) - \cosh\left(\frac{e\psi_m}{k_B T}\right)\right\}}\right] \frac{k\varepsilon_d\varepsilon_{d0} k_B T}{e} \tag{17}$$

Equating the right-hand sides of Eqs. (12) and (17) and solving with Eq. (9), we can get the midplane (ψ_m) and surface potential (ψ_s) in terms of the dissociation constant K_d and the electrolyte concentration c_{el}. It may be noted that for ionic surfactants, the electrolyte concentration c_{el} will depend on the surfactant concentration. Thus, if c_e is the concentration of the monovalent salt (e.g., NaCl) and c_s is the concentration of the monovalent ionic surfactant (e.g., NaDS), we have:

$$c_{el} = c_s + c_e \tag{18}$$

It may be noted that the use of Eq. (17) implies that the effect of the interactions between the adsorbed species has been ignored. Because these interactions provide a positive contribution to the free energy, they decrease the degree of dissociation. These effects have been considered for micelles by Ruckenstein and Beunen [55].

3. The Short-Range Repulsive Force (Π_{SR})

As mentioned earlier, large exponentially decaying repulsive forces can arise due to the steric interaction of long-chained molecules, or due to the organization of water molecules near charged surfaces. We will first discuss the latter. A short discussion of steric forces is provided in a later section.

(a)   Short-Range Repulsion Due to Organization of Water Molecules.   It has been experimentally observed [44] that in many systems, the resistance of a foam to collapse increases sharply at sufficiently high surfactant and salt concentrations. Experimental measurements of the disjoining pressure isotherms of single films containing sodium dodecyl sulfate [50,51] show that qualitatively different isotherms are obtained depending on the concentration of sodium chloride. As the pressure on a film is increased, its thickness decreases and the repulsive disjoining pressure increases. When the thickness is small enough (about 10 nm), black spots appear. These black spots eventually cover the entire film, giving rise to a common black film. A further increase in pressure, however, gives rise to different phenomena depending on the salt concentration. At lower salt concentrations, the common black film ruptures, whereas at higher salt concentrations, there is a sudden transition to a very stable Newton black film. In the latter case, the film rupture predicted by the DLVO theory is prevented by the emergence of a large short-range repulsive force. It is possible to qualitatively explain this force using a simple model proposed by Bhakta and Ruckenstein [39] and Schiby and Ruckenstein [56,57]. The basic idea is that the surface dipoles generate a local electrical field that polarizes the nearby water molecules, which in turn polarize the neighboring water molecules and so on, thus generating a net polarization through water. The surface dipoles are due to the undissociated surfactant molecules present on the interface. When two surfaces approach each other, the polarized layers will overlap, decreasing the dipole moment of the water molecules and thus increasing the free energy of the system. Some details are provided below.

To account for the polarization of the water molecules, it is reasonable to assume that the water is organized in the vicinity of a surface in a layered structure, similar to that of ice with successive layers containing out-of-plane hexagonal rings of water molecules, parallel to the external surface [76].

Assuming that all the water molecules from layer i have the same average polarization m_i normal to the layer, the field E_i^{local}, acting on a site of layer i, due to all the other polarized molecules can be written as:

$$E_i^{local} = (C_0 m_i + C_1(m_{i-1} + m_{i+1}) + \cdots + C_k(m_{i-k} + m_{i+k}) + \cdots) \quad (19)$$

where the coefficient C_k accounts for the contribution of the dipoles of layer $i \pm k$ to the local field at a site of layer i. Note that C_0 corresponds to the field at a site of layer i generated by all the other dipoles of the same layer i.

For a perfect icelike structure of water in the vicinity of the surface, the coefficients C_k are given by [77]:

$$C_0 = -\frac{3.7763}{4\pi\varepsilon_{d0}\varepsilon_d'' l^3}; \quad C_1 = \frac{1.8272}{4\pi\varepsilon_{d0}\varepsilon_d'' l^3} \quad (20)$$

where l is the distance between two adjacent water molecules in the tetragonal coordination of the ice structure, ε_d'' is the local dielectric constant for the interaction between adjacent molecules, and ε_{d0} is the dielectric constant of the vacuum. The field produced by the remote dipoles is screened by the medium, which has a large dielectric constant $\varepsilon_d \cong 80$. In contrast, the screening of the neighboring dipoles is much weaker because there are no intervening water molecules. It can therefore be assumed that E_i^{local} is produced only by the dipoles located within a radius $2l$ from the given site and that the dielectric constant of the medium screening their interactions has a much lower value ε_d''. Only the contributions from the first 26 neighbors (12 from the same layer and 14 from the two adjacent layers) were taken into account in the derivation of an expression for the local field E_i^{local}.

If there is no external electrical field and the molecules of the first water layer have an average dipole moment m_1 oriented perpendicular to the layer, caused by the surface dipoles, the average dipole moment of the molecules of layer i situated at the distance $x = i\Delta$, $m_i = m(x)|_{x=i\Delta}$ is given by:

$$m_i = m(x)|_{x=i\Delta} = \gamma E_i^{local} = \gamma(C_0 m(x) + C_1(m(x - \Delta) + m(x + \Delta)) + \cdots) \quad (21)$$

where Δ is the distance between the centers of two successive layers and γ is the molecular polarizability.

An approximate solution can be obtained by expanding $m(x \pm \Delta)$ in a series:

$$m(x \pm \Delta) = m(x) \pm \frac{dm(x)}{dx}\Delta + \frac{1}{2}\frac{d^2 m(x)}{dx^2}\Delta^2 \pm \cdots \quad (22)$$

The odd derivatives cancel because of symmetry and the terms of order 4 and higher can be neglected; one then obtains:

$$m(x)(1 - \gamma(C_0 + 2C_1 + 2C_2 + \cdots)) = \gamma\Delta^2(C_1 + 4C_2 + \cdots)\frac{d^2m(x)}{dx^2} \quad (23)$$

which is of the form:

$$\frac{d^2m(x)}{dx^2} - \frac{1}{\lambda_h^2}m(x) = 0, \quad (24)$$

with a decay length λ_h given by [77]:

$$\lambda_h = \sqrt{\frac{(C_1 + 4C_2 + 9C_3 + \cdots)\gamma\Delta^2}{1 - (C_0 + 2C_1 + 2C_2 + \cdots)\gamma}} \quad (25)$$

In order to solve Eq. (24), two boundary conditions are needed. The first boundary condition is provided by the symmetry of the system ($m = 0$ at the midway between plates). The second boundary condition is provided by the dipole moment (m_1) of the water molecules at the interface. The solution of Eq. (24) for two planar parallel surfaces separated by the distance x_F (hence for the boundary condition $m(-x_F/2) = m_1, m(x_F/2) = -m_1$) is:

$$m(x) = -m_1\frac{\sinh\left(\dfrac{x}{\lambda_h}\right)}{\sinh\left(\dfrac{x_F}{2\lambda_h}\right)} \quad (26)$$

where x is measured from the midplane.

The dipole moment of the water molecules at the interface (m_1) depends on the electrical field. The computation of the electrical field and the dipole moment is discussed below.

In the traditional electrodynamics of continuum media, a constant surface dipole density on a plane neighboring a continuum medium of constant dielectrical constant ε_d does not induce any electrical field inside the medium because the system is equivalent to two parallel planar sheets of surface charge density $+\sigma_c$ and $-\sigma_c$. Therefore, the corresponding field vanishes everywhere above or below the planar sheets. However, the traditional theory implies the same screening by the neighboring dipoles and the remote dipoles. The assumption that the medium is a continuum with a constant ε_d ceases to be valid at molecular dimensions. It can therefore be assumed that a water molecule from the vicinity of the surface is polarized mainly by the neighboring surface dipoles because the local dielectric constant ε_d' corresponding to this interaction is small. The field generated by the remote surface dipoles is screened by a medium with a large dielectric constant $\varepsilon_d \approx 80$, and therefore can be neglected. The electrical field generated by the surface dipoles in the second water layer is also considered negligible. A detailed analysis of the validity of this assumption has been provided in the literature [78].

It was assumed that the interfaces are located at the external boundaries of the first organized water layers, whereas the surface dipoles are situated at a distance Δ' below the interface. Assuming that the area per surface dipole of the water layer that is polarized by the corresponding dipole is a disk of radius R_d, located at a distance Δ' from the dipole $\vec{p}$ with the component $p_\perp$ normal to the surface, the average local field generated by the surface dipoles is given by [77]:

$$\overline{E} = \frac{p_\perp}{\varepsilon_d'} \frac{1}{2\pi\varepsilon_{d0}\left(R_d^2 + \Delta'^2\right)^{\frac{3}{2}}}. \tag{27}$$

where ε_d' is the local dielectric constant near the interface.

The polarization m_1 is proportional to the total field $E^t = \overline{E} + E_i^{\text{local}}$ (E_i^{local} being generated by the dipoles from the first and second water layers):

$$m_1 = \gamma\left(\overline{E} + E_1^{\text{local}}\right) = \gamma\left(\overline{E} + m_1 C_0 + m_2 C_1 + \cdots\right) \tag{28}$$

Combining the above two equations, we get:

$$m_1 = \frac{\gamma\overline{E}}{\left(1 - \gamma\left(C_0 + C_1 \dfrac{\sinh\left(\dfrac{x_F - 2\Delta}{2\lambda_h}\right)}{\sinh\left(\dfrac{x_F}{2\lambda_h}\right)} + \cdots\right)\right)} \tag{29}$$

where Δ is the distance between the centers of two neighboring water layers ($\Delta = 4/3l$).

The electrostatic energy of a water molecule with the polarizability γ, which acquires the polarization m in the field E_e, is:

$$U = -\int_0^{E_e} m(E)\,\mathrm{d}E = -\int_0^{E_e} \gamma E\,\mathrm{d}E = -\frac{1}{2}\gamma E_e^2 = -\frac{m^2}{2\gamma} \tag{30}$$

and hence the electrostatic interaction energy per unit area between the two parallel plates is given by [77]:

$$U_h(x_F) = \frac{\gamma\overline{E}^2}{\left(1 - \gamma\left(C_0 + C_1 \dfrac{\sinh\left(\dfrac{x_F - 2\Delta}{2\lambda_h}\right)}{\sinh\left(\dfrac{x_F}{2\lambda_h}\right)} + \cdots\right)\right)^2} \frac{x_F - \lambda_h\sinh\left(\dfrac{x_F}{\lambda_h}\right)}{4v\sinh^2\left(\dfrac{x_F}{2\lambda_h}\right)} \tag{31}$$

where $v = (8/3\sqrt{3})l^3$ is the volume corresponding to a water molecule in an icelike structure. It may be noted that it was assumed that the electrostatic energy of the remaining water molecules, located among the head groups of the surfactant molecules adsorbed on the interface, does not depend on the separation distance x_F between plates and hence does not contribute to the disjoining hydration pressure $\Pi_h = -dU_h/dx_F$. Some results are presented in the following sections.

4. Effect of Salt Concentration (NaCl) on the Disjoining Pressure Isotherms of Ionic Surfactants

An increase in the concentration of sodium chloride has the following effects: (a) it compresses the electrical double layer by raising the ionic strength; (b) it decreases the degree of dissociation of the adsorbed surfactants (see Eq. (13)) by causing increased binding of the counterions, thus increasing the surface dipole density and the short-range hydration disjoining pressure; (c) it affects the concentration of R^- ions near the interface via the surface potential and hence the surface excess (Γ). In experiments with single thin liquid films, it has been observed by Exerowa et al. [50] that at high salt concentrations, as the capillary pressure is increased, there is, instead of rupture, a sudden transition to very stable Newton black films. A model calculation of the disjoining pressure [79] using the formulations developed in previous sections is presented in Fig. 6 for three concentrations of sodium chloride [(a) $c_e = 0.150$ mol/dm^3, (b) $c_e = 0.165$ mol/dm^3, and (c) $c_e = 0.180$ mol/dm^3]. The values of the other parameters used were: $\Gamma_\infty = 5 \times 10^{-6}$ mol/m^2, $b_1 = 881$ m^3/mol, $a_1 = -1.53, b = -3.08 \times 10^{-22}, c = -6.28 \times 10^{-14}, d = 8.28 \times 10^7, e = 6.13 \times 10^{15}, q = -9.00 \times 10^{-23}, K_d = 0.050$ mol/dm^3, $c_s = 1 \times 10^{-3}$ mol/dm^3, $l = 2.76$ Å, $\varepsilon_d'' = 9$, and $p_\perp/\varepsilon_d' = 3.45$ D.

Consequently, for the decay length λ_h calculated with Eq. (25), one obtains the value $\lambda_h = 1.0$ Å. For $c_e = 0.150$ mol/dm^3 (Fig. 6a), the disjoining pressure maximum at the higher film thickness (common black film) is larger than that corresponding to a Newton black film. Therefore, the increase of the external pressure above the disjoining pressure corresponding to the common black film results in the rupture of the film instead of its transition to a Newton black film.

At higher concentrations of sodium chloride (see Fig. 6b, $c_e = 0.165$ mol/dm^3), the disjoining pressure maximum corresponding to a Newton black film is higher than that for a common black film. This results in a transition from a common black film to a Newton black film with an increase in external pressure. The value $p_\perp/\varepsilon_{d'} = 3.45$ D was selected to provide about the same height for the peaks of disjoining pressure at $c_e = 0.165$ mol/dm^3, which was the value reported by Exerowa et al. [50] as an upper limit beyond which there is transition to Newton black films instead of rupture. A further increase of the

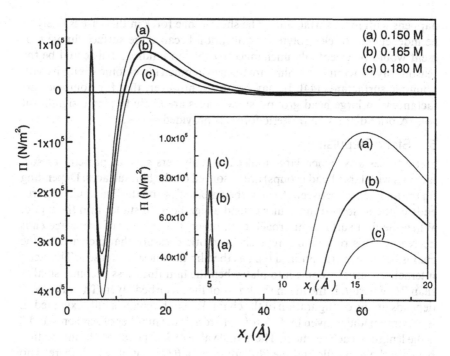

FIG. 6 Disjoining pressure isotherm for $\Gamma_\infty = 5 \times 10^{-6}$ mol/m^2, $a_1 = 881$ m^3/mol, $a_2 = -1.53$, $b_1 = -3.08 \times 10^{-22}$, $b_2 = -6.28 \times 10^{-14}$, $b_3 = 8.28 \times 10^7$, $b_4 = 6.13 \times 10^{15}$, $b_5 = -9.00 \times 10^{-23}$, $K_d = 0.050$ mol/dm^3, $c_s = 1 \times 10^{-3}$ mol/dm^3, $\lambda_1 = 1.0$ Å, $p_\perp/\varepsilon' = 3.45$ D, and (a) $c_e = 0.150$ mol/dm^3, (b) $c_e = 0.165$ mol/dm^3, and (c) $c_e = 0.180$ mol/dm^3. In the inset, the peaks of the disjoining pressure are presented at a smaller scale.

electrolyte concentration ($c_e = 0.18$ mol/dm^3, Fig. 6c) lowers the transition pressure from the common to the Newton black film because of the screening of the double layer and also slightly increases the rupture pressure of the Newton black films because of the increase in the hydration repulsion with increasing ionic strength. (The hydration repulsion increases as the concentration of electrolyte increases because the dipole surface density increases through the reassociation between the R$^-$ surfactant anions present on the interface and the Na$^+$ cations.) At high salt concentrations, the disjoining pressure maximum corresponding to common black films disappears and Newton black films are the only type of stable films formed.

Although the calculations above have been carried out for ionic surfactants, the model is applicable in principle to nonionic surfactants with short head groups as well. In these systems, the surface charge arises due to ion binding rather than the dissociation of the adsorbed surfactant molecules as happens with ionic surfactants. The short-range force in this case is likely to be independent of electrolyte concentration because the surface dipole moment would be essentially unchanged by the ion binding. This could be the reason why Newton black films are formed at much lower ionic strengths with nonionic surfactants [44]. It may be noted, however, that for nonionic surfactants with large head groups, steric forces are likely to play a significant role. A brief discussion of steric forces is provided below.

5. Steric Repulsion

Steric forces arise when large molecules (polymers such as proteins or surfactants with large head groups) are adsorbed on the film surfaces. Depending on the interaction between the adsorbed molecules and the film liquid, the free ends of the adsorbed molecules extend a significant distance into the liquid. When the film surfaces approach each other, the entropy of these free ends decreases. This results in a repulsive osmotic pressure that increases as the chains become more confined (i.e., as the film thickness decreases). The steric interactions usually come into play when the film thickness becomes smaller than $2L$, where L is the mean thickness of the adsorbed layer. The value of L depends on the configuration of the chain. When the chain is fully extended, L is maximum and is given by $L = n_s l_s$, where n_s is the number of segments and l_s is the length of each segment. The actual value of L depends on the interaction between the molecule and the film liquid. In a θ solvent, in which there is no interaction between various segments, we have $L \approx L_0 \equiv l_s \sqrt{n_s}$. For low surface coverage in a θ solvent, Dolan and Edwards [58] have derived a simple expression for the interaction energy per unit area:

$$W(x_F) = \Gamma k_B T\left[\frac{\pi L_0^2}{6x_F^2} + \frac{1}{2}\ln\left[\frac{3x_F^2}{8\pi L_0^2}\right]\right] \quad \text{for } x_F \leq \sqrt{3}L_0$$

$$= 2\Gamma k_B T\left[\exp\left(-\frac{3x_F^2}{2L_0^2}\right)\right] \quad \text{for } x_F \geq \sqrt{3}L_0$$

(32)

The force per unit area is therefore given as:

$$\Pi_{ST} = \Gamma k_B T\left[\frac{2\pi^3 L_0^2}{3x_F^3} - \frac{2}{x_F}\right] \quad \text{for } x_F \leq L_0\sqrt{3}$$

$$= 12\Gamma k_B T\left(\frac{x_F}{L_0^2}\right)\exp\left(-\frac{3x_F^2}{2L_0^2}\right) \quad \text{for } x_F \geq L_0\sqrt{3}$$

(33)

At high surface coverage, the adsorbed molecules form a brush. The thickness of the brush in a good solvent has been given by Alexander [59] as:

$$L = \Gamma^{\frac{1}{2}}R_f^{\frac{5}{2}}$$

(34)

where $R_f = l_s n_s^{3/4}$ is the Flory radius. The repulsive pressure between two brush-bearing surfaces is given by the de Gennes theory [60]:

$$\Pi_{ST} \approx k_B T \Gamma^{\frac{3}{2}} \left[\left(\frac{2L}{x_F} \right)^{\frac{9}{4}} - \left(\frac{x_F}{2L} \right)^{\frac{3}{4}} \right] \quad \text{for } x_F < 2L \tag{35}$$

The first term in Eq. (35) results from the osmotic repulsion whereas the second term arises from the elastic energy that opposes stretching.

D. Flow in a Plateau Border Channel

The flow of the continuous-phase liquid out of a foam takes place through the Plateau border channels. The flow of liquid in a single Plateau border channel must therefore be understood before an attempt to model drainage in a foam is made. An expression for the average velocity (u) of the liquid in a vertical Plateau border channel has been derived for a triangular Plateau border cross-section and is given by [60]:

$$u = \frac{c_v a_p}{20\sqrt{3}\mu} \left(\rho_c g - \frac{\partial p}{\partial z} \right) \tag{36}$$

In Eq. (36), z is the vertical space coordinate (see Fig. 7) that increases in the downward direction and g, ρ_c, μ, and p refer to the gravitational acceleration, density, viscosity, and pressure in the continuous phase, respectively. a_p is the cross-sectional area of a Plateau border channel and can be computed from

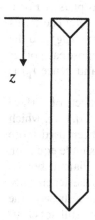

FIG. 7 A Plateau border channel.

the radius of curvature of the walls (r_p) and the film thickness (x_F) using the expression [21]:

$$a_p = \frac{\left(0.322r_p + 1.732x_F\right)^2 - 2.721x_F^2}{0.644} \tag{37}$$

The factor c_v in Eq. (36) accounts for the effect of finite surface viscosity (η_s) and has been computed by Desai and Kumar [61] as a function of the inverse of the dimensionless surface viscosity ($\gamma_s = 0.4387\mu\sqrt{a_p}/\eta_s$). It must be emphasized that their results for the calculation of c_v are valid only for foams because they neglected the viscosity of the dispersed phase. Equation (36) can be used for liquid–liquid concentrated emulsions only when the surfaces are immobile (i.e., when $c_v = 1$). The pressure gradient ($\partial p/dz$) can be computed as follows.

If p_{ref} is the pressure at a position z_{ref} inside a bubble, the p_i at any position z inside a bubble is:

$$p_i = p_{ref} + \rho_D g(z - z_{ref}) \tag{38}$$

where ρ_D is the density of the dispersed phase. Also, if σ is the interfacial tension between the two phases, $p_i - p = \sigma/r_p$. Thus, if we assume that the dispersed phase is not compressed, ρ_D is independent of z and Eq. (36) can be rewritten as:

$$u = \frac{c_v a_p}{20\sqrt{3}\mu}\left((\rho_c - \rho_D)g + \frac{\partial}{\partial z}\left(\frac{\sigma}{r_p}\right)\right) = \frac{c_v a_p}{20\sqrt{3}\mu}\left(\rho g + \frac{\partial}{\partial z}\left(\frac{\sigma}{r_p}\right)\right) \tag{39}$$

where $\rho = \rho_c - \rho_D$ is the density difference between the two phases. For a foam, because the dispersed phase is a gas, $\rho_c \gg \rho_D$ and $\rho \approx \rho_c$. The terms inside the parentheses represent the driving forces due to gravity (ρg) and the gradient of Plateau border suction $\partial/\partial z(\sigma/r_p)$. These forces oppose each other when the latter is negative (i.e., when the liquid fraction (and hence r_p) is smaller at the top).

Leonard and Lemlich [21] were the first to consider the effects of surface viscosity on drainage in a Plateau border channel. However, Eq. (36), which differs from their equation only by a constant factor, has been used more frequently because Desai and Kumar have provided a very simple equation for c_v. Experiments [62] on the flow of fluid in individual Plateau border channels seem to suggest that Eq. (36) appropriately describes the dependence of the velocity on the viscosity and channel dimensions. The usefulness of the factor c_v in describing the effect of surface viscosity is, however, still not clear. In most of our calculations, we have assumed that $c_v = 1$.

E. Theoretical Model for the Drainage and Coalescence in Foam

1. Conservation Equation for Drainage

Modeling of the drainage in a foam essentially involves the formulation of the bulk conservation equations in terms of the expressions for film thinning (Eq. (1)) and Plateau border drainage (Eq. (39)) [39]. There are two ways to approach this problem. In the microscopical approach, each Plateau border channel and film in the foam is considered separately and the liquid content in the foam at any time is computed by summing the liquid content in each Plateau border channel and film in the foam. Typically, this would involve the construction of a detailed polyhedral network in which the position and the orientation of each channel and film in the foam would need to be specified. An attempt was made to formulate a microscopic model for foam drainage using Voronoi polyhedra to generate the network. In the nondegenerate case, these polyhedra possess two important structural properties seen in actual foams: four and only four edges (Plateau border channels) meet at a point, and three and only three faces (films) meet at an edge. The goal was to solve the partial differential equations arising from the mass balances for each channel with appropriate conditions at the junctions and the top and bottom of the foam. The basic balance equation in a single Plateau border channel is:

$$\frac{\partial a_p}{\partial t} = -\frac{\partial u_y a_p}{\partial y} \tag{40}$$

where y is the direction of flow and u_y is the fluid velocity in the y direction. If a channel is inclined at an angle θ to the vertical, u_y is given by:

$$u_y = \frac{c_v a_p}{20\sqrt{3}\mu} \left[\rho g \cos\theta + \frac{\partial}{\partial y} \left(\frac{\sigma}{r_p} \right) \right] \tag{41}$$

Two conditions need to be satisfied at the junctions in the network.

The pressure in the continuous phase is continuous at the junction. Thus, if r_{pi} and r_{pj} are the Plateau border radii of channels i and j meeting at a junction, we have:

$$\frac{\sigma}{r_{pi}} = \frac{\sigma}{r_{pj}} \Rightarrow r_{pi} = r_{pj} \tag{42}$$

Because there is no accumulation at a junction, the net volumetric flow rate of fluid into the junction is 0. Thus, we have:

$$\sum c_i a_{pi} u_{yi} = 0 \tag{43}$$

The summation is over all channels meeting at the junction and c_i is a constant which takes the values $+1$ or -1, depending on whether the channel is above or below the junction. A serious attempt was made to model foam drainage using this microscopic approach. However, we concluded that this approach was not practical given the amount of CPU time that would be required to get results of any value. We therefore decided to use the conventional macroscopic approach in which the foam is treated as a continuous fluid. In this approach, one does not concern oneself with details of individual films and Plateau border channels. Rather, one considers infinitesimal elements of foam, with each element containing a large number of bubbles. The differential element is large enough to contain a large number of bubbles, yet small enough relative to the total volume of the foam for a continuum treatment to be valid. This approach is usually appropriate when the bubble size is much smaller than the length of the foam. Thus, in each element, we consider a mean Plateau border radius $(\overline{r_p})$, a mean film thickness $(\overline{x_F})$, and a mean bubble radius $(\overline{R})$. The ultimate goal of a theoretical model of foam drainage is to obtain these quantities as a function of time and position within the foam. Because all quantities of interest such as the liquid volume fraction (ε) can be expressed in terms of $\overline{r_p}$, $\overline{x_F}$, and $\overline{R}$, the conservation equations are formulated solely in terms of these quantities. The exact distribution of x_F, r_p, and R within an element is important only to the extent that it is required to compute these mean quantities. The Plateau border radius r_p can be assumed to be uniform within a volume element because of the tendency of the capillary pressure (σ/r_p) to equalize due to the flow of liquid from the thicker to thinner channels. This is because the pressure $(p-\sigma/r_p)$ will be smaller in channels with smaller r_p and liquid will flow from regions of higher pressure (larger r_p) to regions of lower pressure (smaller r_p). Thus, a "local" equilibrium of the capillary pressure is established in a very short time and one can assume $r_p = \overline{r_p}$. However, the film thickness is not necessarily uniform. Before we formulate the conservation equations in terms of the mean quantities, it must be emphasized here that we ignore the increase in the bubble size due to Ostwald ripening, which occurs due to the diffusion of gas from smaller to larger bubbles. In monodisperse foams containing gases such as nitrogen, which are sparingly soluble in water, the time scale for Ostwald ripening is likely to be much larger than that for drainage and collapse, so that foam decay can be said to occur in two stages. In the first stage, drainage, coalescence, and collapse are dominant, whereas in the second stage, Ostwald ripening dominates. We consider only the first stage. Ostwald ripening for steady-state foams has been considered by Narsimhan and Ruckenstein [23].

Fig. 8 shows a schematic of a foam when it is freshly formed (a) and when some decay has taken place (b). The space coordinate z increases in the downward direction, and z_1 and z_2 represent the upper and lower boundaries

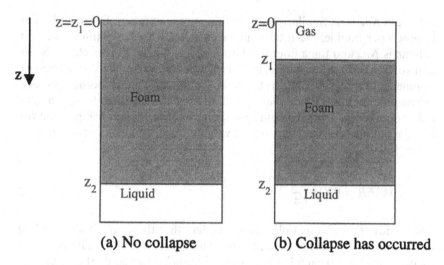

FIG. 8 A foam column.

of the foam, respectively. The origin ($z = 0$) defines the upper boundary of the entire system. In other words, all the gas entering the foam lies between the reference plane ($z = 0$) and the foam/liquid interface ($z = z_2$). Before collapse, all the gas lies within the foam and $z_1 = 0$ (see Fig. 8a). After collapse starts, the foam/gas interface moves down as gas is released from the bubbles at the top due to film rupture and $z_1 > 0$ (see Fig. 8b). The gas released from the ruptured bubbles lies between $z = 0$ and $z = z_1$. The foam volume also decreases due to the loss of liquid and the lower boundary of the foam moves up as the drainage occurs and z_2 decreases with time. It is to be noted that because wall effects are neglected, all quantities are functions of vertical position z only.

Some variables relevant to the macroscopical approach need to be defined [22,23] before we proceed to formulate the conservation equations. As mentioned earlier, we deal with groups of bubbles rather than individual bubbles. Consider a volume element of foam at a level z. Let ε be the volume fraction of liquid in this element. The volume fraction of gas is then $(1-\varepsilon)$. If $\overline{V}$ is the mean bubble volume in this element, the number of bubbles per unit volume (N) in this element is given by:

$$N = \frac{1-\varepsilon}{\overline{V}} \tag{44}$$

Let n_F be the average number of films per bubble. The total number of films per unit volume is then Nn_F and the amount of liquid in the films per unit volume is $Nn_F\overline{A_F x_F}$. Similarly, if n_p is the average number of Plateau border channels per bubble, the total number of Plateau border channels per unit volume is Nn_p and the amount of liquid in the Plateau border channels per unit volume is $Nn_p a_p \bar{l}$, where a_p is the cross-sectional area of a Plateau border channel in this element and $\bar{l}$ is the mean length of a Plateau border channel. We assume that in spite of coalescence, the bubbles maintain their pentagonal dodecahedral geometry and that the geometrical relationships valid for a single bubble (see Appendix A) are valid for mean values as well. Thus, we have:

$$\bar{l} = 0.816\overline{R} = 0.816\left(\frac{3}{4\pi}\overline{V}\right)^{\frac{1}{3}} \tag{45}$$

The values of n_F and n_p follow from the fact that there are 12 films and 30 channels in a pentagonal dodecahedral bubble. Because each film is shared by two bubbles and each Plateau border channel is shared by three bubbles, we have $n_F = 6$ and $n_p = 10$. The liquid volume fraction is given by the sum of liquid per unit volume in the Plateau border channels and the liquid per unit volume in the films as:

$$\varepsilon = Nn_p a_p \bar{l} + Nn_F\overline{A_F x_F} \tag{46}$$

Combining Eqs. (44) and (46), ε can be expressed as:

$$\varepsilon = \frac{n_p a_p \bar{l} + n_F\overline{A_F x_F}}{\overline{V} + n_p a_p \bar{l} + n_F\overline{A_F x_F}} \tag{47}$$

Each of the variables $\bar{l}$, $\overline{A_F x_F}$, $\overline{V}$, and a_p are functions of time and vertical position z.

The variation of ε in any volume element with time depends on the net rate at which liquid flows into the volume element. Hence, the volumetrical flux of liquid at each z is required in order to formulate the bulk conservation equations. This will be considered next.

(a) Calculation of the Volumetrical Flux of Liquid. Liquid flow in a foam takes place in two ways. First, there is gravity drainage through the Plateau border channels. In addition, as the foam becomes drier, there is a net upward movement of bubbles as the volume of the foam decreases (z_2 decreases) due to the outflow of liquid from the foam. As these bubbles move, they carry with them the liquid in the associated films and Plateau border channels. Thus, an upward convective flow of liquid due to bubble movement is superimposed on the downward gravity driven flow through the Plateau border channels. We will discuss each of these components of the flux below.

(b) Gravity Drainage Through the Plateau Border Channels. The flow rate of liquid across any horizontal plane in a foam is given by the sum of the flow rates through all the Plateau border channels that intersect the plane [35]. The flow rate through a Plateau border channel with cross-sectional area a_p will depend on its orientation. It is highest in a vertical channel and lowest in a horizontal channel. Consider a Plateau border channel inclined at an angle θ with the vertical. The average velocity of liquid in this channel is given by:

$$u_\theta = \frac{c_v a_p}{20\sqrt{3}\mu} \left[\rho g \cos\theta + \sigma \frac{d}{d\left(\frac{z}{\cos\theta}\right)} \left(\frac{1}{r_p}\right) \right] \qquad (48)$$

Comparing Eq. (48) with Eq. (39), we get:

$$u_\theta = \cos\theta \left[\frac{c_v a_p}{20\sqrt{3}\mu} \left(\rho g + \sigma \frac{d}{dz}\left(\frac{1}{r_p}\right) \right) \right] = u\cos\theta \qquad (49)$$

Thus, the volumetrical flow rate of liquid through this channel is given by:

$$q_\theta = u_\theta a_p = u a_p \cos\theta \qquad (50)$$

If we assume that the Plateau border channels are oriented randomly, the probability that a channel is oriented between θ and $\theta + d\theta$ is $\sin(\theta)d\theta$ and the average volumetrical flow rate per channel is given by:

$$\bar{q} = \int_0^{\frac{\pi}{2}} q_\theta (\sin\theta)d\theta = \frac{u a_p}{2} \qquad (51)$$

The total volumetrical flux through the Plateau border channels across the plane is therefore given as q_{PB} = (number of PB channels per bubble intersected by a horizontal plane) × (number of bubbles intersected by the horizontal plane per unit area) × (mean flow rate through a PB channel).

For a pentagonal dodecahedral bubble, the number of Plateau border channels per bubble intersected by a horizontal plane is $n_p/5$. The number of bubbles intersected by a horizontal plane can be obtained as shown below using a technique common in dealing with the flow of fluids in porous media [22,23].

Let $F(R)dR$ be the fractional number of bubbles with radii between R and $R + dR$. The number of bubbles with centers at a distance between X and $X + dX$ from the plane is $2NAF(R)dRdX$, where A is the cross-sectional area of the foam. Only those bubbles whose centers are at a distance smaller than their radii ($X < R$) will intersect the plane. Therefore, the number of bubbles of radius R that intersects the plane per unit area is $\int_{X=0}^{X=R} 2NF(R)dRdX = 2NF$

$(R)R dR$, where the factor 2 arises due to the fact that bubbles on both sides of the plane are considered. The total number of spheres per unit area intersected by the plane is then given as $\int_{R=0}^{R=\infty} 2NRF(R) dR = 2N\bar{R}$ and the flux of liquid due to gravity drainage through the Plateau border channels is:

$$q_{PB} = 2N\bar{R}\left(\frac{n_p}{5}\right)\bar{q} = \frac{3}{15} N\bar{R}n_p a_p u \tag{52}$$

(c) Calculation of the Convective Flux. As mentioned earlier, drainage of the liquid causes a decrease in the foam volume, which leads to an upward movement of the bubbles. We need to know the rate at which the liquid associated with the bubbles moves up. Let q_G be the volumetrical flux of gas due to the convection of the bubbles. The number of bubbles moving per unit area per unit time is then given by $q_G/\bar{V}$. Because the amount of liquid per bubble is ε/N, the rate (q_c) at which the liquid is convected along with the bubbles is given as:

q_c = Number of bubbles moving up per unit area per unit time

$\quad\quad \times$ Volume of liquid associated with each bubble

$$= \frac{q_G}{\bar{V}}\frac{\varepsilon}{N} = q_G\frac{\varepsilon}{1-\varepsilon} \tag{53}$$

The total flow rate of liquid per unit area per unit time (q_L) is obtained by summing the contribution due to the Plateau border drainage (q_{PB}) and the convection due to bubble movement (q_c) and is given as:

$$q_L = q_{PB} + q_c = q_{PB} + q_G\left(\frac{\varepsilon}{1-\varepsilon}\right) \tag{54}$$

However, we still do not know q_G. When the foam drains, the convective motion of the bubbles occurs in the upward direction as the foam volume decreases (i.e., q_G is negative). In addition, because this movement occurs due to drainage in the Plateau border channels, it is natural to expect q_G to be related to q_{PB}. The relation between these two quantities is derived as follows.

Because the rate of accumulation of liquid in a volume element equals the net flow rate into the volume element, the conservation equation for the liquid is:

$$\frac{\partial \varepsilon}{\partial t} = -\frac{\partial q_L}{\partial z} \tag{55}$$

If the dispersed gas phase is assumed to be incompressible, a similar conservation equation can be written for the gas phase as:

$$\frac{\partial}{\partial t}(1 - \varepsilon) = -\frac{\partial q_G}{\partial z} \tag{56}$$

that is,

$$\frac{\partial \varepsilon}{\partial t} = \frac{\partial q_G}{\partial z} \tag{57}$$

Comparing Eqs. (55) and (57), we have:

$$-q_L = q_G \tag{58}$$

Combining Eqs. (54) and (58), q_L and q_G can be expressed in terms of q_{PB} as:

$$q_G = -(1 - \varepsilon)q_{PB} \tag{59}$$

$$q_L = (1 - \varepsilon)q_{PB} \tag{60}$$

Thus, the total flux (q_L) is smaller than the flux (q_{PB}) due to Plateau border drainage alone because bubble convection moves some liquid upward.

Equation (55) can therefore be rewritten as:

$$\frac{\partial \varepsilon}{\partial t} = -\frac{\partial (1 - \varepsilon)q_{PB}}{\partial z} \tag{61}$$

Equation (61) is the conservation equation for the liquid in which the quantities ε and q_{PB} can be expressed in terms of r_p, $\overline{A_F x_F}$, and $\overline{R}$ using Eqs. (45) (46) (47). It is to be noted that Eq. (61) is valid whether coalescence occurs or not. Coalescence occurring in a volume element does not, by itself, change the amount of liquid in the element. It only causes a redistribution of liquid from the ruptured films and Plateau border channels to the remaining Plateau border channels within the element. The change in the amount of liquid in the element occurs only due to drainage. The effect of coalescence is felt through $\overline{R}$. When coalescence occurs, the local $\overline{R}$ increases. The two boundary conditions required for Eq. (61) are presented next.

(d) Boundary Conditions. When the foam is collapsing (i.e., when (dz_1/dt) > 0), the liquid released from the collapsed bubbles reenters the foam. The amount of liquid released depends on the rate of foam collapse (dz_1/dt). Equating the rate at which liquid is released from the collapsed bubbles ($\varepsilon dz_1/dt|_{z_1}$) to the flow rate of liquid at the foam gas interface ($q_L|_{z_1}$) gives the boundary condition at the foam/gas interface ($z = z_1$):

$$q_L|_{z_1} = \varepsilon \frac{dz_1}{dt}\bigg|_{z_1} \tag{62}$$

The other boundary condition follows from the fact that the bubbles are spherical at the foam/liquid interface:

$$r_p|_{z_2} = \overline{R}|_{z_2} \tag{63}$$

Because coalescence never occurs at the foam/liquid interface, $\overline{R}\big|_{z_2} = R_0$, where R_0 is the radius of a bubble in the freshly formed monodisperse foam.

It may be noted that the effect of bubble convection was ignored in our earlier papers [35–38]. Although this does not affect the computed liquid profiles much, it cannot be ignored when the conservation equations for the surfactant are formulated because bubble convection produces a significant movement of the surface area and hence of the adsorbed surfactant. The balance equations for the surfactant are considered next.

2. Conservation Equation for Surfactant

Part of the surfactant in the foam is present in dissolved form in the liquid in the films and Plateau border channels, whereas the rest is adsorbed on the film surfaces. Thus, the amount of surfactant per unit volume of the foam is given by $\varepsilon c_s + (2Nn_F\overline{\Gamma A_F})$, where the first term (εc_s) represents the surfactant dissolved in the liquid contained in the films and the Plateau border channels, and the second term corresponds to the adsorbed surfactant on the film surfaces. The term $\overline{\Gamma A_F}$ is the mean value of the surfactant adsorbed on the surface of a film. The mass flux of the surfactant contains three contributions:

(a) From the flow of liquid in the Plateau border channels ($q_L c_s$)
(b) From the dispersion $\left(-\varepsilon D \frac{\partial c_s}{\partial z}\right)$ where D is an apparent diffusivity
(c) From the movement of the adsorbed surfactant due to the convection of bubbles $\left(\frac{q_G}{\overline{V}} 2n_F\overline{\Gamma A_F}\right)$ where $q_G/\overline{V}$ is the number of bubbles per unit area per unit time moving in the downward direction and $2n_F\overline{\Gamma A_F}$ is the surfactant adsorbed per bubble.

It must be emphasized that D is different from the molecular diffusivity and depends on the degree of local mixing. D is actually different in different portions of the foam. We have, however, assumed D to be constant in order to avoid complexity. The net flux of surfactant (q_{cs}) is then given as:

$$q_{cs} = q_L c_s - \varepsilon D \frac{\partial c_s}{\partial z} + \frac{q_G}{\overline{V}} (2n_F\overline{\Gamma A_F}) = q_L c_s - \varepsilon D \frac{\partial c_s}{\partial z} - \frac{q_L}{\overline{V}} (2n_F\overline{\Gamma A_F})$$

(64)

and the conservation equation for the surfactant can be written as:

$$\frac{\partial}{\partial t} (\varepsilon c_s + 2Nn_F\overline{\Gamma A_F}) = -\frac{\partial q_{cs}}{\partial z} = -\frac{\partial}{\partial z}\left[q_L c_s - \varepsilon D \frac{\partial c_s}{\partial z} - \frac{q_L}{\overline{V}} 2n_F\overline{\Gamma A_F}\right]$$

(65)

This equation is also second order in space and we need two boundary conditions for the surfactant concentration c_s.

(a) Boundary Conditions for Surfactants. Before collapse starts, there is no surfactant entering the system at the top. In other words, the surfactant flux at the foam/gas interface ($z = z_1$) is 0. The boundary condition at the top before collapse is therefore:

$$q_{cs}\,|_{z_1} = q_L c_s - \varepsilon D \frac{\partial c_s}{\partial z} - \frac{q_L}{V}(2n_F \overline{\Gamma A_F})\,|_{z_1} = 0 \tag{66}$$

Now, because from Eq. (62) $q_L|_{z_1}$ is 0 before collapse starts, Eq. (66) becomes:

$$\left.\frac{\partial c_s}{\partial z}\right|_{z_1} = 0 \tag{67}$$

On the other hand, once collapse starts, the surfactant in the collapsed bubbles is released into the foam. If A is the cross-sectional area of the foam, the number of bubbles collapsing when the front moves with a velocity dz_1/dt is $(A/\overline{V}|_{z_1})dz_1/dt$. In addition, the surfactant associated with a bubble includes the amount adsorbed on the film surfaces and that dissolved in the liquid in the films and Plateau border channels and is given by $(n_p a_p \overline{l} + n_F \overline{A_F x_v})c_s + 2n_F \overline{\Gamma A_F}|_{z_1}$. Thus, the flux of surfactant into the system due to collapse is given by: (number of bubbles collapsing per unit area per unit time) × (amount of surfactant per bubble) $= (1/\overline{V}|_{z_1})dz_1/dt\left[(n_p a_p \overline{l} + n_F \overline{A_F x_F})c_s + 2n_F \overline{\Gamma A_F}\right]_{z_1}$. Equating this to the flux of surfactant at the foam/gas interface from Eq. (64), we get the boundary condition:

$$-\varepsilon D \left.\frac{\partial c_s}{\partial z}\right|_{z_1} + \left[q_L c_s - \frac{q_L}{V}(2n_F \overline{\Gamma A_F})\right]\Big|_{z_1}$$
$$= \frac{1}{\overline{V}}\frac{dz_1}{dt}\left[(n_p a_p \overline{l} + n_F \overline{A_F x_F})c_s + 2n_F \overline{\Gamma A_F}\right]\big|_{z_1} \tag{68}$$

Using Eq. (62) to eliminate dz_1/dt and simplifying, we get:

$$\varepsilon D \left.\frac{\partial c_s}{\partial z}\right|_{z_1} + q_L \left[\frac{\varepsilon}{1-\varepsilon}c_s + \frac{1+\varepsilon}{\overline{V}\varepsilon}(2n_F \overline{\Gamma A_F})\right]\Big|_{z_1} = 0 \tag{69}$$

The other boundary condition is obtained from the fact that at the foam/liquid interface, the concentration is that of the original foaming solution (c_{s0}):

$$c_s|_{z_2} = c_{s0} \tag{70}$$

The basic conservation equations for the continuous-phase liquid and the surfactant have now been formulated. However, several additional details are required to specify the system completely. Expressions are needed for the movement of the boundaries z_1 and z_2. Furthermore, the mean values $\overline{x_F}$, $\overline{A_F x_F}$, and $\overline{R}$ depend on the degree of coalescence. The movement of the foam/liquid interface will be considered first. This will be followed by a discussion of foam collapse in the absence of coalescence. For simplicity, it is instructive to first discuss foam collapse when coalescence is absent as many of the important phenomena are better understood if the complications arising from the change in bubble size due to coalescence are ignored. In the absence of coalescence, R is essentially a constant parameter and the system of equations will be complete once expressions for dz_1/dt and dz_2/dt are formulated. Detailed expressions for the change of the bubble radius and the rate of collapse in the presence of coalescence will be derived toward the end of the chapter after we have presented the results obtained with the assumption of no coalescence.

3. Movement of the Foam/Liquid Interface

The foam/liquid interface moves up (z_2 decreases) as liquid drains out of the foam at the bottom. This upward movement occurs because the volume of the foam decreases as liquid leaves the foam. Thus, the rate at which this interface moves up is equal to the rate at which the height of the liquid pool at the bottom increases, which in turn depends on the flow rate of liquid through the Plateau border channels at the bottom. Therefore, if A is the cross-sectional area of the foam:

$$-A \frac{dz_2}{dt} = Aq_{PB}|_{z_2}; \quad -\frac{dz_2}{dt} = q_{PB}|_{z_2} \tag{71}$$

4. Movement of the Foam/Gas Interface (Collapse)

When coalescence is neglected, the implicit assumption is that all films at a given level (z) have the same thickness. Because the capillary pressure is usually uniform at a given level, it follows from Eq. (1) that films at the same z drain at the same rate and hence rupture at the same time, leading to the complete disappearance of the foam at that level. Because films are thinnest at the top of the foam, the bubbles at the top will rupture first, followed by the bubbles in the next layer and so on, resulting in a downward movement of the foam/gas interface. Thus, foam collapse starts when the films at the top become critical ($x_F = x_{Fc}$) and the foam/gas interface moves downward (i.e., z_1 increases), with its position at any time being determined by the level (z) at which $x_F = x_{Fc}$. Mathematically, this would be expressed as:

$$x_F = x_{Fc} \quad \text{for} \quad z = z_1 \tag{72}$$

The treatment of this moving boundary problem is simplified considerably if the two boundaries are immobilized by carrying out a transformation from t,z space to t,ξ space, where:

$$\xi = \frac{z - z_1}{z_2 - z_1} \tag{73}$$

With this transformation, the variable ξ lies between 0 and 1 and the Reynolds equation (Eq. (1)) can be written in t,ξ space as:

$$\frac{dx_F}{dt}\bigg|_z = -V_f = \frac{\partial x_F}{\partial t}\bigg|_\xi + \frac{\partial x_F}{\partial \xi}\bigg|_t \frac{\partial \xi}{\partial t}\bigg|_z \tag{74}$$

where:

$$\frac{d\xi}{dt}\bigg|_z = \frac{1}{z_2 - z_1}\left[(\xi - 1)\frac{dz_1}{dt} - \xi\frac{dz_2}{dt}\right] \tag{75}$$

Using Eq. (75), Eq. (74) becomes:

$$-V_f = \frac{\partial x_F}{\partial t}\bigg|_\xi + \frac{\xi - 1}{z_2 - z_1}\frac{\partial x_F}{\partial \xi}\frac{dz_1}{dt} - \frac{\xi}{z_2 - z_1}\frac{dz_2}{dt} \tag{76}$$

Equation (72) can then be written as:

$$x_F|_{\xi=0} = x_{Fc}(c_s|_{\xi=0}) \tag{77}$$

that is,

$$-V_f|_{\xi=0} = -V_f|_{x_{Fc}(c_s|_{\xi=0})} \tag{78}$$

Differentiating Eq. (77) with respect to time, we get:

$$\frac{\partial x_F}{\partial t}\bigg|_{\xi=0} = \frac{dx_{Fc}}{dc_s}\frac{\partial c_s}{\partial t}\bigg|_{\xi=0} \tag{79}$$

Combining Eqs. (76)–(79) for $\xi = 0$, we get:

$$-V_f\bigg|_{\xi=0} = -V_f\bigg|_{x_{Fc}(c_s|_{\xi=0})} = \frac{dx_{Fc}}{dc_s}\frac{\partial c_s}{\partial t}\bigg|_{\xi=0} - \frac{1}{z_2 - z_1}\frac{\partial x_F}{\partial \xi}\bigg|_{\xi=0}\frac{dz_1}{dt} \tag{80}$$

which can be rewritten as:

$$\frac{dz_1}{dt} = \frac{V_f|_{x_{Fc}(c_s|_{\xi=0})} - \dfrac{dx_{Fc}}{dc_s}\dfrac{\partial c_s}{\partial t}\bigg|_{\xi=0}}{\dfrac{1}{z_2 - z_1}\dfrac{\partial x_F}{\partial \xi}\bigg|_{\xi=0}} \tag{81}$$

Before we proceed further, some features of Eq. (81) need to be noted. The rate (dz_1/dt) at which the foam collapses depends strongly on the film velocity (V_f) corresponding to the critical thickness (x_{Fc}) at that level. dz_1/dt increases as the film velocity corresponding to the critical thickness increases. On the other hand, because $dx_{Fc}/dc_s < 0$, the rate of foam collapse will decrease if the local surfactant concentration increases (i.e., if $(\partial c_s/\partial t) > 0$). However, when the critical thickness (x_{Fc}) is independent of the surfactant concentration, as is probably the case with nonionic surfactants, Eq. (81) reduces to:

$$\frac{dz_1}{dt} = \frac{V_f|_{x_{Fc}}}{\dfrac{1}{z_2 - z_1} \left. \dfrac{\partial x_F}{\partial \xi} \right|_{\xi=0}} \tag{82}$$

The system is now completely specified and with the appropriate initial conditions, the conservation equations can be solved with equations for the change of z_1 and z_2. Before we proceed with the formulation of the initial conditions and the solution of the differential equations, we will briefly discuss the phenomenon of drainage equilibrium.

F. Drainage Equilibrium

It has been experimentally observed that a draining foam usually arrives to a metastable mechanical equilibrium in which all processes associated with drainage come to a halt. This equilibrium is characterized by the following properties:

1. There is no flow in the Plateau border channels (i.e., $u = 0$) throughout the foam. This happens when the gradient in the capillary pressure balances gravity, that is, when $\rho g = -\partial/\partial z(\sigma/r_p)$. Indeed, experimental measurements [63,64] of the pressure in a draining foam indicate that after a sufficiently long time, the pressure profile becomes linear with a slope equal to ρg.
2. There is no film thinning (i.e., $dx_F/dt = 0$) throughout the foam. This occurs when the repulsive disjoining pressure (Π) in the films balances the capillary pressure (σ/r_p). In other words, $(\sigma/r_p) = \Pi$ throughout the foam.
3. The surfactant concentration does not change with time. There is no diffusion [$(\partial c_s/\partial z) = 0$] and there is no convection ($u = 0$).
4. There is no movement of the boundaries (i.e., there is no collapse at the top, $dz_1/dt = 0$, and no drainage at the bottom, $dz_2/dt = 0$). This actually follows from conditions (1) and (2). If there is no drainage, r_p will not change and the capillary pressure cannot increase to overcome the maximum disjoining pressure (Π_{max}) and cause collapse. Thus, in an equilibrated foam, $(\sigma/r_p) \leq \Pi_{max}$ and $x_F \geq x_{Fc}$.

The initial conditions are formulated in Section II.G.

G. Initial Conditions

The extent of drainage and collapse that occur by the time equilibrium is established depends on the difference in the amount of liquid in the foam initially and at equilibrium. The larger this difference is, the larger is the amount of drainage that will occur. The state of the foam (distribution of liquid) at the start of the experiment therefore plays an important role in determining its drainage and collapse behavior. The initial distribution of liquid, however, depends on the manner in which the foam is produced. Foams can be divided into two categories depending on the method of generation: (a) homogeneous foams, and (b) pneumatic foams.

Homogeneous foams are usually produced by vigorous agitation of surfactant solutions. As the name suggests, the distribution of liquid is uniform in such foams. Pneumatic foams, on the other hand, are produced by bubbling a gas through a surfactant solution. As the bubbling proceeds and the length of the foam increases, there is drainage of liquid in the foam and a profile of liquid fraction develops. The foam is drier (liquid fraction is lower) higher up in the foam where the bubbles are older and more drainage has occurred.

We will first discuss homogeneous foams in detail and then consider pneumatic foams. Before we proceed further, it must be mentioned that several assumptions were involved in obtaining most of the results that will be presented in the following sections:

1. Coalescence was ignored [35–38]. The films areas were assumed to be the same and the film thickness was only a function of the vertical position z.
2. The surfactant concentration was assumed to be uniform (i.e., the surface potential and surface tension were assumed to be constant).
3. In our first two papers [35,36], film drainage was ignored.

These assumptions were relaxed as we learned more about the system. However, most of our results (which were qualitative in any case) remain valid. In fact, an analysis of the results obtained using these simplifying assumptions actually improves our understanding of the system.

The results and the simplifying assumptions are presented below.

H. Homogeneous Foams

1. Initial and Boundary Conditions for Homogeneous Foams

As mentioned earlier, homogeneous foams are characterized by a uniform initial liquid fraction [36]. The initial condition for homogeneous foams therefore is:

$$At \ t = 0, \ \varepsilon = \varepsilon_0 \ for \ all \ z \tag{83}$$

The other distinguishing property of homogeneous foams is that they do not start draining immediately. Some time usually elapses before some continuous-phase liquid appears at the bottom of the foam. This is in contrast to pneumatic foams, which are always in contact with the continuous-phase liquid at the bottom. It has also been noted [64] that in homogeneous foams, phase separation occurs only when ε at the bottom exceeds a certain minimum (ε_b). (The exact value though has not been specified.) It is reasonable to expect that for a "monodisperse" foam/concentrated emulsion, this value corresponds to the point when the amount of liquid at the bottom becomes large enough for the bubbles at the bottom to become spherical ($r_p = R_0$). The liquid fraction ε at this point corresponds to that for close-packed spheres (i.e., $\varepsilon \approx 0.26 \equiv \varepsilon_b$). The reasoning behind this can be understood as follows: Consider a foam with a uniform initial continuous-phase fraction of $\varepsilon_0 < \varepsilon_b$ (say, $\varepsilon_0 = 0.15$). The bubbles at the bottom will be polyhedral (see Fig. 9a). As time passes, the Plateau border channels at the top drain into those below, so that ε decreases at the top and increases at the bottom. Suppose, at a given point in time, that ε at the bottom just exceeds 0.26. At this instant (see Fig. 9b), the bubbles at the bottom will cease to be in contact and become spherical. (Implicit in this argument is the assumption that the bubble radius is much smaller than the capillary length $R \ll \sqrt{\sigma/\rho g}$). However, due to buoyancy, these bubbles will rise and arrange themselves in a compact fashion ($\varepsilon = 0.26$) (see Fig. 9c), causing the excess continuous phase to accumulate at the bottom as a separate phase. As drainage proceeds, this process continues and the interface rises (z_2 increases), always maintaining $\varepsilon = 0.26$ at the bottom until a drainage equilibrium is established. Thus, there is no flow at the bottom as long as $\varepsilon|_{z=z_2} < \varepsilon_b$ and $\varepsilon|_{z=z_2}$ does not change once it attains the value ε_b. The appropriate boundary condition at the bottom is therefore:

$$t < t_b \Rightarrow u|_{z=z_2} = 0; \quad t > t_b \Rightarrow \varepsilon = \varepsilon_b \qquad (84)$$

where t_b is the time when ε at the bottom just becomes ε_b.

Thus, for a homogeneous foam, there are two critical conditions: one for the separation of the dispersed phase ($x_F = x_{Fc}$) and the other for the sepa-

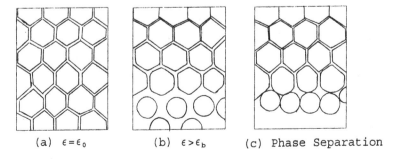

(a) $\epsilon = \epsilon_0$ (b) $\epsilon > \epsilon_b$ (c) Phase Separation

FIG. 9 Separation of the continuous phase.

ration of the continuous phase ($r_p = R_0$ or $\varepsilon = \varepsilon_b = 0.26$). Four scenarios are therefore possible for a foam with a certain initial uniform continuous-phase volume fraction (ε_0) depending on whether one or both these critical conditions are achieved before equilibrium:

(a) Only the continuous phase separates out at the bottom.
(b) Only the disperse phase separates out at the top.
(c) Both phases separate out.
(d) There is no phase separation.

By analyzing the equilibrium state of a foam in relation to its initial state, it is possible to determine, at the outset, which of the above situations arises. In the following sections, we will discuss the equilibrium state in detail and show that by making suitable approximations, it is possible to determine analytically the phase behavior from the initial conditions.

The system can be simplified considerably if films are neglected. In this section, we will discuss the equilibrium in detail using simplified equations that are obtained when films are neglected. Film drainage is important in determining the kinetics of foam collapse (see Eq. (82)). However, because most of the liquid in a foam is stored in the Plateau border channels, films can be ignored when computing the liquid volume fraction. Our main concern in this section is the drainage equilibrium. This assumption is therefore valid. This assumption was made in our first two papers [35,36] and proved to be useful in obtaining insights into many features of foam decay. In addition, in an equilibrated foam, the bubble radius is uniform and equal to the initial bubble radius (R_0) because no coalescence has occurred in the foam that remains at equilibrium. When films are neglected, the liquid fraction ε depends only on the local Plateau border radius r_p and the various macroscopic variables can be expressed as:

$$\varepsilon = N n_p a_p l; \quad N = \frac{1 - \varepsilon}{V}; \quad a_p = \left(\frac{V}{n_p l}\right)\left(\frac{\varepsilon}{1 - \varepsilon}\right);$$

$$r_p = R_0 \sqrt{\frac{4\pi}{3(0.161)n_p(0.816)}} \sqrt{\frac{\varepsilon}{1 - \varepsilon}} \tag{85}$$

Let us reexamine the condition for foam collapse within the framework of this assumption. Because film rupture can occur only if the capillary pressure (σ/r_p) exceeds the maximum disjoining pressure (Π_{max}), it is possible to define a critical Plateau border radius (r_{pc}) given by:

$$r_{pc} = \frac{\sigma}{\Pi_{max}} \tag{86}$$

For film rupture to occur, the Plateau border radius (r_p) must be less than r_{pc}. When a collapsing foam arrives to an equilibrium (i.e., when dz_1/dt

decreases to 0), $r_p = r_{pc}$ at the foam/gas interface. This is because (see Eq. (82)) when collapse comes to a halt:

$$V_f|_{x_{Fc}} = 0; \quad \frac{\sigma}{r_p} = \Pi_{max} \text{ at } z = z_1 \tag{87}$$

that is,

$$\varepsilon|_{z_1} = \varepsilon_c \tag{88}$$

where ε_c is the liquid volume fraction corresponding to $r_p = r_{pc}$. Let us now consider the equilibrium state in more detail.

2.   Drainage Equilibrium in Homogeneous Foams

Because at equilibrium $u = 0$ for all z, we have:

$$\frac{d}{dz}\left(\frac{1}{r_p}\right) = -\frac{\rho g}{\sigma} \tag{89}$$

Using Eq. (85), Eq. (89) can be written in terms of ε at equilibrium (ε_{eq}) as:

$$\sqrt{\frac{3(0.161)n_p 0.816}{4\pi}} \frac{1}{R_0} \frac{d}{dz}\sqrt{\frac{1 - \varepsilon_{eq}}{\varepsilon_{eq}}} = -\frac{\rho g}{\sigma} \tag{90}$$

Now let us define $K \equiv \sqrt{\dfrac{4\pi}{3(0.161)n_p 0.816}}$. Integration of Eq. (90) gives:

$$\sqrt{\frac{1 - \varepsilon_{eq}}{\varepsilon_{eq}}} = -\frac{K\rho g R_0 z}{\sigma} + c \tag{91}$$

where c is a constant of integration. Equation (91) can be rewritten as:

$$\varepsilon_{eq} = \frac{1}{1 + \left(c - \dfrac{K\rho g R_0 z}{\sigma}\right)^2} \tag{92}$$

Let $z_1 = z_{1e}$ and $z_2 = z_{2e}$ at equilibrium. The volume of the continuous phase per unit cross-section at equilibrium is then given by:

$$V_{eq} = \int_{z_{1e}}^{z_{2e}} \varepsilon_{eq} dz \tag{93}$$

Substituting the expression for ε_{eq} from Eq. (92) into Eq. (93) and integrating, we get:

$$V_{eq} = \frac{\sigma}{K\rho g R_0}\left[\tan^{-1}\left(c - \frac{K\rho g R_0 z_{1e}}{\sigma}\right) - \tan^{-1}\left(c - \frac{K\rho g R_0 z_{2e}}{\sigma}\right)\right] \tag{94}$$

If L_0 is the initial foam length, the volume of liquid per unit area initially present in the foam is $L_0\varepsilon_0$. The distance moved by the lower boundary by the time equilibrium is established is equal to the loss of continuous phase per unit cross-section. Thus:

$$L_0 - z_{2e} = L_0\varepsilon_0 - V_{eq} \tag{95}$$

Combining Eqs. (94) and (95) gives:

$$L_0 - z_{2e} = L_0\varepsilon_0 - \frac{\sigma}{K\rho g R_0}\left[\tan^{-1}\left(c - \frac{K\rho g R_0 z_{1e}}{\sigma}\right)\right.$$
$$\left. - \tan^{-1}\left(c - \frac{K\rho g R_0 z_{2e}}{\sigma}\right)\right] \tag{96}$$

To accurately determine the conditions at equilibrium, we need to evaluate the constant of integration c (i.e., we require ε_{eq} at one of the boundaries). There are four possible scenarios for the state of the foam/concentrated emulsion at equilibrium. Each is discussed below.

(a) Case a: The Equilibrated Foam is in Contact with the Continuous Phase at the Bottom.   In this case, because there is no collapse, we have $z_{1e} = 0$ and because the foam is in contact with liquid at the bottom, $\varepsilon_{eq}|_{z=z_{2e}} = \varepsilon_b = 0.26$. Using these conditions in Eq. (91) gives the integration constant for this case:

$$c = \sqrt{\frac{1 - \varepsilon_b}{\varepsilon_b}} + \frac{K\rho g R_0 z_{2e}}{\sigma} \tag{97}$$

Using this expression for c in Eqs. (94) and (96) gives:

$$V_{eq} = \frac{\sigma}{K\rho g R_0}\left[\tan^{-1}\left(\sqrt{\frac{1 - \varepsilon_b}{\varepsilon_b}} + \frac{K\rho g R_0 z_{2e}}{\sigma}\right) - \tan^{-1}\left(\sqrt{\frac{1 - \varepsilon_b}{\varepsilon_b}}\right)\right] \tag{98}$$

$$L_0 - z_{2e} = L_0\varepsilon_0 - \frac{\sigma}{K\rho g R_0}\left[\tan^{-1}\left(\sqrt{\frac{1 - \varepsilon_b}{\varepsilon_b}} + \frac{K\rho g R_0 z_{2e}}{\sigma}\right)\right.$$
$$\left. - \tan^{-1}\left(\sqrt{\frac{1 - \varepsilon_b}{\varepsilon_b}}\right)\right] \tag{99}$$

(b) Case b: The Equilibrated Foam is in Contact with the Separated Disperse Phase at the Top.   In this case, because no drainage of liquid from the foam has taken place, the lower boundary does not move and $z_{2e} = L_0$. In

addition, because a collapsing foam arrives to equilibrium with $r_p = r_{pc}$ at the top, $\varepsilon|_{z=z_{1e}} = \varepsilon_c$. The constant of integration is then given by:

$$c = \sqrt{\frac{1 - \varepsilon_c}{\varepsilon_c}} + \frac{K\rho g R_0 z_{1e}}{\sigma} \tag{100}$$

Substituting the expression for c in Eqs. (94) and (96), we get:

$$L_0\varepsilon_0 - \frac{\sigma}{K\rho g R_0}\left[\tan^{-1}\left(\sqrt{\frac{1 - \varepsilon_c}{\varepsilon_c}}\right)\right.$$
$$\left. - \tan^{-1}\left(\sqrt{\frac{1 - \varepsilon_c}{\varepsilon_c}} - \frac{K\rho g R_0}{\sigma}(L_0 - z_{1e})\right)\right] = 0 \tag{101}$$

$$V_{eq} = \frac{\sigma}{K\rho g R_0}\left[\tan^{-1}\left(\sqrt{\frac{1 - \varepsilon_c}{\varepsilon_c}}\right)\right.$$
$$\left. - \tan^{-1}\left(\sqrt{\frac{1 - \varepsilon_c}{\varepsilon_c}} - \frac{K\rho g R_0}{\sigma}(L_0 - z_{1e})\right)\right] \tag{102}$$

(c)   Case c: Both Phases Separate.   In this case, because both drainage and collapse have taken place, $z_{1e} > 0$ and $z_{2e} < L_0$, and $\varepsilon|_{z=z_{1e}} = \varepsilon_c$ and $\varepsilon|_{z=z_{2e}} = \varepsilon_b$. Thus, from Eq. (91), we have:

$$c = \sqrt{\frac{1 - \varepsilon_b}{\varepsilon_b}} + \frac{K\rho g R_0 z_{2e}}{\sigma} \tag{103}$$

In addition, we have:

$$c = \sqrt{\frac{1 - \varepsilon_c}{\varepsilon_c}} + \frac{K\rho g R_0 z_{1e}}{\sigma} \tag{104}$$

Combining Eqs. (103) and (104), we have:

$$z_{2e} - z_{1e} = \frac{\sigma}{K\rho g R_0}\left[\sqrt{\frac{1 - \varepsilon_c}{\varepsilon_c}} - \sqrt{\frac{1 - \varepsilon_b}{\varepsilon_b}}\right] \equiv L_{max} \tag{105}$$

In Eq. (105), $z_{2e} - z_{1e}$ is the length of the foam at equilibrium. Because the right-hand side of Eq. (105) is a constant for a given system, it is clear that when both phases separate, the final length $(z_{2e} - z_{1e})$ of the equilibrated column is fixed and given by L_{max}.

An important point to be noted is that the length of a foam cannot increase. When one or more phases separate, the length of the foam has to decrease. If there is no loss of liquid or gas, there is no change in the foam length. Thus, the following condition always holds:

$$z_{2e} - z_{1e} \leq L_0 \tag{106}$$

Thus, if $L_0 < L_{max}$, it is not possible to have an equilibrium with both phases separating. In other words, when $L_0 < L_{max}$, only one phase will separate if at all. The initial composition (ε_0) of the system determines which phase separates. If, during drainage, ε at the bottom exceeds ε_b before the films at the top rupture, the continuous phase separates and Eqs. (97)–(99) are valid. On the other hand, if film rupture occurs first, collapse occurs at the top and Eqs. (100)–(102) are valid. If, however, u becomes zero before either of the above occurs, no phase separation takes place. The initial state of the foam (L_0, ε_0) therefore plays a crucial role in determining the phase behavior. The effect of the initial condition on phase separation is discussed next.

3. Effect of Initial Condition on Phase Separation

Based on the initial state of a foam/concentrated emulsion, it is possible to determine, at the outset, which of the two phases separates out. As has been discussed earlier, if $L_0 < L_{max}$, only one phase will separate out. Two situations are possible in this case. Each is discussed separately.

(a)   Case 1: Separation of the Continuous Phase.   Let us define a critical volume:

$$V_c \equiv \frac{\sigma}{K\rho g R_0} \left[\tan^{-1}\left(\sqrt{\frac{1 - \varepsilon_b}{\varepsilon_b}} + \frac{K\rho g R_0 L_0}{\sigma} \right) - \tan^{-1}\left(\sqrt{\frac{1 - \varepsilon_b}{\varepsilon_b}} \right) \right] \tag{107}$$

A comparison with Eq. (98) indicates that V_c is the volume of the continuous phase per unit cross-section in an equilibrated foam when $z_{1e} = 0$ and $z_{2e} = L_0$ (see Fig. 10). Thus, if $\varepsilon_0 L_0 = V_c$, the continuous phase will redistribute itself until an equilibrium is established with $\varepsilon|_{z=z_2} = \varepsilon_b$. However, if $\varepsilon_0 L_0 > V_c$, some continuous phase has to separate out before an equilibrium can be established. To understand the reasoning behind this, consider Fig. 11a. The curve HAFB represents the equilibrium profile of ε in a column of length L_{max}. Consider a column of length $L_0 < L_{max}$ with an initial profile represented by the horizontal line EFG. Thus:

$$L_0 = \text{Length(EG)}; \quad \text{Area(CDEG)} = \varepsilon_0 L_0 \tag{108}$$

Also note that for this system, the volume of continuous phase per unit cross-section in an equilibrated column of length L_0 is given by Area(ABCD). Thus:

$$V_c = \text{Area(ABCD)} \tag{109}$$

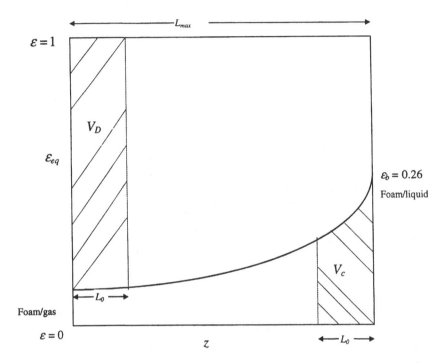

FIG. 10 Profile of a liquid fraction in an equilibrated foam of length L_{max} that has undergone collapse as well as drainage. V_c is the volume of liquid per unit area in an equilibrated foam of length L_0 with $\varepsilon = 0.26$ at the bottom. V_D is the volume gas per unit area in an equilibrated foam of length L_0 with $\varepsilon = \varepsilon_c$ at the top.

It is clear from the figure that the part of the foam from E to F has an excess (compared to equilibrium) of continuous phase given by $V_1 = \text{Area(AEF)}$, whereas the part of the foam from F to G has a deficiency of continuous phase given by $V_2 = \text{Area(BFG)}$. Thus:

$$\varepsilon_0 L_0 - V_c = V_1 - V_2 \tag{110}$$

Now if $V_1 = V_2$ (i.e., if $V_c = \varepsilon_0 L_0$), the amount of continuous phase that the upper part (EF) loses as it attains equilibrium is just equal to the amount the lower part needs. Thus, at equilibrium, the ε profile is given by curve (AFB) and the length of the column remains unchanged. On the other hand, if $V_1 > V_2$ (i.e., if $\varepsilon_0 L_0 > V_c$) (see Fig. 11b), the upper part must lose a greater amount of liquid than the lower part requires (i.e., the lower part will gain more continuous-phase liquid than it can hold). As a result, the excess separates out and the column length decreases. The final profile is now given by

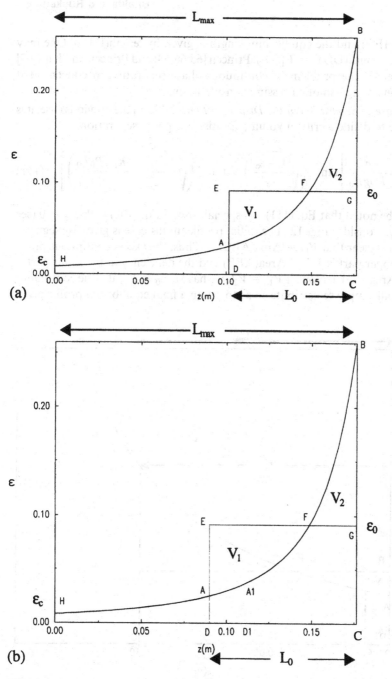

FIG. 11 A pseudo-phase diagram for continuous-phase separation: (a) $V_1 = V_2$, and (b) $V_1 > V_2$.

curve(A1FB) and the equilibrium length is given by length(D1C). One may note that length(DD1) = $V_1 - V_2$. Princen [65,66,68] and Princen and Kiss [67] have treated the problem of continuous-phase separation in concentrated emulsions using osmotic pressure considerations.

(b)   Case 2: Separation of the Dispersed Phase.   As in the previous case, it is possible to define a critical volume for disperse phase separation:

$$V_D \equiv \frac{\sigma}{K\rho g R_0} \left[\tan^{-1} \left(\sqrt{\frac{1-\varepsilon_G}{\varepsilon_c}} \right) - \tan^{-1} \left(\sqrt{\frac{1-\varepsilon_c}{\varepsilon_c}} - \frac{K\rho g R_0 L_0}{\sigma} \right) \right] \quad (111)$$

It may be noted that Eq. (111) is a special case of Eq. (102) with $z_{1e} = 0$ (see Fig. 10). Consider Fig. 12. The initial profile in this case is given by segment EFG. It is clear that V_D = Area(ABCD). Thus, the excess continuous phase in the upper part is $V_1' $ = Area(AEF) and the deficiency in the lower part is V_2' = Area(BGF). Now, if $V_1' = V_2'$, we have $\varepsilon_0 L_0 = V_D$ and the continuous phase will simply redistribute so as to attain a final equilibrium profile given

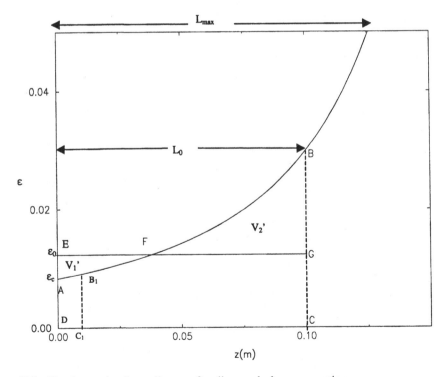

FIG. 12   A pseudo-phase diagram for dispersed phase separation.

by curve(AFB). However, if $V_1' < V_2'$, the lower part (FG) requires more continuous phase than can be provided by the upper part (EF), so that ε at the top falls below ε_c and the disperse phase separates out. This causes the length of the foam to decrease and the final profile is given by curve(BFB1). In this case, the change in length is given by length(DC1).

The discussion in the above sections can be summarized as follows: when $L_0 < L_{max}$:

$L_0 \varepsilon_0 < V_D \Rightarrow$ separation of the dispersed phase gas
$\varepsilon_0 L_0 > V_c \Rightarrow$ separation of the continuous-phase liquid.

If $V_D < \varepsilon_0 L_0 < V_c$, equilibrium is established before ε at either end can reach ε_c or ε_b, and no phase separation takes place.

(c) Separation of Both Phases. As mentioned earlier, when both phases separate, the length of the column at equilibrium is fixed and given by L_{max}. Because the length of a foam column cannot increase, separation of both phases can occur only if $L_0 > L_{max}$. However, $L_0 > L_{max}$ does not necessarily imply that both phases separate. Using a reasoning similar to that used above, it is possible to determine the conditions under which both phases separate.

The volume of continuous phase per unit cross-section in a foam in equilibrium with two phases (V_{ceq}) is given by (see Fig. 13):

$$V_{ceq} = \frac{\sigma}{K\rho g R_0} \left[\tan^{-1}\left(\sqrt{\frac{1 - \varepsilon_c}{\varepsilon_c}} \right) - \tan^{-1}\left(\sqrt{\frac{1 - \varepsilon_b}{\varepsilon_b}} \right) \right] \qquad (112)$$

The volume per unit cross-section of the dispersed phase (V_{Deq}) in this situation is given by:

$$V_{Deq} = L_{max} - V_{ceq} \qquad (113)$$

that is,

$$V_{Deq} = \frac{\sigma}{K\rho g R_0} \left[\sqrt{\frac{1 - \varepsilon_c}{\varepsilon_c}} - \sqrt{\frac{1 - \varepsilon_b}{\varepsilon_b}} - \tan^{-1}\left(\sqrt{\frac{1 - \varepsilon_c}{\varepsilon_c}} \right) \right.$$
$$\left. + \tan^{-1}\left(\sqrt{\frac{1 - \varepsilon_b}{\varepsilon_b}} \right) \right] \qquad (114)$$

Clearly, V_{ceq} and V_{Deq} are fixed for given R_0, σ, and ρ. Thus, if the volumes per unit cross-section of the liquid and gas in the initial foam exceed V_{ceq} and V_{Deq}, respectively (i.e., if $\varepsilon_0 L_0 > V_{ceq}$ and $(1-\varepsilon_0)L_0 > V_{Deq}$), both phases separate out. If any one of these conditions is not satisfied, only one phase will separate. The final length will then be less than L_{max} and the situation is the same as that discussed earlier.

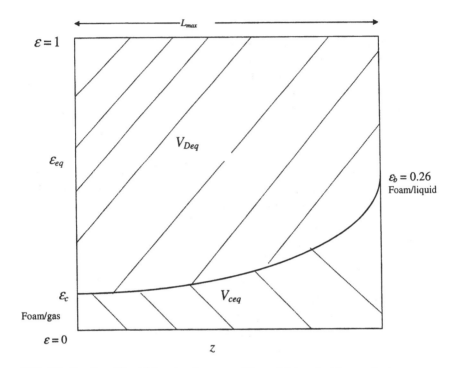

FIG. 13 Profile of liquid fraction in an equilibrated foam that has undergone some collapse and drainage.

(d)   A Generalized "Phase" Diagram.   The condition for the separation of the continuous phase can be written as:

$$\varepsilon_0 L_0 > V_c \text{ for } L_0 < L_{max} \tag{115}$$

$$\varepsilon_0 L_0 > V_{ceq} \text{ for } L_0 > L_{max} \tag{116}$$

Using Eq. (107), Eq. (115) can be written as:

$$\varepsilon_0 > \frac{\sigma}{K\rho g R_0 L_0} \left[\tan^{-1}\left(\sqrt{\frac{1-\varepsilon_b}{\varepsilon_b} + \frac{K\rho g R_0 L_0}{\sigma}} \right) - \tan^{-1}\left(\sqrt{\frac{1-\varepsilon_b}{\varepsilon_b}} \right) \right] \tag{117}$$

and using Eq. (112), Eq. (116) can be written as:

$$\varepsilon_0 > \frac{\sigma}{K\rho g R_0 L_0} \left[\tan^{-1}\left(\sqrt{\frac{1-\varepsilon_c}{\varepsilon_c}} \right) - \tan^{-1}\left(\sqrt{\frac{1-\varepsilon_b}{\varepsilon_b}} \right) \right] \tag{118}$$

Similarly, using Eqs. (111) and (114), the condition for dispersed phase gas separation (collapse) can be written as:

$$\varepsilon_0 < \frac{\sigma}{K\rho g R_0 L_0}\left[\tan^{-1}\left(\sqrt{\frac{1-\varepsilon_c}{\varepsilon_c}}\right) - \tan^{-1}\left(\sqrt{\frac{1-\varepsilon_c}{\varepsilon_c} - \frac{K\rho g R_0 L_0}{\sigma}}\right)\right]$$

for $L_0 < L_{max}$ (119)

$$\varepsilon_0 < 1 - \frac{\sigma}{K\rho g R_0 L_0}\left[\sqrt{\frac{1-\varepsilon_c}{\varepsilon_c}} - \sqrt{\frac{1-\varepsilon_b}{\varepsilon_b}} - \tan^{-1}\left(\sqrt{\frac{1-\varepsilon_c}{\varepsilon_c}}\right)\right.$$

$$\left. + \tan^{-1}\left(\sqrt{\frac{1-\varepsilon_b}{\varepsilon_b}}\right)\right] \text{ for } L_0 > L_{max}$$

(120)

Let us define a dimensionless number $P \equiv \frac{\sigma}{K\rho g R_0 L_0}$. The above conditions can then be written in terms of P as follows.

Continuous-phase liquid separates when:

$$\varepsilon_0 > P\left[\tan^{-1}\left(\sqrt{\frac{1-\varepsilon_b}{\varepsilon_b}} + \frac{1}{P}\right) - \tan^{-1}\left(\sqrt{\frac{1-\varepsilon_b}{\varepsilon_b}}\right)\right] \text{ for } L_0 < L_{max}$$

(121)

$$\varepsilon_0 > P\left[\tan^{-1}\left(\sqrt{\frac{1-\varepsilon_c}{\varepsilon_c}}\right) - \tan^{-1}\left(\sqrt{\frac{1-\varepsilon_b}{\varepsilon_b}}\right)\right] \text{ for } L_0 > L_{max}$$ (122)

Dispersed phase gas separates when:

$$\varepsilon_0 < P\left[\tan^{-1}\left(\sqrt{\frac{1-\varepsilon_c}{\varepsilon_c}}\right) - \tan^{-1}\left(\sqrt{\frac{1-\varepsilon_c}{\varepsilon_c}} - \frac{1}{P}\right)\right] \text{ for } L_0 < L_{max}$$

(123)

$$\varepsilon_0 < 1 - P\left[\sqrt{\frac{1-\varepsilon_c}{\varepsilon_c}} - \sqrt{\frac{1-\varepsilon_b}{\varepsilon_b}} - \tan^{-1}\left(\sqrt{\frac{1-\varepsilon_c}{\varepsilon_c}}\right) + \tan^{-1}\left(\sqrt{\frac{1-\varepsilon_b}{\varepsilon_b}}\right)\right]$$

for $L_0 > L_{max}$ (124)

It can be seen from Eqs. (121)–(124) that for a given ε_0, the occurrence of phase separation is determined entirely by the value of P. Fig. 14 shows the right-hand sides of Eqs. (121)–(124) plotted versus P for $\varepsilon_c = 0.00826$. The curves AB (Eq. (124)) and BC (Eq. (123)) show the lower limit of ε_0, below which foam collapse occurs, and curves DB (Eq. (122)) and BE (Eq. (121)) represent the upper limit for ε_0, beyond which the continuous phase separates. Point B corresponds to $L_0 = L_{max}$. At this point, there is no phase

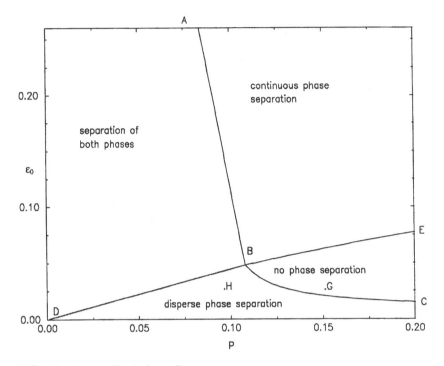

FIG. 14 A generalized phase diagram.

separation and at equilibrium, $\varepsilon = \varepsilon_b$ at the bottom and $\varepsilon = \varepsilon_c$ at the top. The region ABD corresponds to those initial conditions ($L_0 > L_{max}$) for which both phases separate. Fig. 14 is also useful in understanding the effect of initial column height (L_0) and bubble radius (R_0) on the stability of a foam. The influence of these two parameters is determined entirely by their effect on P. For a given pair of fluids, we have $P \propto 1/L_0R_0$. Consider a foam with $\varepsilon_0 = 0.025$. Now if its length and bubble size are such that the initial condition is represented by H, it is unstable. However, if its length and/or bubble size is decreased so that it is characterized by point G, it is stable and no phase separation will take place.

Pneumatic foams will be considered next.

I. Pneumatic Foams

1. Drainage and Collapse During Bubbling

Fig. 15 shows a schematic diagram of a typical experimental setup used to produce foam by bubbling. An inert gas is bubbled at a fixed volumetric flow rate through a porous frit into the surfactant solution. As the foam that is formed moves up, drainage of liquid occurs from the upper to the lower

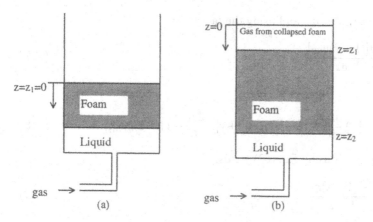

FIG. 15 Schematic of foam being generated by bubbling: (a) no collapse, and (b) collapse occurs during generation.

portions of the foam and by the time the gas supply is stopped, a profile of liquid fraction (ε) develops. The upper portions of the foam are drier (lower ε) because more drainage has taken place there. In order to compute this profile, the drainage equations need to be solved during foam generation. The basic equations are the same as those presented earlier. However, as the foam moves up, an observer at the origin sees the foam/liquid interface moving away from him (i.e., z_2 increases) at a rate that depends on the superficial gas velocity. Thus, Eq. (71) is not valid during bubbling and a new expression needs to be formulated.

Depending on the collapse occurring at the top of the foam, part of the gas resides in the foam, whereas the rest escapes from the top due to collapse. Because we have defined the reference plane to be such that all the escaped gas resides between it and the foam/gas interface (i.e., between $z = 0$ and $z = z_1$), a mass balance for the bubbling gas can be written as:

$$GA = \frac{d}{dt} \int_{z_1}^{z_2} A(1 - \varepsilon)dz + A \frac{dz_1}{dt} \tag{125}$$

In Eq. (125), A is the cross-sectional area and G is the superficial gas velocity. The first term represents the gas in the foam, whereas the second term corresponds to the gas escaped from the collapsed bubbles. Equation (125) can be simplified to obtain:

$$G = \frac{dz_2}{dt} - \frac{d}{dt} \int_{z_1}^{z_2} \varepsilon dz \tag{126}$$

Applying Leibnitz's rule and using the fact that:

$$\frac{\partial \varepsilon}{\partial t} = -\frac{\partial[(1-\varepsilon)q_{PB}]}{\partial z} \tag{127}$$

we have:

$$G = (1-\varepsilon|_{z_2})\frac{dz_2}{dt} + (1-\varepsilon)q_{PB}|_{z_2} + \varepsilon|_{z_1}\frac{dz_1}{dt} - (1-\varepsilon)q_{PB}|_{z_1} \tag{128}$$

Eliminating dz_1/dt using Eq. (62) yields:

$$\frac{dz_2}{dt} = \frac{G - (1-\varepsilon)q_{PB}}{1-\varepsilon}\bigg|_{z_2} \tag{129}$$

When the gas supply is shut off, $G = 0$ and Eq. (129) reduces to $dz_2/dt = -q_{PB}|_{z_2}$ which is the same as Eq. (71). As long as there is no collapse at the top during generation, the motion of the foam/gas interface is determined by the bulk movement of the foam. However, as soon as the thickness of the liquid films at the top decreases to a critical value, the foam begins to collapse and a downward component is superimposed on the bulk movement of the foam. As a result, the net upward velocity of the foam/gas interface decreases. With less stable foams, it is possible to arrive at a steady state when the rate of foam collapse at the top becomes equal to the rate of foam generation (i.e., $dz_1/dt = dz_2/dt$) and the foam length (z_2-z_1) does not change with time. This steady state has been used to characterize the stability of short-lived foams.

In the sections that follow, some results are presented for pneumatic foams. These results are reproduced from our papers and reflect the improvements with time. We first present results for foam drainage where there is no collapse. Here [35] the films are neglected and a comparison is made with some experimental data [69] and the results from the quasi-steady-state model for foam drainage proposed by Narsimhan [32]. Results are then presented for the model in which film drainage and rupture are accounted for but coalescence and the change in surfactant concentration are ignored. The effect of various parameters on the steady-state height is examined.

2. Simulation Results for Pneumatic Foams

Narsimhan [32] has used a quasi-steady-state model to compute the initial liquid fraction in a pneumatic foam. Instead of solving the drainage equations using moving boundaries as we have done, he computed the distribution of liquid by assuming that the downward flow rate of liquid per unit cross-section due to gravity drainage (q_{PB}) is compensated by the liquid entrained by the rising bubbles ($q_B = Gn_pa_pl/V$), so that quasi-steady-state material balances can be written along the length of the foam. The quasi-steady-state balance is written as:

$$\frac{d}{dz}\left(\frac{Gn_p a_p l}{V}\right) = \frac{d}{dz}(q_{PB})$$

(130)

This ordinary differential equation is solved using the boundary conditions:

$$u = 0 \text{ for } z = z_1 = 0; \text{ at } \varepsilon = 0.26 \text{ for } z = z_2$$

(131)

Presented below are comparisons of our results with those from the quasi-steady-state model.

Fig. 16 shows a comparison of the profiles of the initial liquid fraction (profile of ε when the gas supply is shut off) obtained from the two models for $L_0 = 5$ cm and $G = 10^{-3}$ m/s. It is clear that the profiles are qualitatively different. The total amount of liquid in the foam per unit cross-section (V_0) when the gas supply is just shut off (i.e., when $z_2 - z_1 = L_0$) is the area under the ε-versus-z curve and is given by:

$$V_0 = \int_0^{L_0} \varepsilon dz$$

(132)

It is clear that the quasi-steady-state model predicts a higher V_0. Fig. 17 shows the evolution of the fractional change in foam height with time t' for the above system. Here t' is the time elapsed after the gas supply is shut off. Note that the curves flatten out, indicating that a drainage equilibrium is established. When the quasi-steady-state model is used to compute the initial distribution, the change in the height is about 70% greater than that obtained from our model. Because the two models differ only in the way the initial distribution is computed, it is evident that the initial distribution strongly affects the drainage process. Figs. 18–20 show comparisons of the theoretical predictions of the two models with the experimental data of Germick et al. [69]. Fig. 18 shows the effect of bubble size, whereas Fig. 19 shows the results for three viscosities. It is clear that the unsteady-state model shows better agreement with the data. Fig. 20 shows the effect of initial foam height (L_0) on the drainage. Our model shows reasonable agreement for the smallest height ($L_0 = 10.61$ cm). However, the theoretical drainage curves do not change to the same extent with height as the experimental curves. Before we proceed further, some qualitative features need to be noted in the figures. The extent and the rate of drainage are greater when the bubble size is smaller and the viscosity is higher. With smaller bubble sizes (i.e., smaller PB cross-sectional area) and higher viscosities, the downward fluid velocity is smaller. This means that a smaller amount of liquid drains out of the foam during bubbling. A greater amount of liquid is therefore present in the foam when the gas

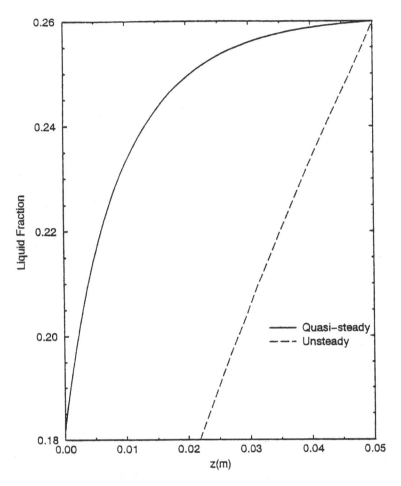

FIG. 16 Initial liquid fraction profiles for $L_0 = 5$ cm, $\mu = 1$ cP, $R = 0.2$ mm, $G = 0.001$ m/s, $\rho = 10^3$ kg/m^3, $\mu_s = \infty$, and $\sigma = 50$ mN/m.

supply is shut off. This in turn means that the foam is farther from equilibrium when the gas supply is shut off. The change in foam height is therefore larger. It must be emphasized here that film drainage was ignored [35] in obtaining the curves shown in Figs. 18–20.

Film drainage was incorporated into the model in a subsequent article [37]. This, however, brought to the fore a new issue regarding the boundary conditions at the foam/liquid interface during foam formation.

In foams formed by bubbling, new bubbles are continuously introduced into the system at the foam/liquid interface. Thus, when the foam is being generated, the boundary condition at the bottom is a statement about the

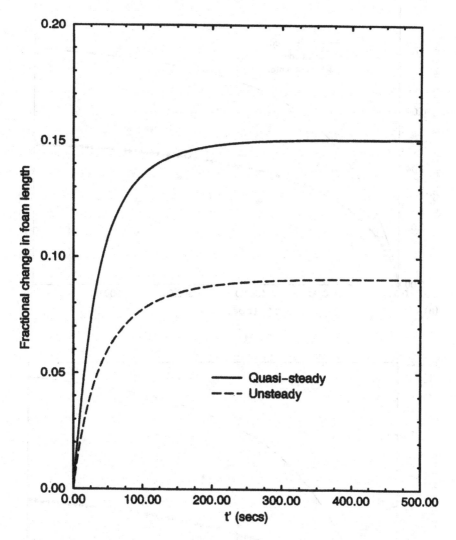

FIG. 17 Variation of the fractional change in foam length with time (t') for the system of Fig. 16.

condition of the bubbles as they enter the foam. When films were neglected [35,36], the boundary condition at the bottom was simply the condition for close-packed spheres ($\varepsilon = 0.26$). This was sufficient because there was only one variable (ε). However, with the inclusion of films into the model, there are two independent variables (r_p and x_F) and information is needed about the distribution of liquid between the films and the Plateau border channels. Because the differential equation for x_F is first order in time, an initial con-

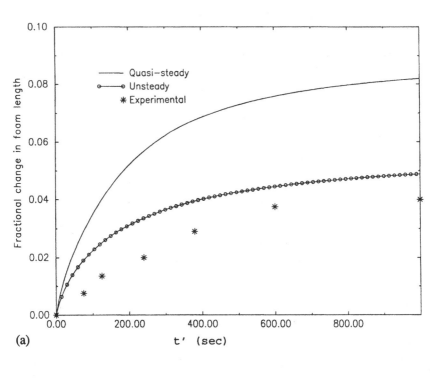

(a)

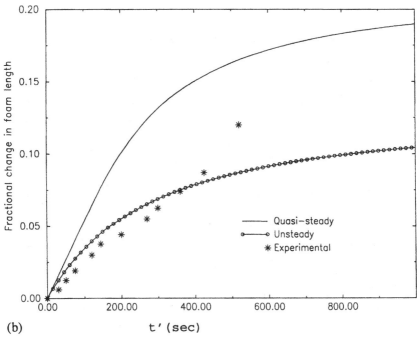

(b)

dition is needed. In other words, the value of x_F at the foam/liquid interface during bubbling (x_{F0}) is required. We have no physical arguments to specify the film thickness. In order to identify the effect of its value, we carried out simulations with several values of x_{F0}. It was found that this value has practically no effect on the drainage [37] because, for thicker films, most of the drainage takes place within a very short distance from the bottom. Thus, as long as the foam is being generated (gas is being pumped in), the condition at the bottom is:

$$x_F|_{z_2} = x_{F0} = \text{constant} \tag{133}$$

Once the gas supply is stopped, the films at the bottom simply obey the Reynolds equation, with Eq. (133) as the initial condition. For the present calculations, x_{F0} was taken to be 500 nm.

The results of some simulations done to examine the effect of various parameters on foam collapse are discussed below.

(a) Effect of Superficial Gas Velocity on the Steady-State Height. It is practical to use the steady-state height as a measure of foam stability only for relatively short-lived foams [37]. For more stable foams, it is possible that the height of the foam when collapse starts will be too large to be conveniently measured in a laboratory. It is also evident that the superficial gas velocity used to generate the foam will have a significant impact on the steady-state height. Fig. 21 shows the steady-state height (H_0) as a function of the superficial gas velocity (G). It is clear that the steady-state height increases with G. The slope of the curve increases and seems to become unbounded for $G = 0.00035$ m/s, indicating that there is an upper limit on the superficial gas velocity beyond which a steady-state foam height will never be achieved. Some explanation for this can be provided as follows. It can be inferred from Eqs. (60) and (62) that the rate of foam collapse is determined by $q_{PB}|_{z_1}$. Because the film thickness at the collapsing front is fixed, q_{PB} is determined primarily by the value of the Plateau border radius at the top. However, because collapse can occur only if the capillary pressure at the top exceeds the maximum disjoining pressure, there is an upper limit to the Plateau border radius at the top given by:

$$r_{pmax} = \frac{\sigma}{\Pi_{max}} \tag{134}$$

FIG. 18 Fractional change in foam length $[(L_0 - L(t'))/L_0]$ versus time (t') for two bubble sizes with $G = 0.0578$ cm/s, $L_0 = 19.8$ cm, $\mu = 1$ cP, $\rho = 1000$ kg/m^3, and $\mu_s = \infty$: (a) $R = 0.153$ mm, and (b) $R = 0.203$ mm.

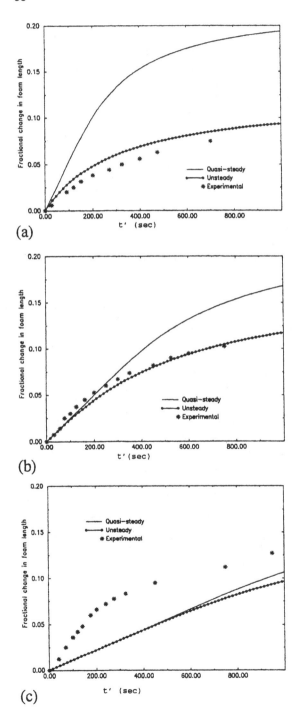

(a)

(b)

(c)

In other words, there is an upper limit on the value of $q_{PB}|_{z_1}$, given by:

$$q_{PBmax} = \frac{3}{15} Nn_p a_p R_0 \left. \frac{a_p \rho g}{20\sqrt{3}\mu} \right|_{r_p = r_{pmax}, x_F = x_{Fc}} \tag{135}$$

This means that there is an upper limit to the value of dz_1/dt, given by:

$$\frac{dz_1}{dt} = \frac{q_{PB \; max}\left(1 - \varepsilon|_{r_p = r_{pmax}, x_F = x_{Fc}}\right)}{\varepsilon|_{r_p = r_{pmax}, x_F = x_{Fc}}} \tag{136}$$

Thus, for values of G for which dz_2/dt (Eq. (129)) is larger than the maximum value given by Eq. (136), dz_2/dt will always be larger than dz_1/dt and a steady-state height [for which $(dz_1/dt) = (dz_2/dt)$] will never be achieved.

(b) *Effect of Electrolyte Concentration on Foam Collapse.*   The concentration of electrolytes in the surfactant solution plays a crucial role in determining the stability of a foam due to its effect on the repulsive electrical double-layer forces in a film [37]. The primary effect of an increase in electrolyte concentration is to compress the electrical double layer (i.e., the Debye length decreases). The lower the value of Π_{max} is, the smaller is the capillary pressure required to cause collapse. On the other hand, the smaller the critical thickness is, the larger is the residence time required for the film to become critical. Figs. 22 and 23 show the variation of Π_{max} and the critical thickness (x_{Fc}) with electrolyte concentration. At small concentrations, the double layer is not too compressed. The maximum of the disjoining pressure therefore corresponds to large film thicknesses for which the van der Waals forces are not very large. An increase in the electrolyte concentration therefore raises Π_{DL} without affecting Π_{VDW} much and raises Π_{max}. At higher electrolyte concentrations, the double layer is compressed and Π_{max} corresponds to very small thicknesses at which the van der Waals force is significant. Π_{max} now decreases with an increase in electrolyte concentration because Π_{VDW} rises sharply with a decrease in the film thickness. Π_{max} therefore increases, attains a maximum, and then decreases as the electrolyte concentration increases. The position of this maximum, however, moves to smaller film thicknesses. This means that the critical thickness decreases with an increase in electrolyte concentration. It is therefore reasonable to expect that as the electrolyte con-

---

FIG. 19   Fractional change in foam height $[(L_0 - L(t'))/L_0]$ for three viscosities with $L_0 = 18.67$ cm, $G = 0.0578$ cm/s, $\rho = 1000$ kg/m^3, and $\mu_s = \infty$: (a) $\mu = 1.3$ cP, $\sigma = 57.6$ dyn/cm, and $R = 0.171$ mm; (b) $\mu = 3.02$ cP, $\sigma = 55.8$ dyn/cm, and $R = 0.179$ mm; and (c) $\mu = 6.28$ cP, $\sigma = 54.2$ dyn/cm, and $R = 0.203$ mm.

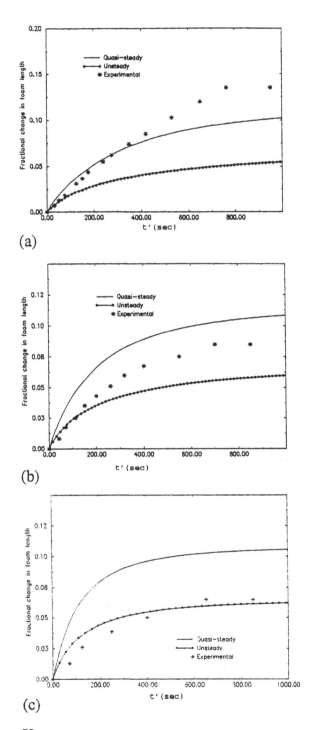

(a)

(b)

(c)

58

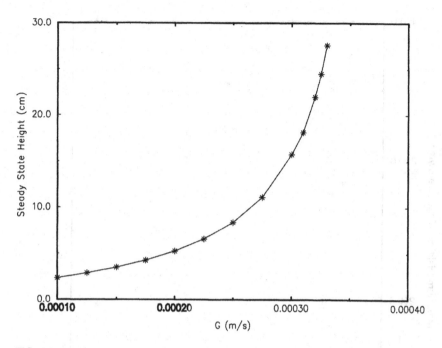

FIG. 21 Effect of superficial gas velocity (G) on the steady-state height. The values of the parameters used are: $\rho = 1000 \, \text{kg/m}^3$, $\mu = 1 \, \text{cP}$, $\psi_s = 18 \, \text{mV}$, $R = 0.2 \, \text{mm}$, $T = 298$ K, $A_h = 3.7 \times 10^{-20} \text{J}$, $c_{el} = 0.001$ M, $\Pi_{max} = 227.23 \, \text{N/m}^2$, and $\sigma = 40 \, \text{mN/m}$.

centration increases, the stability of the foam as quantified by the steady-state height (H_0) increases at first, achieves a maximum, and then decreases. Fig. 24, which presents the effect of electrolyte concentration on H_0, shows that this, indeed, is the case. Fig. 25 shows the effect of electrolyte concentration on the time required for a foam to collapse to half the steady-state height. It is clear that the trend is the same as that for the steady-state height. It has been observed experimentally [70] that plots of $(z_2 - z_1)/H_0$ versus $\log(t/t_{1/2})$ practically coincide and, at lower salt concentrations, are linear. To check if our model predicts this feature (see Fig. 26), such plots were generated for different

FIG. 20 Fractional change in foam length $[(L_0 - L(t'))/L_0]$ for three values of L_0 with $\mu = 0.9632 \, \text{cP}$, $\sigma = 59.4 \, \text{dyn/cm}$, $G = 0.0375 \, \text{cm/s}$, $\rho = 1000 \, \text{kg/m}^3$, and $\mu_s = \infty$: (a) $L_0 = 22.78$ cm and $R = 0.203$ mm; (b) $L_0 = 16.27$ cm and $R = 0.198$ mm; and (c) $L_0 = 10.61$ cm and $R = 0.193$ mm.

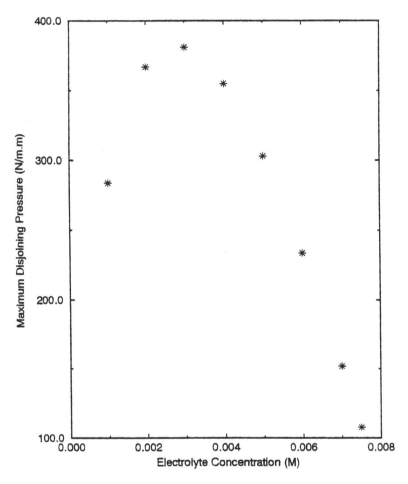

FIG. 22 Effect of electrolyte concentration (c_{el}) on the maximum disjoining pressure (Π_{max}). The values of the other parameters used are: $\psi_s = 19$ mV, $T = 298$ K, and $A_h = 3.7 \times 10^{-20}$ J.

values of the bubble radius. Although the curves are not linear, they seem to coincide quite well. The shape of the curve is more similar to the data provided in the above paper at higher concentrations (labeled by these authors as "not-so-nice data"). The slope of the linear portion of the theoretical curves is reasonably close to the universal experimental line provided by these authors. A more direct comparison with the above experiments is not possible because there is no information on the bubble size used in the above experiments.

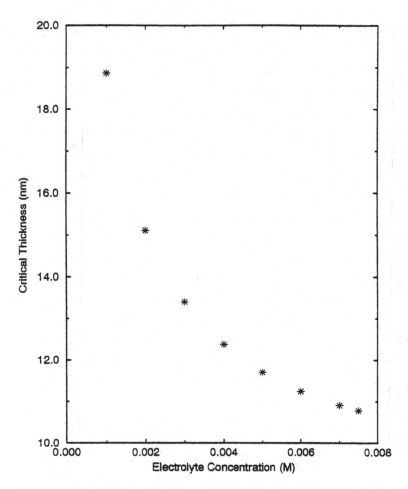

FIG. 23 Effect of electrolyte concentration (c_{el}) on critical thickness (x_{Fc}) for the system of Fig. 22.

The concentration of the electrolyte also plays an important role in determining whether the foam collapses completely, or arrives at a drainage equilibrium. Fig. 27 shows the variation in foam height with time for two systems that differ only in the electrolyte concentration. The initial foam height is 4 cm and the values of the other parameters used are: $\rho = 1000$ kg/m^3, $\mu = 1$ cP, $\psi_s = 19$ mV, $R_0 = 0.2$ mm, $A_h = 3.7 \times 10^{-20}$, $\sigma = 40$ mN/m, and $G = 0.0005$ m/s. The foam with $c_{el} = 0.003$ M equilibrates at a height of about 1.85 cm, whereas the foam with $c_{el} = 0.007$ M collapses completely. This can be ex-

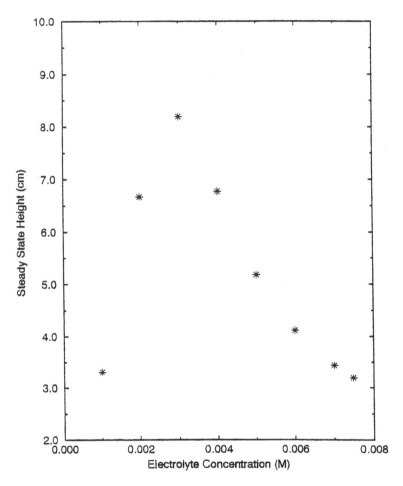

FIG. 24 Effect of the electrolyte concentration on the steady-state height. The values of the parameters used are: $\rho = 1000\ \mathrm{kg/m^3}$, $\mu = 1\ \mathrm{cP}$, $\psi_s = 19\ \mathrm{mV}$, $R = 0.2\ \mathrm{mm}$, $T = 298\ \mathrm{K}$, $A_h = 3.7\times10^{-20}\ \mathrm{J}$, $\sigma = 40\ \mathrm{mN/m}$, and $G = 0.0001\ \mathrm{m/s}$.

plained as follows: The capillary pressure has its smallest value (σ/R_0) at the bottom of the foam, which in this case is 200 $\mathrm{N/m^2}$. For $c_{el} = 0.003$ M, $\Pi_{max} = 380.85\ \mathrm{N/m^2}$, whereas for $c_{el} = 0.007$ M, it is about 150 $\mathrm{N/m^2}$. Because in the latter case the smallest capillary pressure possible in the system exceeds the maximum disjoining pressure, the foam collapses completely.

(c) Effect of Bubble Size. Fig. 28 shows the effect of bubble radius (R_0) on H_0 for a system with the following parameters: $\psi_s = 19\ \mathrm{mV}$, $\sigma = 40\ \mathrm{mN/m}$,

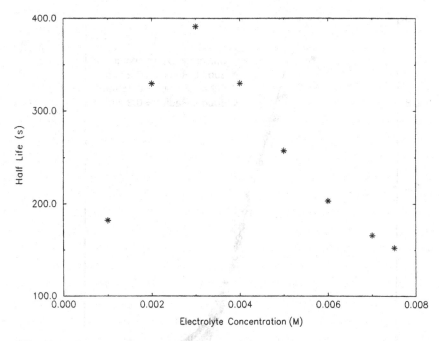

FIG. 25 Effect of electrolyte concentration on collapse half-life for the system of Fig. 24.

$G = 0.0001$ m/s, $c_{el} = 0.005$ M, and $T = 298$ K. Clearly, H_0 increases with bubble size. This can be explained as follows: In general, in a foam column draining under gravity, the curvature of the Plateau border channels (r_p) is the largest and the capillary pressure is the smallest at the foam/liquid interface where $r_p = R_0$. For a given Π_{max}, the smaller the bubble size is, the smaller is the decrease in r_p required for the capillary pressure to rise to Π_{max}. This means that collapse starts at smaller heights, resulting in smaller values of H_0. Fig. 29 shows a plot of $z_2 - z_1$ versus time for two bubble sizes. It is clear that there is a qualitative difference because the foam with $R_0 = 0.3$ mm collapses to reach an equilibrium height, whereas the foam with $R_0 = 0.125$ mm collapses completely. This is because (σ/R_0) is less than Π_{max} for the former and greater than Π_{max} for the latter.

J. Coalescence in Foam

In earlier sections, bubble coalescence was ignored under the assumption that all films at a given level drain at the same rate and hence rupture at the same time [38]. The reality, however, is that there is a time lapse between the onset of

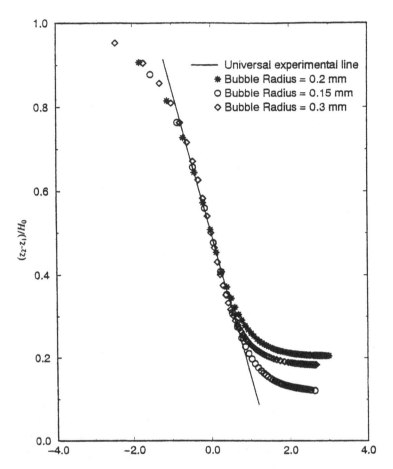

FIG. 26 Effect of bubble size on the dimensionless plots of $(z_2 - z_1)/H_0$ versus $\log(t/t_{1/2})$. The values of parameters are the same as in Fig. 24.

film rupture and the beginning of foam collapse. Some films at a given level rupture earlier than others. This triggers the coalescence of neighboring bubbles sharing these films and leads to an increase in bubble size. Foam collapse occurs only when all films at the top of the foam rupture. Thus, earlier expressions for foam collapse are likely to be valid only when the time interval between the start of film rupture and the onset of foam collapse is short. It is reasonable to expect that coalescence is a consequence of the lack of homogeneity of the films at any level. This could be due to a difference in initial thicknesses and the amount of surfactant adsorbed at the film surfaces or

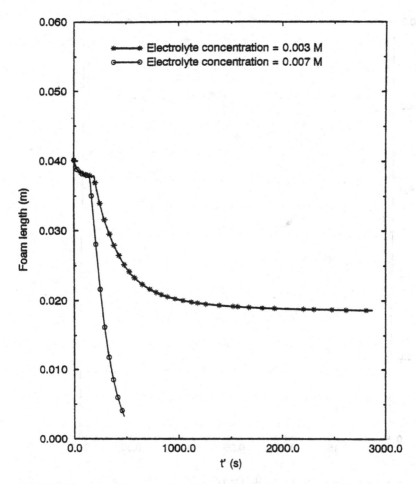

FIG. 27 Variation of foam height with time for a foam with an initial length of 4 cm and $G = 0.0005$ m/s. With $c = 0.007$ M, complete collapse occurs, whereas with $c_{el} = 0.003$ M, the foam achieves an equilibrium height. The other parameters are the same as in Fig. 24.

film size. We hypothesize that the main cause of coalescence is nonuniformity of film areas. Even in a monodisperse foam, the faces of the polyhedral bubbles are not identical [13] and there is a nonuniformity in the film areas ($A_F = \pi R_F^2$). This causes a nonuniformity of film thicknesses because the rate of film thinning depends on R_F (see Eq. (1)). For a given capillary pressure (σ/r_p), films with smaller R_F drain faster and are therefore thinner. Thus, if $R_{F\,min} \leq R_F \leq R_{F\,max}$ and $f dR_F$ is the fraction of films in the element that

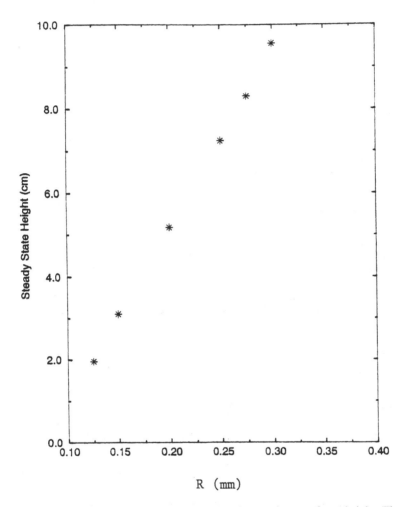

FIG. 28 Effect of the bubble radius R on the steady-state foam height. The values of the parameters used are: $\rho = 1000 \, \text{kg/m}^3$, $\mu = 1 \, \text{cP}$, $\psi_s = 19 \, \text{mV}$, $T = 298 \, \text{K}$, $A_h = 3.7 \times 10^{-20} \, \text{J}$, $\sigma = 40 \, \text{mN/m}$, $c_{el} = 0.005 \, \text{M}$, and $G = 0.0001 \, \text{m/s}$.

have radii between R_F and $R_F + dR_F$, the mean film thickness ($\overline{x_F}$) and the mean film area ($\overline{A_F}$) are given by [39]:

$$\overline{x_F} = \int_{R_{Fmin}}^{R_{Fmax}} x_F f \, dR_F; \quad \overline{A_F} = \int_{R_{Fmin}}^{R_{Fmax}} A_F f \, dR_F = \int_{R_{Fmin}}^{R_{Fmax}} \pi R_F^2 f \, dR_F \quad (137)$$

In an initially monodispersed foam, the bubble size is uniform within an element (i.e., all bubbles within the element have the same radius (R) and

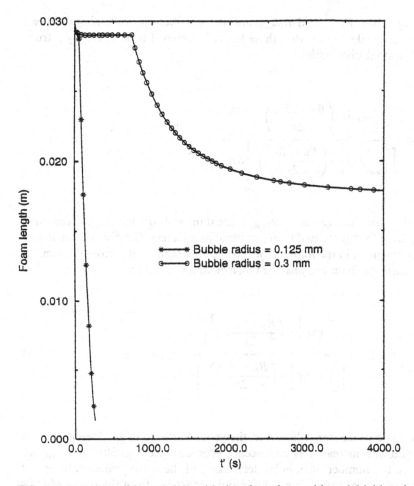

FIG. 29 Variation of foam height with time for a foam with an initial length of 3 cm. With $R = 0.125$ mm, complete collapse occurs, whereas with $R = 0.3$ mm, an equilibrium height is attained. The values of the parameters used are: $\rho = 1000$ kg/m^3, $\mu = 1$ cP, $\psi_s = 19$ mV, $T = 298$ K, $A_h = 3.7 \times 10^{-20}$ J, $\sigma = 40$ mN/m, $c_{el} = 0.005$ M, and $G = 0.0005$ m/s.

$\overline{R} = R$). However, once films start rupturing, coalescence starts and gives rise to a size distribution within the element (i.e., $\overline{R} \neq R$). The mean bubble radius ($\overline{R}$) increases because neighboring bubbles coalesce to form a larger bubble. The rate at which $\overline{R}$ increases with time, which in turn depends on the number of films rupturing per unit time, hence depends on the distribution of film areas. The film radii (R_F) are assumed to lie between two limiting values R_{Fmin} and R_{Fmax}, and the distribution of R_F about a mean value $R_{F0} = 0.606R_0$ within these limits is assumed to be given by a truncated normal distribution:

$$f = \frac{\exp\left[-\left(\dfrac{R_F - R_{F0}}{2s}\right)^2\right]}{\displaystyle\int_{R_{Fmin}}^{R_{Fmax}} \exp\left[-\left(\dfrac{R_F - R_{F0}}{2s}\right)^2\right] dR_F} \tag{138}$$

Before coalescence starts, R_{Fmin} is constant and equal to R_{Fmin0} (the value of R_{Fmin} in freshly formed foam; an input parameter). The distribution in any volume element in the foam before coalescence starts is therefore the same as that when the foam has just been formed and is given by:

$$f_0 = f|_{t=0} = \frac{\exp\left[-\left(\dfrac{R_F - R_{F0}}{2s}\right)^2\right]}{\displaystyle\int_{R_{Fmin0}}^{R_{Fmax}} \exp\left[-\left(\dfrac{R_F - R_{F0}}{2s}\right)^2\right] dR_F} \tag{139}$$

Once coalescence starts, however, the value of $R_{F\,min}$ in an element is not a constant and increases as coalescence proceeds (see Fig. 30). As R_{Fmin} increases, the number of bubbles decreases and the mean bubble volume ($\overline{V}$) increases. At the same time, liquids and surfactants from the ruptured films are distributed among the remaining Plateau border channels. It may be noted that we assume R_{Fmax} to be a constant. This is not strictly true. It is possible that geometrical rearrangements occurring during coalescence will give rise to films larger than R_{Fmax}. It may also be noted that f and R_{Fmin}, and hence the mean bubble volume, are functions of only the vertical position (z) and time (t). This is because if wall effects are neglected, the capillary pressure (σ/r_p) at a given level (z) is independent of the radial position and any volume element at a given level will have the same distribution of film radii. On the other hand, at any time t and at each z, we need to know the film thickness x_F corresponding to each R_F so that we can determine when coalescence starts at each level and how quickly coalescence proceeds. Thus, x_F is now a function of R_F in addition to t and z. Thus, coalescence gives rise to an additional moving boundary

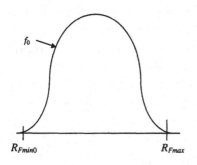

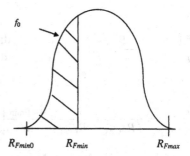

(a) Before coalescence

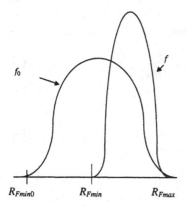

(b) After coalescence

FIG. 30 Distribution of film radii before and after coalescence.

in R_F space. The boundary in R_F space is simplified and can be immobilized by carrying out a transformation from R_F space to r space where:

$$r = \frac{R_F - R_{Fmin}}{R_{Fmax} - R_{Fmin}} \tag{140}$$

With this transformation, the variables ξ and r lie between 0 and 1, and the Reynolds equation (Eq. (1)) can be written in t,r,ξ space as:

$$\left.\frac{dx_F}{dt}\right|_{R_F,z} = -V_f = \left.\frac{\partial x_F}{\partial t}\right|_{r,\xi} + \left.\frac{\partial x_F}{\partial r}\right|_{t,\xi}\left.\frac{dr}{dt}\right|_{R_F,z} + \left.\frac{\partial x_F}{\partial \xi}\right|_{t,r}\left.\frac{d\xi}{dt}\right|_{R_F,z} \tag{141}$$

where:

$$\left.\frac{d\xi}{dt}\right|_{R_F,z} = \frac{1}{z_2 - z_1}\left[(\xi - 1)\frac{dz_1}{dt} - \xi\frac{dz_2}{dt}\right] \tag{142}$$

and

$$\left.\frac{dr}{dt}\right|_{R_F,z} = \frac{(r-1)}{(R_{Fmax} - R_{Fmin})}\left.\frac{dR_{Fmin}}{dt}\right|_z$$

$$= \frac{(r-1)}{(R_{Fmax} - R_{Fmin})}\left[\left.\frac{\partial R_{Fmin}}{\partial t}\right|_\xi + \frac{\partial R_{Fmin}}{\partial \xi}\frac{d\xi}{dt}\right] \tag{143}$$

Using Eqs. (142) and (143), Eq. (141) becomes:

$$-V_f = \left.\frac{\partial x_F}{\partial t}\right|_{r,\xi} + b_1\left.\frac{\partial R_{Fmin}}{\partial t}\right|_\xi + b_2\frac{dz_1}{dt} + b_3\frac{dz_2}{dt} \tag{144}$$

where:

$$b_1 = \frac{\partial x_F}{\partial r}\frac{(r-1)}{(R_{Fmax} - R_{Fmin})} \tag{145}$$

$$b_2 = \frac{\xi - 1}{z_2 - z_1}\left[\frac{\partial x_F}{\partial \xi} + \frac{(r-1)}{(R_{Fmax} - R_{Fmin})}\frac{\partial R_{Fmin}}{\partial \xi}\right] \tag{146}$$

and

$$b_3 = \frac{-\xi}{z_2 - z_1}\left[\frac{\partial x_F}{\partial \xi} + \frac{(r-1)}{(R_{Fmax} - R_{Fmin})}\frac{\partial R_{Fmin}}{\partial \xi}\right] \tag{147}$$

It may be noted that Eq. (144) reduces to Eq. (76) when there is no coalescence (i.e., when $\partial R_F/\partial \xi = 0$ and $dR_{Fmin}/dt = 0$). The expression for dR_{Fmin}/dt is derived below.

1. Expression for dR_{Fmin}/dt

Before coalescence starts, the value of R_{Fmin} is constant throughout the foam and is given by R_{Fmin0}, which is an input parameter. Coalescence begins at a given z when the thinnest films at that z (those with $R_F = R_{Fmin0}$) become critical ($x_F = x_{Fc}$) and rupture. Once this happens, R_{Fmin} starts increasing. Because the capillary pressure and the disjoining pressure isotherm are different in different parts of the foam, the degree of coalescence and hence the value of R_{Fmin} will be different in different portions of the foam. Thus, R_{Fmin} is a function of time t and vertical position z. As long as coalescence occurs, R_{Fmin} increases and the rate of increase is obtained by recognizing that x_F $|_{z,R_{Fmin}}$ (i.e., $x_F|_{\xi,r=0}$) always corresponds to the critical thickness (x_{Fc}), which in turn is a function of the surfactant concentration (c_s) at that ξ. Thus, coalescence at a given ξ is characterized by the following conditions:

$$x_F|_{\xi, r=0} = x_{Fc}(c_s|_\xi) \tag{148}$$

that is,

$$-V_f|_{r=0,\xi} = -V_f|_{x_{Fc}(c_s|_\xi)} \tag{149}$$

Differentiating Eq. (148) with respect to time, we get:

$$\left.\frac{\partial x_F}{\partial t}\right|_{\xi,r=0} = \left.\frac{\partial x_{Fc}}{\partial t}\right|_\xi = \left.\frac{dx_{Fc}}{dc_s}\frac{\partial c_s}{\partial t}\right|_\xi \tag{150}$$

Combining Eqs. (144) and (150) for $r = 0$, we get:

$$
\begin{aligned}
-V_f\bigg|_{r=0,\xi} &= -V_f\bigg|_{x_{Fc}(c_s|_\xi)} = \left.\frac{dx_{Fc}}{dc_s}\frac{\partial c_s}{\partial t}\right|_\xi \\
&\quad + b_1\bigg|_{r=0,\xi}\left.\frac{\partial R_{Fmin}}{\partial t}\right|_\xi + b_2\bigg|_{r=0,\xi}\frac{dz_1}{dt} + b_3\bigg|_{r=0,\xi}\frac{dz_2}{dt}
\end{aligned}
\tag{151}
$$

which can be rewritten as:

$$\left.\frac{dR_{Fmin}}{dt}\right|_\xi = \frac{-V_f|_{x_{Fc}(c_s|_\xi)} - \left.\dfrac{dx_{Fc}}{dc_s}\dfrac{\partial c_s}{\partial t}\right|_\xi - b_2|_{r=0,\xi}\dfrac{dz_1}{dt} - b_3|_{r=0,\xi}\dfrac{dz_2}{dt}}{b_1|_{r=0,\xi}} \tag{152}$$

The following features of Eq. (152) need to be noted. The rate (dR_{Fmin}/dt) at which coalescence proceeds at any level depends strongly on the film velocity (V_f) corresponding to the critical thickness (x_{Fc}) at that level. Because $b_1 < 0$, dR_{Fmin}/dt increases as the film velocity corresponding to the critical thickness increases. On the other hand, because (dx_{Fc}/dc_s) < 0, the coalescence rate will decrease if the local surfactant concentration increases (i.e., if ($\partial c_s/\partial t$) > 0).

As mentioned earlier, coalescence gives rise to a bubble size distribution in an initially monodispersed foam. The number of bubbles decreases and the mean bubble volume increases. Owing to the complexity involved in computing the evolution of the bubble size distribution, we restrict our treatment to the calculation of the mean bubble volume $\overline{V}$, which is required in the formulation of the bulk conservation equations. This is considered next.

2. Calculation of the Mean Bubble Volume

Before coalescence begins, $\overline{V}$ is the same throughout the foam and is equal to the volume of the bubble as it is first formed ($V_0 = 4/3\pi R_0^3$). As coalescence starts (i.e., as R_{Fmin} starts increasing), $\overline{V}$ increases because the rupture of a film causes coalescence of the neighboring bubbles that share the film, and the same volume of gas is distributed among a smaller number of bubbles. Consider a certain volume element. If f_0 is the distribution of film radii in this element before coalescence starts (see Eq. (139)), the following relationship is always satisfied:

Number of films in the element at any time t

$$= \text{(Initial number of films in the element)} \int_{R_{Fmin}(t)}^{R_{Fmax}} f_0 dR_F \qquad (153)$$

It may be noted that $\int_{R_{Fmin}(t=0)}^{R_{Fmax}} f_0 dR_F = 1$ and f_0 is only a function of R_F and is independent of time. As films rupture and disappear, $R_{Fmin}(t)$ increases and $\int_{R_{Fmin}(t)}^{R_{Fmax}} f_0 dR_F$ decreases (see Fig. 30). Because the number of bubbles in this element is proportional to the number of films, we have from Eq. (153):

Number of bubbles in the element at any time t

$$= \text{(Initial number of bubbles in the element)} \int_{R_{Fmin}(t)}^{R_{Fmax}} f_0 dR_F. \qquad (154)$$

Also, by definition:

$$\text{Mean bubble volume in element} = \frac{\text{Volume of gas in element}}{\text{Number of bubbles in element}} \qquad (155)$$

Now because the volume of gas in this element remains unchanged, the mean bubble volume in the element at any time is inversely proportional to the number of bubbles in the element. In other words:

$$\frac{\text{Number of bubbles in element before coalescence}}{\text{Number of bubbles in element after coalescence}}$$

$$= \frac{\text{Mean bubble volume after coalescence}}{\text{Initial bubble volume}} \qquad (156)$$

Thus, we have:

$$\frac{\overline{V}(t)}{\overline{V}(t=0)} = \frac{\int_{R_{Fmin}(t=0)}^{R_{Fmax}} f_0 dR_F}{\int_{R_{Fmin}(t)}^{R_{Fmax}} f_0 dR_F} = \frac{1}{\int_{R_{Fmin}(t)}^{R_{Fmax}} f_0 dR_F} \qquad (157)$$

All the equations dealing with coalescence have now been formulated. However, it is clear that the expression for foam collapse (Eq. (81)) derived earlier is no longer valid because it is based on the premise that all films at a given level rupture simultaneously. The expression for foam collapse in the context of coalescence is derived below.

3. Movement of the Foam/Gas Interface in the Presence of Coalescence

When R_{Fmin} at any z becomes equal to R_{Fmax}, all the films at this z are critical and rupture simultaneously. When this happens at the foam/gas interface ($z = z_1$), all the bubbles at the top of the foam collapse and the interface moves downward (i.e., z_1 starts increasing). In other words, the boundary starts moving when the distribution function (f) at the top becomes a Dirac delta function. The moving interface is, therefore, always characterized by $R_{Fmin} = R_{Fmax} = $ constant, that is, $\partial R_{Fmin}/\partial t|_{\xi=0} = 0$. Using this condition in Eq. (152) with $\xi = 0$, we get:

$$0 = \frac{-V_f|_{x_{Fc}(c_s|_{\xi=0})} - \dfrac{dx_{Fc}}{dc_s}\dfrac{\partial c_s}{\partial t}\bigg|_{\xi=0} - b_2|_{r=0,\xi=0}\dfrac{dz_1}{dt} - b_3|_{r=0,\xi=0}\dfrac{dz_2}{dt}}{b_1|_{r=,\xi=0}}$$

(158)

Because $b_3|_{\xi=0} = 0$, Eq. (158) can be rewritten as:

$$\frac{dz_1}{dt} = \frac{-V_f|_{x_{Fc}(c_s|_{\xi=0})} - \dfrac{dx_{Fc}}{dc_s}\dfrac{\partial c_s}{\partial t}\bigg|_{\xi=0}}{b_2|_{r=0,\xi=0}}$$

(159)

Because it is not possible numerically to deal with the Dirac delta function, collapse was deemed to occur when the distribution was narrow enough (i.e., when R_{Fmin} became equal to R_{Fc}, where R_{Fc} is slightly smaller than R_{Fmax} and is given by $R_{Fc} = R_{Fmax}/1.001$).

Some simulation results for foam coalescence and collapse are presented below.

4. Results for Coalescence and Collapse

In this section, results of simulations done to examine the effect of various parameters such as the apparent diffusivity (D), the concentration of surfactant (c_{s0}) and salt (c_e) in the foaming solution, and the initial distribution of film radii (f) will be discussed.

The mean film radius was taken to be $R_{F0} = 0.606R_0$, where R_0 is the bubble radius (taken to be 0.2 mm) in the monodispersed foam before coalescence starts. The standard deviation s was taken to be $0.5R_{F0}$, and R_{Fmin0} and R_{Fc} were taken to be $0.99R_{F0}$ and $1.01R_{F0}$, respectively, except when their effects on the results were examined. This represents a deviation of about 1% from the mean value. These values are arbitrary as we have no way of identifying what the exact distribution is. Unless otherwise specified, the apparent diffusivity is taken as $D = 5 \times 10^{-5}\,\mathrm{m^2/s}$. This value is apparent and is much higher than the molecular diffusivity, which has a value of $10^{-10}\,\mathrm{m^2/s}$. In reality, D will vary throughout the foam depending on the fluid velocity and will be higher in regions where significant coalescence (which results in greater mixing) occurs. The values of the other parameters used were $G = 0.0001\,\mathrm{m/s}$, $c_e = 8\,\mathrm{mol/m^3}$, $R_0 = 0.2\,\mathrm{mm}$, $\mu = 1\,\mathrm{cP}$, $T = 298\,\mathrm{K}$, $c_{s0} = 0.012\,\mathrm{mol/m^3}$, and $K_d = 0.156\,\mathrm{mol/m^3}$.

5. Effect of Diffusivity

To examine the effect of the apparent diffusivity (D) on foam collapse, simulations were carried for two values of D viz. $D = 5 \times 10^{-5}\,\mathrm{m^2/s}$ and $D = 5 \times 10^{-6}\,\mathrm{m^2/s}$. Fig. 31 shows the variation in foam length with time for these two cases. Points A and B indicate the onset of foam collapse. The foam with the higher diffusivity starts collapsing earlier and collapses at a much higher rate. This can be explained as follows: Coalescence leads to a release of surfactant from the ruptured films and hence leads to a local accumulation of surfactant. This local increase of the surfactant concentration tends to stabilize the remaining films because the maximum disjoining pressure increases with surfactant concentration (i.e., a film becomes more stable as the surfactant concentration increases). Further coalescence depends strongly on the rate at which the surfactant is transported away. With low diffusivity, the transport of surfactant away from the region is slower and the local accumulation is higher. The films are therefore stabilized to a greater extent and the rate of coalescence is much smaller. Thus, it takes longer for R_{Fmin} to approach R_{Fmax} and foam collapse starts much later. A similar explanation applies to the rate of foam collapse. A higher D means that the surfactant released due to collapse quickly diffuses away without significantly stabilizing the films and the foam collapses at a much faster rate. Fig. 32 shows a comparison of the surfactant concentration profiles at different times. The rise in surfactant concentration in the upper levels of the foam is primarily due to coalescence and collapse. However, the surfactant concentration throughout the foam is always higher when the diffusivity is smaller. Fig. 33 shows the profiles of the dimensionless bubble radius $\bar{R}/R_0$ at three different times: when the gas supply has just been turned off ($t' = 0$), 10 min later ($t' = 600\,\mathrm{s}$), and 30 min later ($t' = 1800\,\mathrm{s}$). Here t' represents the time elapsed after the gas supply is shut off. At $t' = 0$, there is no coalescence and the bubble size throughout the foam is uniform and equal to the initial bubble radius (i.e., $\bar{R}/R_0 = 1$). At other times, however, there is a sharp increase in the bubble radius in the upper portion of the foam. This can be explained as follows: Due to drainage, the liquid content is lowest at the top of the foam. In other words, at $z = z_1 = 0$, r_p is the smallest and the capillary pressure (σ/r_p) is the largest. This

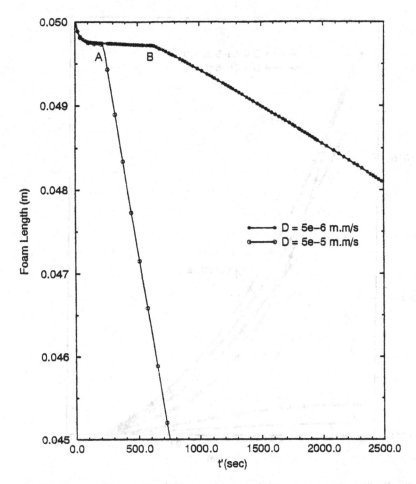

FIG. 31 Variation of foam length with time for two values of the apparent diffusion coefficient D. The values of the other parameters used are $c_{s0} = 0.012$ mol/m^3, $c_e = 8$ mol/m^3, $R = 0.2$ mm, $R_{F0} = 0.606R$, $R_{Fmin0} = 0.99R_{F0}$, and $R_{Fc} = 1.01R_{F0}$.

means that for a given R_F, films at the top of the foam drain most rapidly and rupture the earliest. Coalescence therefore starts first at the top of the foam and propagates down the foam with the passage of time. It is clear from the figure that for both values of D, there are regions in which coalescence has taken place at $t' = 600$ s and $t' = 1800$ s. However, for larger D, significantly more coalescence has occurred and the foam/gas interface has moved farther for reasons already outlined earlier in this section.

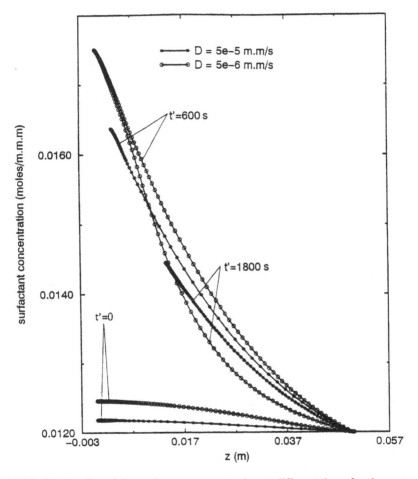

FIG. 32 Profiles of the surfactant concentration at different times for the system of Fig. 31.

In Section III, we will discuss briefly the phenomenon of interbubble gas diffusion (Ostwald ripening), which has been ignored in our model.

III. INTERBUBBLE GAS DIFFUSION (OSTWALD RIPENING)

When the pressure in the foam bubbles varies significantly, there is a transfer of gas from bubbles with higher pressure to those with lower pressure. This transfer occurs by diffusion through the liquid films separating the bubbles

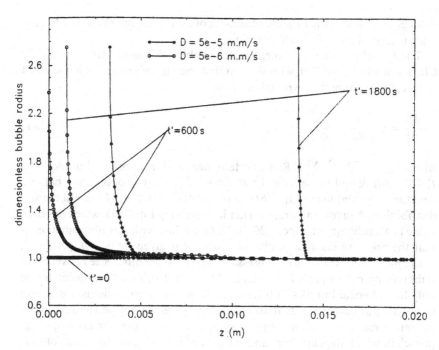

FIG. 33 Profiles of the dimensionless bubble radius ($\overline{R}/R$) at different times of the system of Fig. 31.

and is driven by the difference in the concentration of the dissolved gas at the two film surfaces. Because these concentrations are proportional to the pressure at the interfaces (Henry's law), the transport of gas is faster when the pressure difference is higher. The other important factor is the solubility of the gas in the liquid. Interbubble diffusion increases dramatically as the solubility of the gas increases. This is the main reason why CO_2 foams are less stable than N_2 foams. The pressure inside a bubble is determined by its volume when it is just formed. Because the bubble is initially spherical, the pressure inside the bubble is given by:

$$p_i = p_{pool} + \frac{2\sigma}{R} \tag{160}$$

where R is the bubble radius and p_{pool} is the pressure in the liquid pool. Thus, the pressure is higher in the smaller bubbles. This means that gas is transferred from smaller to larger bubbles, causing the smaller bubbles to get smaller and the larger bubbles to get larger. This is very similar to the phenomenon of Ostwald ripening observed in supersaturated solutions, where the larger particles grow in size at the expense of smaller ones. Hence, interbubble diffusion is also referred to as Ostwald ripening.

The first theoretical treatment of interbubble diffusion was provided by Clark and Blackman [71], who showed that the rate of growth of a bubble of radius R is given by the expression:

$$\frac{dR}{dt} = C\left(\frac{1}{R_m} - \frac{1}{R}\right) \tag{161}$$

where $R_m = \sum N_i R_i^2 / \sum N_i R_i$ is a certain mean radius such that bubbles with $R < R_m$ shrink and those with $R > R_m$ grow. Typically, a population balance equation is solved using Eq. (161) to obtain the evolution of the bubble size distribution. A pivotal assumption made in deriving Eq. (161), which has been made in all subsequent papers [25,71,72,73] dealing with this phenomenon, is that the pressure inside a bubble of radius R is given by $p_i = p_{atm} + 2\sigma/R$, where p_{atm} is the atmospherical pressure. The pressure difference between two bubbles with radii R_1 and R_2 is then $2\sigma(1/R_1 - 1/R_2)$. This relationship is true only for spherical bubbles. It is not clear how the pressure changes with size in polyhedral bubbles. Equation (161) may therefore not be valid for polyhedral foams. One option for polyhedral foams is to use a microscopic approach in which the curvature and the area of each film are functions of time and change as the pressure in the bubbles sharing the film changes. This approach has been used to simulate the evolution of a two-dimensional foam [17–19]. Using this approach for three-dimensional foam and integrating it with drainage and coalescence are likely to be extremely difficult. A model that incorporates drainage, coalescence, and interbubble gas diffusion has been proposed by Narsimhan and Ruckenstein [22,23] for steady-state foams. However, drainage is treated by assuming pentagonal dodecahedral geometry, and interbubble gas diffusion is treated by assuming spherical bubbles. Given the complexity of the system, such assumptions are probably inevitable. Even with these assumptions, extending this model to standing foams would be very difficult. The results of our model are therefore valid only for initially monodispersed foams composed of gases that are sparingly soluble in the continuous-phase liquid.

We shall conclude this article with a brief discussion of a new approach used to study foam drainage viz. the wetting of dry foam.

IV. WETTING OF DRY FOAM

A novel approach [40,74,75] has been used recently by some investigators to study foam drainage. One starts with a stable dry foam that has stopped draining. Liquid is then continuously fed at the top of the foam. It is seen that a well-defined moving front is established between the wet and dry regions of the foam. Interestingly, this front moves with a velocity that is proportional to the square root of the flow rate at which the liquid is fed into the foam. A brief discussion of the theory for this system is provided below. The interested reader is advised to refer to an excellent discussion provided by Verbist et al. [40].

Consider the drainage equation for a single vertical PB channel:

$$\frac{\partial a_p}{\partial t} = -\frac{\partial}{\partial z}(u a_p) = -\frac{\partial}{\partial z}\left(\frac{1}{20\sqrt{3}\mu}a_p^2\left(\rho g + \sigma\frac{\partial}{\partial z}\left(\frac{1}{r_p}\right)\right)\right) \tag{162}$$

If the liquid in the films is neglected, the PB cross-sectional area (a_p) is proportional to the square of the PB radius (r_p) and we have:

$$a_p = C r_p^2 \tag{163}$$

where C is a constant. Equation (164) can be scaled and rearranged into the following form:

$$\frac{\partial \alpha}{\partial t^*} = \frac{\partial}{\partial z^*}\left(\alpha^2 - \frac{\sqrt{\alpha}}{2}\frac{\partial \alpha}{\partial z^*}\right) \tag{164}$$

where α, t^*, and z^* are dimensionless quantities obtained by scaling PB area (a_p), distance (z), and time (t) quantities that are obtained using scaling factors x_0 and τ in space and time. Thus:

$$\alpha = \frac{a_p}{x_0^2}; \quad z^* = \frac{z}{x_0}; \quad t^* = \frac{t}{\tau} \tag{165}$$

The scaling factors x_0 and τ are given by:

$$x_0 = \sqrt{\frac{\sigma C}{\rho g}} \quad and \quad \tau = \frac{20\sqrt{3}\mu}{\sqrt{\sigma \rho g C}} \tag{166}$$

It has been shown that Eq. (164) has an analytical solution of the form:

$$\begin{aligned}
\alpha(z^*, t^*) &= v_w \tanh^2[\sqrt{v_w}(z^* - v_w t^*)] && \text{for } z^* \leq v_w t^* \\
&= 0 && \text{for } z^* \geq v_w t^*
\end{aligned} \tag{167}$$

which describes the evolution of the liquid fraction profile in a foam that is fed with liquid at the top. v_w is the velocity of the moving front that can be measured experimentally. The relation between the front velocity and the feed flow rate can be obtained as follows.

Let ε_1 be the liquid fraction above the front. Because most of the variation in ε is restricted to a small region near the boundary of the dry and wet regions of the foam, we can consider the foam to consist of a wet portion ($\varepsilon = \varepsilon_1$) and a dry portion ($\varepsilon = \varepsilon_0$) demarcated by a sharp boundary that moves downward at a rate v_w. Let Q be the volumetrical flow rate of the feed per unit cross-sectional area at the top of the foam. A mass balance gives:

$$Q = \varepsilon_1 v_w \tag{168}$$

The velocity of the front (v_w) is proportional to the fluid velocity (u_1) in the PB channels. Because:

$$u_1 \propto \varepsilon_1 \tag{169}$$

it follows that:

$$v_w \propto \varepsilon_1 \tag{170}$$

and one obtains:

$$v_w \propto Q^{\frac{1}{2}} \tag{171}$$

The relationship described in Eq. (171) is really a consequence of the fact that the fluid velocity in a PB channel given by the Hagen–Poiseuille law and is proportional to the liquid fraction. This premise has been disputed in a recent publication, which contends that the exponent in Eq. (171) is really 3/2 [75]. However, more verification is required before this can be accepted as fact.

NOMENCLATURE

a_1	Empirical parameter in the Frumkin adsorption equation
a_p	Average cross-sectional area of a Plateau border channel
A	Foam cross-sectional area
A_1, A_2, A_3	Functions used in the calculation of the electrical double-layer force
A_h	Hamaker constant
A_F	Area of the surface of a film
$\overline{A_F} = \int_{R_{Fmin}}^{R_{Fmax}} f A_F dR_F$	Mean surface area of a film at any level
b	Empirical constant used in the calculation of the van der Waals force
b_1	Empirical parameter in the Frumkin adsorption equation
b_1, b_2, b_3	Variables defined in Eqs. (145–147)

c	Empirical constant used in the calculation of the van der Waals force; constant of integration
c_e	Concentration of salt
c_{el}	Electrolyte concentration
c_f	Correction factor for Reynolds equation accounting for the mobility of the film surfaces
c_{Na^+}	Concentration of Na^+ ions near the surface
c_{R^-}	Concentration of surfactant anions near the surface
c_s	Concentration of dissolved surfactant
c_{s0}	Surfactant concentration in the foaming solution
c_v	Coefficient accounting for the mobility of the walls of a Plateau border channel
D	Apparent diffusivity: diffusivity
D_s	Surface diffusivity
e	Protonic charge; also empirical constant used in the calculation of the van der Waals force
E	Electrical field
E_i^{local}	Local field
$\overline{E}$	Average local field generated by the surface dipoles
f	Distribution of film radii
f_0	Initial distribution of film radii
g	gravity
G	Superficial gas velocity
k_B	Boltzmann constant
$K \equiv \sqrt{\dfrac{4\pi}{2(0.161)n_P 0.816}}$	
K_d	Dissociation constant of the adsorbed surfactant molecules
l	Distance between two adjacent water molecules in the ice structure; edge length

$\bar{l}$	Average length of a Plateau border channel
l_s	Length of a segment
L	Thickness of the brush
L_{max}	Maximum length of an equilibrated foam
L_0	Initial length of foam; chain length in a theta solvent
m	polarization
n	Number of water molecules per unit volume
n_F	Number of films per bubble
n_p	Number of PB channels per bubble
n_s	Number segments in molecule
N	Number of bubbles per unit volume
N_A	Avogadro's number
p	Pressure in liquid
p_{atm}	Atmospherical pressure
p_i	Pressure inside bubble/droplet
p_{pool}	Pressure in the liquid pool
p_{ref}	Reference pressure in liquid
$p_\perp$	Normal component of the dipole moment
P	Dimensionless number given by $\dfrac{\sigma}{K\rho g R_0 L_0}$
ΔP	Pressure difference driving film drainage
q	Empirical constant used in the calculation of the van der Waals force
q_c	Volumetric flux of liquid due to bulk movement of bubbles
q_{cs}	Flux of surfactant
q_G	Volumetric flux of gas
q_L	Total flux of liquid
q_{PB}	Volumetric flux of liquid due to drainage of liquid through the PB channels

q_θ	Volumetric flow rate through a Plateau border channel inclined at an angle θ to the vertical
Q	Volumetric flow of feed per unit cross-sectional area at the top of the foam
$r = (R_F - R_{Fmin})/(R_{Fmax} - R_{Fmin})$	Dimensionless coordinate in R_F space
r_p	Plateau border radius
R	Bubble radius
R_d	Radius of the area polarized by a surface dipole
R_F	Radius of a film
R_{Fc}	The upper limit for $R_{F\ min}$ beyond which the distribution is assumed to be Dirac delta
R_{Fmax}	Largest value of the film radius
R_{Fmin}	Smallest value of the film radius at a given level
R_G	Molar gas constant
R_m	A mean radius used in models for interbubble gas diffusion
R_0	Bubble radius in the initially monodispersed foam
$\bar{R} = \left(\frac{3}{4\pi}\bar{V}\right)^{\frac{1}{3}}$	Mean bubble radius
s	Standard deviation
t	Time
t'	Time elapsed after gas supply is shut off
T	Absolute temperature
u	Average velocity in a PB channel
u_θ	Average velocity in a PB channel inclined at an angle θ to the vertical
v_w	Velocity of the boundary separating the dry and wet regions of a foam
V_c	Volume of continuous phase per unit area in an equilibrated foam of length L_0, which is in contact with continuous phase at the bottom

V_{ceq}	Volume of continuous phase per unit area in an equilibrated foam of length L_{max}
V_{eq}	Volume of continuous phase per unit area at equilibrium
V_D	Volume of dispersed phase per unit area in an equilibrated foam of length L_0 in which the films at the top are critical
V_{Deq}	Volume of dispersed phase per unit area in an equilibrated foam of length L_{max}
V_f	Rate of film thinning
$\bar{V}$	Mean bubble volume
W	Interaction energy
x	Distance from the midplane
x_0	Scaling factor for the vertical coordinate z
x_F	Film thickness
x_{Fc}	Critical thickness of film rupture
x_{Fm}	Film thickness corresponding to the maximum disjoining pressure
$x_{Fm,dl}$	Film thickness corresponding to the maximum disjoining pressure when only electrical double-layer forces are present
$\overline{x_F} = \int_{R_{Fmin}}^{R_{Fmax}} f x_F \, dR_F$	Mean film thickness at a given level
z	Space coordinate
z_1	Coordinate of the foam/gas interface
z_{1e}	Coordinate of the foam/gas interface at equilibrium
z_2	Coordinate of the foam/liquid interface
z_{2e}	Coordinate of the foam/liquid interface at equilibrium

Greek letters

α_d	Degree of dissociation of the adsorbed surfactant molecules

$\xi = (z-z_1)/(z_2-z_1)$	Dimensionless coordinate in z-space
Δ	Distance between the centers of two neighboring water layers
Δ'	Distance indicating the location of the surface dipoles
ε	Liquid fraction
ε_b	Liquid fraction at the foam/liquid interface $= 0.26$
ε_c	Liquid fraction at the foam/gas interface when collapse occurs
ε_d	Dielectric constant; liquid fraction
ε_d', ε_d'', ε_{d0}	Dielectric constants
ε_{eq}	Liquid fraction at equilibrium
ε_0	Initial liquid fraction in a foam; liquid fraction in the dry portion of a foam
γ	Polarizability of water
γ_s	Inverse of dimensionless surface viscosity
η_s	Surface viscosity
κ	Reciprocal Debye length
λ_h	Decay length in Eq. (24)
λ_n	nth root of the equation $J_0(\lambda_n) = 0$
μ	Bulk viscosity
μ_s	Dipole moment of an adsorbed surfactant molecule
μ_w	Dipole moment of a water molecule
v	Volume of a water molecule in an icelike structure
ρ	Density difference between the continuous phase and the dispersed phase
ρ_c	Density of the continuous phase
ρ_D	Density of the dispersed phase
σ	Surface tension
σ_c	Surface charge per unit area
σ_0	Surface tension of pure water

ψ	Potential
ψ_m	Midplane potential
ψ_s	Surface potential
Γ	Surface density of surfactant
Γ_∞	Surface density at saturation
$\overline{\Gamma}$	Mean surface density of surfactant at a given level taken to be a function of the mean film thickness and the surfactant concentration
Π	Disjoining pressure
Π_h	Contribution to the disjoining pressure due to organization of water molecules
Π_{max}	Maximum disjoining pressure
$\Pi_{max,dl}$	The maximum disjoining pressure when no short-range repulsive forces are present
Π_{DL}	Electrical double-layer force per unit area
Π_{SR}	Short-range repulsive force per unit area
Π_{ST}	Steric repulsive force per unit area
Π_{VDW}	van der Waals force per unit area

APPENDIX A

The size of a bubble is expressed in the model in terms of the bubble radius R, which is the radius of a sphere having the same volume as that of the bubble. All other parameters that depend on the bubble size, such as the area of the films (faces) and the length of the Plateau border channels, are expressed in terms of R as shown below. The volume of a regular pentagonal dodecahedron with edge length l is:

$$V = 7.7l^3 \tag{A1}$$

and the area (A_F) of each pentagonal film is:

$$A_F = 1.72l^2 \tag{A2}$$

Because, by definition, $V = (4\pi R^3)/3$, we have $l = 0.816R$.

In the model, the pentagonal films are represented by circular disks, which have the same surface area. If R_F is the radius of these circular disks, we have:

$$A_F = \pi R_F^2 \Rightarrow R_F = 0.606R \tag{A3}$$

APPENDIX B

A. Calculation of Velocity Coefficient

Desai and Kumar [61] have expressed the velocity coefficient (c_v) as a function of the dimensionless surface viscosity $\left(\gamma_s = 0.4387\mu\sqrt{a_p}/\eta_s 2 \right)$ as:

$$c_v = b_{i0} + b_{i1}(\gamma_s - \gamma_i) + b_{i2}(\gamma_s - \gamma_i)^2 + b_{i3}(\gamma_s - \gamma_i)^3 \tag{B1}$$

where $\gamma_i \le \gamma_s \le \gamma_{i+1} = 3$ are constants. The values of the coefficients $\{b_{ij} = 0\text{--}3\}$ can be found elsewhere in the literature [61].

REFERENCES

1. Uraizee, F.; Narsimhan, G. Enzyme Microb. Technol. 1990, *12*, 232.
2. Uraizee, F.; Narsimhan, G. Enzyme Microb. Technol. 1990, *12*, 315.
3. Brown, L.; Narsimhan, G.; Wankat, P.C. Biotechnol. Bioeng. 1990, *36*, 947.
4. Lemlich, R. In: *Adsorptive Bubble Separation Techniques;* Lemlich, R., Ed.; Academic Press: New York, 1972; 33 pp.
5. Davis, W.; Haas, P.A. In: *Adsorptive Bubble Separation Techniques;* Lemlich, R., Ed.; Academic Press: New York, 1972; 279 pp.
6. Aubert, J.H.; Kraynik, A.M.; Rand, P.B. Sci. Am. 1986, *254*, 74.
7. Rossen, W.R. In: *Foams: Theory, Measurements and Applications;* Prud'homme, R.K. Khan, S.A., Eds.; Marcel Dekker: New York, 1995; 413 pp.
8. Kulkarni, R.D.; Goddard, E.D.; Chandar, P. In: *Foams: Theory, Measurements and Applications;* Prud'homme, R.K. Khan, S.A., Eds.; Marcel Dekker: New York, 1995; 555 pp.
9. Garret, P.R. In: *Defoaming, Theory and Industrial Applications;* Garrett, P.R., Ed.; Marcel Dekker: New York, 1993; 1 pp.
10. Ruckenstein, E.; Sun, F. Ind. Eng. Chem. Res. 1995, *34*, 3581.
11. Park, J.S.; Ruckenstein, E. J. Appl. Polym. Sci. 1989, *38*, 453.
12. Bikerman, J.J. *Foams*; Springer-Verlag: Berlin, 1973.
13. Matzke, E.B. Am. J. Bot. 1946, *33*, 58.
14. Williams, R.E. Science 1968, *161*, 276.
15. Ross, S.; Prest, H.F. Colloids Surf. 1986, *21*, 179.
16. Princen, H.M.; Levinson, P. J. Colloid Interface Sci. 1987, *120*, 172.
17. Weaire, D.; Rivier, N. Contemp. Phys. 1984, *25*, 59.
18. Weaire, D.; Kermode, J.P. Philos. Mag. B 1983, *48*, 245.
19. Weaire, D.; Kermode, J.P. Philos. Mag., B 1984, *50*, 379.

20. Thompson, D.W. *On Growth and Form*; Cambridge University Press: New York, 1963.
21. Leonard, R.A.; Lemlich, R. AIChE J. 1965, *11*, 18.
22. Narsimhan, G.; Ruckenstein, E. Langmuir 1986, *2*, 230.
23. Narsimhan, G.; Ruckenstein, E. Langmuir 1986, *2*, 494.
24. Narsimhan, G.; Ruckenstein, E. In: *Foams: Theory, Measurements and Applications;* Prud'homme, R.K. Khan, S.A., Eds.; Marcel Dekker: New York, 1995; 99 pp.
25. Hartland, S.; Bourne, J.R.; Ramaswami, S. Chem. Eng. Sci. 1993, *48*, 1723.
26. Jeelani, S.A.K.; Ramaswami, S.; Hartland, S. Trans. IChemE 1990, *68*, 271.
27. Hartland, S.; Barber, A.D. Trans. IchemE 1974, *52*, 43.
28. Miles, G.D.; Sheklovsky, L.; Ross, T. J. Phys. Chem. 1945, *49*, 93.
29. Jacobi, W.H.; Woodcock, K.E.; Grove, C.S. Ind. Eng. Chem. 1956, *48*, 9046.
30. Haas, P.A.; Johnson, H.F. Ind. Eng. Chem. Fundam. 1967, *6*, 225.
31. Krotov, V.V. Colloid J. USSR 1981, *43*, 33; English translation.
32. Narsimhan, G. J. Food Eng. 1991, *14*, 139.
33. Bhakta, A.; Khilar, K.C. Langmuir 1991, *7*, 1827.
34. Ramani, M.V.; Kumar, R.; Gandhi, K.S. Chem. Eng. Sci. 1993, *48*, 455.
35. Bhakta, A.R.; Ruckenstein, E. Langmuir 1995, *11*, 1486.
36. Bhakta, A.R.; Ruckenstein, E. Langmuir 1995, *11*, 4642.
37. Bhakta, A.R.; Ruckenstein, E. Langmuir 1996, *12*, 3089.
38. Ruckenstein, E.; Bhakta, A.R. Langmuir 1996, *12*, 4134.
39. Bhakta, A.R.; Ruckenstein, E. J. Colloid Interface Sci. 1997, *191*, 184.
40. Verbist, G.; Weaire, D.; Kraynik, A.M. J. Phys. Condens. Matter 1996, *8*, 3715.
41. Reynolds, O. Philos. Trans. R. Soc. Lond. A 1886, *177*, 157.
42. Ivanov, I.B.; Dimitrov, D.S. Colloid Polym. Sci. 1974, *252*, 982.
43. Vrij, A. Discuss. Faraday Soc. 1966, *42*, 23.
44. Scheludko, A. Adv. Colloid Interface Sci. 1967, *1*, 391.
45. Ivanov, I.B.; Radoev, B.; Manev, E.; Scheludko, A. Trans. Faraday Soc. 1970, *66*, 1262.
46. Ruckenstein, E.; Jain, R.K. Faraday Trans. II 1974, *70*, 132.
47. Sharma, A.; Ruckenstein, E. Colloid Polym. Sci. 1988, *266*, 60.
48. Lucassen, J. In: *Anionic Surfactants—Physical Chemistry of Surfactant Action;* Lucassen-Reynders, E.H., Ed.; Marcel Dekker: New York, 1981; 27 pp.
49. Donners, W.A.B.; Rijnbout, J.B.; Vrij, A. J. Colloid Interface Sci. 1977, *60*, 540.
50. Exerowa, D.; Kolarov, T.; Khristov, K. Colloids Surf. 1987, *22*, 171.
51. Kolarov, T.; Khristov, K.; Exerowa, D. Colloids Surf. 1989, *42*, 49.
52. Oshima, H.; Kondo, T. J. Colloid Interface Sci. 1988, *122*, 591.
53. Narsimhan, G. Colloids Surf. 1992, *62*, 41.
54. Fainerman, V.D. Colloids Surf. 1991, *57*, 249.
55. Ruckenstein, E.; Beunen, J.A. Langmuir 1988, *4*, 77.
56. Schiby, D.; Ruckenstein, E. Chem. Phys. Lett. 1983, *95*, 435.
57. Schiby, D.; Ruckenstein, E. Chem. Phys. Lett. 1983, *100*, 277.
58. Dolan, A.K.; Edwards, S.F. Proc. R. Soc. Lond. A 1974, *337*, 509.
59. Alexander, S.J. Physique 1977, *38*, 983.

60. de Gennes, P.G. Adv. Colloid Interface Sci. 1987, *27*, 189.
61. Desai, D.; Kumar, R. Chem. Eng. Sci. 1982, *37*, 1361.
62. Koczo, K.; Racz, G. Colloids Surf. 1987, *22*, 97.
63. Racz, G.; Erdos, E.; Koczo, K. Coll. Polym. Sci. 1982, *260*, 720.
64. Kann, K.B. Colloid J. USSR 1979, *41*, 714; English translation.
65. Princen, H.M. J. Colloid Interface Sci. 1979, *71*, 55.
66. Princen, H.M. Langmuir 1986, *2*, 519.
67. Princen, H.M.; Kiss, A.D. Langmuir 1987, *3*, 36.
68. Princen, H.M. J. Colloid Interface Sci. 1990, *134*, 188.
69. Germick, R.J.; Rehill, A.S.; Narsimhan, G. J. Food Eng. 1994, *23*, 555.
70. Iglesias, E.; Anderez, J.; Forgiarini, A.; Salager, J. Colloids Surf. A Physicochem. Eng. Asp. 1995, *98*, 167.
71. Clark, N.O.; Blackman, M. Trans. Faraday Soc. 1948, *44*, 1.
72. Lemlich, R. Ind. Eng. Chem. Fundam. 1978, *17*, 89.
73. Ramaswami, S.; Hartland, S.; Bourne, J.R. Chem. Eng. Sci. 1993, *48*, 1709.
74. Goldfarb, I.I.; Kann, K.B.; Shreiber, I.R. Fluid Dyn. 1988, *23*, 244.
75. Koehler, S.A.; Hilgenfeldt, S.; Stone, H.A. Phys. Rev. Lett. 1998, *82*, 4232.
76. Eisenberg, D.; Kauzmann, W. *The Structure and Properties of Water*; Oxford University Press, 1969.
77. Manciu, M.; Ruckenstein, E. Langmuir 2001, *17*, 7061.
78. Manciu, M.; Ruckenstein, E. Langmuir 2001, *17*, 7582.
79. Ruckenstein, E.; Manciu, M. Langmuir 2002, *18*, 2727.

2
Foam Film Stability in Aqueous Systems

TAKAMITSU TAMURA and YUKIHIRO KANEKO Lion Corporation, Tokyo, Japan

I. ASSESSMENT OF FOAM STABILITY

The ability to control foam stability or coalescence rate of bubbles is important in many industrial applications. Foams can persist for a few minutes to several days depending on storage conditions. To effectively utilize foams in any of these situations, it is important to have some control over their stability. Therefore it is very important to deepen our understanding of the mechanisms involved in foam persistence and decay. Most works in this area have been rather empirical and many experimental data are rendered useless because important parameters such as bubble size have not been measured. In this chapter we attempt to summarize the quantitative analysis on foam film stability in aqueous systems in terms of surface tension measurements.

A. Standard Test Methods

1. Static Methods

Foam is a dispersion of gas bubbles in a liquid, in which each bubble dimension falls within the range of colloid size. Surfactant solutions are often characterized by their foaminess, which belongs to an important characteristic of surfactant solutions. In this chapter we review the existing test methods for measuring the foaming and antifoaming properties of liquids. A number of methods used in evaluating the aspects of foam formation and stabilization have been reported [1–5]. According to the principle of foam generation, foaming methods are classified into two groups: static and dynamic. The dynamic test method uses foam volume measurement under a state of dynamic equilibrium between rates of foam formation and decay. The static test method usually consists of forming a foam and then foam volume measurements are performed when the rate of foam formation is zero. The typical measurement of a dynamic foam is the volume of foam at equilibrium; the typical measurement of a static foam is the rate of foam collapse. The foaminess of the solutions is generally defined as the foam volume that can be obtained from a unit volume of solution. Evaluation of the foaminess of materials very often depends on physical conditions such as pressure or temperature and also strongly depends on the test methods. The standard test methods are summarized in Table 1. As many factors influence the foaming of solutions in practical situations, it is not possible to establish specific units to define the foaming power and foam stability of surfactant solutions.

The static methods are classified into four methods depending on their principles: the pouring method, shaking method, beating method, and stirring method. The pouring method was developed by Ross and Miles in 1941 as a foaming power evaluation for synthetic detergents [6], then approved as ASTM [7], JIS [8], and ISO standards [9]. A test solution (200 mL) under test passes through a nozzle and drips from a height of 90 cm through a vertical thermostatted tube of 50 mm wide into 50 mL of the same solution. This method has been widely accepted to give accurate and reproducible results among laboratories. The amount of foam obtained by this method is limited by the volume of the measuring tube, and therefore adjustment of concentration or selection of suit surfactants is necessary for the measurements. The shaking method, in which the surfactant solution moves quickly up and down, is very simple and inexpensive but subject to many sources of error [10,11]. This method is also limited by the volume of the reservoir. The increase in volume is determined by the increase in the total height of the test solution after vigorous shaking as defined in the ASTM standard [10]. This method covers the measurement of the increase in volume of a low-viscosity

TABLE 1 Standard Methods for Foaming Assessment

Principle	Classification	Method	Standard
Static methods	Pouring	Ross and Miles Test	ASTM Standard D 1173
		Modified Ross and Miles Test	ISO Standard 696-1975(E)
	Shaking	Bottle Test	ASTM Standard D 3601
	Beating	Perforated Disk Test	DIN Standard 53902 Part 1
	Stirring	Blender Test	ASTM Standard D 3519
Dynamic methods	Air injection	Diffuser Stone Test	ASTM Standard D 892
			ASTM Standard D 1881
		Gas Bubble Separation Test	ASTM Standard D 3427
	Circulation	Recycling and Fall Test	AFNOR Draft T73-421

aqueous liquid due to its tendency to foam under low-shear conditions. The beating method was developed as an alternative to the shaking method, eliminating some of the latter's problems [12–14]. Foam is produced by passing perforated disks through each test surfactant solution. A DIN-standard electric drive system is employed which varies the speed of the pistons in order to eliminate the errors of manual shaking or beating [11]. The stirring method, being very simple, is widely accepted, but different results tend to be obtained depending on how the mechanical stirring method is implemented in scientific literatures [15–19]. The increase in volume is determined by the increase in total height of a test solution after vigorous shaking in the blender as defined by the ASTM standard [15]. This method covers the measurement of the increase in volume of a low-viscosity aqueous liquid due to its tendency to foam under high-shear conditions.

2. Dynamic Methods

The dynamic methods are classified into two tests depending on their principles: air injection method and circulating method. The first air injection method was developed by Bikerman in 1938 [20]. As the operating conditions are often closer to their practical conditions, air injection methods offer many advantages over other foaming methods. The air injection method, designated the ASTM standard, is specific for the determination of foaming characteristics of lubricating oils [21]. The foam is produced by passing air through a diffuser stone attached to the top of an air-inlet tube. If the method is used, however, to determine the foaming power of an aqueous surfactant solution, the air-flow rate and the surfactant concentration must be adjusted in order to obtain sensible results. The similar injection test method is accepted as ASTM standard as a simple beaker test for evaluating the tendency of an engine coolant to foam [22]. The circulating method is suitable for the determination of the foaming characteristics of a surfactant solution and also for the study of the performance of antifoaming agents [23]. Foams are produced by circulating the surfactant solution with a pump and spraying the solution through a nozzle onto the surface of the solution. The advantage of this method is that it operates in conditions very close to the practical ones, as foam produced in practice is, in many cases, caused by the circulation of a liquid capable of foaming.

B. Characterization of Foam

1. Foam Lifetime

As liquid foams are dynamic, once generated, foams disappear gradually by drainage, coalescence, and disproportion. A certain volume of foam is prepared and the changes occurring in it are monitored as a function of time. To improve our understanding of foam stability, it should be characterized as a foam feature. The characterization of foams must take into consideration three features: foam lifetime, distribution of foam, and water contents in foam.

Szekrenyesy et al. described models that correspond to several types of bubbles and they investigated stability under different circumstances in each model [24]. The lifetimes of single floating bubbles are rather long because of their low probability for foam film rupture. The lifetime of bubbles has a minimum value arising from the duration of thinning (τ_{th}). The residual number of single foams generated from the chip of a capillary is measured in the apparatus as shown in Fig. 1 and the number of bubbles left after time τ ($N\tau$) is related to the initial number (N_0) as:

$$N_\tau = N_0 \exp[-\kappa - \tau_{th}] \tag{1}$$

where κ is a kinetic constant related to the probability of breakage. By plotting experimental data, corresponding to the logarithmic form of Eq. (1), the τ_{th} value can be determined.

The lifetime of bubbles in a foam column can be defined as a significant physical parameter. Bikerman proposed the ratio of steady-state foam volume to gas flow rate as a unit of foaminess, denoted Σ. A glass frit is used to generate the bubbles in a jacketed glass column [20]. The steady state of

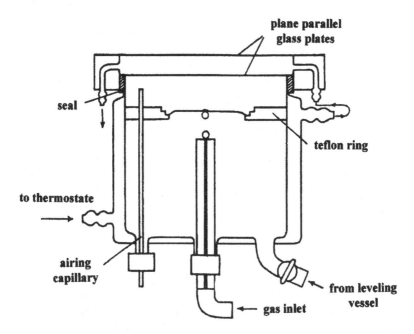

FIG. 1 Apparatus for the investigation of single bubbles and foams. (From Ref. 24.)

foam volume (V_f) is read at the rate of gas flow (U_g). The ratio $\Sigma = V_f \cdot t / U_g$ is almost independent of U_g and is applicable to the lifetime of the bubble in the foam column. Furthermore, Marysa et al. examined the dependence of foam volume on the rate of gas flow for aqueous alcohol and fatty acid solutions [25,26]. They found a linear relation between total gas volume (V_g) and flow rate (U_g). The slope of the line is defined as retention time t_r as:

$$t_r = \Delta V_g / \Delta U_g \qquad (2)$$

where ΔV_g is the change in foam volume, and ΔU_g is the change in gas flow. This parameter is an improved Bikerman's Σ by considering the volume of liquid contained within the foam. This parameter is almost independent of U_g and container shape.

2. Bubble-Size Distribution

Foam properties vary with time as a result of changing the distribution of gas and liquid in the foam. A number of theoretical and experimental research on bubble-size distribution have been reported [3,27,28]. Hartland described drainage and coalescence times in steady-state dispersions and related these processes to a foam column [28]. The effect of bubble-size distribution in the foam is discussed with theoretical treatments. Direct observation of bubble-size distribution has been performed in terms of photographing the foam through a glass wall [29,30], an electron microscopy and video microscopy [3], a capillary technique [31,32], pressure measurement [33], and an optical-fiber technique [34]. In the photographic methods, the decay of static foams is measured by changing the size distribution of bubbles in a foam with time. Video microscopy is a technique involving the connection of a video camera to a microscope objective lens using a fiber-optic light pipe. Preparing foams for electron microscopy by freezing confirms mechanical stability by immobilizing the liquid state. Prins et al. described a method where a thin-glass fiber is introduced into a foam, the diameter of the fiber is 200 μm, and the diameter of the tip is only 20 μm [34]. Light is reflected at the end of the tip and its intensity depends on the difference in the reflective indexes of the glass and the surrounding medium. The returning beam is received by a light-sensitive cell then converted into an electronic signal. Fig. 2 shows a schematic diagram of the apparatus. Foam structure is also clarified with the aid of magnetic resonance imaging (MRI) [3,35–37]. The principal advantage of using x-rays for structural investigations of foams is that the foam can be investigated under atmospheric pressure in its natural wet state. In the late 1980s, MRI tomography was applied to observe the profile in various draining foams. Experimental development was hampered by the lack of adequate theoretical insights, and MRI has only recently been taken up again for this purpose. Magnetic resonance imaging extends the ideas of nuclear magnetic resonance (NMR) by adding to the applied homogeneous external magnetic field a

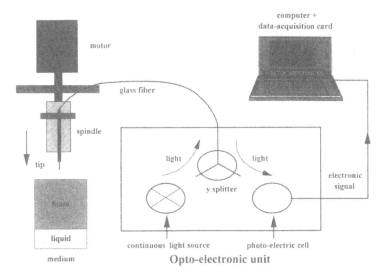

FIG. 2 A schematic presentation of the foam analyzer. (From Ref. 34.)

pulsed linear magnetic field gradient. Electrical measurements, capacitance [38–40], and conductance [38,41,42] have been applied to pursue foam drainage. A perspex tube surrounded by a capacitance sensor is placed in a bath of detergent solution. Foam is created inside the tube by blowing air through a fine nozzle located beneath it. In order to perform capacitance measurements, a nonionic detergent solution is required. The capacitance of a foam is dependent on its liquid content, on account of the different dielectric constants of its constituents. In the conductance apparatus, the electrodes are in direct contact with the foam, which reduces the parasitic effects because of the influence of specific geometry of the capacitance apparatus. The easiest way to determine the local liquid fraction is by inserting two stripped wire electrodes into a tube filled with foam and measuring the resistance as a function of time. It was used to measure the profile of a solitary wave and to monitor free drainage.

II. DYNAMIC SURFACE PROPERTIES OF FOAM FILM

A. Dynamic Surface Tension

1. Dynamic Surface Tension Measurements

Interpretation of the dynamic adsorption process at the gas–liquid interface is significant for analyzing foam stabilization processes. Although a number of techniques for measuring an equilibrium surface tension exist, only a few techniques can be applied to monitor the dynamic surface tension according to the time scale. Several experimental methods are described for studying adsorption dynamics: the oscillating jet [44–46], pendant drop designation [47], drop volume or drop weight technique [48–51], Wilhelmy plate method [52–54], maximum bubble pressure [55–79], and the overflow cylinder technique [80,81]. As shown in Fig. 3, these methods have different time windows [43]. At longer adsorption times conventional techniques for measuring static surface tension exist, reaching into the time domain of hours and even days. At shorter adsorption times the maximum bubble pressure technique covers a wide range of time window, where the oscillating jet, incline plate, and growing drop methods are used. Thus the maximum bubble pressure technique is commonly utilized for measurement of less than 1 sec. Although this technique belongs to a classical method in interfacial science, a large number of researchers are using it in recent years, because of the development of new devices for measuring very short time regions.

2. Surface Adsorption Mechanisms

The equilibrium and dynamic aspects of surface tension and adsorption of surfactants at the air–water interface are important factors in foam film stability [82]. Dynamic adsorption models with the diffusion-controlled and mixed-kinetic mechanisms are discussed in some surfactant solution litera-

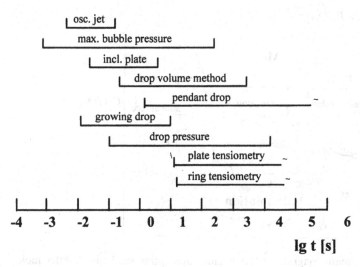

FIG. 3 Overlap of time window of different dynamic experimental technique. (From Ref. 43.)

tures [83–85]. When a new surface of a surfactant solution is created, a finite
time is required to reach an equilibrium between the surface concentration
and the bulk concentration. The dynamic surface tension is a function of time
and surface composition. For solutions of soluble surfactants, the dynamic
adsorption behavior is governed by a two-step process (Fig. 4). The first step is
an adsorption process and the second step is a bulk mass transfer process.
When surfactants are adsorbed from the solution to the new interface, the
concentration decreases in the subsurface. Diffusion then tends to restore the
initial concentration in the subsurface by bringing surfactants from the bulk
solution.

In diffusion-controlled adsorption models, one assumes that there is no
activation energy barrier to the transfer of surfactant molecules between the
subsurface and the surface [85]. Thus diffusion is the only mechanism needed
in establishing adsorption equilibrium. The time required for the molecules to
transfer from the bulk to the subsurface is much longer than the time
required for equilibration between the surface and the subsurface. On the
contrary, if the adsorption or desorption rate at the interface is slow or
comparable to the diffusion rate, the adsorption process is significant. This
model is called the mixed-kinetic adsorption model. This condition may
depend not only on the properties of the system but also on the diffusion
length and possibly on convection conditions. The diffusion-controlled model
of Eqs. (3) and (4) have been given by Fainerman et al. [86,87].

At short times

$$\gamma_t = \gamma_0 - 2RT_c(D_s t/\pi)^{1/2} \tag{3}$$

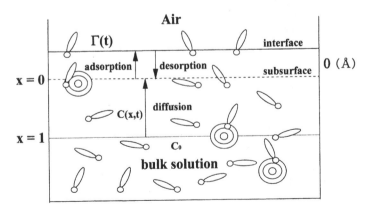

FIG. 4 A schematic diagram of the dynamic adsorption mechanism. After molecules adsorb from the subphase ($x = 0$), diffusion ensues from a diffusion length I.
(From Ref. 85.)

and at long times

$$\gamma_t = \gamma_{eq} - (RT\Gamma^2/2c)(D_L t/\pi)^{-1/2}. \tag{4}$$

The parameters c, Γ, D_s, and D_L represent the bulk concentration, equilibrium surface excess, monomer diffusion coefficient of surfactant for short times, and monomer diffusion coefficient of surfactant for long times, respectively.

According to the study of Rosen et al., the typical dynamic surface tension curve can be divided into four stages: induction region, rapid fall region, mesoequilibrium region, and equilibrium region [61–64]. The maximum rate of surface tension decrease $(d\gamma_t/dt)_{max}$ was calculated from dynamic surface tension data. The dynamic surface tension at a constant surfactant concentration can be represented by the following relaxation function.

$$\gamma_t = \gamma_m + (\gamma_0 - \gamma_m)/\{1 + (t/t^*)^n\} \tag{5}$$

where γ_m is the mesoequilibrium surface tension, which means the change in γ in 30 sec was attained in less than 1 mN m^{-1}, γ_0 is the surface tension of solvent (water), t^* is the time when γ_t attained the intermediate value between γ_0 and γ_m, and n is constant. The values of t^* and n are calculated by utilizing a successive approximation technique to Eq. (5). A differential form of Eq. (5) represents Eq. (6).

$$d\gamma_t/dt = -(\gamma_0 - \gamma_m)[n(t/t^*)^n - 1/t^*]/[1 + (t/t^*)^n]^2 \tag{6}$$

where the maximum rate of decrease in the surface tension, $(d\gamma_t/dt)_{max}$, can be expressed by Eq. (7), which is obtained by substituting t^* for t in Eq. (6).

$$(d\gamma_t/dt)_{max} = n(\gamma_0 - \gamma_m)/4t^* \tag{7}$$

Fig. 5 shows the γ_t decay data and predictions using Eqs. (3), (4), and (5). The calculated values using Eq. (5) are in good agreement with the observed values. Until 0.1 sec the observed tension drop from γ_0 is about 20 mN m^{-1}, and clearly Eq. (3) describes these early stages of the decay well. Turning to the end of the decay after 1 sec, there is a good agreement with the approximation given by Eq. (4). Although the dynamic surface tension curves can be fitted at short times and long times, respectively, by using the diffusion-limited model treatment, the calculated curves using the relaxation function treatment are widely fitted for observed time scale. The data treatment by the relaxation function has not yet been connected to models of mass transfer to the interface. However, the calculated values obtained by this treatment are in good agreement with the observed values. The implication of this empirical approach would be good when considered from the standpoint of the interpretation of the phenomena.

The effect of hydrocarbon chain length on $(d\gamma_t/dt)_{max}$ for 1.0 mM CmE8 ($m = 10$ to 16) solutions at different temperatures is shown in Fig. 6 [79]. The

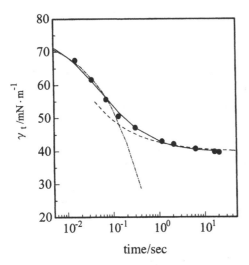

FIG. 5 Dynamic surface tension for 1 mM C12E8 at 2°C. The lines are calculated using Eq. (1): —, Eq. (4): —·—, and Eq. (5):– –, respectively.

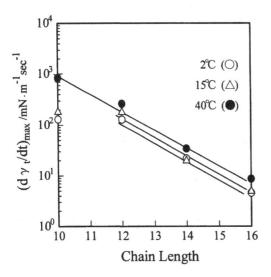

FIG. 6 Effect of hydrocarbon chain length on $(d\gamma_t/dt)_{max}$ for 1 mM CmE8 ($m = 10$, 12, 14, and 16) at different temperatures ($T = 2$, 15, and 40°C).

rapid fall region of γ_t shifted to a longer measuring time region with increasing hydrocarbon chain length and was only slightly dependent on temperature. Thus, except for the C10E8 system, only a slight temperature dependence on the adsorption curves was observed, and also all curves were parallel to the lower γ_t value with increasing temperature. At a low bulk concentration region, diffusion is expected to be the dominant factor during the adsorption process. Colin et al. have shown that the typical dynamic surface tension curves of aqueous C10E10 solutions are in good accordance with the penetration depth theory [88], indicating that the adsorption of C10E10 is then limited by the surfactant diffusion toward the surface. And the monomer concentration depends on the critical micellar concentration (CMC). The CMC values tend to decrease with temperature and the hydrocarbon chain length. Our results also show that the shorter the length of the hydrocarbon chain the faster the adsorption rate, which may be due to a higher monomer concentration.

B. Foamability and Foam Stability

1. Gibbs–Marangoni Effect

Surface elasticity is a dominating parameter imparting stability of thin liquid films [89,90]. Actual foams contain many bubbles which are separated by these liquid films and are continuously enforced by the dynamic change of liquid, such as liquid drainage or bubble motions, etc. [91,92]. In surfactant-stabilized aqueous films, any stretching force in the film causes a local decrease in the surface concentration of the adsorbed surfactant. This decrease causes the local surface tension increase (Gibbs effect), which in turn acts in opposition to the original stretching force. With time the original surface concentration of the surfactant is restored. The time-dependent restoring force existing in thin liquid films is referred to as the Gibbs–Marangoni (Marangoni) effect [94]. This concept has been strongly supported by many researchers. It is possible to generate a surface tension gradient in a foam film by stretching various elements of the film to some extent. This property of foam film is usually referred to as the Gibbs elasticity. The Gibbs elasticity has been related to foam stability experimentally using the wire cage technique for polyoxyethylene ($n = 14$) nonylphenol solutions by Prins [95]. Furthermore, the plot of the Gibbs elasticity against SDS concentration has clearly revealed a maximum point below the CMC [93]. Prins et al. revealed the relationship between the surface tension gradient and the resulting surface expansion rate by the overflow cylinder technique [80,81]. The rate of film drainage can be slowed down by the generation of a surface tension gradient along the film surface, which opposes the flow of liquid out of the film. Fig. 7 shows the relationship

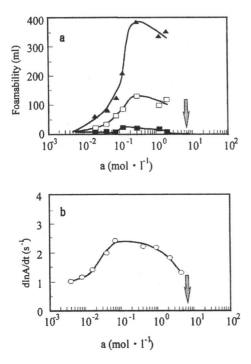

FIG. 7 Foamability (a) as measured by means of the closed loop foam generator at three different gas flow rates and surface expansion rates (b) as measured by the overflowing cylinder technique as a function of the alcohol activity for butanol solutions. The arrow indicates the limit of solubility. (From Ref. 80.)

between foamability and surface expansion rate ($d\ln A/dt$) as a function of concentration (activity) for aqueous butanol solutions. Foamability exceeds the optimum value in a butanol activity. The butanol concentration at which an optimum foamability is found coincides with the concentration at which an optimum surface tension gradient can be generated.

2. Dynamic Surface Tension and Elongation of Lamellae

Our results also proved the correlation between foamability and surface tension gradient for aqueous nonionic surfactant solutions. Foam formation was estimated from a dynamic surface tension using the maximum bubble pressure method, and foam stability was estimated from a transfer distance of lamella using a laminometer. Laminometer measurements were made using the Du Noüy ring method [1,78,96]. Force profile during the expansion of lamella was monitored using an electronic-balance with expansion rate at 0.1 mm sec^{-1}. The degree of elongation of lamella, $L_{lamellae}$, was defined as follows.

$$L_{lamellae} = L_{rupture} - L_{max} \tag{8}$$

where L_{max} is the transfer distance from the height representing maximum force to that representing zero force. The γ_t of 1.0 mM C12En solutions as a function of time is shown in Fig. 8 [78]. In the relatively longer measuring time region (>1 sec), γ_m decreases with the decrease in EO units. Scott et al. have shown the relation between the cross section of the polyoxyethylene dodecyl ether molecules and the equilibrium surface tension of their aqueous solutions [97]. The equilibrium surface tension showed a monotonous increase in EO units from 7 to 30, as well as in the molecular cross section. Although the surfactant molecules in the mesoequilibrium lamella seem to be oriented completely at the air–water interface, low γ_t would be observed with nonionic surfactants having a smaller cross section. In the relatively shorter measuring time region (< 0.1 sec), however, γ_t decreases with increasing EO units. An initial adsorption process of surfactants to air–water interface would take place; nonionic surfactants having a large CMC would be much more effective in increasing the rate of adsorption at the air–water interface. Furthermore, nonionic surfactants having a large hydrophilic area/molecule would also be much more effective in breaking hydrogen bondings among water molecules.

The Ross–Miles foam behavior of 1.0 mM aqueous C12En solutions is shown in Fig. 9 [78]. It is seen that the initial foam height increases linearly with the increase in EO units, n (up to 22), but then it slightly falls for longer EO units. A similar behavior has been observed in several types of polyoxy-

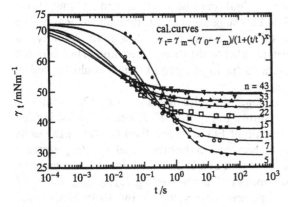

FIG. 8 Effects of the number of EO units on the dynamic surface tension, γ_t, vs. bubble surface lifetime, t, curve for 1 mM aqueous C12En solutions at 26°C.

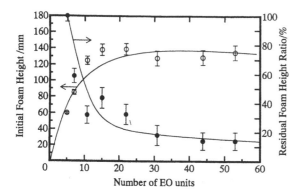

FIG. 9 Effects of the number of EO units on the Ross–Miles foam behavior for 1 mM aqueous C12En solutions at 26°C.

ethylene adducts. Stirton et al. have shown that polyoxyethylene alkanols exhibit a maximum foam height at $n = 17$–19 in hard water [98]. Schick et al. have shown the effect of EO units on foam formation and foam stability of several distilled polyoxyethylene alkylethers. In dodecyl ether homologous, the initial foam height decreases with the increase in EO units after attaining a maximum at $n = 7$ [99]. However, the residual foam height ratio, which reflects the foam stability of the solutions, shows that stability strongly depends upon the number of EO units at a given concentration as shown in Fig. 9. The residual foam height ratio steeply decreases with the increase in EO units until it reaches 31, and then it gradually falls for the longer EO units. Accordingly, the relationship between the initial foam height and EO units is contrary to the relationship between the residual foam height ratio and EO units for a homologous series of C12En. This suggests that it is difficult to measure the initial foam height of the solution having a very low foam stability, as the defoaming would happen parallel to the foam generation. Therefore the initial foam height of surfactants having longer EO units could be underestimated in these conditions.

Both the $(d\gamma_t/dt)_{max}$ and $L_{lamellae}$ of the 1.0 mM aqueous C12En solutions as a function of EO units are shown in Fig. 10 [78]. A linear increase in $(d\gamma_t/dt)_{max}$ was observed in the region of EO units less than 40. When EO units exceed 40, $(d\gamma_t/dt)_{max}$ seems unchanged. On the other hand, the steep decrease in $L_{lamellae}$ was observed in the region of EO units less than 10. Particularly, when EO units reach 40, the lamella is difficult to elongate ($L_{lamellae}$ is only 1.7 mm). Compared with the experimental results for the Ross–Miles foam behavior as shown in Fig. 9, $(d\gamma_t/dt)_{max}$ seems to correlate to the initial foam height, while $L_{lamellae}$ seems to correlate to the residual foam height after 5

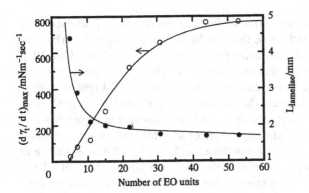

FIG. 10 Effects of the number of EO units on the dynamic parameters for 1 mM aqueous C12En solutions at 26°C.

min. This coincidence indicates that the solutions having a large $(d\gamma_t/dt)_{max}$ value have a potential for making a large total surface area of bubbles, because $(d\gamma_t/dt)_{max}$ reflects the maximum rate of the decrease in the surface tension under a constant value in the work of foam generation. Therefore the initial foam height for the Ross–Miles technique is in good concordance with the value of $(d\gamma_t/dt)_{max}$. On the other hand, the residual foam height ratio for this method is in good concordance with $L_{lamellae}$, because $L_{lamellae}$ reflects the durability of lamella under elongation. We believe that the increase in EO units on initial foam height has a tendency to generate bubbles easily. In fact, the initial foam height of surfactants having EO units within 5 to 15 increases from 60 to 140 mm. These increases correlate to the increase in the values of $(d\gamma_t/dt)_{max}$ from 28 to 278 mN m^{-1} sec^{-1}. Despite the increase in the value of $(d\gamma_t/dt)_{max}$ from 519 to 770 mN m^{-1} sec^{-1}, the decrease in initial foam height is observed when EO units are from 22 to 53. This indicates that excessive decrease in surface tension generated by surfactants having longer EO units would weaken a restoring force in resisting the thinning of the lamella. On the contrary, the higher initial foam film stability, in the shorter EO region, would be caused by the Marangoni effect.

III. STABILIZATION OF FOAM FILM

A. Thin Film Measurements

The presence of surfactants plays an important role in the drainage behavior of foam films, depending on the surface forces (the disjoining pressure) and hydrodynamic forces in their films [100]. The simultaneous action of disjoining pressure and hydrodynamic forces determines the stage of the formation and evolution of a liquid film of fluid surfaces during drainage. Because of film surface corrugations, which cause thermal fluctuations or other disturbances, the film either ruptures or transforms into a thinner Newton black film [101]. For the thicker foam films, film rupture is a result of the local fluctuation thinning of the film due to the existence of surface waves. Light scattering from thin liquid films can reveal the existence of the surface waves [102–104]. There are two types of countermodes for a foam film system: squeezing and bending modes. The light-scattering curves of soap film for the squeezing and bending modes are very different functions of the film thickness as shown in Fig. 11. The bending-mode curve is shown just opposite to that of the squeezing mode. The bending mode is not effective in causing film rupture, thus the investigation of squeezing mode instability is sufficient to determine the conditions of film stability [100].

If the short-range repulsive disjoining pressure is large enough, the black foam films are stable. There are two types of black foam films: common and Newtonian. While the common black films are the thicker type of black films (from about 5 to 20 nm in thickness), the Newtonian black (NB) films are bimolecular thin films (less than 5 nm in thickness). A mechanism of rupture of NB films is considered as a process of new phase nucleation in a two-dimensional system [105–108]. There exist in the film elementary vacancies (unoccupied positions of surfactant molecules) moving randomly, which associate to form clusters of vacancies called holes. A hole can grow up by fluctuations to a critical size and become a nucleus of a hypothetical two-dimensional phase of vacancies. Further spontaneous growth of the nucleus leads irreversibly to the rupture of the film. When the rupture of NB film is due to formation of holes in it by a nucleation mechanism, it has been shown that the mean film lifetime τ depends on the monomer surfactant concentration C as:

$$\tau = A(r)\exp[B/\ln(Ce/C)], \tag{9}$$

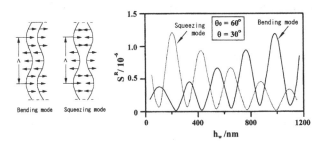

FIG. 11 Surface scattering ratio (S^R) as a function of film thickness h_w. (From Ref. 104.)

where A is the kinetic factor, B is proportional to the work of hole formation, C is the surfactant concentration in the solution, and Ce is the C-value at which the NB film is in thermodynamic equilibrium with a sufficiently large hole in it. The constants A, B, and Ce are parameters of the theory independent of the surfactant concentration. The NB foam stability can be estimated from this theory experimentally based on the dependence of τ on C. The following parameters can thus be determined: work for nucleus formation, number of elemental vacancies in the nucleus, critical surfactant concentration below which the NB film cannot be formed.

A detailed review of theoretical and experimental techniques on foam film stability or drainage was reported [109–116]. Simple observation techniques for observing thinning behavior of foam film photographed their interference fringes on liquid films as a function of time. Soap bubbles can be easily generated at the top of a capillary after dipping it into a surfactant solution by blowing air. Vertical foam film can be also generated in a rectangular Pt and features of interference fringes on the film are observed. However, quantitative studies on film drainage have been carried out by using horizontal [117–147] and vertical [148–168] film apparatus. Horizontal circular films are commonly nursing adaptations of the techniques pioneered by Derjaguin et al. [116]. Direct measurement of the thickness of the aqueous film can be determined by microreflectance methods. We utilize the porous-plate technique, first developed by Mysels [153] and later refined by Bulgarian research groups [224]. Single and thin-liquid films are formed in a hole drilled through a fritted glass disk as shown in Fig. 12. The disjoining pressure is

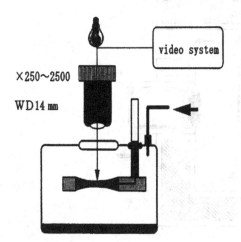

FIG. 12 Schematic diagram of the cell for formation of films in the capillary of a porous plate.

calculated from the pressure balance between an applied gas pressure and the bulk liquid pressure derived from Laplace pressure. On the other hand, foam film can be produced by pulling a vertical frame out of a reservoir solution. Precise measurements of the aqueous core thickness of a foam film (d_2) are carried out by FT-IR measurement, and the measurement cell used in our laboratories is shown in Fig. 13 [79,161]. A measuring cell having two CaF$_2$ windows and a frame is used for measurements in the cell. The cell consists of a double wall glass container with flowing water for temperature control. The rectangular frame is hollowed out in a sintered glass plate and an infrared beam passes through the center of the frame. The d_2 value is determined from the absorbance of the adsorption band at 3450 cm^{-1}, which is assigned to the OH stretching vibration of a water molecule with a molar absorption coefficient.

Many physical factors are involved in the control of surfactant foam film stability. This means that foam film stability can be determined by the following factors: surface rheology, the Marangoni effect, disjoining pressure, and hydrophobic interaction. A dominant factor for maintaining the foam varies with the thickness of the liquid films as shown in Fig. 14 [162]. The rate of thinning (drainage) of liquid films is drastically influenced by the rheolog-

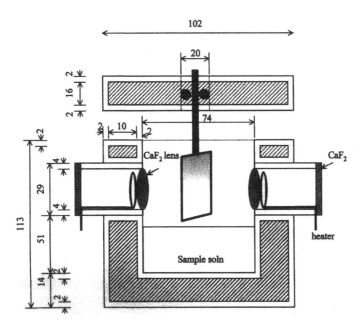

FIG. 13 Schematic diagram of the cell for formation of films formed by a frame pulled out of a solution. The sizes of the apparatus are shown in mm.

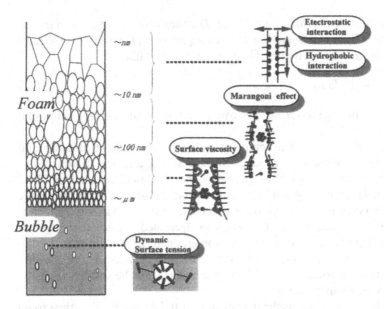

FIG. 14 Schematic diagram of foam stabilizing factors depending on the lamellae thickness.

ical properties of the related adsorption layer. Surface rheology factors affect the rate of drainage in a vertical foam film that proceeds immediately after generation (d_2 a few hundred nanometers). The most effective contribution of the Marangoni effect on the drainage would occur for d_2 values from tens to hundreds of nanometers. Furthermore, a disjoining pressure or hydrophobic interaction between hydrocarbon chains in a surfactant effect is a major factor in the stability of the foam film, which has d_2 values below tens of nanometers. From the following sections, we describe the relationship between foam stabilizing factors and foamability by monitoring the vertical foam film thickness.

B. Surface Viscosity

Investigation of the influence of surface viscosity on the rate of film thinning is performed by numerical and technical approaches [110,169–173]. The surface tension gradient in the thinning film, which is created by the efflux of liquid from the film and sweeping of surfactant along the film surfaces from the film and the sweeping of surfactant along the film surfaces to the plateau borders, can be characterized by the dimensionless elasticity number, E_s, as:

$$E_s = E_0 R_f / \mu D \tag{10}$$

where μ is the bulk viscosity of the liquid, D is the diffusivity of the surfactant, and E_0 is the Gibbs elasticity. The Boussinesq number (B_o) measures the ratio of interfacial and bulk viscous effects in the draining film as:

$$B_o = (\mu^s + \kappa^s)/\mu R_f \tag{11}$$

where μ^s is the surface-shear viscosity, and κ^s is the surface dilational viscosity.

During drainage, the surfactant monolayer may undergo dilating and shearing deformation that produce surface stress, and the drainage also depends on these rheological properties of the film interface. The combined effects of surface tension gradient and interfacial viscosity on the film drainage time are depicted in Fig. 15 [110]. An increase in surface viscosity, which is characterized by increasing B_o, results in decreased surface mobility and higher drainage time. If the surface tension gradient, E_s, is high, then the film drainage will be slow even at low values of surface viscosity. In the film drainage of these solutions the Marangoni effect, not the surface viscosity, appears to play a major role.

The influence of surface-shear viscosity on the drainage of vertical foam film is also investigated in four different model foam films as shown in Table 2 [161]. The $(d\gamma_t/dt)_{max}$ values for these four systems were adjusted in the range from 7200 to 7600 mN m^{-1} sec^{-1}. Under these conditions, we can correctly estimate the effect of surface shear viscosity on foam film stability, because the

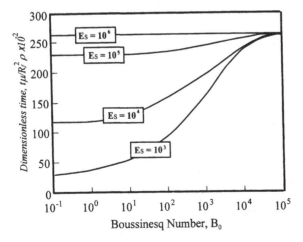

FIG. 15 Dimensionless drainage time for the film to drain from a dimensionless thickness, h_i, to the thickness, h_f, vs. Boussinesq number, at various values of the dimensionless interfacial elasticity number. (From Ref. 110.)

TABLE 2 Effects of Glycerin and NaCl on $(d\gamma_t/dt)_{max}$ and Surface Shear Viscosity (η_s) of SDS 60 mM Aqueous Solutions at 25°C

Additions to 60 mM aqueous solutions	$(d\gamma_t/dt)_{max}$ mN m^{-1} sec^{-1}	$\Delta\lambda/-$		$T/$sec		η_s µN sec m^{-1}
		SDS aqueous solutions	Water	SDS aqueous solutions	Water	
Blank	7600	0.125	0.118	16.2	16.2	3.16
Glycerin 4%	7500	0.130	0.121	16.2	16.2	3.88
Glycerin 4%/ NaCl 0.4 M	7300	0.141	0.129	16.2	16.2	5.27
NaCl 0.4 M	7200	0.121	0.120	16.2	16.2	0.55

surface tension gradient has little contribution to the thinning processes. Furthermore, there are no differences in the period of oscillation among these samples, as shown in Table 2, which means these films have extremely low elasticity. The η_s values of the SDS solutions with glycerin went up by about 20% compared with those of the solutions without glycerin. The η_s values of the SDS solutions with NaCl were reduced to one-sixth. However, the η_s values of the SDS/glycerin solutions with NaCl went up about 10 times compared with those of the solutions without NaCl. It is well known that by adding glycerin the bulk viscosity of the aqueous solution will increase. Our data have also shown that the presence of glycerin markedly increased the η_s values of the SDS solutions. Clark et al. demonstrated that the diffusion of surface-adsorbed fluorophore in SDS-stabilized films was dependent on the glycerin concentration present in the solution by using fluorescence recovery after the photobleaching technique [143]. The diffusion coefficient of surface-adsorbed fluorophore decreased with increasing viscosity, confirming the presence of an interaction between the surface film and the interlamellar liquid. Thus it would be understandable if the surface shear viscosity is affected by the addition of glycerin. The result of the thinning behavior of vertical films is shown in Fig. 16. As shown in the thinning pattern, the transition point from silver film to black film is shifted to a longer time region in the presence of glycerin, despite the little differences in the initial thick films. The effects of surface shear viscosity on foam film stability were observed in fairly thin film condition for the films stabilized by an electrostatic repulsion. On the contrary, when glycerin was added to SDS/0.4 M NaCl solution, the thinning pattern serves to slow down the rate of thinning in the initial thick film. This delay of the thinning is caused by an increase in surface shear viscosity. However, for both cases, foam film thickness at transition to black film and breakdown of the black film is not affected by surface viscosity.

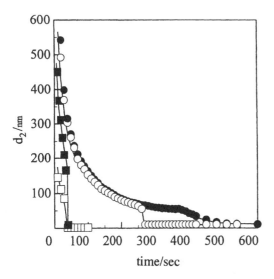

FIG. 16 Aqueous core thickness (d_2) of foam films of aqueous solutions of 60 mM SDS (○), 60 mM SDS/4 wt.% glycerin (●), 60 mM SDS/0.4 M NaCl (□), and 60 mM SDS/4 wt.% glycerin/0.4 M NaCl (■) as a function of time at 26°C.

C. Marangoni Stabilization

The Marangoni effect can be explained schematically in Fig. 17, which shows that if the solution is too dilute to stabilize the foam film, the differential tension will be relatively small and little foaming will occur. In the case of the solution being too concentrated, the differential tension relaxes too rapidly because of the supply of surfactant, which diffuses to the surface. This causes the restoring force to have time to contact the disturbing forces, producing a dangerously thinner film and poor foaming. The maximum foam volume can be observed at the intermediate surfactant concentration range at which the Marangoni effect enhances foam stability.

The thinning behavior of vertical thin films as a function of concentration for the C12E7 solution is also shown in Fig. 18 [161]. It is seen that the drainage for every system gradually continued just before rupture. The drainage time up to rupture decreased in the order of concentration, 0.5 > 0.1 > 10 mM. Further evidence of the effect of Marangoni stabilization on the drainage curves was provided by monitoring the $L_{lamellae}$ by laminometer measurement. As shown in the inserted figure in Fig. 18, the drainage speed coincided with the order of the $L_{lamellae}$ for each solution, indicating a reflection of the stabilization with the Marangoni effect.

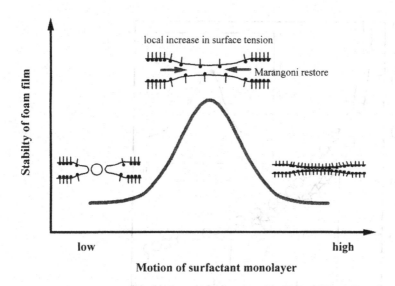

local increase in surface tension

Marangoni restore

low   high

Motion of surfactant monolayer

FIG. 17   A schematic diagram of the Marangoni effect.

Furthermore, the foaming behavior of nonionic surfactants has become measurable using both $(d\gamma_t/dt)_{max}$ and $L_{lamellae}$. The values of $(d\gamma_t/dt)_{max}$ for the series of aqueous C12En solutions are plotted against the surfactant concentration shown in Fig. 19a [161]. The line represents a linear relationship between $(d\gamma_t/dt)_{max}$ and the concentration, respectively. The value of $L_{lamellae}$ from the solutions is plotted against the log of concentration shown in Fig. 19b. In each homologous series, $L_{lamellae}$ reaches a maximum at the specific concentration. These maxima have shifted to a lower concentration with the increase in EO units. Furthermore, there is a significant decrease in the maximum value of $L_{lamellae}$ with the increase in EO units, i.e., $L_{lamellae}$ for 4.3 mm ($n = 5$) and 1.9 mm ($n = 31$). In order to make clear the relationship between $(d\gamma_t/dt)_{max}$ and $L_{lamellae}$, the relations shown in Fig. 19a and b were transferred to Fig. 20 [161]. In Fig. 20, $(d\gamma_t/dt)_{max}$ is given as a function of $L_{lamellae}$. The maxima of plots for all homologous series coinciding with the $(d\gamma_t/dt)_{max}$ value are nearly 20 mN m^{-1} sec^{-1}. From these results, we confirm that foam stability is closely related to the maximum rate of decrease in surface tension. The optimum surface tension gradient to keep foam stability, accomplished according to the Marangoni effect, would exist if the magnitude of $(d\gamma_t/dt)_{max}$ is nearly 20 mN m^{-1} sec^{-1} regardless of the number of EO units. Therefore if the surface tension gradient were too low, the foam films would tend to be dynamically unstable because the rate of surface tension decrease is too small. The consequent surface tension gradient is insufficient to prevent

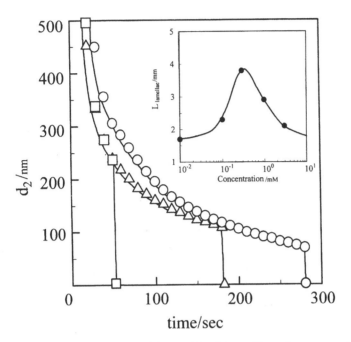

FIG. 18 Aqueous core thickness (d_2) of foam films of aqueous solutions of C12E7 as a function of time at 26°C. Concentrations of C12E7 are 0.1 mM ($\triangle$), 0.5 mM (O), and 10 mM ($\square$). Inserted figure: The L_{lamellae} vs. concentration for aqueous solutions of C12E7 at 25°C.

further thinning and may result in foam film rupture. Whereas, at too high surface tension gradient, the foam films are also dynamically unstable, because the change in surface tension with time will be too large to prevent foam film rupture and the films can become very thin. It is concluded from the findings of the conditions that the most remarkable foam stability of the solutions would be obtained under 20 mN m^{-1} sec^{-1} of $(d\gamma_t/dt)_{\text{max}}$, because L_{lamellae} measured by a laminometer is related to the foam stability measured by the Ross–Miles technique.

The difference in the value of L_{lamellae} among C12En homologs cannot be explained on the basis of the time-dependent restoring force described above. Therefore the effects of the number of EO units on the surface viscosity were measured in the concentration, where L_{lamellae} attains maximum for each homolog. The surface viscosity values increase with the increase in EO units remarkably. This suggests that the environment of the internal liquid in lamella for small EO units is more mobile, resulting in lower surface viscosity.

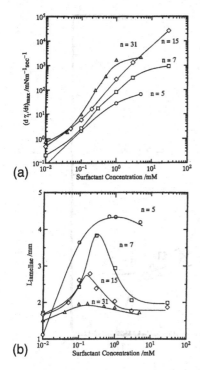

FIG. 19 Effect of surfactant concentration on $(d\gamma_t/dt)_{max}$ (a) and $L_{lamellae}$ (b) for the series of C12En at 26°C.

D. Disjoining Pressure

Typically, a certain volume of foam is prepared and the changes occurring in it are monitored as a function of time. Plateau border channels form a complex interconnected network through which liquid flows out of the foam under the action of gravity. At the same time, the liquid in the film is sucked into the PB channels. As a result, they become thinner and finally rupture. In the early stage of formation, foam film drains under the action of gravitation or capillary force. At this stage, a positive disjoining pressure may slow down and prevent drainage. It has been well established that, in addition to the Laplace capillary pressure, three additional types of force can operate in aqueous film layers: the electrostatic double-layer repulsion, the van der Waals interaction, and the short-range hydration or structural repulsive force. Our summary in this section closely follows the review by Bhakta and Ruckenstein [114].

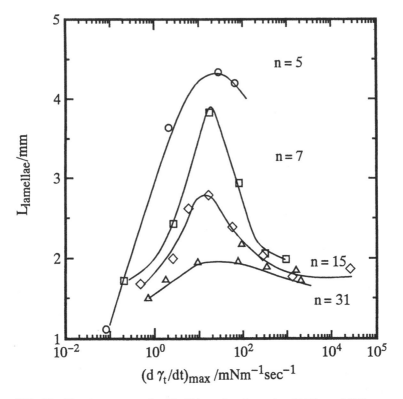

FIG. 20 The $L_{lamellae}$ vs. the $(d\gamma_t/dt)_{max}$ for the series C12En at 26°C.

As drainage of the foam film proceeds, numerous forces other than simple capillary forces come into play depending on the film thickness. These forces involve the Marangoni effect and the disjoining pressure. The effects of disjoining pressure (Π) seen in the films can be considered with the balance of van der Waals and electrostatic forces. The driving force is a net result of suction pressure in the adjacent PB channels and the disjoining pressure in the film. The Π refers to repulsive forces that arise to interact with each other when the film surfaces are close enough ($x < 100$ nm). Rupture of a film occurs when waves generated on the surface due to mechanical and thermal perturbation grow in an unbounded fashion. The repulsive disjoining force increases in response to the local thinning due to the disturbance, the wave is damped, and no rupture occurs. On the contrary, the disjoining pressure decreases, local thinning is accelerated and leads to rupture.

Π is the sum of an attractive van der Waals force (Π_{VDW}), a repulsive force (Π_{DL}) due to the interaction of electrical double layers on the two surfaces, and a short-range repulsive force (Π_{SR}) due to the steric interaction of long-chained molecules or due to organization of water molecules near charged surfaces.

$$\Pi = \Pi_{VDW} + \Pi_{DL} + \Pi_{SR} \tag{12}$$

The plot of Π vs. x_F is referred to as the disjoining pressure isotherm and plays a crucial role in determining the rate of film thinning and its stability against rupture.

Π_{VDW} tends to thin the film and its contribution to the disjoining pressure can be calculated approximately as,

$$P_d = -A/(6\pi h^3) \tag{13}$$

where h is the film thickness and A is the Hamaker constant for the liquid film. If the surface charge of foam film is carried from ionic surfactants, the electrostatic disjoining pressure repulses the surfaces and tends to thicken the film. An electrostatic contribution to the disjoining pressure Π_{DL} can be calculated as,

$$P_{el} = 2C_{el}RT \left[\cosh(zF\psi_d/RT) - 1\right] \tag{14}$$

where C_{el} is the electrolyte concentration, z is the ion valence, F is Faraday's constant, and ψ_d is the electrical potential midway between the film surfaces. For nonionic surfactants, the surface charge due to the binding ions is present in the water and is probably independent of surfactant concentration. For ionic surfactants, the surface charge results from the dissociation of the adsorbed surfactant molecules and is likely to be dependent on the surfactant concentration. The Π_{SR} due to organization of water molecules constructs the collapse of a foam film. As the pressure on a film is increased, its thickness decreases and the repulsive disjoining pressure increases. When the thickness is small enough (about 10 nm), black spots appear which eventually cover the entire film giving rise to a common black film. Further increase in pressure, however, gives rise to different phenomena depending on the salt concentration. At lower salt concentration, a common black film ruptures. At higher salt concentration, there is a sudden transition to a very stable Newton black film. An electric field is generated by undissociated surfactant molecules via their dipoles. The electric field organizes the water molecules near the surface and gives rise to the repulsive force. Thus increased counterion binding at high salt concentration increases the number of surface dipoles and hence the electric field and makes the short-range repulsive force stronger. With no salt, there is just one peak corresponding to the electrical double layer. As the salt concentration is increased, the curve changes first to one in which there are two maxima and then to one having only a short-range repulsion. An increase in the salt concentration increases the binding of sodium ions and raises the number of dipoles on the surface. The short-range repulsive force therefore increases. At the same time, the double layer is compressed. At intermediate concentrations, the double layer is not too compressed and the increased short-range repulsive force gives rise to an additional strongly repulsive force at a smaller thickness. In this case, there is a jump transition when the capillary pressure is raised above the value at point A. At higher concentrations, the electrical double layer is so compressed that Π_{DL} has a range comparable to that of the short-range polarization force and the two combine to give a single-peaked short-range repulsion. In nonionic surfactant systems, the surface charge arises because of ion binding rather than because of the dissociation of adsorbed surfactant molecules as happens with ionic surfactants. The short-range force in this case is likely to be independent of electrolyte concentration, because the surface dipole moment would be essentially unchanged by ion binding. This could be the reason why Newton black films are formed at much lower ionic strength with nonionic surfactants.

The thickness reaches a width at which the capillary pressure is counter-balanced by disjoining pressure. Films in this situation are referred to as common black films and these films do not rupture in this metastable state for a long time. Although the stability provided by repulsive electrostatic force would be the most effective for the foam film under fluctuation, the thickness is controlled mainly by the electrolyte strength of the surfactant solution. Evidence for the effect of disjoining pressure on foam film stability is provided by monitoring the film thickness by FT-IR. The thinning behavior of vertical thin films as a function of time for 60 mM SDS with different NaCl concentrations is shown in Fig. 21 [161]. Depending on the salt concentration, two different types of black films for SDS solutions can be observed. The common black film is known to be thick and exists at low salt concentrations (<0.1 M), whereas the second or the Newtonian black film can be observed at higher salt concentrations. It is seen that each film is thinning due to drainage until it undergoes no further thinning. Drainage time in the foam film for a pure SDS solution required for the first black film formation was 250 sec, while the SDS/0.4 M NaCl system required a drainage time of only about 30 sec to form the second black film, indicating a quick drainage. Furthermore, the aqueous core thickness of the black film from SDS solution was 10 nm, while that from SDS/0.4 M SDS solution reached as low as 2 nm. The results clearly show that the thickness of the SDS black film decreases constantly with NaCl concentration, because the surface charge was neutralized by counter-ions in high salt concentrations and the minimum in the interaction energy between the surface will be deeper. Therefore the drainage of SDS films with a high salt concentration would be faster due to a stronger squeezing-out effect.

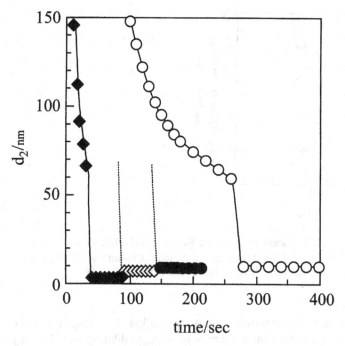

FIG. 21 Aqueous core thickness (d_2) of foam films for aqueous solutions of 60 mM SDS with different NaCl concentration as a function of time at 26°C. Concentrations of NaCl are 0 M (○), 0.04 M (●), 0.2 M (◇), and 0.4 M (◆).

Although the electrostatic force affects the rate of drainage in a wide range of film thicknesses, little change in the D_{rup} value was observed. Thus both the Ross and Miles foam stability and the D_{rup} values as a function of NaCl concentration are shown in Fig. 22 [162]. A linear decrease in the D_{rup} value was observed with increasing NaCl concentration. But the residual foam height ratio showed very little dependence on NaCl concentration, indicating a relatively small contribution of the electrostatic force to defoaming effect.

E. Hydrophobic Interaction

Dilational elasticity gives the surface tension variation of a liquid surface with respect to the unit fraction area change, and it is a measure of the ability of the surface to adjust its surface tension to an instantaneous stress. As the dilational elasticity measures the ability of the surface to develop surface tension gradient, this property characterizes the film and foam stability. The elasticity of the film is proportional to the dilational modulus from which the film was formed. Films from solutions with higher elasticity have higher

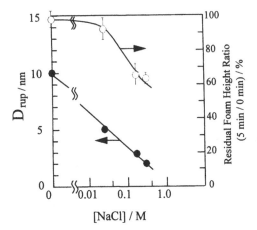

FIG. 22 Effects of NaCl concentration on the Ross and Miles foam stability (5 min) and aqueous core thickness of rupture (D_{rup}) for aqueous solutions of 60 mM SDS at 26°C. Symbols are residual foam height after 5 min (O) and D_{rup} (●).

drainage time, and both properties increase with alkyl chain length. During film drainage, the surfactant monolayer may undergo dilating and shearing deformation that also produce surface stress, and the drainage also depends on these rheological properties of the film interfaces.

Thus we have examined the effect of variations in the structure of the hydrophobic group for a constant oxyethylene chain length on foaming and vertical foam film stability. The $(d\gamma_t/dt)_{\mathrm{max}}$ values of the nonionics for each temperature as a function of hydrocarbon chain length are already shown in Fig. 6. It was found that log $(d\gamma_t/dt)_{\mathrm{max}}$ decreased linearly with increasing length of the hydrophobic chain for each temperature, except for the C10E8 system results. Only a slight temperature dependence on adsorption rate was observed for the C12E8, C14E8, and C16E8 surfactants. These results clearly show that the interaction between the hydrocarbon chain of surfactant molecules is one of the dominant factors for their adsorption processes. On the other hand, large temperature dependencies on adsorption rate for C10E8 and C12E8 surfactants are caused by hydration around the hydrophilic moiety of these nonionics. Usually, the adsorption rate of many surfactants tends to decrease below the CMC. Thus the sudden decrease in the $(d\gamma_t/dt)_{\mathrm{max}}$ values for C10E8 at 2°C and 15°C results is attributed to a reduction in the concentration below the CMC.

A number of articles have reported about the foamability of aqueous POE alkyl ether solutions [174–177]. Cox et al. reported the initial foamability measured by the Ross and Miles test, although it depends on the EO moiety contents and tends to decrease with increasing hydrocarbon chain length in the order C10 > C12 > C14 > C16 > C18 at 100°F [176]. Also, Satkowski et al. reported that the initial foamability tends to decrease in the order C10 > C12 = C14 > C18 at 50°C [177]. Based on the results in Fig. 6, the effects of hydrocarbon chain length on the initial foamability strongly depend on the value of $(d\gamma_t/dt)_{max}$.

The thinning behavior of vertical foam films as a function of time for 1 mM CmE8 (m = 10 to 16) solution at different temperatures is shown in Fig. 23 [79]. The increase in temperature caused a rapid drainage in the films for every surfactant system. These increases in the drainage suggest that they might be caused by a decrease in the bulk viscosity of the solution, an increase in the adsorption rate of the surfactants, and a release of hydration around the hydrophobic moiety of the surfactants adsorbed at the surface. When we compare the drainage pattern at 2°C and 15°C, despite the fact that a similar drainage pattern was observed for the C10E8 and C12E8 systems, large

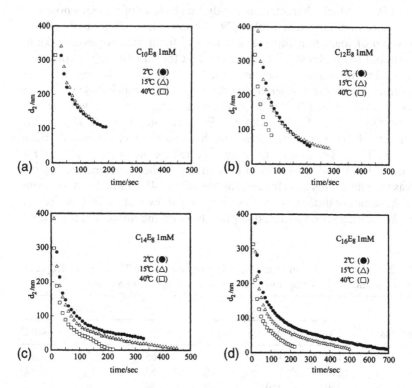

FIG. 23 Aqueous core thickness (d_2) of vertical foam films for 1 mM CmE8 (m = 10, 12, 14, and 16) solutions as a function of time at different temperatures.

differences in the pattern were observed in the C14E8 and C16E8 systems. These results will be discussed from two different points of view, namely, the Marangoni effect and the hydrophobic interaction. We believe that the appearance of the optimum $(d\gamma_t/dt)_{max}$ value (ca. 20 mN m^{-1} sec^{-1}) to maintain foam stability in nonionic solutions, under a constant speed of elongation by using a ring-type laminoneter, indicates the Marangoni effect would act to enhance stability on film drainage (Sec. III-C). In Fig. 6, it is seen that the values of $(d\gamma_t/dt)_{max}$ of the C10E8 and C12E8 solutions were observed around 10^2 to 10^3 mN m^{-1} sec^{-1} at each temperature. As the Marangoni effect would be a small contribution to the retardation of the drainage under these conditions, similar drainage took place in both films. On the other hand, the values of $(d\gamma_t/dt)_{max}$ of the C14E8 and C16E8 solutions were observed around 10 mN m^{-1} sec^{-1} at each temperature. The increment in the drainage time up to rupture also suggests that the Marangoni effect may have a contribution to the stabilization in both films.

The lifetime of the vertical film (T_{rup}) and the aqueous core thickness of rupture (D_{rup}), which characterizes the drainage curve of aqueous nonionic solutions, are summarized in Table 3 [79]. Although it is difficult to determine the position of foam film rupture in a vertical foam film, we would like to discuss about foam film stability assuming that the film drainage parameters (T_{rup}, D_{rup}) monitored at the center of the frame reflect the film properties at the rupture. There was a difference in the T_{rup} values, which depends on the hydrocarbon chain length. It was found that the D_{rup} increased with temperature in the case of C10E8 and C12E8 solutions and the D_{rup} value remarkably decreased with increasing hydrocarbon chain length from C12E8 to C16E8. The foam films were ruptured on the order of 100 nm for both the C10E8 and C12E8 solutions in the higher temperature region, whereas the black films were formed for both the C14E8 and C16E8 solutions under the same conditions. These results clearly show that the intermolecular hydrophobic interaction in the adsorption layer on the surfaces of foam films

TABLE 3 Lifetime of Foam Film (T_{rup}) and Aqueous Core Thickness of Rupture (D_{rup}) as a Function of Temperature for Aqueous 1 mM CmE8 Solutions

Surfactants	T_{rup}/sec			D_{rup}/nm		
	2°C	15°C	40°C	2°C	15°C	40°C
C10E8	190	140	10	105	127	315
C12E8	210	280	70	52	47	84
C14E8	330	450	220	34	6	2
C16E8	690	500	220	11	12	16

was a significant factor in the stability of the vertical thick foam films during drainage. The two surfactants (C10E8, C12E8), which occupy an area of 0.655 and 0.589 nm^2, respectively, form unstable foam films with a D_{rup} of 315 and 84 nm at 40°C, whereas the other two surfactants (C14E8, C16E8), which occupy an area of 0.496 and 0.381 nm^2, respectively, form stable foam films with a D_{rup} of 2 and 16 nm, respectively. However, the foam film of the C14E8 solution was ruptured at a silver film with a D_{rup} of 34 nm at 2°C, whereas the Newton black film was observed at 15°C and 40°C. These results may reflect the relative differences in the contribution of the two factors: liberation of restricted water around the hydrophobic moiety of nonionics with increasing temperature and the strengthening of the hydrophobic interaction which prevents fluctuation in the vertical thick foam films during drainage. Thus the black film was not maintained at 2°C because the hydrophobic interaction could be reduced by an increase in the HLB of the surfactant with decreasing temperature. The CMC of the POE alkyl ether solution decreases with increasing temperature [178]. It can be assumed that the black film formation of the systems, which act as an effective hydrophobic interaction, was performed with a decrease in the CMC around the high temperature region. Furthermore, a Newton black film was observed in the C16E8 solution even when the temperature was lowered to 2°C. This is in line with the decrease in the CMC with an increase in hydrocarbon chain length. With the increase in the hydrophobic interaction, the lifetime of the vertical films was prolonged and a sufficient drainage of the films resulted.

We have also examined the effects of variation on the structure of the hydrophilic group with a constant hydrocarbon chain length in foaming and film drainage parameters. The thinning behavior of the vertical films as a function of time for 1 mM ethoxylate surfactants (C12En: N = 8, 22, and 53) and hydrophobically modified ones (C12E8Et and C12E8Bu) is shown in Fig. 24 [79]. The T_{rup} and D_{rup} values for these surfactant solutions are shown in Table 4. A large difference in the D_{rup} values was found, while the same drainage curves were observed for every nonionic solution. Thus the NRE8 solution forms the most stable vertical film with a large T_{rup} value and this value decreased with increasing EO units. Furthermore, adding a hydrophobic section at the end of the polar head leads to a decrease in the T_{rup} values and to an increase in the D_{rup} values in the order C12E8 > C12E8Et > C12E8Bu. It is suggested that the decrease in the T_{rup} value from 210 to 90 sec would be related to the increase in the A values from 0.604 to 0.924 nm^2. Further evidence for the relationship between the D_{rup} value and the surface coverage is proved by the observation using the Ross and Miles test. Both the Ross and Miles foam stability and the D_{rup} values as a function of the A values of nonionics are shown in Fig. 25 [79]. A linear increase in the D_{rup} values was observed with increasing A value. It was found that the D_{rup} value reaches its

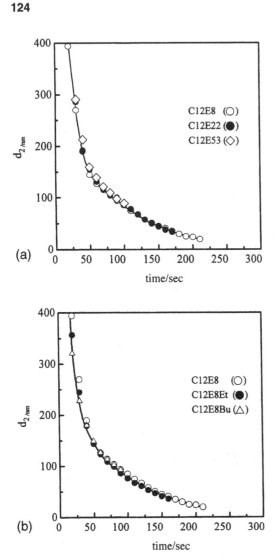

FIG. 24 Aqueous core thickness (d_2) of vertical foam films for 1 mM ethoxylated surfactants as a function of time at 25°C. Results are shown for C12En: n = 8, 22, and 53 (a); C12E8Et and C12E8Bu (b).

TABLE 4 Film Drainage Parameters (T_{rup}, D_{rup}) and $(d\gamma_t/dt)_{max}$ for 1 mM Ethoxylate Surfactants

Surfactants	T_{rup}/sec	D_{rup}/nm	$(d\gamma_t/dt)_{max}/\text{mN m}^{-1}\text{ sec}^{-1}$
C12E8	210	20	141
C12E22	170	34	510
C12E53	110	77	790
C12E8Et	160	36	236
C12E8Bu	90	93	133

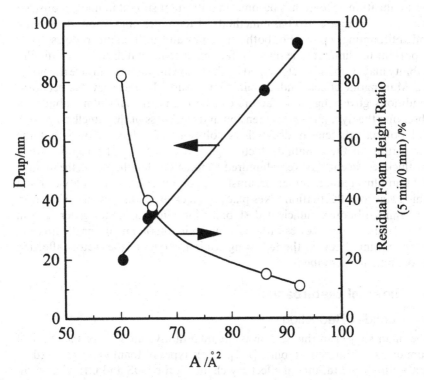

FIG. 25 Effects of area per molecule (A) on the Ross and Miles foam stability (5 min) and aqueous core thickness of rupture (D_{rup}) for 1 mM nonionics at 25°C.

maximum at five times (about 100 nm) the value of the C12E8 solution, where the modified nonionics have an area per molecule of 1.1 to 1.5 times as large as the nonmodified one. The residual foam height ratio sharply decreased with increasing A value. Thus the residual foam height ratio for the modified nonionics was reduced to one-tenth of that for C12E8, while the foam height was maintained at about 80% after 5 min in the Ross and Miles test. These studies suggest that the decease in the hydrophobic interaction between the hydrocarbon chain of the nonionics contributes to decreasing foam film rupture, because the $(d\gamma_t/dt)_{max}$ values for all the nonionics were larger than $100 \text{ mN m}^{-1} \text{ sec}^{-1}$, and no differences in the drainage curves were observed, where the contribution of the Marangoni stabilization was negligibly small in the foam film. It is concluded that the hydrophobic interaction would also be a significant factor in the vertical film stabilization.

IV. DESTABILIZATION OF FOAM FILM

Destabilization of foam has become an important subject in many industrial processes. There are two basic methods of foam destabilization, defoaming and antifoaming actions. For both defoaming and antifoaming processes, it is important to eliminate the physical factors of foam stabilization. Foam film stability may be mainly determined by the following factors: surface rheology, the Marangoni effect, and disjoining pressure. In the previous sections, conditions governing these factors can be found in foam film stability by observing the dynamic surface tension and thickness of the lamellae for SDS and polyoxyethylene dodecylethers solutions. We have shown that the elimination of these stability factors gave rise to foam film rupture when the thickness becomes several hundred nanometers. Furthermore, the lifetime of foam films reduces under circumstances of low relative humidity (RH) at which vigorous fluctuation takes place as an external disturbance. However, we cannot obtain a sufficient destabilization effect on foam generation in several industrial processes without using antifoams as an internal fluctuation on foam film. Thus, in the following sections, some of the factors affecting defoaming are described.

A. External Disturbance

1. Humidity Reduction

The main subject in this section is the quantitative analysis of the effect of humidity on foam film stability [161]. Three types of foam film were used by controlling the Marangoni effect, by changing the SDS and C12E7 concentration, and by disjoining pressure with the addition of electrolytes. In the low humidity conditions with regulating dry nitrogen gas, water evaporation from the surface takes place to reduce foam film stability. We believe that, under these external disturbances, surface coverage on the foam may affect one of the significant factors of foam stability. Thus foam film stability is compared at the surface concentration with and without the Marangoni stabilization on film surfaces. As a result, foam film stability was remarkably decreased at a humidity below 60% RH. However, the stability of SDS solution persisted because of the electrostatic repulsion force between the two surfaces under humidity reduction. The aim of the present study is to clarify the most effective factors in the foam film stability of aqueous SDS solutions where the disjoining pressure occurs, compared with that of aqueous nonionic solutions. When a frame is withdrawn from an aqueous SDS solution, a fairly thick film can be produced in the frame. If the air around the frame is not saturated with the vapor of water, evaporation may occur in the film surface. As humidity is difficult to control experimentally, many studies on thin liquid films were usually conducted in an enclosed cell that allows saturated condition. Three different foam film models were used by controlling the Marangoni effect, by changing the SDS and C12E7 concentration, and by disjoining pressure with the addition of electrolytes. The lifetime of these films was measured under circumstances of low relative humidity at which vigorous fluctuation took place to elucidate the contribution of the abovementioned factors to overall foam film stability. The differences among these foam models will be elucidated from the viewpoint of dynamic surface tension and elongation of the lamellae using a laminometer.

The γ_t values of SDS, SDS/0.4 M NaCl, and C12E7 solutions as a function of time are shown in Fig. 26a–c [151]. The rapid fall region of γ_t shifted to a shorter measuring time region with increasing SDS concentration. On the other hand, no differences were observed in these curves above 10 mM (Fig. 26a). For the SDS/0.4 M NaCl solution, lower γ_t values were observed compared with that for the SDS solution, indicating that the adsorption rate increased with the addition of electrolytes (Fig. 26b). Similar curves were observed for the C12E7 solution shown in Fig. 26c. The $L_{lamellae}$ values of the surfactant solutions described are shown in Fig. 27 [161]. Although there was little change in the $L_{lamellae}$ value of the SDS solution over a wide range of $(d\gamma_t/dt)_{max}$, 2 to 5×10^2 mN m^{-1} sec^{-1}, the $L_{lamellae}$ exceeded the optimum value at a particular gradient in SDS solution with the addition of NaCl. The maximum value of the $L_{lamellae}$ of the SDS/0.4 M NaCl solution was reached when the magnitude of $(d\gamma_t/dt)_{max}$ was nearly 20 mN m^{-1} sec^{-1}, which coincided with the magnitude where the maximum value of the $L_{lamellae}$ of the $C_{12}E_7$ solution was reached. We believe that the appearance of the optimum rate of change of the surface tension to maintain foam stability in nonionic surfactant solution suggests that the Marangoni effect contributes to foam film stability [78]. If the $(d\gamma_t/dt)_{max}$ value is too low, the foam film tends

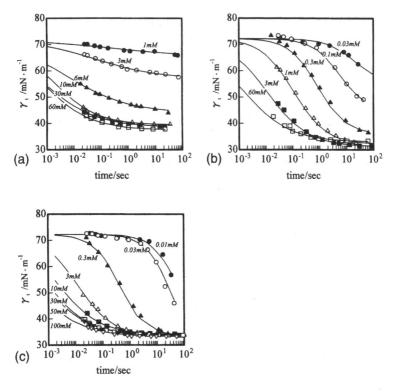

FIG. 26 Dynamic surface tension of aqueous surfactants solutions as a function of time. Results are shown for solutions of SDS (a), SDS/0.4 M NaCl (b), and C12E7 (c) at 25°C.

to be dynamically unstable because the surfactant concentration is too low to stabilize the foam film. The consequent surface tension gradient is insufficient to elongate the lamellae in the foam film and will result in foam rupture. In contrast, if the $(d\gamma_t/dt)_{max}$ value is too high, the foam film is also dynamically unstable, because the rate of surface tension gradient leads to excess lamellae thinning in the foam and will also result in foam film rupture. Our results strongly suggest that the Marangoni effect would be a dominant factor in foam film stability for both SDS/0.4 M NaCl and C12E7 solutions, because the optimum $L_{lamellae}$ values are observed at the same $(d\gamma_t/dt)_{max}$ for both systems. In addition, the $L_{lamellae}$ values for SDS and SDS/0.2 M NaCl solutions were smaller than those for the other two solutions. The low $L_{lamellae}$ values might be due to the existence of an electrostatic force of repulsion between the two ionic surfaces in which the SDS molecules were adsorbed.

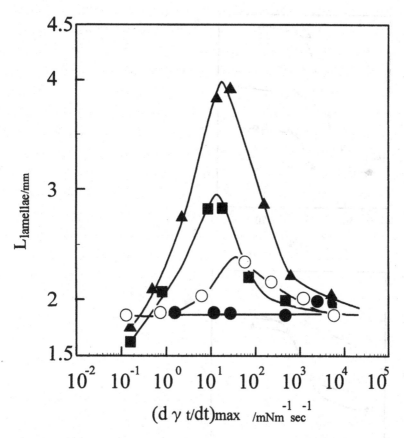

FIG. 27 $L_{lamellae}$ vs. $(d\gamma_t/dt)_{max}$ for aqueous solutions of SDS (●), SDS/0.2 M NaCl(○), SDS/0.4 M NaCl (■), and C12E7 (▲) at 26°C. The rate of expansion was kept constant at 0.1 mm sec^{-1}.

Because the increase in the electrostatic force of repulsion is inversely proportional to the distance between the two surfaces, the force was gradually increased with the thinning of the foam film. This repulsion force would prevent further thinning when a certain thickness was reached in the foam film, and the rupture would occur with further elongation. We described the relationship between the magnitude of this repulsive force and the thickness of the foam film in the previous section (Sec. III-D).

The change in the $L_{lamellae}$ of the surfactant concentration with the optimum Marangoni effect was measured as a function of relative humidity as shown in Fig. 28a [161]. The condition where the optimum Marangoni effect exists in the foam film can be determined by measuring the maximum

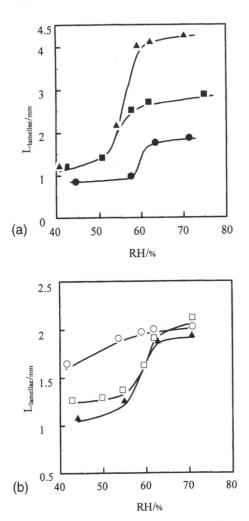

FIG. 28 Effect of relative humidity (RH) on $L_{lamellae}$ for aqueous solutions (a) of 2 mM SDS (●), 0.3 mM SDS/0.4 M NaCl (■), and 0.3 mM C12E7 (▲); for aqueous solutions (b) of 60 mM SDS (○), 60 mM SDS/0.4 M NaCl (□), and 10 mM C12E7 (▲) at 26°C. The rate of expansion was kept constant at 0.1 mm sec^{-1}.

with different $(d\gamma_t/dt)_{max}$. In Fig. 27, there was a maximum point in the $L_{lamellae}$ at 20 mN m^{-1} sec^{-1} in $(d\gamma_t/dt)_{max}$, and the concentrations of surfactants were 2 mM for the SDS solution and 0.3 mM for both the SDS/0.4 M NaCl and C12E7 solutions. The $L_{lamellae}$ values remained constant at 60% to 75% RH for these solutions. However, there was a rapid drop in $L_{lamellae}$ for both SDS/0.4 M NaCl and C12E7 solutions. This means that the reduction of RH caused a considerable reduction in foam film stability. The fluctuation is believed to be caused by surface cooling accompanied with water evaporation from the surfaces of foam films at low RH condition. The fluctuation caused by the reduction of RH is one of the known causes of destruction of foam films [179]. Our results clearly show that if the RH of the surrounding is less than 60%, foam film stability tends to decrease suddenly when the surfactant concentrations are at the optimum Marangoni effect. The change in $L_{lamellae}$ as a function of RH at a high surfactant concentration where the Marangoni effect completely disappears is shown in Fig. 28b. The $(d\gamma_t/dt)_{max}$ value, shown in Fig. 27, reaches 750 mN m^{-1} sec^{-1} where the decrease in $L_{lamellae}$ ends. This means that the Marangoni effect does not exist at these concentrations, therefore the adsorption rate of the surfactant at the surface is too fast to restore film thinning. The appropriate surfactant concentrations for comparison are 60 mM for both SDS and SDS/0.4 M NaCl solutions and 10 mM for C12E7 solution. In Fig. 28b, $L_{lamellae}$ values are found to be about 2 mm above 60% RH. However, a large difference in foam stability appeared below the 60% RH region. A sudden decrease in the $L_{lamellae}$ value was found in both SDS/0.4 M NaCl and C12E7 solutions under these conditions, and the values at 45% RH were reduced to half of those at 65% RH. On the other hand, the $L_{lamellae}$ values remained unchanged for the SDS system, indicating that the value at 45% RH fell to only 80% of that at 65% RH. These studies suggest that the difference in foam stability strongly depends on the electrostatic force of repulsion carried by the ionic charge of adsorbed surfactants between the two surfaces. However, the surface charge was neutralized by counterions, such as that in SDS/0.4 M NaCl solution, and, consequently, the foam behavior was similar to that for C12E7 which possesses no dissociation function in the molecule. These studies suggest that the stability of foam films may be a consequence of electrostatic repulsion which resists the rupture induced by fluctuation under low RH condition.

B. Internal Disturbance

1. Antifoaming Mechanisms

The main subject in this paragraph is foam breakdown mechanisms by antifoam particles. Various mechanisms proposed for the action of antifoams are described in many superior reviews [112,180–184]. A number of investigations about the breakdown of foams by dispersed insoluble oils [185–202], hydrophobic solid particles [203–214], and oils and particles in combination [215–220] were reported as antifoams. The mixed-type antifoams exhibit excellent defoaming performance; thus they are widely used in many industries. Several theories have been presented in the literatures on the mechanisms of defoaming through the use of three types of antifoams [216,217]. A widely accepted mechanism for antifoaming action separates the process into two steps: first, an oil drop enters the air–water interface and in the second step the oil begins to spread over the foam film causing rupture. Ross and McBain suggested that the oil breaks the foam film by spreading on both sides of the foam film [221]. Garrett suggested that oil spreading is not a necessary condition of antifoaming action by oils [181].

In most antifoaming mechanisms, the last step in foam film rupture by oil drop is the formation of an oil bridge in the foam film. The oil lens-bridge theory was suggested by Garrett [181] and Frye and Berg [205]. The oil drop first enters one of the foam film surfaces and forms a lens (Fig. 29a). On further thinning of the film, the lens enters the opposite film surface and an oil bridge is formed. The bridge is unstable because capillary forces dewet the film from the bridge and the film then ruptures. These mechanisms were based on a direct observation of horizontal film rupture using cinematographic techniques proposed by Dippenaar [206,207]. The theory of foam film bridging by a solid particle was also suggested by Garrett [208]. When the hydrophobic particle bridges the film, the capillary force dewets the particle, and the film

(a) Hydrophobic Oil

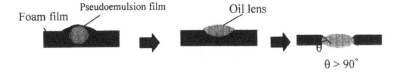

(b) Solid-particle

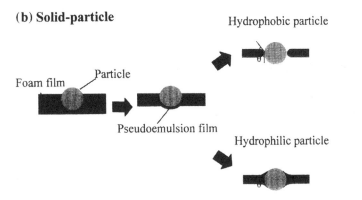

FIG. 29 Antifoaming mechanisms for the oil-type antifoams (a) and the solid-type antifoams (b). (From Refs. 205 and 208.)

then ruptures (Fig. 29b). The mechanism for the mixed-type antifoam was also suggested by Wasan et al. [182,217,218]. They pointed out that antifoaming efficiency strongly depends on the stability of a pseudoemulsion film (formed between the air–water interface and approaching oil droplet). After breaking the pseudoemulsion film, the antifoam particles enter the other foam film and form the lens bridge (Fig. 30).

2. Antifoaming Efficiency

Based on the above background, we have confirmed these mechanisms by direct observation of foam film rupture using a scanning laser microscope (SLM). Decay curves of foam volume as a function of time with and without using four types of antifoams are shown in Fig. 31 [222,223]. It is seen that fairly high initial foam volumes were observed for both the silicone oil and the silica particles. Although the silicone oil has a slightly higher antifoaming effect than the silica particle, both antifoams have only weak antifoaming efficiency. Compared with these antifoams, initial foam volumes were reduced to one-half their blank solution for the mixed-type antifoam. Effective antifoaming performance was observed within a short time region, in particular a slight foam volume could be observed in the mixed-type system. These defoaming efficiencies were consistent with the results reported by Wasan et al. [182,218]. They also pointed out that the oil alone had a low foam-breaking efficiency, the hydrophobic particles alone were little more effective, and the mixed-type agent was a much better antifoam.

3. Rupture of Pseudoemulsion Foam Film

We have observed some actions of the four different types of antifoams in two-dimensional thin liquid film, which reproduced the phenomena of foam film breakdown at the three-dimensional plateau border, by using SLM techniques. We have also investigated the preparation of a novel solid antifoamer by observing the defoaming performance and the physical properties of the solid particles. Hydrophobic silicone resins, which have the required physical properties for optimum defoaming conditions, were prepared by interfacial polymerization via the hydrolysis reaction of trichloromethylsilane. In this method, polymerization proceeded at the interface of the oil and water phases. This method can obtain silicone resin particles having various surface shapes. Scanning laser microscopy was used to observe the phenomena during foam film breakdown by the antifoams. A single He–Ne laser beam ($\lambda = 633$ nm) was directed on the sample situated on a slide glass plate, and the reflective interference fringes that appeared around the sample were observed by a CCD camera. A proper quantity of sample solution was poured on a slight inclined slide glass fixed on the movable stage. Microscopic observation was carried out with a spontaneous drainage

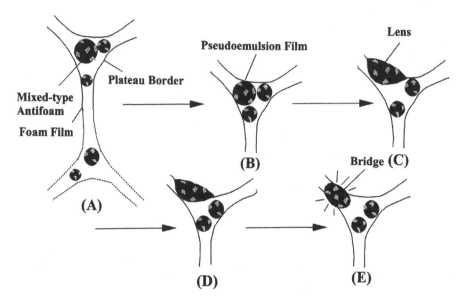

FIG. 30 Antifoaming mechanisms for the mixed-type antifoams. (From Ref. 218.)

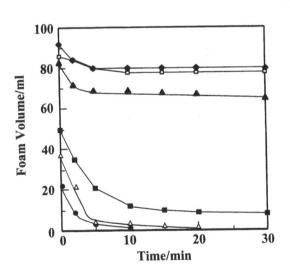

FIG. 31 Effects of different types of antifoams (200 ppm) on the rate of breakdown of foams from the model foaming solution (60 mM SDS/0.5 wt.% PVA). Blank (♦), silicone oil: KF96 (▲), hydrophobic silica: H51 (□), mixed-type antifoam: F-16 (△), silicone resin particle (■).

in a liquid sample. Reflective fringe patterns around the antifoam particles were recorded during the spontaneous drainage. The interval of each interference fringe is always regular and corresponds to $\lambda/2$ in the direction of thickness when using monochromic light. Furthermore, by focusing on the top and bottom of the sample, any change in the particle form can be observed with a high resolution (> 0.1 μm). We could not see any differences in the droplet shape between the silicone oil and the mixed-type antifoam in the model foaming solution. The diameter (d) of the dispersed silicone oil was about 10 μm, and a similar size of silicone oil droplets, in which some silica particles ($d < 0.1$ μm) existed as aggregates ($d = 2$ to 3 μm), was observed as the mixed-type antifoams. Thus antifoaming efficiency of the mixed-type is clearly due to the presence of the solid particles. A large difference in antifoaming efficiency was also observed between the hydrophobic silica and the silicone resin, although there was a slight difference in their particle diameters. Because the silica particle has a large specific gravity, it might be readily swept from the foam film and hence reduce the defoaming efficiency. However, Frye and Berg reported the effects of silica particle geometry on foam stability [204]. The polyhedral silica particles had a more effective antifoaming efficiency than the spherical silica particles. Thus our results suggest that the significant factor for foam breaking could be derived from the surface roughness of the solid particles.

Several antifoaming mechanisms may be proposed in the literature and the most suggested mechanism may be bridging mechanisms by several types of antifoams. In the last step of foam breaking, the capillary force dewets the film from the oil or solid bridge and the film then ruptures. Dippenaar performed a direct observation of solid particles by a cinematographic approach [206,207]. Garrett [208], Frye and Berg [205], and Aveyard et al. [211] also did discuss the mechanism of rupture in view of Dippenaar's work and the thermodynamics of the systems from the viewpoint of contact angle. The antifoaming efficiency strongly depends on the stability of the pseudoemulsion film. The stability of the pseudoemulsion film was directly studied by Wasan et al., forming such a film from a surfactant solution on the tip of a capillary [182,189]. They pointed out that the edge of the particles penetrating into an aqueous phase can pierce and break the pseudoemulsion film. Furthermore, the particle concentration required to break the foam film is very low, because it is sufficient that the film be pierced at only one point to rupture. We have tried to directly clarify the mechanism of breakdown of such films and correlate it to the defoaming efficiency of different types of antifoaming particles using laser microscopic techniques.

Fig. 32 illustrates the SLM reflection features of the pseudoemulsion films formed on the top of the silicone oil antifoams (A series) and the mixed-type antifoams (B series). When the interval of interference fringe in the image is

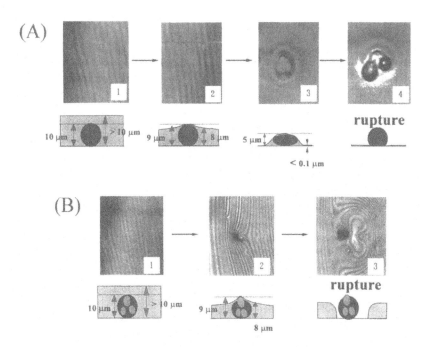

FIG. 32 Scanning laser microscopic reflection features of pseudoemulsion films from the model foaming solution (60 mM SDS/0.5 wt.% PVA) for the silicone oil antifoams (A series) and the mixed-type antifoams (B series). Estimated images of vertical sectional views are shown under these features.

wide, the liquid film has a gentle surface incline on the other hand, when the interval is narrow, it has a steep incline. The estimated images of the vertical sectional views are shown under these SLM features. It is seen that widespread fringes were observed for both antifoams when the thickness of the surfactant solutions is greater than the diameter of the antifoam droplets, indicating gentle surface inclines (A-1, B-1). Considerable differences in the features appeared as the drainage of the solution progressed. In the case of the silicone oil, the pseudoemulsion film was formed around the top of the droplets, which was confirmed from the appearance of circular fringes (A-2). Although changes in the shape of the droplets and aggregation of each particles occurred with additional drainage, the pseudoemulsion film persisted until the film thickness became less than 0.1 μm (A-3). The foam film rupture occurred around the droplets after the removal of most of the surfactant solution (A-4). As the dimensions of the thinning film become less than the original oil droplet diameter, distortion of the particle takes place. Solid particles existing on the surface of oil droplet form marked projection, which may be wetted by oil film. These projections give rise to distortion of the thinning film (B-2). As soon as the distortion of the thinning film took place on the top of the antifoam droplet, the pseudoemulsion film can be instantaneously ruptured (B-3). A large area of the slide glass surface can be seen in the direction of the thinner part of the liquid film after the rupture. Thus the liquid film rupture occurred when the film thickness decreased to a diameter comparable to the antifoam particle (ca. 8 μm).

Fig. 33 illustrates the SLM reflection features of the pseudoemulsion films from the model foaming solution formed on the top of the hydrophobic silica antifoams (C series) and the silicone resin antifoams (D series). Estimated images of the vertical sectional views are also shown under these SLM features. It is seen that the pseudoemulsion films were confirmed on the top of their particles when the thickness of the surfactant solutions was slightly

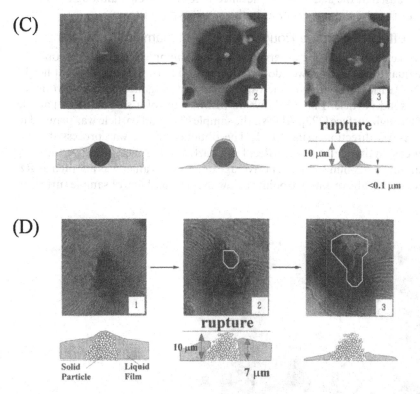

FIG. 33 Scanning laser microscopic reflection features of pseudoemulsion films from the model foaming solution (60 mM SDS/0.5 wt.% PVA) for the hydrophobic silica antifoams (C series) and the silicon resin antifoams (D series).

thinner than the diameter of the antifoam particles (C-1, D-1). Considerable differences in the features appeared as the drainage of the solution progressed. In the case of the hydrophobic silica particles, the pseudoemulsion film completely surrounded the particles with drainage and the film persisted until the film thickness became less than 0.1 μm (C-2). The foam film rupture occurred on the surface of the particles after the removal of most of the surfactant solution (C-3). On the other hand, when the liquid film thickness decreased and became slightly less than the diameter of the silicone resin particles, the acute top of solid particles distorted the pseudoemulsion film. As soon as the distortion took place on the top of the antifoam droplet, the pseudoemulsion film can be momentarily ruptured and part of the particle appeared from the liquid film (D-2: surrounded by the white line). The surface of the particle gradually appeared depending on the thinning of the liquid film. These results clearly show that the rupture of the pseudoemulsion film was caused at the acute top of the silicone resin antifoam although instantaneous rupture did not occur as in the mixed-type antifoam.

4. Effect of Surface Roughness on Antifoaming Properties

Surface shape could be observed using a scanning electron microscope. As the quantitative value, we adopted the method known as the 10-point height irregularities (Rz). It gives the value of surface roughness of the SLM image of a solid particle. Fig. 34 shows an SLM image of a silicone resin particle with rough surface [223]. At first, the sample of a solid particle was scanned in the z-axis direction using SLM. The obtained image was processed three dimensionally, and then the Rz of the particles was measured five times by scanning a 1.5-mm square. The average of these 25 values was adopted as Rz. Five points above and five points below the standard line of sample surface in

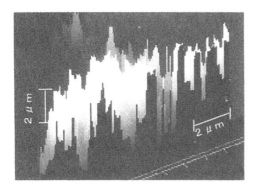

FIG. 34 Scanning laser microscope image of the silicone resin particle having a rough surface.

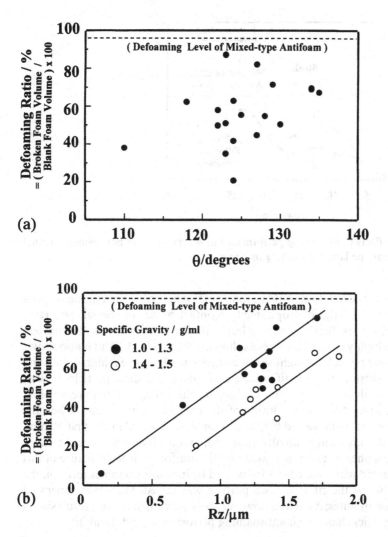

FIG. 35 Relationship between defoaming performance evaluated by the cylinder shaking test and the contact angle = θ (a), and the 10-point height irregularities = Rz of various kinds of developed silicone resin antifoams (b).

order of distance are chosen. Rz is defined by dividing the sum of the absolute value of these numbers by 5. By using the Rz, the surface roughness of a solid particle can be quantitatively determined as a physical property.

Various kinds of silicon resin particles with different surface properties for the Rz (0.82–1.87) and the contact angle (110–135°) were prepared with different reaction conditions (temperature, oil type, and silylchloride type).

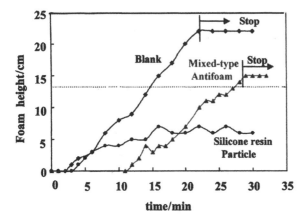

FIG. 36 Effects of defoaming performance of several types of antifoams evaluated by the drum-type laundry washing machine.

No significant correlation could be found in this figure between the defoaming performance and the hydrophobicity (contact angle) of the silicone resin particle. However, the defoaming effect evaluated by the cylinder shaking test correlated highly to the Rz values as shown in Fig. 35 [223]. And it also shows that the solid particles of light specific gravity tend to have a high defoaming effect. Furthermore, the antifoaming performance of these particles is confirmed in the drum-type laundry washing machine (Fig. 36). In the case of no antifoam, foam filled up the inside of the machine after 20 min. When the mixed-type antifoam was added, little foam was observed in the first 10 min, but after 10 min foam gradually appeared, and finally increased to 15 cm in height. This phenomenon suggested that the antifoaming performance of this type decreased with time or was influenced by increasing temperature. On the contrary, when the silicone resin particle was added, foam was generated within a few minutes; it did not, however, increase with time. This result shows that this particle has a high antifoaming performance and durability.

REFERENCES

1. Bikerman, J.J. In *Foams*; Springer-Verlag: New York, 1973; 65 pp.
2. Sosis, P. In *Detergency*; Cutler, W.G.; Davis, R.C., Eds.; Dekker: New York, 1975; 625.
3. Wilson, A.J. In *Foams: Theory, Measurements, and Applications*; Purd'homme; Khan, S.A., Eds.; Dekker: New York, 1996; 243 pp.
4. Domingo, X.; Fiquet, L.; Meijer, H. Tenside Surfactants Deterg. 1992, *29*, 16.

5. Zocchi, G. In *Handbook of Detergents, Part A: Properties*; Broze, G., Ed.; Dekker: New York, 1999; 419 pp.
6. Ross, J.; Miles, G.D. Oil Soap 1941, *18*, 99.
7. ASTM Standard D-1173-53.
8. JIS K-3362.
9. ISO Standard 696.
10. ASTM Standard D-3601.
11. JIS Standard 2234.
12. Schlachter, A.; Dierkes, H. Seifen, Ole, Fette, Waches 1951, *53*, 207.
13. DIN Standard D 3601.
14. Cox, M.F. J. Am. Oil Chem. Soc. 1989, *66*, 1637.
15. ASTM Standard D 3519.
16. Hart, J.R.; Degeorge, M.T. J. Soc. Cosmet. Chem. September/October 1980, *31*, 223.
17. JIS K-2241.
18. Weeks, L.E.; Harris, J.C.; Brown, E.L. J. Am. Oil Chem. Soc. 1954, *31*, 254.
19. Bhat, G.R.; Harper, D.L. In *Surfactants in Solution*; Mittal, K.L., Ed.; Plenum: New York, 1989; Vol. 10, 381.
20. Bikerman, J.J. Trans. Faraday Soc. 1938, *34*, 634.
21. ASTM Standard D 892.
22. ASTM Standard D 1881.
23. AFNOR Draft T 73-412.
24. Szekrenyesy, T.; Liktor, K.; Sandor, N. Colloids Surf. 1992, *68*, 267.
25. Malysa, K.; Miller, R.; Lunkenheimer Colloids Surf. 1991, *53*, 47.
26. Malysa, K. Adv. Colloid Interface Sci. 1992, *40*, 37.
27. Bisperink, C.G.J.; Ronteltap, A.D.; Prins, A. Adv. Colloid Interface Sci. 1992, *38*, 13.
28. Hartland, S. In *Thin Liquid Films: Fundamentals and Applications*; Ivanov, I.B., Ed.; Dekker: New York, 1988; 663 pp.
29. Rosen, M.J.; Solash, J. J. Am. Oil Chem. Soc. 1969, *46*, 399.
30. Sasaki, H.; Matsukawa, H.; Usui, S.; Matijevic, E. J. Colloid Interface Sci. 1986, *113*, 500.
31. Selecki, A.; Wasiaki, R. J. Colloid Interface Sci. 1984, *102*, 557.
32. Besio, G.J.; Oyler, G.; Prud'homme Rev. Sci. Instrum. 1985, *55*, 746.
33. Monsalve, A.; Schechter, R.S. J. Colloid Interface Sci. 1984, *97*, 327.
34. Bisperink, C.G.J.; Ronteltap, A.D.; Prins, A. Adv. Colloid Interface Sci. 1992, *38*, 13.
35. Assink, R.A.; Caprihan, A.; Fukushima, E. AIChEJ 1988, *34*, 2077.
36. German, J.B.; McCarty, M.J. J. Agric. Food Chem. 1989, *37*, 1321.
37. McCarthy, M.J. AIChEJ 1990, *36*, 287.
38. Weaire, D.; Hutzler, S.; Verbist, G.; Peters, E. Adv. Chem. Phys. 1997, *102*, 315.
39. Prause, B.A.; Glazier, J.A.; Gravina, S.J.; Montemagno, C.D. J. Phys. Condens. Matter. 1995, *7*, L511.
40. Ohya, M.; Minagawa, H. J. Jpn. Oil Chem. Soc. 1993, *42*, 302.

41. Phelan, R.; Weaire, D.; Peters, E.A.J.F.; Verbist, G. J. Phys. Condens. Matter. 1996, *8*, L475.
42. Kao, A.; Takahashi, A.; Matsudomi, N.; Kobayashi, K. J. Agric. Food Chem. 1989, *37*, 1321.
43. Dukhin, S.S.; Kretzschmar, G.; Miller, R. In *Dynamics of Adsorption at Liquid Interfaces*; Mobius, D., Miller, R., Eds.; Elsevier: Amsterdam, 1995; 140 pp.
44. Owens, D.K. J. Colloid Interface Sci. 1969, *29*, 469.
45. Abe, Y.; Matsumura, S. Tenside Deterg. 1983, *20*, 218.
46. Coltharp, K.A.; Franses, E.I. Colloids Surf. A 1996, *108*, 225.
47. Lin, S.-Y.; McKeigue, K.; Maldarelli, C. AIChEJ 1990, *36*, 1785.
48. Svitova, T.; Smirnova, Y.; Yakubov, G. Colloids Surf. 1995, *101*, 251.
49. Kragel, J.; Wustneck, R.; Clark, D.; Wilde, P.; Miller, R. Colloids Surf. A 1995, *98*, 127.
50. Svitova, T.; Hoffmann, H.; Hill, M. Langmuir 1996, *12*, 1712.
51. Miller, R.; Hoffmann, R.; Hartomann, R.; Schano, K.-H.; Halbig, A. Adv. Mater. 1992, *4*, 370.
52. Lange, H. J. Colloid Sci. 1965, *20*, 50.
53. Lucassen, J.; Giles, D.J. Chem. Soc., Fraday Trans. 1 1975, *71*, 217.
54. Patino, J.M.R.; Nino, M.R.R. Colloids Surf. 1995, *103*, 91.
55. Austin, M.; Bright, B.B.; Simpson, E.A. J. Colloid Interface Sci. 1967, *23*, 108.
56. Bendure, R.L. J. Colloid Interface Sci. 1971, *35*, 238.
57. Kloubek, J. J. Colloid Interface Sci. 1972, *41*, 17.
58. Mysels, K.J. Langmuir 1986, *2*, 423.
59. Mysels, K.J. Langmuir 1986, *2*, 428.
60. Woolfrey, S.G.; Banzon, G.M.; Groves, M.J. J. Colloid Interface Sci. 1986, *112*, 583.
61. Hua, X.Y.; Rosen, M.J. J. Colloid Interface Sci. 1988, *124*, 652.
62. Rosen, M.J.; Hua, XI.Y. J. Colloid Interface Sci. 1990, *139*, 397.
63. Hua, X.Y.; Rosen, M.J. J. Colloid Interface Sci. 1991, *141*, 180.
64. Rosen, M.J.; Hua, X.Y.; Zhu, Z.H. In *Surfactants in Solution*, Vol. 11; Mittal, K.L., Sha, D.O., Eds.;. Plenum: New York, 1991; 315 pp.
65. Garrett, P.R.; Ward, D.R. J. Colloid Interface Sci. 1989, *132*, 475.
66. Hirt, D.E.; Prud'homme, R.K.; Miller, B.; Rebenfeld, L. Colloids Surf. 1990, *44*, 101.
67. Mysels, K.J. Colloids Surf. 1990, *43*, 241.
68. Fainerman, V.B. Colloids Surf. 1991, *57*, 249.
69. Fainerman, V.B.; Makievski, A.V.; Miller, R. Colloids Surf. A 1993, *75*, 229.
70. Miller, R.; Joos, P.; Fainerman, V.B. Progr. Colloid Polym. Sci. 1994, *97*, 188.
71. Fainerman, V.B.; Miller, R. J. Colloid Interface Sci. 1995, *175*, 118.
72. Miller, R.; Fainerman, V.B.; Schano, K.-H.; Hofmann, A.; Heyer, W. Tenside Surfactants Deterg. 1997, *34*, 357.
73. Iliev, T.Z.; Dushkin, C.D. Colloid Polym. Sci. 1992, *270*, 370.
74. Dushkin, C.D.; Iliev, T.H. Colloid Polym. Sci. 1994, *272*, 1157.
75. Joos, P.; Fang, J.P.; Serrien, G. J. Colloid Interface Sci. 1992, *151*, 144.
76. Garrett, P.R.; Gratton, P.L. Colloids Surf. A 1995, *103*, 127.

77. Kao, R.L.; Edwards, D.A.; Wasan, D.T.; Chen, E. J. Colloid Interface Sci. 1992, *148*, 247.
78. Tamura, T.; Kaneko, Y.; Ohyama, M. J. Colloid Interface Sci. 1995, *173*, 493.
79. Tamura, T.; Takuuchi, Y.; Kaneko, Y. J. Colloid Interface Sci. 1998, *206*, 112.
80. Bergink-Martena, D.J.M.; Bos, H.J.; Prins, A. J. Colloid Interface Sci. 1994, *165*, 221.
81. Tunier, R.; Bisperink, J.; Berg, V.D.C.; Prins, A. J. Colloid Interface Sci. 1996, *179*, 327.
82. Danov, K.D.; Kralchevsky, P.A.; Ivanov, I.B. In *Handbook of Detergents: Part A. Properties*; Broze, G., Ed.; Dekker: New York, 1999; 303 pp.
83. Chang, C.H.; Franses, E.I. Colloids Surf. 1992, *69*, 189.
84. Chang, C.H.; Wang, N.-H.L.; Franses, E.I. Colloids Surf. 1992, *62*, 321.
85. Chang, C.H.; Franses, E.I. Colloids Surf. A 1995, *100*, 1.
86. Fainerman, V.D.; Makievski, A.V.; Miller, R. Colloids Surf. A 1994, *87*, 61.
87. Eastoe, J.; Dalton, J.S.; Rougueda, P.G.A.; Crooks, E.R.; Pitt, A.R.; Simister, E.A. J. Colloid Interface Sci. 1997, *188*, 423.
88. Colin, A.; Giermanska-Kahn, J.; Langevin, D. Langmuir 1997, *13*, 2953.
89. Durian, D.J.; Weitz, D.A. In *Encyclopedia of Chemical Technology*; 4th Ed.; Howe-Grant, M., Kroschwitz, J.I., Eds.; Wiley: New York, 1992; Vol. 11, 783.
90. Malhotra, A.K.; Wasan, D.T. In *Thin Liquid Films: Fundamentals and Applications*; Inanov, I.B., Ed.; Dekker: New York, 1988; 829 pp.
91. Bhakta, A.; Ruckenstein, E. Adv. Colloid Interface Sci. 1997, *70*, 1.
92. Lucassen, J. In *Anionic Surfactants*; Lucassen-Reynders, E.H., Ed.; Dekker: New York, 1981; 217 pp.
93. Schick, M.J.; Schmolka, I.R. In *Nonionic Surfactants: Physical Chemistry*; Dekker: New York, 1987; 835 pp.
94. Marangoni, G.M. Nuovo Cim. 1872, *2*, 239.
95. Prins, A. In *Foams*; Akers, R.J., Ed.; Academic Press: New York, 1977; 51 pp.
96. Izumi, Y.; Joke, K. J. Jpn. Oil Chem. Soc. 1993, *42*, 811.
97. Schott, H. J. Pharm. Sci. 1969, *58*, 1521.
98. Wrigley, A.N.; Smith, F.D.; Stirton, A.J. J. Am. Oil Chem. Soc. 1957, *34*, 39.
99. Schick, M.J.; Beyer, E.A. J. Am. Oil Chem. Soc. 1963, *40*, 66.
100. Danov, K.D.; Kralchevsky, P.A.; Ivanov, I.B. In *Handbook of Detergents: Part A. Properties*; Dekker: New York, 1999; 303 pp.
101. Ivanov, I.B.; Dimitrov, D.S. In *Thin Liquid Films: Fundamentals and Applications*; Ivanov, I.B., Ed.; Dekker: New York, 1988; 379 pp.
102. Vrij, A. J. Colloid Sci. 1964, *19*, 1.
103. Mann, J.A., Jr.; Caufield, K., Jr.; Gulden, G. J Opt. Soc. Am. 1971, *61*, 76.
104. Joosten, J.G.H. In *Thin Liquid Films: Fundamentals and Applications*; Ivanov, I.B., Ed.; Dekker: New York, 1988; 569 pp.
105. Kashiev, D.; Exerowa, D. J. Colloid Interface Sci. 1980, *77*, 501.
106. Exerowa, D.; Balinov, B.; Kashiev, D. J. Colloid Interface Sci. 1983, *94*, 45.
107. Exerowa, D.; Kashiev, D. Contemp. Phys. 1986, *27*, 429.
108. Muller, H.J.; Balinov, B.B.; Exerowa, D.R. Colloid Polym. Sci. 1988, *266*, 921.
109. Sheludko, A. Adv. Colloid Interface Sci. 1967, *1*, 391.

110. Malhotra, A.K.; Wasan, D.T. Chem. Eng. Commun. 1987, *55*, 95.
111. Exerowa, D.; Kashchiev; Platokanov, D. Adv. Colloid Interface Sci. 1992, *40*, 201.
112. Pugh, R.J. Adv. Colloid Interface Sci. 1996, *64*, 67.
113. Kralchevsky, P.A.; Danov, K.D.; Ivanov, I.B. In *Foams: Theory, Measurements, and Applications*; Prud'homme, R.K., Khan, S.A., Eds.; Dekker: New York, 1996; 1 pp.
114. Bhakta, A.; Ruckenstein, E. Adv. Colloid Interface Sci. 1997, *70*, 1.
115. Weaire, D.; Hutzler, S.; Verbist, G.; Peters, E. In *Adv. Chem. Phys.*; Prigogine, I., Rice, S.A., Eds.; Wiley: New York, 1997; Vol. 102, 315.
116. Kruglyakov, P.M. In *Thin Liquid Films: Fundamentals and Applications*; Ivanov, I.B., Ed.; Dekker: New York, 1988; 767 pp.; Derijaguin, B.V.; Titievskaya, A.S. Kolloid Zh., 1953; *15*, 416.
117. Exerowa, D.; Kolarov, T.; Khristov, K.H.R. Colloids Surf. 1987, *22*, 171.
118. Kolarov, T.; Cohen, R.; Exerowa, D. Colloids Surf. 1989, *22*, 49.
119. Cohen, R.; Koynova, R.; Technov, B.; Exerowa, D. Eur. Biophys. J. 1991, *20*, 203.
120. Exerowa, D.; Nikolova, A. Langmuir 1992, *8*, 3102.
121. Nikolov, A.D.; Wasan, D.T. J. Colloid Interface Sci. 1989, *133*, 1.
122. Rao, A.A.; Wasan, D.T.; Manev, E.D. Chem. Eng. Commun. 1982, *15*, 63.
123. Kralchevsky, P.A.; Nikolov, A.D.; Wasan, D.T.; Ivanov, I.B. Langmuir 1990, *6*, 1180.
124. Ivanov, I.B.; Dimitrov, A.S.; Nikolov, A.D.; Denkov, N.D.; Wasan, D.T. J. Colloid Interface Sci. 1992, *151*, 446.
125. Dimitrov, A.S.; Nikolov, A.D.; Kralchesky, P.A.; Ivanov, I.B. J. Colloid Interface Sci. 1992, *151*, 462.
126. de Feijter, J.A.; Vrij, A. J. Colloid Interface Sci. 1978, *64*, 269.
127. de Feijter, J.A.; Vrij, A. J. Colloid Interface Sci. 1979, *70*, 456.
128. Johansson, G.; Pugh, R.J. Int. J. Miner. Process 1992, *34*, 1.
129. Waltermo, A.; Manev, E.; Pugh, R.; Claesson, P. J. Disp. Sci. Technol. 1994, *15*, 273.
130. Manev, E.; Pugh, R.J. Langmuir 1991, *7*, 2253.
131. Manev, E.; Pugh, R.J. J. Colloid Interface Sci. 1992, *151*, 505.
132. Manev, E.; Pugh, R.J. Colloids Surf. A 1993, *70*, 289.
133. Bergeron, V.; Radke, C.J. Langmuir 1992, *8*, 3020.
134. Aronson, A.S.; Bergeron, V.; Fagan, M.E.; Radke, C.J. Colloids Surf. A 1994, *83*, 109.
135. Bergeron, V.; Radke, C.J. Colloid Polym. Sci. 1995, *273*, 165.
136. Bergeron, V.; Langevin, D.; Asnacios, A. Langmuir 1996, *12*, 1550.
137. Bergeron, V. Langmuir 1997, *13*, 3474.
138. Swayne, E.N.; Newman, J.; Radke, C.J. J. Colloid Interface Sci. 1998, *203*, 69.
139. Vassilieff, C.S.; Manev, E.D. Colloid Polym. Sci. 1995, *273*, 512.
140. Velev, O.C.; Constantinides, G.N.; Avraam, D.G.; Payatakes, A.C.; Borwankar, R.P. J. Colloid Interface Sci. 1995, *175*, 68.

141. Koczo, K.; Nikolov, A.D.; Wasan, D.T.; Borwankar, R.P.; Gonsalves, A. J. Colloid Interface Sci. 1996, *178*, 694.

142. Sharma, A.; Ruckenstein, E. J. Colloid Interface Sci. 1985, *106*, 12.

143. Clark, D.C.; Dann, R.; Mackie, A.R.; Mingins, J.; Pinder, A.C.; Purdy, P.W.; Russell, E.J.; Smith, L.J.; Wilson, D.R. J. Colloid Interface Sci. 1990, *138*, 195.

144. Joye, J.-L.; Miller, C.A. Langmuir 1992, *8*, 3083.

145. Jones, M.N.; Mysels, K.J.; Scholten, P.C. Trans. Faraday Soc. 1966, *62*, 1336.

146. Müller, H.J.; Rheinländer, Th. Langmuir 1996, *12*, 2334.

147. Rutland, M.W. Colloids Surf. A 1994, *83*, 121.

148. Prins, A.; Arcuri, C.; Van Den Tempel, M. J. Colloid Interface Sci. 1967, *24*, 84.

149. Yamanaka, T. Bull. Chem. Soc. Jpn. 1970, *43*, 633.

150. Yamanaka, T. Bull. Chem. Soc. Jpn. 1975, *48*, 1755.

151. Yamanaka, T.; Hatashi, M.; Matsuura, R. J. Colloid Interface Sci. 1982, *88*, 458.

152. Yamanaka, T.; Tano, T.; Kamegaya, O.; Exerowa, D.; Cohen, R.D. Langmuir 1994, *10*, 1871.

153. Mysels, K.J. J. Phys. Chem. 1964, *68*, 3441.

154. Lyklema, J.; Scholten, P.C.; Mysels, K.J. J. Phys. Chem. 1965, *69*, 116.

155. Vikingstad, E.; Skauge, A.; Hoiland, H. J. Colloid Interface Sci. 1978, *66*, 240.

156. Lai, J.; Di Megrio, J.-M. J. Colloid Interface Sci. 1994, *164*, 506.

157. Zhang, Z.; Liang, Y.; Zhang, Y. Chem. Phys. Lett. 1995, *246*, 101.

158. Basheva, E.S.; Danov, K.D.; Kralchevsky, P.A. Langmuir 1997, *13*, 4342.

159. Schalchli, A.; Sentenac, D.; Benattar, J.J. J. Chem. Soc., Faraday Trans. 1996, *92*, 553.

160. Baets, P.J.M.; Stein, H.N. Langmuir 1992, *8*, 3099.

161. Tamura, T.; Kaneko, Y.; Nikaido, M. J. Colloid Interface Sci. 1997, *190*, 61.

162. Tamura, T., Ph.D. Dissertation, Science University of Tokyo, 2000.

163. McEntee, W.R.; Mysels, K.J. J. Phys. Chem. 1969, *73*, 3018.

164. Umemura, J.; Matsumoto, M.; Kawai, T.; Takenaka, T. Can. J. Chem. 1985, *63*, 1713.

165. Umemura, J.; Kawai, T. Can. J. Chem. 1990, *94*, 62.

166. Benattar, J.J.; Schalchili, A.; Belorgey, O. J. Phys., I (France) 1992, *2*, 955.

167. Stein, H.N. Adv. Colloid Interface Sci. 1991, *34*, 175.

168. Hudales, J.B.; Stein, H.N. J. Colloid Interface Sci. 1990, *138*, 354.

169. Kimizuka, H.; Sasaki, T. Bull. Chem. Soc. 1951, *24*, 230.

170. Ruckenstein, E.; Sharma, A. J. Colloid Interface Sci. 1987, *119*, 1.

171. Wasan, D.T.; Nikolov, A.D.; Lobo, L.A.; Koczo, K.; Edwards, D.A. Progr. Surf. Sci. 1992, *39*, 119.

172. Shah, D.O.; Djabbarab, N.F.; Wasan, D.T. Colloid Polym. Sci. 1978, *256*, 1002.

173. Brown, A.G.; Thuman, W.C.; Mc Bain, J.W. J. Colloid Interface Sci. 1953, *8*, 491.

174. Schick, M.J.; Schmolka, I.R. In *Nonionic Surfactants: Physical Chemistry*; Schick, M.J., Ed.; Dekker: New York, 1987; 835 pp.

175. Rosen, M.J. In *Surfactants and Interfacial Phenomena*; 2nd Ed.; Wiley: New York, 1989; 276 pp.

176. Cox, M.F. J. Am. Oil Chem. Soc. 1989, *66*, 367.

177. Satkowski, W.B.; Huang, S.K.; Liss, R.L. In *Nonionic Surfactants*; Schick, M.J., Ed.; Dekker: New York, 1967; 108 pp.
178. Meguro, K.; Ueno, M.; Esumi, K. In *Nonionic Surfactants: Physical Chemistry*; Schick, M.J., Ed.; Dekker: New York, 1987; 109 pp.
179. Overbeek, J.Th.G. J. Phys. Chem. 1960, *64*, 1178.
180. Owen, M.J. In *Encyclopedia of Chemical Technology*; Vol. 7; Kroscwitz, H.P.R., Ed.; Dekker: New York, 1993; 928 pp.
181. Garett, P.R. In *Defoaming: Theory and Industrial Applications*; Garrett, P.R., Ed.; Dekker: New York, 1993; 1 pp.
182. Wasan, D.T.; Koczo, K.; Nikolov, A.D. In *Foams: Fundamentals and Applications in Petroleum Industry*; Schramm, L.L., Ed.; ACS Press: Washington, 1994; 47 pp.
183. Aveyard, R.; Clint, J.H. Curr. Opin. Colloid Interface Sci. 1996, *1*, 764.
184. Pelton, R. Pulp. Pap. Can. 1989, *90*(2), T61.
185. Ross, S.; Haak, R.M. J. Am. Chem. Soc. 1958, *62*, 1260.
186. Lobo, L.A.; Nikolov, A.D.; Wasan, D.T. J. Disp. Sci. Technol. 1989, *10*, 143.
187. Koczo, K.; Lobo, L.A.; Wasan, D.T. J. Colloid Interface Sci. 1992, *150*, 150.
188. Lobo, L.; Wasan, D.T. Langmuir 1993, *9*, 1668.
189. Rácz, G.; Koczo, K.; Wasan, D.T. J. Colloid Interface Sci. 1996, *181*, 124.
190. Kim, Y.-H.; Koczo, K.; Wasan, D.T. J. Colloid Interface Sci. 1997, *187*, 29.
191. Nñmeth, Zs.; Rácz, Gy.; Koczo, K. Colloids Surf. A 1997, *127*, 151.
192. Binks, B.P.; Dong, J. J. Chem. Soc., Faraday Trans. 1998, *94*, 401.
193. Garrett, P.R. J. Colloid Interface Sci. 1980, *76*, 587.
194. Garrett, P.; Moore, P. J. Colloid Interface Sci. 1993, *159*, 214.
195. Garrett, P.R. Langmuir 1995, *11*, 3576.
196. Aveyard, R.; Binks, B.P.; Fletcher, P.D.I.; Garrett, P.R. J. Chem. Soc. Faraday Trans. 1993, *89*, 4313.
197. Aveyard, R.; Clint, J.H. J. Chem. Soc., Faraday Trans. 1995, *91*, 2681.
198. Aveyard, R.; Clint, J.H. J. Chem. Soc., Faraday Trans. 1996, *92*, 85.
199. Roberts, K.; Axberg, C.; Österlund, R. J. Colloid Interface Sci. 1977, *62*, 264.
200. Kulkarni, K.; Goddard, E.D.; Kanner, B. J. Colloid Interface Sci. 1977, *59*, 468.
201. Bergeron, V.; Fagan, M.E.; Radke, C.J. Langmuir 1993, *9*, 1704.
202. Bergeron, V.; Cooper, P.; Fischer, C.; Giermanska-Kahn, J.; Langevin, D.; Pouchelon, A. Colloids Surf. 1997, *122*, 103.
203. Kulkarni, R.D.; Goddard, E.D.; Kanner, B. Ind. Eng. Chem., Fundam. 1977, *16*, 427.
204. Frye, G.C.; Berg, J.C. J. Colloid Interface Sci. 1989, *127*, 222.
205. Frye, G.C.; Berg, J.C. J. Colloid Interface Sci. 1989, *130*, 54.
206. Dippenaar, A. Int. J. Miner. Process. 1982, *9*, 1.
207. Dippenaar, A. Int. J. Miner. Process. 1982, *9*, 15.
208. Garrett, P.R. J. Colloid Interface Sci. 1979, *69*, 107.
209. Johansson, G.; Pugh, R.J. Int. J. Miner. Process. 1992, *34*, 1.
210. Aveyard, R.; Cooper, P.; Fletcher, P.D.I.; Rutherford, C.E. Langmuir 1993, *9*, 604.

211. Aveyard, R.; Binks, B.P.; Fletcher, P.D.I.; Rutherford, C.E. J. Disp. Sci. Technol. 1994, *15*, 251.

212. Aveyard, R.; Beake, B.D.; Clint, J.H. J. Chem. Soc., Faraday Trans. 1996, *92*, 4271.

213. Tang, F.-Q.; Xiao, Z.; Tang, J.-A.; Jiang, L. J. Colloid Interface Sci. 1989, *131*, 498.

214. Aronson, M. Langmuir 1986, *2*, 653.

215. Frye, G.C.; Berg, J.C. J. Colloid Interface Sci. 1989, *130*, 54.

216. Garrett, P.R.; Davis, J.; Rendall, H.M. Colloids Surf. A 1994, *85*, 159.

217. Rácz, G.; Koczo, K.; Wasan, D.T. J. Colloid Interface Sci. 1996, *181*, 124.

218. Koczo, K.; Koczone, J.K.; Wasan, D.T. J. Colloid Interface Sci. 1994, *166*, 225.

219. Kulkarni, R.D.; Goddard, E.D.; Kanner, B. J. Colloid Interface Sci. 1977, *59*, 468.

220. Aveyard, R.; Cooper, P.; Fletcher, P.D.I.; Rutherford, C. Langmuir 1993, *9*, 604.

221. Ross, S.; McBain, W. Ind. Eng. Chem. 1944, *36*, 570.

222. Tamura, T.; Kageyama, M.; Kaneko, Y.; Kishino, T.; Nikaido, M. J. Colloid Interface Sci. 1999, *213*, 179.

223. Tamura, T., Kageyama, M., Kaneko, and Nikaido, M., In *Emulsions, Foams, and Thin Films*; Mittal, K.L., Kumar, P., Eds.; Dekker: New York, 2000; 161 pp.

224. Excerowa, D.; Scheludko, A. Comptes Rendus Bulg. Acad. Sci. 1971, *24*, 47.

3

Liquid Drops at Surfaces

EDGAR M. BLOKHUIS Leiden Institute of Chemistry, Leiden University, Leiden, The Netherlands

I. INTRODUCTION

Any review on the shape of a liquid droplet on top of a solid surface has to start with the pioneering work by P. S. Laplace and Sir Thomas Young almost two centuries ago [1,2]. Young and Laplace set out to describe the phenomenon of "capillary action" in which the liquid inside a small capillary tube may rise several centimeters above the liquid outside the tube [3]. To understand this effect, two fundamental equations were derived by Young and Laplace. The first equation, known as the Laplace or Young–Laplace equation [1], relates the curvature at a certain point of the liquid surface to the pressure difference between both sides of the surface, and we consider it next in more detail. The second equation is Young's equation [2], which relates the contact angle to the surface tensions involved.

A. Laplace Law

When we consider the interface between a liquid and its vapor, it has been recognized since the work by Gibbs [4] that it is convenient to introduce the concept of the *dividing surface*. This is the hypothetical surface to which we assign all the properties of the surface, such as surface tension. On a molecular scale, the precise location of the dividing surface cannot be uniquely determined and a certain choice has to be made for it. According to Gibbs, it has to be chosen "sensibly coincident" with the interfacial region, that is, the region in which all the properties of the liquid, such as the density (see Fig. 1), smoothly cross over to those of the vapor. A very common choice for the location of the dividing surface, especially for a one component system, is the so-called *equimolar* surface, which is defined as the surface for which the excess number density of a certain component vanishes—the excess of a quantity being defined as the integrated difference between that quantity and its

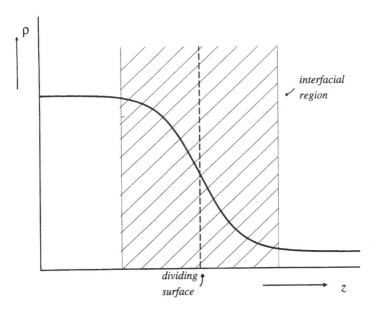

FIG. 1 Density profile of a liquid–vapor interface as a function of height z showing the interfacial region with the location of the dividing surface.

value extrapolated to the dividing surface (see Fig. 2). In this chapter we assume that a choice for the location of the dividing surface has been made so that we can treat the liquid–vapor surface as a mathematical surface and investigate its properties.

A very important feature of the surface is its *curvature*. Any point on the surface is characterized by two radii of curvature, R_1 and R_2. The radius of curvature at a certain point along a certain direction is defined as the radius of the sphere with which the surface can be locally approximated such that the first and second derivatives in that direction are equal to those of the sphere. Together with this definition, one needs to determine the *sign* of the radius of curvature, since the curvature can be directed toward either the liquid or the vapor phase. The usual convention, the one that we will also adopt here, is to define curvature toward the liquid phase (or in general, toward the denser phase) as positive.

In terms of the two radii of curvature the total curvature, J, and Gaussian curvature, K, are defined by

$$J \equiv \frac{1}{R_1} + \frac{1}{R_2},$$

$$K \equiv \frac{1}{R_1 R_2}.$$

$$(1)$$

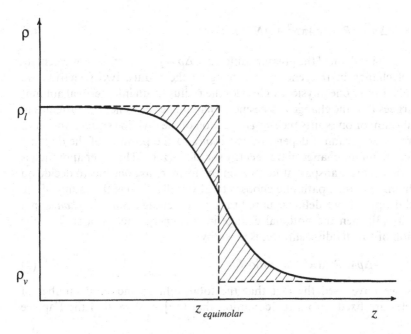

FIG. 2 Density profile in the interfacial region with the definition of the equimolar surface as the location where the excess density vanishes.

The important advantage of characterizing a point on the surface by J and K instead of by R_1 and R_2 is that while R_1 and R_2 are only defined in certain directions, J and K are *direction independent* as long as R_1 and R_2 are chosen perpendicular to each other.

Having defined what we mean by the surface of a liquid and its curvature, we now consider the derivation of Laplace's law for a spherical drop of liquid with radius R. In this case the radii of curvature at every point on the surface are simply equal to R. The usual derivation of the Laplace equation follows from a consideration of the net change in free energy of the liquid droplet resulting from a change in the radius R. Consider the Helmholtz free energy of a spherical drop of liquid consisting of only one type of molecule

$$F = -p_1 V_1 - p_v V_v + \sigma A + \mu N. \tag{2}$$

In this expression, $p_{1,v}$ and $V_{1,v}$ are the pressure and volume of the liquid and vapor phase, respectively, σ is the surface tension, and A is the surface area. Furthermore, μ is the chemical potential and N the total number of particles. If V is the total volume of the system, then we can write Eq. (2) in terms of the radius of the droplet

$$F = -\Delta p \frac{4}{3} \pi R^3 + \sigma 4\pi R^2 + \mu N - p_v V, \qquad (3)$$

where we have defined the pressure difference $\Delta p \equiv p_l - p_v$. If we now calculate the net change in free energy by changing the radius, two scenarios are possible. Either one physically changes the radius by an infinitesimal amount and argues that the change in free energy should be zero since the system is in equilibrium, or one shifts the *chosen* location of the dividing surface and since the free energy cannot depend on the *choice* of the position of the dividing surface, again the change in free energy should be zero. The latter approach is somewhat more transparent because in the former case one has to decide on the thermodynamic path one chooses to physically change the radius of the liquid droplet. If we define square brackets to denote a *notional change* in a quantity [3], then the notional change in free energy due to a shift in the location of the dividing surface, is given by

$$[dF] = -\Delta p 4\pi R^2 [dR] + \sigma 8\pi R[dR], \qquad (4)$$

where we have used the fact that the total volume and total number of particles are fixed. From the condition that $[dF] = 0$ we find the Laplace equation [1]

$$\Delta p = \frac{2\sigma}{R}. \qquad (5)$$

The Laplace equation relates the difference between the pressures of the liquid and vapor phase across a *curved* surface to the surface tension of the *planar* interface at coexistence, and in this sense it is a remarkable relation. It should be realized that the Laplace equation is an asymptotic equation, only valid in the thermodynamic limit of very large liquid droplets. For small radii, corrections do exist, and the form of the Laplace equation in Eq. (5) has to be seen as the first term in an expansion in $1/R$. The leading order correction in $1/R$ to the Laplace equation defines the so-called Tolman length δ [5]

$$\Delta p = \frac{2\sigma}{R} - 2\delta\sigma \frac{1}{R^2} + \cdots. \qquad (6)$$

To uniquely define δ via the above equation, a certain choice for the location of the dividing surface, i.e., the value of R, has to be made. This can be deduced by the fact that if one shifts the radius R by a microscopic distance Δ, the coefficient of the $1/R^2$ term is affected. The convention is that in the above equation R corresponds to the equimolar radius, $R = R_e$.

The typical value of the Tolman length is expected to be microscopic, i.e., of the order of a molecular diameter. However, the Tolman length is also a measure of the asymmetry between the profile on the liquid side and the vapor side of the planar interface. For a simple liquid–vapor interface, the density profile is expected to be nearly antisymmetric and the Tolman length might only be a fraction of a molecular diameter. The smallness of the Tolman length for a simple, one-component system makes it difficult to obtain reliable values from experiments or even from simulations. The most reliable estimates from Molecular Dynamics (MD) simulation of a Lennard–Jones liquid–vapor interface [6] indicate that the Tolman length is not very temperature dependent and is about two tenths of the Lennard–Jones radius in magnitude.

We have now derived the Laplace equation for the case of a spherical liquid droplet. In a more general form, the Laplace equation relates the total curvature at a point of the surface to the difference in pressure on both sides of the surface at that point.

$$\Delta p = \sigma J. \tag{7}$$

One may easily verify that the above equation reduces to the Laplace equation in Eq. (5) for a spherical surface when one inserts $J = 2/R$. When we consider the shape of a droplet in a gravitational field in the next section, the above expression is used to describe the shape. Furthermore, we present in Section II.C a detailed derivation of the Laplace equation in Eq. (7) as it pertains to the nonspherical shape of a droplet in a gravitational field. First, we need to consider the interaction of the liquid surface in contact with the solid substrate on which the droplet resides.

B. Young's Law

In 1805, Thomas Young wrote on page 66 of his essay on the "Cohesion of Fluids" [2]

> But it is necessary to premise one observation, which appears to be new, and which is equally consistent with theory and experiment; that is that for each combination of a solid and a fluid, there is an appropriate angle of contact between the surfaces of the fluid, exposed to the air, and to the solid.

Young then proceeded to describe in words—Young took pride in not using any formulas—the equation that we now know as "Young's equation" [2]

$$\sigma_{sv} = \sigma_{sl} + \sigma_{lv} \cos \theta, \tag{8}$$

where σ_{sv}, σ_{sl}, and σ_{lv} are the surface tensions of the solid–vapor, solid–liquid, and liquid–vapor surfaces, respectively. The angle θ is the contact angle at which the liquid–vapor surface meets the solid substrate (see Fig. 3).

Young derived the above equation by considering the balance of surface forces acting on the triple line where the solid, liquid, and vapor phases meet;

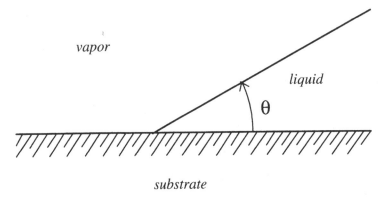

FIG. 3 Definition of the contact angle θ.

the surface force (per unit of length of the triple line) of the solid–vapor surface is equal to the sum of the surface force of the solid–liquid surface and the component of the surface force of the liquid–vapor surface along the surface of the substrate.

An alternative derivation of Young's equation follows the same route as the derivation of the Laplace equation using a notional change of the location of the dividing surface. Consider the surface free energy of the system depicted in Fig. 4. Around the line of three-phase contact a cylinder is drawn with length L and radius R, and implicitly we assume that R and L approach

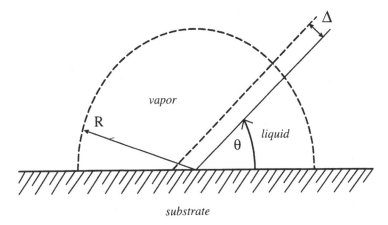

FIG. 4 Cylindrical region with radius R around the line of three-phase contact. The liquid–vapor surface is shifted over a distance Δ.

infinity. The contribution to the surface free energy associated with the three surfaces reads in this geometry

$$F_s = \sigma_{sv} RL + \sigma_{sl} RL + \sigma_{lv} RL. \tag{9}$$

If the location of the dividing surface is now shifted by a distance Δ (dashed line in Fig. 4), the notional change in surface free energy associated with the notional change in surface area is given by

$$[dF_s] = \sigma_{sv} \frac{\Delta}{\sin\theta} L + \sigma_{sl} \frac{\Delta}{\sin\theta} L + \sigma_{lv} \frac{\Delta}{\tan\theta} L. \tag{10}$$

Again, from the requirement that $[dF_s] = 0$, we recover Young's law.

Even though alternative and more rigorous derivations of Young's law have appeared in the literature [1,3,7–11], the validity of Young's law has been a continued subject of scrutiny [12–17]. Young's equation has been questioned as a general rule [12,13] and in the presence of a gravitational field [12–15]. Its validity has been investigated using integral relations and arguments have been given for the introduction of a microscopic contact angle [16,17]. As we will discuss in Section III, it is now recognized that for very small liquid droplets on a substrate, Young's law indeed has to be modified to account for the presence of the line tension of the triple line [18–21]. For macroscopically large droplets and for macroscopically large distances from the triple line, however, the contact angle is given by Young's law.

C. Wetting and Antonow's Rule

Young's equation describes the situation depicted in Fig. 3 in which three phases meet at a common line of contact [22]. This is the situation of *partial wetting*, and it is characterized by the existence of a contact angle as shown in the figure. In general, if all three phases are fluidlike, three contact angles, of which only two are independent, describe the situation of partial wetting, but in this chapter we continue to assume that one of the three phases is a solid substrate.

The situation as depicted in Fig. 3 is only one of two possible configurations. The second situation is the one of *complete wetting*, and it is characterized by the presence of a macroscopically thick layer of one of the phases between the other two (see Fig. 5). Whereas in the case of partial wetting the three surface tensions involved obey Young's law, in the case of complete wetting the surface tensions are related through Antonow's rule [22,23], which states that the surface tension of the solid–vapor interface, when it is intruded by a macroscopically thick liquid layer, is given by the sum of the surface tensions of the solid–liquid and liquid–vapor interfaces

$$\sigma_{sv} = \sigma_{sl} + \sigma_{lv} \tag{11}$$

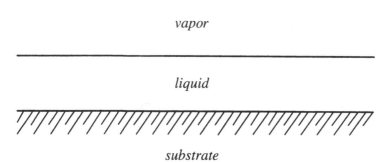

substrate

FIG. 5 Situation corresponding to the complete wetting regime in which the liquid phase intrudes between the substrate and vapor phase.

Antonow discovered this rule empirically after measuring the surface tensions of different three-phase systems. At the time, Antonow's rule was the subject of intense scrutiny in the literature partly because Antonow claimed the validity of the rule in *any* three-phase system with deviations being caused only by supposed nonequilibrium effects [23]. We now know that the rule holds exactly, but only in the complete wetting regime.

    We have seen that a three-phase system is in a state of either partial wetting or complete wetting. It is then not difficult to imagine the situation in which a thermodynamic variable of the system is changed, say the temperature, so that the system undergoes a transition from the partially wet state to the completely wet state. Such a transition is termed the *wetting transition*. One convenient way to monitor the wetting transition experimentally is to measure the contact angle as a function of temperature, for instance. On approach to the wetting transition, the contact angle decreases until at the wetting transition it is zero and a thick liquid layer spreads at the solid–vapor interface.

    Theoretically, the wetting transition can occur in two distinct fashions. To elaborate this point, we consider the possible ways the surface tensions involved may depend on temperature. In Fig. 6, we have plotted σ_{sv} and the sum $\sigma_{sl} + \sigma_{lv}$ as a function of temperature. These two curves represent the surface free energies of the two competing surface phases, i.e., the *partially wet* phase (σ_{sv}) versus the *completely wet* phase ($\sigma_{sl} + \sigma_{lv}$) that the system can "choose" from. Below the *wetting temperature* T_w, the system is partially wet since the curve σ_{sv} is below the curve $\sigma_{sl} + \sigma_{lv}$, while above $T = T_w$ the system is completely wet. If the temperature dependence is that of Fig. 6a, the transition from one state to the other occurs abruptly; the surface tension is continuous but the first derivative of the surface tension jumps from one value to the other on passing the wetting transition. Here the wetting transition is

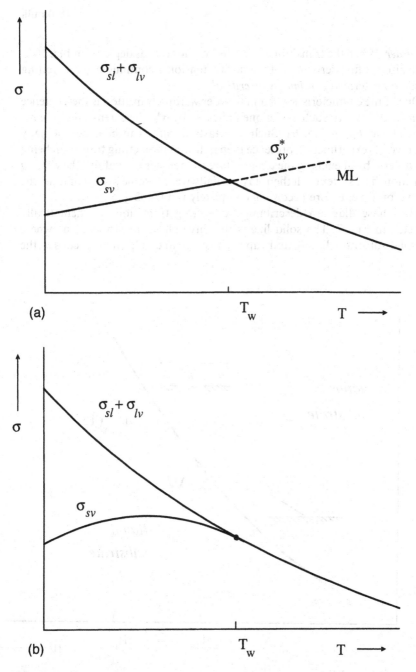

FIG. 6 Surface tensions σ_{sv} and $\sigma_{sl} + \sigma_{lv}$ as a function of temperature T. In (a) is the wetting transition ($T = T_W$), a first-order wetting transition with a mestastable continuation of the partially wet phase above T_W (σ_{sv}^*) until the metastable limit (ML). In (b) the wetting transition is continuous.

first order. When the transition occurs more smoothly as depicted in Fig. 6b—
when also the first derivative of the surface tension is continuous—the wetting
transition is termed *continuous* or *critical*.

One of the characteristics of a first-order wetting transition is the existence
of *metastable states* such as the one indicated by σ_{sv}^*, the extension of the σ_{sv}
curve above T_w, in Fig. 6a. Such a metastable state can be experimentally
observed; its experimental presence then indicates the wetting transition being
first order—by quenching the temperature to a temperature above the wetting
transition. The system will then stay partially wet for some period of time, the
nucleation time, before becoming completely wet [24–27].

The phase diagram describing the wetting transition is schematically
depicted in Fig. 7. The solid line is the three-phase coexistence line where
the solid substrate, liquid, and vapor phase are in equilibrium. It ends at the

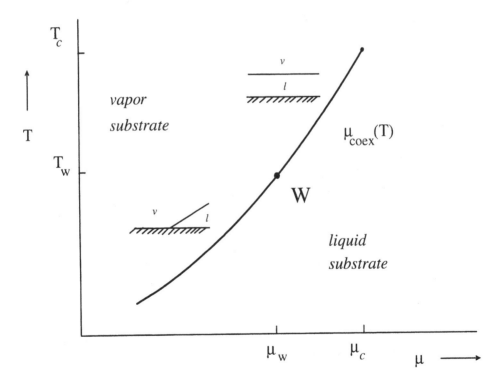

FIG. 7 Phase diagram. Below the wetting transition (W), along the line of three-
phase coexistence [$\mu = \mu_{coex}(T)$], the system is partially wet, while it is completely wet
above the wetting transition.

critical point of the liquid and vapor phase. Somewhere along the three-phase coexistence line, a wetting transition takes place (W). That a wetting transition must occur on approach to the critical point was first argued by Cahn [22,28]. Cahn's argument is based on the fact that at the critical point, the surface tension of the liquid–vapor vanishes with a critical exponent μ, which is argued to be larger than the critical exponent with which $\sigma_{sv} - \sigma_{sl}$ vanishes. The result is that close enough to the critical point $\sigma_{sv} - \sigma_{sl}$ *must* be larger then σ_{lv}, which means that $\sigma_{sv} > \sigma_{sl} + \sigma_{lv}$, and a wetting transition occurs.

Below the wetting transition the system is partially wet, whereas above the wetting transition the system is completely wet. We therefore can see that while it seems that Antonow's rule is nothing but a very special case ($\theta = 0$) of Young's law, this is not correct. From the phase diagram in Fig. 7, it is clear that Antonow's rule is valid for the *entire* temperature range above the wetting transition and not just at the wetting transition itself [23].

D. Outline

After this brief introduction into the necessary background of wetting phenomena, we now continue the description of liquid drops on a substrate. In Section II, we discuss the shape of large droplets in a gravitational field. This analysis dates back to the work in 1883 by Bashforth and Adams [29], who supplied numerical tables for the shape of the liquid droplet. In Section III, we consider very small droplets for which gravity can be disregarded but for which the line tension connected to the line of three-phase contact becomes important. In particular, we address the influence of the presence of line tension on the wetting transition, and the experimental determination of the magnitude of line tension. In Section IV, we discuss systems that are large enough to discard line tension and small enough to discard gravity, but for which the shape is not described by surface tension alone. These are systems such as vesicles, micelles, and microemulsion droplets with surfactant-like molecules present at the surface of the adhered droplet. Finally, in Section V, we provide a brief summary and discussion.

II. SHAPE OF A LIQUID DROP IN A GRAVITATIONAL FIELD

In this section, we consider the influence of a gravitational field on the shape of a liquid droplet residing on a solid substrate (see Fig. 8). This topic was already addressed some 100 years ago by Bashforth and Adams [29], who supplied numerical tables for the shape of the liquid droplet. Their analysis is based on two equations: the Laplace equation to describe the shape of the droplet, and Young's equation to determine the contact angle, Young's angle,

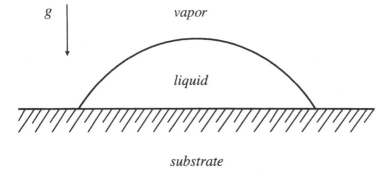

substrate

FIG. 8 Liquid droplet on a solid substrate in a gravitational field.

with which the droplet meets the substrate on which it resides. The way the gravitational field enters the analysis is through the (linear) dependence of the Laplace pressure difference on height. The gravitational field, however, does not influence the value of the contact angle, which remains to be given by Young's Law.

In the following, we first derive the shape equations used by Bashforth and Adams [29] to calculate (numerically) the shape of the liquid droplet. After having rewritten the shape equations somewhat, we then supply a schematic program to solve the shape equations numerically using the Runge–Kutta method. Finally, in this section, we show from a functional minimization of the free energy that the profile thus determined from the Laplace equation and Young's law corresponds to a minimum in the free energy [11].

A. Shape Equations

To derive the shape equations, consider the system depicted in Fig. 9. The origin of our coordinate system is chosen at the apex of the droplet with the direction of the coordinate z chosen opposite to the direction of the gravitational field. Because the droplet is axisymmetric around the z-axis, the shape of the droplet is completely determined by the function $z(r)$, with r the radial distance to the z-axis (see Fig. 9).

The Laplace equation given in the previous section relates the pressure difference between the inside and the outside of a droplet (p_l and p_v, respectively) at any point P on the surface of a droplet to the surface tension and the radii of curvature (R_1 and R_2) at that point

$$\Delta p = \sigma_{lv}\left(\frac{1}{R_1} + \frac{1}{R_2}\right). \tag{12}$$

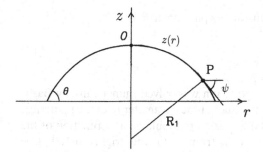

FIG. 9 Shape of the liquid droplet on a substrate in terms of $z(r)$. The origin (O) is located at the apex. The point P is an arbitrary point on the liquid–vapor surface with radii of curvature R_1 (shown) and R_2 (perpendicular to the plane of the figure).

The radii of curvature, R_1 and R_2, can be expressed in terms of the previously defined shape function $z(r)$

$$\frac{1}{R_1} = \frac{z'(r)}{r\left[1 + z'(r)^2\right]^{1/2}},$$

$$\frac{1}{R_2} = \frac{z''(r)}{\left[1 + z'(r)^2\right]^{3/2}}, \tag{13}$$

where R_1 is the radius of curvature in the plane of Fig. 9 and R_2 is the radius of curvature in the plane perpendicular to Fig. 9. In a gravitational field with strength g, the pressure varies linearly with the height z

$$p_1 = p_{10} - \rho_1 g z, \quad p_v = p_{v0} - \rho_v g z, \tag{14}$$

where we have defined ρ_1 and ρ_v as the (constant) densities in the liquid and vapor phase, respectively, and where we have chosen the origin (O) at the apex as our point of reference. At the apex of the droplet, the radii of curvature are equal, say equal to b, so that the Laplace equation at this point reads

$$p_{10} - p_{v0} = \frac{2\sigma}{b}. \tag{15}$$

When we combine Eqs. (12)–(15), we arrive at the following differential equation for $z(r)$

$$\frac{b}{r} \frac{z'(r)}{\left[1 + z'(r)^2\right]^{1/2}} + \frac{bz''(r)}{\left[1 + z'(r)^2\right]^{3/2}} = 2 - \beta\frac{z(r)}{b}, \tag{16}$$

where we have defined the dimensionless parameter β as

$$\beta \equiv \frac{\Delta\rho g b^2}{\sigma}, \tag{17}$$

and the density difference $\Delta\rho \equiv \rho_l - \rho_v$.

This second-order differential equation was solved numerically by Bashforth and Adams [29] for different values of the parameter β. In the numerical tables they supplied, $R \equiv r/b$ and $Z \equiv z/b$ are tabulated as a function of the angle ψ (defined in Fig. 9) which runs from $0°$ (at the origin) to $180°$ (see Fig. 10). The analysis of Bashforth and Adams is completed by locating the substrate such that the angle ψ is equal to the Young angle θ (see Fig. 10). For given θ and fixed value of the parameter β, the volume of the droplet is completely determined and numerical values were also tabulated by Bashforth and Adams [29]. Alternatively, it is thus possible to consider the volume of the droplet as the parameter that can still be varied instead of β (or b). Small values of the parameter β then indicate a small gravitational field or, equivalently, small droplet volumes. When $\beta = 0$, there is no gravitational field and the droplet shape is spherical.

Values for the liquid–vapor surface tension can and have been obtained experimentally [30] by fitting the measured profile to the profile calculated by Bashforth and Adams [29]. In this manner one obtains β, and thus σ, if the volume of the droplet and the liquid and vapor densities are also known.

B. Solving the Shape Equations Numerically

In this subsection, we describe a schematic program to explicitly solve the shape equations in Eq. (16) numerically using the Runge–Kutta method. We

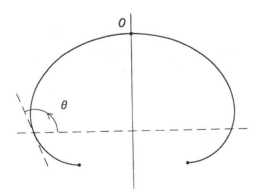

FIG. 10 Shape of the liquid droplet from the numerical tables of Bashforth and Adams [29] for $\beta = 1$. The location of the substrate (the broken horizontal line) is such that the liquid–vapor surface meets the substrate with angle θ.

start by rewriting the differential equations in Eq. (16) in a somewhat more convenient form.

Introduce the arc length s as the distance along the surface from the apex in units of b, and $\psi(s)$ as the angle at the surface defined previously in Fig. 9. The derivative with respect to s of the reduced height $Z(s) \equiv z(s)/b$ and radial distance $R(s) \equiv r(s)/b$ are then related to $\psi(s)$ via

$$R'(s) = \cos\psi(s),$$
$$Z'(s) = -\sin\psi(s), \tag{18}$$

where the prime indicates a differentiation with respect to the argument s. In terms of the functions $\psi(s)$ and $R(s)$ the radii of curvature, R_1 and R_2 are given by

$$\frac{1}{R_1} = \psi'(s),$$
$$\frac{1}{R_2} = \frac{\sin\psi(s)}{R(s)}. \tag{19}$$

The differential equation in Eq. (16) thus reduces to

$$\psi'(s) = -\frac{\sin\psi(s)}{R(s)} + 2 - \beta Z(s). \tag{20}$$

We see that instead of one second-order differential equation, we are now left with a set of coupled first-order differential equations [Eqs. (18) and (20)]. For given value of β these can be solved numerically as a function of s until the angle ψ reaches the contact angle at some s_0, $\psi(s_0) = \theta$. A schematic version of such a program using the Runge–Kutta method for solving differential equations is given in Table 1. As a function of β, stepsize h and contact angle θ, this program determines the shape (R, Z) parametrized in terms of s as well as the droplet volume

$$\frac{V}{b^3} = \int_0^{s_0} ds\, \pi R(s)^2 \sin\psi(s). \tag{21}$$

If instead of β the droplet volume is given, the parameter β needs to be varied iteratively such that the volume is equal to that required.

One final remark concerns the form of the differential equation at the apex $(s = 0)$. There both $\psi(0) = 0$ and $R(0) = 0$ and the ratio to determine $\psi'(0)$ is not well defined. However, we know that at the apex the radii of curvature are equal so that $\psi'(0) = 1$. This result we have used in the *if* statement in Table 1.

As an example, a number of droplet shapes are shown in Fig. 11 keeping the volume of the droplet fixed and increasing the strength of the gravitational field. The Young angle has been chosen to be 180° for the illustration.

TABLE 1 For Given Contact Angle θ and Gravitational Parameter β, This Program Determines the Profile (R, Z) and Volume V in Units of b

Step 1	Input parameters
	stepsize h, contact angle θ, gravitational parameter β
Step 2	Initiation
	$i = 0,\ R_i = 0,\ Z_i = 0,\ \psi_i = 0$
	$f(\psi, R, Z) \equiv -\sin(\psi)/R + 2 - \beta Z$
Step 3	Loop: *repeat* step 3 *until* $\psi_i > \theta$
	$i = i + 1$
	$kR_1 = h\cos(\psi_{i-1})$
	$kZ_1 = -h\sin(\psi_{i-1})$
	if $i = 1$ *then*
	$\qquad k\psi_1 = h$
	else
	$\qquad k\psi_1 = hf(\psi_{i-1}, R_{i-1}, Z_{i-1})$
	end if
	$kR_2 = h\cos(\psi_{i-1} + k\psi_1/2)$
	$kZ_2 = -h\sin(\psi_{i-1} + k\psi_1/2)$
	$k\psi_2 = hf(\psi_{i-1} + k\psi_1/2, R_{i-1} + kR_1/2, Z_{i-1} + kZ_1/2)$
	$kR_3 = h\cos(\psi_{i-1} + k\psi_2/2)$
	$kZ_3 = -h\sin(\psi_{i-1} + k\psi_2/2)$
	$k\psi_3 = hf(\psi_{i-1} + k\psi_2/2, R_{i-1} + kR_2/2, Z_{i-1} + kZ_2/2)$
	$kR_4 = h\cos(\psi_{i-1} + k\psi_3)$
	$kZ_4 = -h\sin(\psi_{i-1} + k\psi_3)$
	$k\psi_4 = hf(\psi_{i-1} + k\psi_3, R_{i-1} + kR_3, Z_{i-1} + kZ_3)$
	$R_i = R_{i-1} + (kR_1 + 2kR_2 + 2kR_3 + kR_4)/6$
	$Z_i = Z_{i-1} + (kZ_1 + 2kZ_2 + 2kZ_3 + kZ_4)/6$
	$\psi_i = \psi_{i-1} + (k\psi_1 + 2k\psi_2 + 2k\psi_3 + k\psi_4)/6$
Step 4	End correction
	$\Delta s = h(\theta - \psi_{i-1})/(\psi_i - \psi_{i-1})$
	$R_i = R_{i-1} + \Delta s\cos(\psi_{i-1})$
	$Z_i = Z_{i-1} - \Delta s\sin(\psi_{i-1})$
Step 5	Volume calculation
	total $= hR_{i-1}^2\sin(\psi_{i-1})/2$
	for $j = 2$ *to* $j = i - 2$ *do*
	$\qquad$ total $=$ total $+ hR_j^2\sin(\psi_j)$
	end do
	total $=$ total $+ \Delta s[R_{i-1}^2\sin(\psi_{i-1}) + R_i^2\sin(\theta)]/2$
	Volume $= \pi$ total

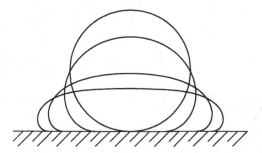

FIG. 11 Different droplet shapes at fixed volume and varying strength of the gravitational field. The Young angle is 180° here.

C. Derivation of the Shape Equations from the Free Energy

In the last part of this section, we show that the profile determined in the previous subsection from the Laplace equation and Young's law indeed corresponds to a minimum in the free energy. A convenient method of proving this fact is that of functional differentiation. This derivation [11] of Young's law and Laplace's law in the presence of gravity is closely related to the derivation first outlined by Gibbs [4] and later given by Johnson [8] (see also Ref. 9).

Consider the free energy, which is the sum of three contributions

$$F = \int dV \Delta\rho gz + \int dA\,\sigma_{lv} + \int dO[\sigma_{sl} - \sigma_{sv}]. \tag{22}$$

The first term is the gravitational energy of the liquid drop, while the second and third terms describe the contributions to the free energy due to surface tension. We have now chosen the origin at the base of the droplet—hence *different* from our previous convention where it was chosen at the apex of the drop (see Fig. 12). The integral V in Eq. (22) is over the volume of the drop, the integral A is over the surface area between the liquid and the vapor phase, and the integral O is over the surface area of the base of the droplet.

If we use the function $r = r(z)$ to describe the shape of the droplet (see Fig. 12), the integration domains in the above expression for the free energy can be written more explicitly as

$$F[r(z)] = \int_0^h dz\,\Delta\rho gz\,\pi r(z)^2 + \int_0^h dz\,\sigma_{lv}\,2\pi r(z)\left[1 + r'(z)^2\right]^{1/2}$$

$$+ \int_0^h dz[\sigma_{sl} - \sigma_{sv}]\pi r(z)^2 \delta(z), \tag{23}$$

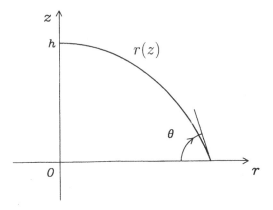

FIG. 12 Half of the liquid droplet on the substrate in terms of $r(z)$; the other half (not shown) is its mirror image reflected in the z-axis. The height h is the distance between the substrate and the apex of the droplet.

where we have introduced h as the height of the apex of the droplet above the substrate (see Fig. 12) and used the rotational symmetry of the system.

The free energy is a *functional* of the function $r(z)$: for every profile $r(z)$ the free energy can be calculated using the above formula. We are now interested in finding *that* profile $r(z)$ that minimizes the above free energy functional under the constraint that the total volume is constant. This is achieved by introducing Δp as the Lagrange multiplier to fix the volume and add to the above free energy a term

$$-\Delta p \int dV = -\Delta p \int_0^h dz\, \pi r(z)^2. \tag{24}$$

Taking the functional derivative of $F - \Delta p V$ yields the following equation to determine $r(z)$

$$
\begin{aligned}
\frac{\delta[F - \Delta p V]}{\delta r(z)} &= 2\pi r(z) \Bigg\{ -\Delta p + g\Delta\rho z + \sigma_{lv} \Bigg[\frac{1}{r(z)[1 + r'(z)^2]^{1/2}} \\
&\quad - \frac{r''(z)}{[1 + r'(z)^2]^{3/2}} \Bigg] \Bigg\} + 2\pi r(0)\delta(z) \\
&\quad \times \Bigg\{ \sigma_{sl} - \sigma_{sv} - \sigma_{lv} \frac{r'(0)}{[1 + r'(0)^2]^{1/2}} \Bigg\}.
\end{aligned}
\tag{25}
$$

The expression above comprises two terms, the first involving the complete profile $r(z)$ and the second, which is multiplied by the delta function $\delta(z)$, involving only the profile and its derivative, at the substrate. In order for the above expression to be zero, both terms in the curly brackets have to be *zero separately*. Two conditions therefore result, the first of which can be written as

$$\Delta p - \Delta \rho g \, z = \sigma_{lv} \left(\frac{1}{R_1} + \frac{1}{R_2} \right), \tag{26}$$

where we have used the explicit expressions for the radii of curvature in terms of the function $r(z)$

$$\frac{1}{R_1} = \frac{1}{r(z) \left[1 + r'(z)^2 \right]^{1/2}},$$

$$\frac{1}{R_2} = -\frac{r''(z)}{\left[1 + r'(z)^2 \right]^{3/2}}. \tag{27}$$

The Lagrange multiplier Δp is determined by relating it to the pressure difference at the apex of the droplet ($z = h$); see Eq. (15). Eq. (26) then becomes

$$\frac{2\sigma_{lv}}{b} + \Delta \rho g (h - z) = \sigma_{lv} \left(\frac{1}{R_1} + \frac{1}{R_2} \right). \tag{28}$$

Using Eqs. (14) and (15), we now see that the above equation is in fact the Laplace equation as given in Eq. (12), the only difference coming from the fact that a different convention for the location of the origin has been used.

Next we examine the second term in curly brackets in Eq. (25). We can relate the term containing the derivative of $r(z)$ at $z = 0$ to the contact angle θ

$$-\frac{r'(0)}{\left[1 + r'(0)^2 \right]^{1/2}} = \cos \theta. \tag{29}$$

We thus rederive Young's law

$$\sigma_{sv} = \sigma_{sl} + \sigma_{lv} \cos \theta. \tag{30}$$

We have thus shown that the profile $r(z)$ that minimizes the free energy in Eq. (22) has to obey the Laplace equation with Young's law as a boundary condition.

III. SMALL LIQUID DROPS: INFLUENCE OF LINE TENSION

We now discuss the situation in which the droplets are so small that the influence of gravity can safely be discarded but large enough so that curvature corrections to the surface tension are still not important. The shape of the droplet is then determined by the usual Laplace equation without gravity with the result that the shape is that of a spherical cap or truncated sphere. However, when the droplets are small enough, Young's equation for the contact angle is modified to account for the presence of the *line tension*, the concept of which dates back to the work of Gibbs [4], associated with the line of three-phase contact.

A. Line Tension

We start by supplying a thermodynamic definition of line tension [3]. Consider the three-phase system as shown in Fig. 13. Around the line of three-phase contact a cylinder is drawn with length L and radius R, and implicitly we assume that R and L approach infinity. The total free energy inside the cylinder comprises terms of the form pressure times volume, surface tension times surface area, and, finally, line tension times line length

$$F = -p_l \theta R^2 L - p_v (\pi - \theta) R^2 L + \sigma_{sv} RL + \sigma_{sl} RL + \sigma_{lv} RL + \tau L. \qquad (31)$$

The line tension can thus be defined by subtracting from the total free energy the contributions due to pressure and surface tension, dividing the result

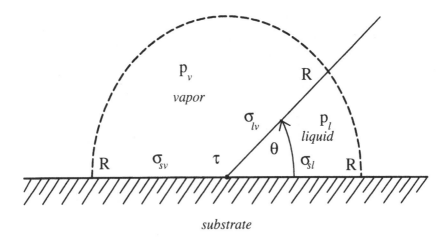

FIG. 13 Different contributions to the total free energy within a cylindrical region with radius R around the line of three-phase contact. p_l and p_v are the (equal) pressures in the liquid and vapor phase, respectively, σ_{sv}, σ_{sl}, σ_{lv}, the three surface tensions, and τ the line tension.

by the length L of the contact line, and taking the thermodynamic limit (R, $L \to \infty$)

$$\tau \equiv \lim_{R,L \to \infty} \frac{1}{L}[F + pV - (\sigma_{sv} + \sigma_{sl} + \sigma_{lv})RL].\tag{32}$$

Since the pressure is equal in both the liquid and vapor phase, we have replaced the separate contributions from the liquid and vapor pressures by the overall pressure, p, times the total volume V.

Before investigating the consequences of the presence of line tension on the wetting transition, we first discuss the physical origin of the magnitude of the line tension. Just as the surface tension originates from the interaction between molecules, the magnitude of the line tension is determined by the interaction between the molecules and the substrate [22]. This interaction can be described by the *surface potential, V(l)*, which is the integrated interaction energy between a liquid layer with thickness l and the substrate. In general $V(l)$ exhibits a strong repulsion at short distances, a minimum at some characteristic distance l_{min}, and decays to its asymptotic value at large distances (see Fig. 14). The value of the interaction potential at the equilibrium thickness, $l = l_{min}$, is the surface tension of the partially wetted substrate, $V(l_{min}) = \sigma_{sv}$, while the surface potential at $l = \infty$ is equal to the surface

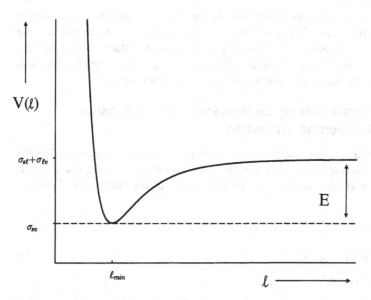

FIG. 14 Characteristic shape of the surface potential $V(l)$ as a function of liquid layer thickness l. At the minimum, $l = l_{min}$, we have $V(l_{min}) = \sigma_{sv}$ and at $l = \infty$ we have $V(\infty) = \sigma_{sl} + \sigma_{lv}$.

tension associated with an infinitely thick layer, which, according to Anto-now's rule [Eq. (11)], is given by $V(\infty) = \sigma_{sl} + \sigma_{lv}$. The difference between the two defines the energy $E \equiv V(\infty) - V(l_{min}) = \sigma_{sl} + \sigma_{lv} - \sigma_{sv}$.

It was shown by de Feijter and Vrij [31] that the line tension is determined by the shape of the interaction potential between $l = l_{min}$ and $l = \infty$ via

$$\tau = \int_{l_{min}}^{\infty} dl \left[\left\{ 2\sigma_{lv}\Delta V(l) - [\Delta V(l)]^2 \right\}^{1/2} - \left\{ 2\sigma_{lv}E - E^2 \right\}^{1/2} \right] \tag{33}$$

where we have defined $\Delta V(l) \equiv V(l) - V(l_{min})$.

The expression above clearly shows that the line tension is determined by the shape of the surface potential $V(l)$. Furthermore, we have seen that the surface tensions involved, and, through Young's Law, then also the contact angle, are related to the value of the surface potential at $l = l_{min}$ and $l = \infty$. So, by varying the shape of the surface potential one varies the value of the line tension and contact angle, so that, in principle, one could then plot the line tension as a function of contact angle. However, in some thermodynamic treatments of the three-phase system [32] this *implicit* dependence of the line tension on contact angle has been made *explicit*, leading to incorrect thermodynamic results in particular for the influence of line tension on the wetting properties.

It is also clear that since Eq. (33) is the difference between two positive quantities, the value of the line tension can be positive or negative depending on the precise shape of $V(l)$. Contrary to the case of surface tension where a negative surface tension destabilizes the interface, a negative line tension does not necessarily destabilize the line of three-phase contact [33].

B. Consequences of the Presence of Line Tension on the Wetting Transition

When the shape of the liquid droplet is that of a spherical cap (see Fig. 15), the volume V, surface area A in contact with the vapor phase, and surface area O in contact with the substrate are given in terms of the radius R and contact angle θ by

$$V = \frac{\pi}{3} R^3 (2 + 3\cos\theta - \cos^3\theta),$$

$$A = 2\pi R^2 (1 + \cos\theta), \tag{34}$$

$$O = \pi R^2 \sin^2\theta.$$

Note that in Fig. 15 we have chosen to define the contact angle somewhat differently than in the previous sections. Whereas first θ was defined as the contact angle of the liquid phase, now it is the contact angle of the vapor phase

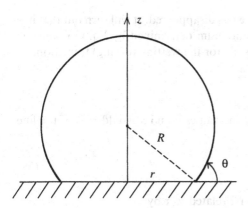

FIG. 15 Spherical cap with radius R and contact angle θ. The radius of the circle in which the three phases meet is r.

(compare Figs. 13 and 15). The only consequence is that now $\theta = 0$ corresponds to a "drying" transition rather than a "wetting" transition.

For fixed volume, the free energy is the sum of the surface tension contributions and the contribution due to the line tension of the three-phase line. Again, it is expressed in terms of R and θ

$$F(R, \theta) = \sigma_{lv} 2\pi R^2 (1 + \cos\theta) + (\sigma_{sl} - \sigma_{sv})\pi R^2 \sin^2\theta + \tau 2\pi R \sin\theta. \quad (35)$$

The above free energy is to be minimized with respect to R and θ adding a Lagrange multiplier, Δp, to fix the volume of the droplet. One finds that the equilibrium contact angle thus determined has to obey the following relation [19–21]

$$-\sigma_{lv}\cos\theta + \sigma_{sl} - \sigma_{sv} + \frac{\tau}{r} = 0, \quad (36)$$

where $r \equiv R \sin\theta$ is the radius of the circle of three-phase contact. The above equation is the balance of forces on the line (circle) of three-phase contact. Without the line tension term τ/r it would be the classical Young's law.

Note, incidentally, that if we would have assumed an explicit contact angle dependence of the line tension $\tau(\theta)$ in the free energy, unphysical additional terms in the minimization of the free energy related to derivative of the line tension with respect to θ, $d\tau(\theta)/d\theta$, would show up, in particular into Eq. (36).

The angle θ determined by Eq. (36) does not always correspond to the angle at which the free energy is minimal; the solution of Eq. (36) might correspond to a local maximum rather than a minimum of F, or even when it is a local minimum, the global minimum might occur at one of the extremes $\theta = 0$ or

$\theta = \pi$, at which the three-phase line has disappeared. It will turn out that it is the competition between a local minimum determined by Eq. (36) and the value of F at $\theta = 0$ that is responsible for the wetting (drying) transition.

Let us define

$$x \equiv \cos\theta, \tag{37}$$

and denote by θ_0 and $x_0 = \cos\theta_0$ the values θ and x would have if the line tension were 0; i.e., from Eq. (36)

$$x_0 = \cos\theta_0 = \frac{\sigma_{sl} - \sigma_{sv}}{\sigma_{lv}}. \tag{38}$$

Define a dimensionless line tension $\bar{\tau}$ related to τ by

$$\bar{\tau} = \frac{\tau}{(3V/\pi)^{1/3}\sigma_{lv}}. \tag{39}$$

Note that this becomes negligibly small in the limit of a macroscopically large droplet volume V. That, ultimately, is why line tension significantly affects the contact angle only of small droplets.

Finally, define a dimensionless free energy Φ related to F by

$$\Phi = \frac{F}{2\pi(3V/\pi)^{2/3}\sigma_{lv}}. \tag{40}$$

Then from Eqs.(32)–(34) and Eqs. (37)–(39), the expression in Eq. (35) for the free energy becomes

$$\begin{aligned}\Phi &= (1+x)^{-1/3}(2-x)^{-2/3}\left[1 + \frac{x_0}{2}(1-x)\right] \\ &\quad + \bar{\tau}(1+x)^{-1/6}(1-x)^{-1/2}(2-x)^{-1/3.}\end{aligned} \tag{41}$$

while the force–balance relation in Eq. (36) becomes

$$\bar{\tau} = (x - x_0)(1+x)^{-1/6}(2-x)^{-1/3}(1-x)^{1/2.} \tag{42}$$

This relation [21] between $\bar{\tau}$ and x is shown for various fixed x_0 as the solid curves in Fig. 16.

The relation Eq. (42) between line tension and contact angle at given x_0 is the condition for a local extremum of F. From Eqs. (41) and (42), one has for the value of the reduced free energy at this local extremum

$$\Phi_{\text{loc extr}} = (1+x)^{-1/3}(2-x)^{-2/3}\left[1 + \left(x - \frac{x_0}{2}\right)(1-x)\right], \tag{43}$$

with x implicitly given as a function of $\bar{\tau}$ by Eq. (42).

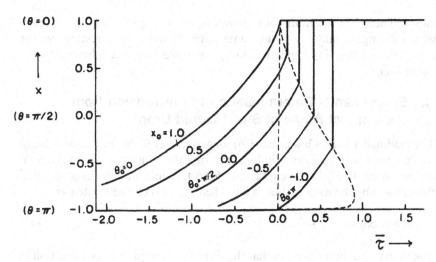

FIG. 16 Equilibrium $x = \cos\theta$ versus τ for various $x_0 = \cos\theta_0$ (see also Ref. 21).

To find the global minimum, one must compare Eq. (43) with the values of Φ at the limits $x = 1$ ($\theta = 0$) and $x = -1$ ($\theta = \pi$). It turns out that only the value of Φ at $x = 1$, $\Phi_{x=1} = 2^{-1/3}$, competes with the local extremum in Eq. (43). The competing free energies are equal when

$$(1+x)^{-1/3}(2-x)^{-2/3}\left[1 + \left(x - \frac{x_0}{2}\right)(1-x)\right] = 2^{-1/3}. \tag{44}$$

This and Eq. (42) are a pair of parametric equations (parameter x_0) for a locus of wetting (drying) transitions in the (x, τ) plane. This locus is shown as the dashed curve in Fig. 16.

The solid curves of x versus τ in Fig. 16, which display the force–balance relation Eq. (42) for various x_0, give the value of the contact angle at which, for given x_0 and τ, the free energy has a local extremum. Everywhere along any of the solid curves in Fig. 16, up to the dashed wetting transition locus (and even for some way beyond, on the curve's analytic continuation), the local extremum is a local minimum, as one may verify. Furthermore, up to the transition locus, $\Phi_{\text{loc extr}} < \Phi_{x=1}$, so that the solid curves correspond to global minima of the free energy.

To the high-τ side of the dashed curve in Fig. 16, we have that $\Phi_{\text{loc extr}} > \Phi_{x=1}$, so that the minimum free energy is that corresponding to $x = 1$. Therefore, when, with increasing τ, the solid curve in Fig. 16, for any given x_0, intersects the dashed curve, the equilibrium x *jumps discontinuously* to 1 (the equilibrium contact angle jumps discontinuously to 0), as shown in the figure. This is a sudden wetting (drying) of the solid–liquid interface by the vapor. There is a discontinuous change in the shape of the liquid droplet, which changes suddenly from being a truncated sphere resting on the substrate to being a full sphere entirely separated from the substrate by the vapor phase.

C. Experimental Determination of Line Tension from the Contact Angle of Small Liquid Drops

The modified Young's law in Eq. (36) suggests a way to determine the value of the line tension experimentally by measuring the contact angle for different droplet sizes. If θ_∞ is the contact angle in the limit of very large droplets determined by the usual Young's law, then Eq. (36) can be written as

$$\cos \theta = \cos \theta_\infty - \frac{\tau}{\sigma_{lv}} \frac{1}{r}. \tag{45}$$

The above equation indicates that the slope of $\cos \theta$, plotted as a function of $1/r$, near $1/r = 0$ determines the line tension. As an example, we show recent results by Wang et al. [34] in Fig. 17 for the contact angle as a function of $1/r$

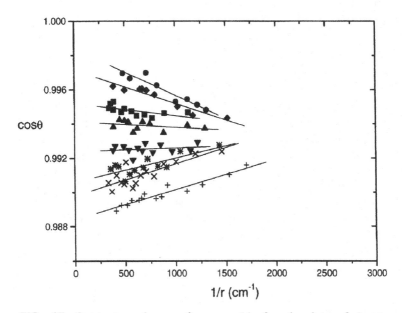

FIG. 17 Contact angle $\cos \theta$ versus $1/r$ for droplets of 1-octene on a hexadecyltrichlorosilane-coated Si wafer at various temperatures: $T = 40.7$ (plusses), 42.1, 43.1, 44.0, 46.0, 47.3, 48.6, and 50.0°C (solid circles) [34]. The solid lines are a linear fit to the data.

for *n*-octane and 1-octene droplets on a hexadecyltrichlorosilane-coated Si wafer for several temperatures. Also shown [34] in Fig. 18, are the associated line tensions calculated from the slope near $1/r = 0$.

How small should the droplets be in order for the line tension to appreciatingly influence the value of the contact angle? The de Feijter–Vrij formula for the line tension, Eq. (33), indicates that a rough estimate for the value of the line tension is given by $\tau \approx \sigma_{lv}d$ with d the interaction range of $V(l)$. This leads to a theoretical estimate [3] for the value of the line tension $\tau_{theor} \approx 10^{-10}$ to 10^{-12} N. One therefore expects that only for very small droplets with droplet radii of the order of 10–100 nm or less, would one, experimentally, be able to observe any influence of the presence of line tension on contact angle.

The experimental observations are different, however [34–41]. For a large range of droplet radii from micrometers up to millimeters, a deviation of cos θ linear near $1/r = 0$ is observed by a variety of experimental groups. Several recent experimental results [34–41] for the determination of the line tension of liquids on a solid substrate are collected in Table 2.

The experimental results roughly subdivide into two groups: those for which the droplet radii are in the millimeter range [34–37] and those for which

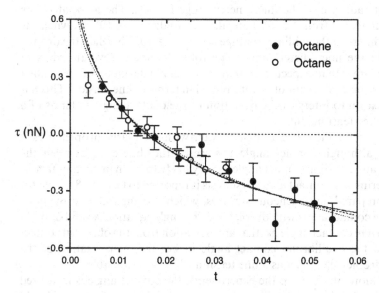

FIG. 18 Line tension as a function of reduced temperature $t \equiv (T_W - T)/T_W$ for octane and octene [34]. The lines are fits to the data for different models. (See Ref. 34 for details.)

TABLE 2 Survey of Some Recent Experimental Studies for the Determination of the Line Tension

First author	System	Line tension $(10^{-9}$ N$)$	Diameter (μm)	Hysteresis	Method
Wang [34]	n-Octane/1-octene on SAM/Si	−0.5 to 0.3	6–30	1°	S
Stöckelhuber [35]	Water on mica	7.6	6–50	None	S
Drelich [36]	Water on quartz	80–3300	30–100	No report	S
Nguyen [37]	Water on glass	1160–5540	10–50	No report	D
Amirfazli [38]	Organic liquids on SAM/Au	810–4960	1000–5000	7°	D
Gu [39]	n-Alkanes on FC725	1700–3700	1000–6000	5°–7°	D
Duncan [40]	Organic liquids on FC721	1020–5620	1000–6000	5°	D
Li [41]	Organic liquids on FC721	2030–6440	1000–5000	6°–8°	D

The method refers to whether the contact angle was measured statically (S) or dynamically (D). Details of the experiments are found in the respective references.

the droplet radii are in the micrometer range [38–41]. The associated line tensions calculated from Eq. (45) are also twofold; $\tau_{exp} \approx 10^{-6}$ N when the droplet radii are in the millimeter range and $\tau_{exp} \approx 10^{-9}$ N when the droplet radii are in the micrometer range. The former value is several orders of magnitude larger than expected, so that the observed deviation of the contact angle with droplet size cannot be interpreted in terms of line tension. This has been the reason to interpret the deviation of the contact angle in terms of a "pseudo"-line tension [36].

The reason for the contact angle to vary with such large droplet sizes is unknown, although contact angle hysteresis, the difference between the advancing and receding contact angle, is very likely to be an important factor. In the experiments, contact angle hysteresis is reported to be up to 8°, even for carefully prepared homogeneous surfaces, which is comparable to the deviation of the contact angle with droplet size. Young's equation is valid only for the *equilibrium* contact angle, so that any deviation from the observed contact angle from the equilibrium contact angle makes the interpretation of the observed droplet size effect as a line tension effect, questionable.

Furthermore, in many of the experiments, the contact angle is measured *dynamically* (the three-phase line advances or recedes), so that one should worry whether true equilibrium conditions have been reached [19,42]. If this is the case, the measured contact angle is not equal to the equilibrium contact angle but rather the advancing or receding contact angle. Arguments have been provided by Amirfazli et al. [38], Duncan et al. [40], and Li and Neumann [41] for the use of the *advancing* contact angle in Eq. (45). In the context of a simple model [43] in which the substrate is thought to be made of strips of material with two distinct surface tensions, it was shown that the advancing contact angle corresponds to the contact angle of the hypothetical substrate consisting of the material with the lowest of the two surface tensions. However, as also shown in the context of this simple model [43], the advancing contact angle differs from the equilibrium contact angle, so that the use of Eq. (45) remains questionable.

Even the determined line tensions with $\tau_{exp} \approx 10^{-9}$ N [34,35] are large compared to what is expected on theoretical grounds and the discussion above may also hold to some extent for these results. What is clear is that no general agreement as to the origin of the disagreement between the different experimental results and between the experimental results and theoretical expectations has been established thus far, and further research is necessary.

Only very recently, computer calculations were performed by Bresme and Quirke [44] to determine the magnitude of the line tension. Using molecular dynamics of small solid particles at the interface between a Lennard–Jones liquid and its vapor phase, they showed the line tension to be of the order $\tau_{comp} \approx 10^{-12}$ N, which is consistent with the theoretical estimate given above.

IV. VESICLES AT SURFACES

After having discussed the influence of gravity on the shape of large droplets in Section II and the influence of line tension on small droplets in Section III, we now turn to the discussion of systems that are large enough to discard line tension and small enough to discard gravity, but for which the shape is not described by surface tension alone. These are systems in which, besides surface tension, the free energy is complemented by the free energy connected with *bending* the surface. The expression for the free energy to take into account the bending energy is due to Helfrich [45], and it has been used to describe membranes, vesicles, and microemulsion systems and to calculate their respective phase diagrams [46].

Seifert and Lipowsky [47,48] were the first to apply the Helfrich free energy for the description of the shape and free energy of such systems adhered to a solid substrate (see Fig. 19). In their analysis they calculated the phase diagram for the unbinding of a vesicle adsorbed to a substrate. In this section, we discuss the analysis by Seifert and Lipowsky [47]. First, we introduce the expression for the curvature free energy by Helfrich and derive the shape equations that minimize the Helfrich free energy for a vesicle in contact with a solid substrate. Since these shape equations cannot, in general, be solved

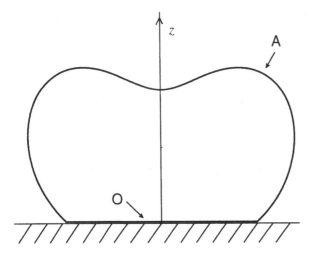

FIG. 19 Vesicle adhered to a substrate located at $z = 0$, A is the surface area of the whole droplet including the surface area O that is in contact with the substrate. The vesicle shape is assumed to be axisymmetric around the z-axis.

analytically a schematic program is provided to solve the shape equations numerically. The large number of parameters involved makes numerical work rather tedious, however, and the need arises for limiting analytical results. In the last subsection, therefore, we derive such a limiting solution by calculating the shape and free energy of a vesicle adhered to a substrate under the condition that the energy cost of bending is small [49].

A. Shape Equations

The Helfrich free energy [45] describes the surface free energy associated with *bending* in terms of the radius of spontaneous curvature, R_0, the rigidity constant associated with bending, k, and the rigidity constant associated with Gaussian curvature, $\bar{k}$,

$$F_{\mathrm{H}} = \int \mathrm{d}A \left[\sigma - \frac{2k}{R_0} J + \frac{k}{2} J^2 + \bar{k} K \right], \tag{46}$$

where σ is the surface tension of the planar membrane surface. The above free energy features an integral over the whole surface area, A, of the total curvature, $J = 1/R_1 + 1/R_2$ and Gaussian curvature, $K = 1/(R_1 R_2)$. The above form for the free energy is the most general form in an expansion up to second order in curvature, and can be viewed as defining the four coefficients $\sigma, k/R_0, k$, and $\bar{k}$. The rigidity constant associated with Gaussian curvature, $\bar{k}$, is a measure of the energy cost for topological changes of the surface. In our discussion we only consider systems where the topology is fixed so that the term proportional to $\bar{k}$ can be dropped.

When we consider the case of vesicle adsorption, part of the vesicle surface is in direct contact with the substrate. As the dimensions of the vesicle are large compared to the range of the interaction between the substrate and the vesicle, the interaction can be described by an adhesion energy integrated over the area of contact [47]

$$F_s = \int dO[\Delta\sigma - \sigma], \tag{47}$$

where O is the area of substrate–vesicle contact (see Fig. 19) and where $\Delta\sigma \equiv \sigma_{sv} - \sigma_s$ is the difference in surface tension of the substrate–vesicle surface and the bare substrate.

The total free energy is the sum of the Helfrich free energy F_H and substrate interaction energy F_s

$$F = F_H + F_s = \int dA \left[\sigma - \frac{2k}{R_0}J + \frac{k}{2}J^2\right] + \int dO[\Delta\sigma - \sigma]. \tag{48}$$

The above free energy is to be minimized with the condition that the total vesicle volume is fixed. Again, this is achieved by adding to the free energy a Lagrange multiplier $-\Delta p$ times the volume V of the vesicle. The functional minimization of $\Omega \equiv F - \Delta p V$ with respect to the shape then leads to the shape equation

$$\Delta p = \sigma J - \frac{4k}{R_0}K - \frac{k}{2}J^3 + 2kJK - k\Delta_s J, \tag{49}$$

where Δ_s denotes the *surface* Laplacian. This equation might be termed the *generalized* Laplace equation because it reduces to the Laplace equation $\Delta p = \sigma J$ [Eq. (7)] when one sets the coefficients k/R_0 and k equal to zero. The shape equation (49) is the same as the shape equation describing the shape of a *free* vesicle. The vesicle adhesion energy is only present in the *boundary conditions* for the curvature at the substrate. They are given by

$$\left.\frac{1}{R_1}\right|_{\text{substrate}} = 0, \quad \left.\frac{1}{R_2}\right|_{\text{substrate}} = [2(\sigma - \Delta\sigma)/k]^{1/2}, \tag{50}$$

where R_2 is the radius of curvature along the meridians of the vesicle and R_1 is the radius of curvature perpendicular to it. The first equation indicates that when $k \neq 0$, the contact angle with which the vesicle meets the substrate is *always equal to zero*. Through the second equation, first derived by Seifert and Lipowsky [47], the adhesion energy, $\Delta\sigma$, enters the description of the vesicle shape.

Unfortunately, the shape equations cannot be solved analytically in general. In practice one solves the shape equations numerically for given values of σ, k/R_0, k, $\Delta\sigma$, and Δp. With the shape of the vesicle thus obtained one is then able to calculate the volume V and free energy of the vesicle. In this way, one can compare the free energy with the free energy of the unbound vesicle and locate unbinding transitions [48]. In the next subsection, an explicit program is given to solve the shape equations numerically with the assumption that the vesicle shape is axisymmetric around the z-axis.

B. Solving the Shape Equations Numerically

In this subsection, we describe a schematic program to explicitly solve the shape equations in Eq. (49) numerically using the Runge–Kutta method. We start by rewriting the differential equations in Eq. (49) similarly to the way the shape equations of a liquid drop in a gravitational field were rewritten in Section II.

Again, introduce s, the arc length, as the distance along the surface from the apex (although now not in units of b), and $\psi(s)$ as the angle at the surface (see Fig. 20). The height $Z(s)$ and radial distance $R(s)$ (see Fig. 20) are given by the previous Eq. (18)

$$
\begin{aligned}
R'(s) &= \cos\psi(s), \\
Z'(s) &= -\sin\psi(s),
\end{aligned}
\tag{51}
$$

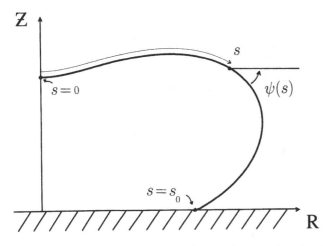

FIG. 20 Half of the vesicle adhered to a substrate; the other half (not shown) is its mirror image reflected in the Z-axis. A certain point on the surface is described by the radial distance $R(s)$ and height $Z(s)$ with the arc length s defined as the distance along the surface from the apex ($s = 0$). At the substrate, $s = s_0$.

and the radii of curvature, R_1 and R_2, by the previous Eq. (19)

$$\frac{1}{R_1} = \psi'(s),$$

$$\frac{1}{R_2} = \frac{\sin\psi(s)}{R(s)}. \tag{52}$$

The surface Laplacian of the total curvature expressed in terms of $R(s)$ and $\psi(s)$ reads

$$\Delta_s J = \psi''' + \frac{2}{R}\cos(\psi)\psi''' + \frac{1}{R^2}[1 - 2\cos^2(\psi)]\psi' - \frac{1}{R}\sin(\psi)[\psi']^2$$

$$+ \frac{1}{R^3}\sin(\psi)\cos^2(\psi), \tag{53}$$

so that the shape equations in Eq. (49) are given by

$$\psi''' = -\frac{\Delta p}{k} + \frac{\sigma}{k}\left(\frac{\sin(\psi)}{R} + \psi'\right) - \frac{4}{R_0}\frac{\sin(\psi)\psi'}{R} - \frac{2}{R}\cos(\psi)\psi''$$

$$- \frac{1}{2}[\psi']^3 + \frac{1}{2}\frac{1}{R^2}[3\cos^2(\psi) - 1]\psi' + \frac{3}{2}\frac{1}{R}\sin(\psi)[\psi']^2 \tag{54}$$

$$- \frac{1}{2}\frac{1}{R^3}\sin(\psi)[\cos^2(\psi) + 1],$$

where we have not written the explicit s dependence of the functions $\psi(s)$ and $R(s)$. The above third-order differential equation is to be solved numerically with the boundary conditions [Eq. (50)]

$$\psi(s_0) = \pi, \tag{55}$$

$$\psi'(s_0) = [2(\sigma - \Delta\sigma)/k]^{1/2},$$

where $s = s_0$ is the value of s at the substrate (see Fig. 20). The profile is obtained in the following way. At the apex ($s = 0$), $R(0) = 0$, $Z(0) = 0$, $\psi(0) = 0$, $\psi'(0) = \psi_0'$, and $\psi''(0) = 0$, where ψ_0' is the unknown initial value of the first derivative $\psi'(0)$. From the shape equations in Eq. (54) it is clear that only the ratios of the parameters $(\Delta p)/k$, σ/k, and $1/R_0$ enter the description so that the number of parameters is reduced by one. For a given value of these ratios and initial value ψ_0', the profile can be determined numerically as a function of s until the angle $\psi(s = s_0) = \pi$. With the profile calculated in this manner, one is then able to determine the corresponding adhesion energy, $\Delta\sigma$, using Eq. (55)

$$\Delta\sigma = \sigma - \frac{k}{2}[\psi'(s_0)]^2 \tag{56}$$

the volume of the adhered vesicle [see also Eq. (21)]

TABLE 3 For Given Pressure Difference, Surface Tension, and Radius of Spontaneous Curvature, This Program Determines the Profile (R, Z), Free Energy $F_k \equiv F/k$, Volume V, and Substrate Interaction $\Delta\sigma_k \equiv (\Delta\sigma)/k$

Step 1   Input parameters
stepsize h, pressure $\Delta p_k \equiv \Delta p/k$, surface tension $\sigma_k \equiv \sigma/k$,
spontaneous curvature $1/R_0$, value ψ' at apex ψ_0'

Step 2   Initiation
$i = 0$, $R_i = 0$, $Z_i = 0$, $\psi_i = 0$, $\psi_i' = \psi_0'$, $\psi_i'' = 0$
$f(\psi, \psi', \psi'', R) \equiv -\Delta p_k + \frac{\sigma_k}{R}\sin(\psi') + \sigma_k\psi' - \frac{4}{R_0 R}\sin(\psi)\psi'$
$\qquad - \frac{2}{R}\cos(\psi)\psi'' - \frac{1}{2}(\psi')^3 + \frac{1}{2R^2}[3\cos^2(\psi) - 1]\psi'$
$\qquad + \frac{3}{2R}\sin(\psi)(\psi')^2 - \frac{1}{2R^3}\sin(\psi)[\cos^2(\psi) + 1]$

Step 3   Loop: repeat step 3 until $\psi_i > \pi$
$i = i + 1$
$kR_1 = h\cos(\psi_{i-1})$
$kZ_1 = -h\sin(\psi_{i-1})$
$k\psi_1 = h\psi_{i-1}'$
$k\psi_1' = h\psi_{i-1}''$
$if\ i = 1\ then$
$\qquad k\psi_1'' = 3/8h[\Delta p_k + 2\sigma_k k\psi_0' - 4/R_0(\psi_0')^2]$
$else$
$\qquad k\psi_1'' = hf(\psi_{i-1}, \psi_{i-1}', \psi_{i-1}'', R_{i-1})$
$end\ if$
$kR_2 = h\cos(\psi_{i-1} + \psi k_1/2)$
$kZ_2 = -h\sin(\psi_{i-1} + k\psi_1/2)$
$k\psi_2 = h[\psi_{i-1}' + k\psi_i'/2]$
$k\psi_2' = h[\psi_{i-1}'' + k\psi_i''/2]$
$k\psi_2'' = h\,f(\psi_{i-1} + k\psi_i/2,\ \psi_{i-1}' + k\psi_1'/2,\ \psi_{i-1}'' + k\psi_1/2,\ R_{i-1} + kR_1/2)$

$kR_3 = h\cos(\psi_{i-1} + k\psi_2/2)$
$kZ_3 = -h\sin(\psi_{i-1} + k\psi_2/2)$
$k\psi_3 = h[\psi_{i-1}' + k\psi_2'/2]$
$k\psi_3' = h[\psi_{i-1}'' + k\psi_2''/2]$
$k\psi_3'' = h\,f(\psi_{i-1} + k\psi_2/2,\ \psi_{i-1}' + k\psi_2'/2,\ \psi_{i-1}'' + k\psi_2''/2,\ R_{i-1} + kR_2/2)$

$kR_4 = h\cos(\psi_{i-1} + k\psi_3)$
$kZ_4 = -h\sin(\psi_{i-1} + k\psi_3)$
$k\psi_4 = h[\psi_{i-1}' + k\psi_3']$
$k\psi_4' = h[\psi_{i-1}' + k\psi_3']$
$k\psi_4'' = h\,f(\psi_{i-1} + k\psi_3,\ \psi_{i-1}' + k\psi_3',\ \psi_{i-1}'' + k\psi_3'',\ R_{i-1} + kR_3)$

TABLE 3 Continued

$$R_i = R_{i-1} + (kR_1 + 2kR_2 + 2kR_3 + kR_4)/6$$
$$Z_i = Z_{i-1} + (kZ_1 + 2kZ_2 + 2kZ_3 + kZ_4)/6$$
$$\psi_i = \psi_{i-1} + (k\psi_1 + 2k\psi_2 + 2k\psi_3 + k\psi_4)/6$$
$$\psi'_i = \psi'_{i-1} + (k\psi'_1 + 2k\psi'_2 + 2k\psi'_3 + k\psi'_4)/6$$
$$\psi''_i = \psi''_{i-1} + (k\psi''_1 + 2k\psi''_2 + 2k\psi''_3 + k\psi''_4)/6$$

Step 4 End correction
$$\Delta s = h(\pi - \psi_{i-1})/(\psi_i - \psi_{i-1})$$
$$R_i = R_{i-1} + \Delta s \cos(\psi_{i-1})$$
$$Z_i = Z_{i-1} - \Delta s \sin(\psi_{i-1})$$
$$\psi_i = \psi_{i-1} + \Delta s\, \psi'_{i-1}$$
$$\psi'_i = \psi'_{i-1} + \Delta s\, \psi''_{i-1}$$
$$\psi''_i = \psi''_{i-1} + \Delta s\, f(\psi_{i-1}, \psi'_{i-1}, \psi''_{i-1}, R_{i-1})$$

Step 5 Substrate energy calculation
$$\Delta\sigma_k = \sigma_k - 1/2(\psi'_i)^2$$

Step 6 Free energy calculation

$$\text{total} = \frac{1}{2} h R_{i-1}\left[\sigma_k - \frac{2}{R_0}\left(\psi'_{i-1} + \frac{\sin(\psi_{i-1})}{R_{i-1}}\right) + \frac{1}{2}\left(\psi'_{i-1} + \frac{\sin(\psi_{i-1})}{R_{i-1}}\right)^2\right]$$

for $j = 2$ to $j = i - 2$ do

$$\text{total} = \text{total} + h R_j\left[\sigma_k - \frac{2}{R_0}\left(\psi'_j + \frac{\sin(\psi_j)}{R_j}\right) + \frac{1}{2}\left(\psi'_j + \frac{\sin(\psi_j)}{R_j}\right)^2\right]$$

end do

$$\text{total} = \text{total} + \frac{1}{2}\Delta s\left\{R_i\left[\sigma_k - \frac{2}{R_0}\left(\psi'_i + \frac{\sin(\psi_i)}{R_i}\right) + \frac{1}{2}\left(\psi'_i + \frac{\sin(\psi_i)}{R_i}\right)^2\right]\right.$$
$$\left. + R_{i-1}\left[\sigma_k - \frac{2}{R_0}\left(\psi'_{i-1} + \frac{\sin(\psi_{i-1})}{R_{i-1}}\right) + \frac{1}{2}\left(\psi'_{i-1} + \frac{\sin(\psi_{i-1})}{R_{i-1}}\right)^2\right]\right\}$$

$$F_k = 2\pi\,\text{total} + \pi\Delta\sigma_k R_i^2$$

Step 7 Volume calculation
$$\text{total} = h R_{i-1}^2 \sin(\psi_{i-1})/2$$
for $j = 2$ to $j = i - 2$ do
$$\text{total} = \text{total} + h R_j^2 \sin(\psi_j)$$
end do
$$\text{total} = \text{total} + \Delta s\, [R_{i-1}^2 \sin(\psi_{i-1}) + R_i^2 \sin(\psi_i)]/2$$
$$V = \pi\,\text{total}$$

$$V = \int_0^{s_0} ds\, \pi R^2 \sin(\psi), \tag{57}$$

and the free energy, Eq. (48),

$$F = 2\pi \int_0^{s_0} ds\, R \left[\sigma - \frac{2k}{R_0} \left(\psi' + \frac{\sin(\psi)}{R} \right) + \frac{k}{2} \left(\psi' + \frac{\sin(\psi)}{R} \right)^2 \right]$$
$$+ \pi \Delta\sigma [R(s_0)]^2. \tag{58}$$

A schematic program to solve the shape equations using the Runge–Kutta method and calculate $\Delta\sigma$, V, and F, is given in Table 3.

Usually, the substrate adhesion energy, $\Delta\sigma$, and vesicle volume, V, are known instead of ψ_0' and $(\Delta p)/k$. The program in Table 3 should then be used to vary ψ_0' and $(\Delta p)/k$ iteratively until the desired values of $\Delta\sigma$ and V have been reached.

One final remark with regard to the program in Table 3 concerns the form of the differential equation at the apex ($s = 0$). Again, there both $\psi(0) = 0$ and $R(0) = 0$ so that Eq. (54) cannot straightforwardly be used to determine the value $\psi(0)$. One therefore should expand ψ around $s = 0$, $\psi(s) = \psi_0's + \psi'''(0) s^3/8 + \cdots$, and take the limit $s = 0$ in Eq. (54). One can show that

$$\psi'''(0) = -\frac{3}{8} \frac{\Delta p}{k} + \frac{3}{4} \frac{\sigma}{k} \psi'(0) - \frac{3}{2} \frac{1}{R_0} [\psi'(0)]^2. \tag{59}$$

This result is used in the *if* statement in Table 3.

Seifert and Lipowsky [47] used the above scheme to calculate phase diagrams for the unbinding of a vesicle adsorbed to a substrate. In Fig. 21, an example of such a phase diagram as calculated by Seifert and Lipowsky is shown [47,48]. In this example, $1/R_0 = 0$, the volume V and surface area A are kept constant, and the unbinding transition (thick solid lines and thick dashed line) is located as a function of the reduced volume $v \equiv [(36\pi V^2)/A^3]^{1/2}$ and reduced substrate adhesion energy, $w \equiv [(\sigma - \Delta\sigma)A]/(4\pi k)$.

The large number of parameters makes numerical work rather tedious. Furthermore, it is difficult to gain physical insight into the role played by the different parameters in vesicle unbinding. To be able to proceed analytically it might be useful to make certain physically reasonable assumptions. One such assumption, which we describe next, is to assume that the rigidity constant of bending is small [49]. In particular, the length $(k/\sigma)^{1/2}$ will be assumed small compared to the typical size of the vesicle.

C. Small Bending Rigidity Limit

To make a systematic expansion in k, we first investigate the case when $k = 0$ and then $k \neq 0$ but small.

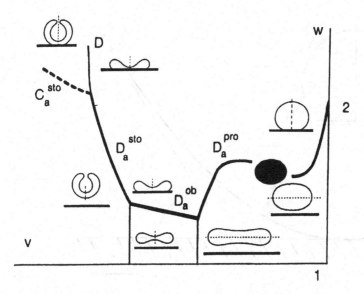

FIG. 21 Schematic phase diagram for the vesicle unbinding transition (thick solid lines and thick dashed line) as calculated by Seifert and Lipowsky [47,48] as a function of the reduced volume, v, and reduced substrate adhesion energy, w. In the solid region, nonaxisymmetric vesicle shapes are relevant. The solid lines are shape transitions of the vesicle within the bound or unbound region. (See Ref. 48 for further details.)

1. No Rigidity

In the absence of bending rigidity, the minimization of the free energy is done in two steps. First, we note that when $k = 0$ the solution of the shape equation in Eq. (49) is that of a spherical cap (see inset, Fig. 22). We know this to be the case when also k/R_0 is taken to be zero so that the shape equation reduces to the well-known Laplace equation, but also with the spontaneous radius of curvature term present, the shape is that of a spherical cap. Second, we insert the spherical-cap profile, which is fully described by the radius R and contact angle θ, into the expression for the free energy, and then minimize with respect to R and θ.

For the spherical-cap profile the integration of the free energy over the surface areas A and O and over the volume V in Eq. (48) can be carried out to yield

$$
F_0(R, x) = 2\pi\sigma R^2(1 + x) + \Delta\sigma\pi R^2(1 - x^2) - 8\pi\frac{k}{R_0}R(1 + x)
$$
$$
- 4\pi\frac{k}{R_0}R(1 - x^2)^{1/2}\arccos(x),
$$

(60)

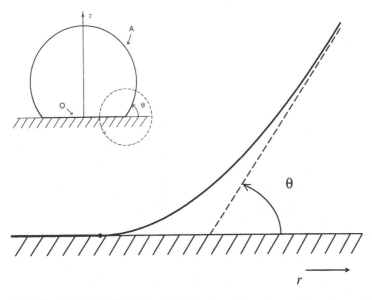

FIG. 22 Height profile of the vesicle in the region close to the substrate (see dashed circle in the inset). The asymptotic spherical-cap profile (dashed line and inset) meets the substrate with contact angle θ.

where we have defined $x \equiv \cos\theta$, and where the subscript 0 stands for the fact that we have set $k = 0$. The last term in the expression for $F_0(R,x)$ stems from an integration of J across the line where the spherical-cap profile meets the substrate [49]. Again, define $\Omega_0 \equiv F_0 - \Delta pV$ to fix the volume $V = (\pi/3)R^3(2 + 3x - x^3)$ of the vesicle. Minimization of Ω_0 with respect to R and x then fixes R and x. From $\partial\Omega_0/\partial x = 0$ and $\partial\Omega_0/\partial R = 0$ one finds the following set of equations [49]

$$\Delta p = \frac{2\sigma}{R} - \frac{4k}{R_0}\frac{1}{R^2}, \tag{61}$$

$$\Delta\sigma = \sigma x - \frac{2k}{R_0}\frac{1}{R}x + \frac{2k}{R_0}\frac{1}{R}\frac{\arccos(x)}{(1-x^2)^{1/2}}. \tag{62}$$

The first equation is the well-known Laplace equation with a finite size correction originally due to Tolman [5]. In the work by Tolman, the Laplace equation is written as $\Delta p = 2\sigma/R - 2\sigma\delta/R^2$ [see Eq. (6)], so that the Tolman length δ is related to the radius of spontaneous curvature via $\sigma\delta = 2k/R_0$. The second equation determines the value of the contact angle. It reduces to Young's equation $\sigma_{sv} = \sigma_s + \sigma\cos\theta$ when one inserts $k/R_0 = 0$.

With R and x given by Eqs. (61) and (62), the free energy F_0 in the absence of rigidity is now fully determined. Next, we expand around this solution in small k.

2. Small Rigidity

To calculate the leading order contribution to the free energy of the adhered vesicle in small k, we need to determine the shape of the vesicle. As shown in the previous section, the shape of the profile is that of a spherical cap when $k = 0$ (see inset, Fig. 22). In the region where the spherical-cap profile meets the substrate, the first derivative of the height profile is *discontinuous* so that the curvature J, which is related to the *second* derivative of the height profile, contains a delta function. Thus, when one integrates J^2, one finds that the curvature energy is *infinite*. Therefore, for any finite value of k the surface profile *has* to meet the substrate with zero contact angle as described by the first boundary condition in Eq. (50). The result is that deviations of the spherical-cap profile to leading order in k are located in the region where the spherical-cap profile meets the substrate. The precise shape of the profile near the kink (see Fig. 22) can be determined [49] by minimizing the curvature free energy with the condition that far away from the kink it smoothly crosses over to the spherical-cap profile with radius R and contact angle x. One can show that the resulting first-order contribution to the free energy, still expressed in terms of the radius R and contact angle x of the asymptotic spherical-cap profile, is given by [49]:

$$F(R,x) = F_0(R,x) + 2\pi R(1-x)^{1/2}(k\sigma)^{1/2}2^{1/2}\left[2^{3/2}(1+x)^{1/2}\right.$$

$$\left. -(3+x)\right] - 2\pi R(1-x)^{1/2}\left(\frac{k}{\sigma}\right)^{1/2}2^{1/2}(\Delta\sigma - \sigma). \tag{63}$$

One should keep in mind that x is now not the actual contact angle, which should be zero for any finite value of k, but rather the *asymptotic* contact angle describing the shape of the spherical cap far from the substrate (see Fig. 22).

It is noted from the expression for the free energy in Eq. (63) that the leading contribution to the free energy is proportional to $k^{1/2}$ rather than k. Therefore, if we limit our analysis to the *leading* contribution in k, which is $k^{1/2}$, all contributions to the free energy proportional to k can be neglected. In particular the contribution arising from the integration of the bending energy term $(k/2)J^2$ [see Eq. (48)] over regions where the surface is not strongly curved, i.e., far away from the substrate, have been discarded.

We now have all the contributions to the free energy of a single droplet adhered to the substrate expressed in terms of two parameters, the radius R and contact angle x. As in the previous subsection, we are only left with the determination of R and x itself via a minimization of the free energy with respect to these parameters. This gives the following set of equations [49]

$$\Delta p = \frac{2\sigma}{R} - \frac{4k}{R_0}\frac{1}{R^2}, \tag{64}$$

$$\Delta\sigma = \sigma x - \frac{2k}{R_0}\frac{1}{R}x + \frac{2k}{R_0}\frac{1}{R}\frac{\arccos(x)}{(1-x^2)^{1/2}}$$

$$- \frac{(k\sigma)^{1/2}}{R}\left[4 - 2^{3/2}(1+x)^{1/2}\right], \tag{65}$$

where we have kept terms only to leading order in $k^{1/2}$. In comparison with the expressions in Eqs. (61) and (62) we note that the rigidity constant only appears through the presence of the last term in Eq. (65). The Laplace equation with the Tolman correction is therefore *unaffected* to leading order in $k^{1/2}$ by the presence of rigidity.

Next, R and x are to be solved from the above set of equations and inserted into the expression for $F(R,x)$ in Eq. (63). This is done perturbatively in an expansion in $(k/\sigma)^{1/2} \ll R$. At the same time we assume that also $k/(\sigma R_0) \ll R$ so that to zeroth order the radius and asymptotic angle are given by $R_v = [3V/\pi(2 + 3x_0 - x_0{}^3)]^{1/3}$ and $x_0 \equiv \Delta\sigma/\sigma$, respectively. In terms of x_0 and R_v the free energy in the constant volume ensemble can then be shown to be equal to [49]

$$F(V) = \pi\sigma R_V^2\left(2 + 3x_0 - x_0^3\right) - 4\pi\frac{k}{R_0}R_V\left(1 - x_0^2\right)^{1/2}\arccos(x_0)$$

$$- 8\pi\frac{k}{R_0}R_V(1 + x_0) + 4\pi R_V(k\sigma)^{1/2} \tag{66}$$

$$\times \left(1 - x_0^2\right)^{1/2}\left[2 - 2^{1/2}(1 + x_0)^{1/2}\right].$$

One can now also investigate other ensembles for the single vesicle adhered to a substrate such as the ensemble in which besides the volume also the surface area A is kept constant. As we discussed previously, this is the ensemble for which Seifert and Lipowsky constructed their phase diagrams of the vesicle unbinding transition. Another ensemble one might wish to consider is the constant pressure ensemble, for which Ω is the appropriate free energy, in which instead of the vesicle volume the Laplace pressure difference Δp is fixed. Analogously to the derivation of $F(V)$ one can show that $\Omega(\Delta p)$ is given by [49]

$$\Omega(\Delta p) = \frac{\pi}{3}\sigma R_p^2 (2 + 3x_0 - x_0^3) - 4\pi \frac{k}{R_0} R_p (1 - x_0^2)^{1/2} \arccos(x_0)$$

$$- 8\pi \frac{k}{R_0} R_p (1 + x_0) + 4\pi R_p (k\sigma)^{1/2} (1 - x_0^2)^{1/2} \tag{67}$$

$$\times \left[2 - 2^{1/2}(1 + x_0)^{1/2} \right],$$

where we have defined $R_p \equiv 2\sigma/\Delta p$ and $x_0 \equiv \Delta\sigma/\sigma$.

As an example, we compare the above free energy to the exact free energy obtained by solving the shape equations numerically. We fix $\Delta p/\sigma$ such that $R_p = 2$, large compared to $(k/\sigma)^{1/2}$ for which we take $(k/\sigma)^{1/2} = 0.1$, in some arbitrary length unit. Furthermore, $\Delta\sigma$ is varied such that $-1 < x_0 < 1$ for two different values of the spontaneous radius of curvature, $k/(\sigma R_0) = 0$ and $k/(\sigma R_0) = 0.05$. The result [49] is shown in Fig. 23. The circles and squares are the numerical results for $k/(\sigma R_0) = 0$ and $k/(\sigma R_0) = 0.05$, respectively. The dashed curve $(k/(\sigma R_0) = 0)$ and the dot–dashed curve $(k/(\sigma R_0) = 0.05)$ are the

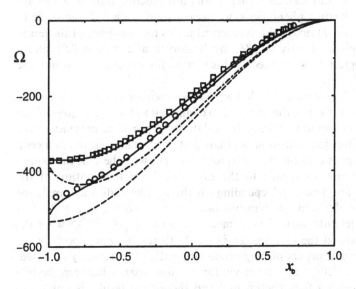

FIG. 23 Free energy, in arbitrary units, as a function of $x_0 = \Delta\sigma/\sigma$ for fixed pressure difference [49]. The expansion parameter $(k/\sigma)^{1/2} = 0.1$ in some arbitrary length unit. Circles and squares are the results obtained by numerical solution of the profile for $k/(\sigma R_0) = 0$ and $k/(\sigma R_0) = 0.05$, respectively. The dashed and the dot–dashed curve are Ω in Eq. (67) with $k = 0$ for $k/(\sigma R_0) = 0$ and $k/(\sigma R_0) = 0.05$, respectively. The solid curves are the full free energy Ω in Eq. (67).

free energy Ω_0 found by setting $k = 0$ in Eq. (67). The solid curve is the full free energy Ω in Eq. (67) for both $k/(\sigma R_0) = 0$ and $k/(\sigma R_0) = 0.05$. As can be seen, it agrees well with the numerically obtained free energy.

Yet other ensembles can be considered. Recently, the onset of droplet aggregation in microemulsions was discussed [50] in the context of the above formulas for the free energy. The appropriate ensemble in a microemulsion system is determined by the two constraints of fixed total amount of surfactant and of fixed total amount of the component *inside* the micro-emulsion droplet, e.g., the amount of water when we consider water-in-oil microemulsions. These two constraints determine the total amount of surface area available to the surfactant and the total amount of droplet volume available to the internal phase. In this analysis, instead of a vesicle adhered to a substrate, the formation of a microemulsion dimer was considered and the phase diagram for droplet–dimer coexistence was constructed [50].

V. SUMMARY AND DISCUSSION

We have presented an overview of the different shapes that liquid droplets may have on a solid substrate. After a brief introduction into the necessary background of wetting phenomena, we discussed, in Section II, the shape of droplets in a gravitational field. Numerical tables for the shape of the liquid droplet were given already in 1883 by Bashforth and Adams [29], and a schematic program was supplied in this section for the calculation of these numerical tables.

In Section III, very small droplets were considered for which gravity can be disregarded but for which the line tension connected to the line of three-phase contact becomes important. We addressed the influence of the presence of line tension on the wetting transition, and saw that positive line tension induces a wetting (drying) transition. Furthermore, we reviewed the current experimental situation with regard to the experimental determination of the magnitude of line tension. Depending on the droplet radii measured, the experimentally obtained line tensions fall into two groups: $\tau_{exp} \approx 10^{-6}$ N when the droplet radii are in the millimeter range and $\tau_{exp} \approx 10^{-9}$ N when the droplet radii are in the micrometer range. Both results, but especially the former, are several orders of magnitude larger than theoretically expected, $\tau_{theor} \approx 10^{-10}$ to 10^{-12} N. The origin for this discrepancy between the two groups and between the experimental and theoretical results is unknown, although contact angle hysteresis and nonequilibrium effects in general might be an important factor. It is clear that much further research is necessary.

In Section IV, we discussed the adhesion of more complex systems that have surfactant-like molecules present at the surface, such as vesicles, micelles, and microemulsion droplets. The adhesion of membranes and vesicles has attracted considerable experimental and theoretical interest because of its importance to biological processes [51]. Experimentally, these systems have been investigated using several techniques such as surface force apparatus, osmotic stress, optical microscopy, micropipette aspiration, reflection interference contrast microscopy, etc. [52]. Theoretical determination of the shapes of vesicles [47,48] has relied mainly on the numerical solution of the shape equations derived from the Helfrich expression [45] for the free energy. However, the large number of parameters involved makes numerical work rather tedious. Furthermore, it is difficult to gain physical insight into the role played by the different parameters in vesicle unbinding or of microemulsion droplet aggregation. To gain this insight and explore the rich experimental phase diagrams, further theoretical research is needed.

REFERENCES

1. Laplace, P.S. *Traité de Mécanique Céleste; Supplément au dixième livre, Sur l'action Capillaire*; Courcier: Paris, 1806.
2. Young, T. On the cohesion of fluids. Philos. Trans. R. Soc. London 1805, *95*, 65.
3. Rowlinson, J.S.; Widom, B. *Molecular Theory of Capillarity*; Clarendon: Oxford, 1982.
4. Gibbs, J.W. *Collected Works*; Dover: New York, 1961.
5. Tolman, R.C. J. Chem. Phys. 1949, *17*, 333.
6. Haye, M.J.; Bruin, C. J. Chem. Phys. 1994, *100*, 556; van Giessen, A.E.; Blokhuis, E.M. J. Chem. Phys. 2002, *116*, 302.
7. Gauss, C.F. Comm. Soc. Reg. Sci. Gött. Rec. 1830, *7*.
8. Johnson, R.E. J. Phys. Chem. 1959, *63*, 1655.
9. Finn, R. Pacific J. Math. 1980, *88*, 541.
10. Avron, J.E.; Taylor, J.E.; Zia, R.K.P. J. Stat. Phys. 1983, *33*, 493.
11. Blokhuis, E.M.; Shilkrot, Y.; Widom, B. Mol. Phys. 1995, *86*, 891.
12. Bikerman, J.J. *Physical Surfaces*; Academic Press: New York, 1970.
13. Bikerman, J.J. Schulman, J.H. Ed.; Second International Congress of Surface Activity 1957, *Vol. III,*. Academic Press: New York, 1957, 125–130.
14. Pethica, B.A.; Pethica, T.J.P. Schulman, J.H. Ed.; Second International Congress of Surface Activity 1957, *Vol. III,*. Academic Press: New York, 1957, 131–136.
15. Pethica, B.A. J. Colloid Interface Sci. 1977, *62*, 567.
16. Ruckenstein, E.; Lee, P.S. Surf. Sci. 1975, *52*, 298.
17. Merchant, G.J.; Keller, J.B. Phys. Fluids A 1992, *4*, 477.
18. Vesselovsky, W.S.; Pertzov, W.N. J. Phys. Chem. U.S.S.R. 1936, *8*, 5.
19. Toshev, B.V.; Platikanov, D.; Scheludko, A. Langmuir 1988, *4*, 489.
20. Toshev, B.V.; Avramov, M.Z. Colloids Surf. A 1993, *75*, 33.
21. Widom, B. J. Phys. Chem. 1995, *99*, 2803.
22. For reviews on Wetting see: Schick, M. in *Liquids at Interfaces*, pp. 415–497; Les Houches XLVIII (1988)., Chavolin, J., Joanny, J.F., Zinn-Justin, J., Eds.; North-

Holland, Amsterdam, 1998; Dietrich, S., in *Phase Transitions and Critical Phenomena*, Vol. 12.; Domb, C., Lebowitz, J., Eds.; Academic Press: London, 1988, p. 1; Sullivan, D.E., Telo da Gama, M.M. in *Fluid Interfacial Phenomena*; Croxton, C.A., Ed.; Wiley, New York, 1986, p. 45; Lipowsky, R., Fisher, M.E., Phys. Rev. B 1987, *36*, 2126; de Gennes, P.G. Rev. Mod Phys. 1985, *57*, 827.

23. Winter, A. Heterogen. Chem. Rev. 1995, *2*, 269.
24. Bausch, R.; Blossey, R. Phys. Rev. E 1994, *50*, 1759; Bausch, R.; Blossey, R.; Burschka, M.A. J. Phys. A 1994, *27*, 1405.
25. Law, B.M. Phys. Rev. Lett. 1994, *72*, 1698.
26. Bonn, D.; Indekeu, J.O. Phys. Rev. Lett. 1995, *74*, 3844.
27. Blokhuis, E.M. Phys. Rev. E 1995, *51*, 4642.
28. Cahn, J.W. J. Chem. Phys. 1977, *66*, 3667.
29. Bashforth, F.; Adams, J.C. *An Attempt to Test the Theories of Capillary Action*; Cambridge University Press: London, 1883.
30. See, e.g., Meunier, J. In *Liquids at Interfaces, Les Houches XLVIII*. Charvolin, J., Joany, J.F., Zinn-Justin, J., Eds.; Elsevier: Amsterdam, 1990; and references therein.
31. de Feijter, J.A.; Vrij, A. J. Electronanal. Chem. 1972, *37*, 9.
32.. Marmur, A. J. Colloid Interface Sci. 1997, *186*, 462.
33. van Giessen, A.E.; Bukman, D.J.; Widom, B. Mol. Phys. 1999, *96*, 1335.
34. Wang, J.Y.; Betelu, S.; Law, B.M. Phys. Rev. Lett. 1999, *83*, 3677.
35. Stöckelhuber, K.W.; Radoev, B.; Schulze, H.J. Colloids Surf. A 1999, *156*, 323.
36. Drelich, J.; Miller, J.D. Colloids Surf. 1992, *69*, 35.
37. Nguyen, A.V.; Stechemesser, H.; Schulze, H.J. J. Colloid Interface Sci. 1997, *187*, 547.
38. Amirfazli, A.; Kwok, D.Y.; Neumann, A.W. J. Colloid Interface Sci. 1998, *205*, 1.
39. Gu, Y.; Li, D.; Cheng, P. J. Colloid Interface Sci. 1996, *180*, 212.
40. Duncan, D.; Li, D.; Gaydos, J.; Neumann, A.W. J. Colloid Interface Sci. 1995, *169*, 256.
41. Li, D.; Neumann, A.W. Colloids Surf. 1990, *43*, 195.
42. Kralchevsky, P.A.; Nikolov, A.D.; Ivanov, I.B. J. Colloid Interface Sci. 1986, *112*, 132.
43. Neumann, A.W.; Good, R.J. J. Colloid Interface Sci. 1972, *38*, 341.
44. Bresme, F.; Quirke, N. J. Chem. Phys. 1999, *110*, 3536.
45. Helfrich, W. Z. Naturforsch. C 1973, *28*, 693.
46. For reviews see, Gelbart, W.M., Ben-Shaul, A., Roux, D., *Micelles, Membranes, Microemulsions, and Monolayers*; Springer: New York, 1994; Nelson, D., Piran, T., Weinberg, S., Eds.; *Statistical Mechanics of Membranes and Surfaces*; World Scientific: Singapore, 1988 Wortis, M.; Seifert, U.; Berndl, K.; Fourcade, B.; Rao, M.; Zia, R. Beysens, D. Boccara, N., Forgacs, G., Eds.; *Dynamical Phenomena at Interfaces, Surfaces and Membranes*; Nova Science: New York, 1993, Safran, S.A. *Statistical Thermodynamics of Surfaces, Interfaces, and Membranes*; Addison-Wesley: Reading, 1994.
47. Seifert, U., Lipowsky, R., Phys. Rev. A 42, 4768; Seifert, U., Z. Phys. B 1995, *97*,

299; Lipowsky R.; Seifert, U., Langmuir 1991, *7*, 1867; Seifert, U. Phys. Rev. Lett. 1995, *74*, 5060.

48. Seifert, U. Adv. in Phys. 1997, *46*, 13.

49. Blokhuis, E.M.; Sager, W.F.C. J. Chem. Phys. 1999, *110*, 3148; Blokhuis, E.M.; Sager, W.F.C. J. Chem. Phys. 1999, *110*, 7062.

50. Sager, W.F.C.; Blokhuis, E.M. Prog. Colloid Polym. Sci. 1998, *110*, 258.

51. Alberts, B.; Bray, D.; Lewis, J.; Raff, M.; Roberts, K.; Watson, J.D. *Molecular Biology of the Cell*; Garland: New York, 1994.

52. For reviews see Lipowsky, R. 521–602 pp. Parsegian, V.A.; Rand, R.P. 643–690 pp. Helfrich, W. 691–722 pp. In *Structure and Dynamics of Membranes*; Lipowsky, R., Sackmann, E., Eds.; Elsevier: Amsterdam, 1995.

4

Simulation of Bubble Motion in Liquids

YING LIAO and JOHN B. MCLAUGHLIN   Clarkson University, Potsdam, New York, U.S.A.

I. INTRODUCTION

Bubbles play an important role in many industrial and natural systems. A few examples include chemical reactors, mass transfer operations, froth flotation, and aeration systems. In some situations, buoyancy is the main cause of bubble motion. However, in low-gravity environments, more subtle effects such as surface tension gradients may be the primary mechanism for bubble motion. In turbulent flows, bubbles tend to follow the motion of the surrounding liquid. These examples may give some impression of the scope of the subject that we will attempt to address in this chapter.

Experimental and theoretical research on the motion of bubbles dates well back into the nineteenth century. However, computational research on bubble motion has become feasible only in the last few decades. This chapter will attempt to provide an overview of this work and to provide the reader with some understanding of the tools that are used. It will become evident that the subject is large and varied. In many situations, there are alternative procedures that are currently in use, and, when possible, we offer our own opinions about the advantages and disadvantages of various options.

It is probably reasonable to say that simulations at the present time are primarily useful in providing basic physical insights that the engineer or scientist can use in developing models of complex processes. Although computer power and algorithms limit the complexity of systems that presently can be treated, simulations can provide useful information about the details of processes that are difficult to obtain by experiments or by theoretical analysis. Phenomena that occur on very small length or timescales and phenomena involving complex interfacial motions are examples. Theoretical analyses are generally limited to relatively simple shapes such as spheres, spherical caps, or spheroids. The motion of a bubble is often assumed to be steady or quasi steady. Processes involving large accelerations, rapid rates of dissolution, coalescence, or breakup are not easily treated without significant modeling simplifications and simulations can provide valuable tests of such approaches.

This chapter will discuss simulation techniques in the context of physical phenomena and processes. Two of the most basic issues of interest are the velocity and the shape of bubbles. One would also like to be able to predict the velocity field in the surrounding liquid and, in many cases, the internal circulation of the bubble. In polar liquids, surface-active molecules tend to concentrate on bubble surfaces, and this can have a significant effect on the rise velocity, the rate of dissolution, and the tendency of bubbles to coalesce. In terrestrial applications, the buoyancy force is often the most important issue in determining the motion of a bubble. However, in low-gravity environments, temperature gradients may play an important role in causing bubble motion through their effect on the surface tension of the interface. In many industrial reactors, turbulence and fluid shear control a bubble's motion and the possibility that it may break up or coalesce with another bubble. In such flows, it is important to predict the motion of bubbles because they contain either a reactant or a product or byproduct of the reaction and their location and concentration affects the efficiency of operation of the reactor. Forces such as the centrifugal force or the lift force can play an important role in determining the trajectory of a bubble.

We will attempt to describe a representative set of applications of bubble motion simulations. The techniques used to perform the simulations are somewhat different for small Reynolds numbers than for large Reynolds numbers. Therefore the applications will be discussed in separate sections devoted to different Reynolds number regimes. We will first briefly summarize some general background on bubble motion and then present the equations that govern bubble motion. We will then describe some of the more common numerical techniques that have been used to perform simulations of bubble motion. Our goal is not to provide a detailed explanation of how to implement these methods. We hope to provide an overview of the basic approach, the advantages and disadvantages of a technique, and a discussion of the situations to which it applies. References to published papers providing more detailed descriptions will be provided.

There are a number of review articles and books that deal with either simulations of bubble motion or general background on bubble motion. The book by Clift et al. [1] provides an extremely useful discussion of the older literature on the subject of bubble motion. References to more recent books and review articles are given in subsequent sections.

II. BACKGROUND

Although the focus of this chapter is on numerical simulation of bubble motion, we will first provide a short review of experimental and theoretical work on bubble motion. In keeping with the overall organization of the chapter, this section will begin the simplest case: single bubbles translating through motionless Newtonian liquids. A more detailed review of the subject may be found in the book by Clift et al. [1].

A. Buoyancy-Driven Bubble Motion

Early progress in understanding the motion of bubbles was mainly concentrated in the extreme cases of Stokes flow and potential flow, corresponding to $Re \ll 1$ and $Re \gg 1$, where Re denotes a Reynolds number based on the size and velocity of the bubble and viscosity of the liquid. In the Stokes flow regime, Hadamard [2] and Rybczynski [3] independently derived exact solutions for a fluid sphere translating through a motionless liquid. When the motion of the fluid sphere is driven by gravity, the general result for the sphere's terminal velocity, U_t, is

$$U_t = \frac{2}{3} \frac{gR^2 \Delta \rho}{\mu} \frac{1 + k_\rho}{2 + 3k_\rho}. \tag{1}$$

where g, R, $\Delta \rho$, μ, and k_ρ denote the acceleration of gravity, the sphere's radius, the density difference between the sphere's phase and the continuous phase, the viscosity of the continuous phase, and the ratio of the densities of the sphere's phase and the continuous phase. Usually, the density of a gas is much smaller than the density of a liquid so that $k_\rho \ll 1$ for a bubble. In this case, Eq. (1) reduces to

$$U_t = -\frac{1}{3} \frac{gR^2 \rho}{\mu}, \tag{2}$$

where ρ denotes the liquid density. The above expression for the bubble velocity is larger by a factor $3/2$ than the expression for the velocity of a massless rigid sphere according to the Stokes drag law. The explanation for this difference is that the viscous contribution to the drag from the shear stress acting on a rigid sphere is absent for a bubble in the low-density approximation.

Although the Hadamard–Rybcynski result is an exact solution of the Stokes equation, it is inconsistent with many experimental results for small bubbles or drops. Bond [4] and Bond and Newton [5] showed that sufficiently small bubbles or drops behave like rigid spheres. This phenomenon was eventually explained by Frumkin and Levich [6] (see Ref. 7) as being due to surface active impurities. For example, consider a bubble or a nonpolar droplet moving through a polar liquid. The surfactant molecules adsorb on the surface of the bubble or nonpolar drop and are swept to the rear by convection. This leads to a surface tension gradient that balances the shear stress at the interface and retards the motion of liquid at the interface. Fig. 1 gives a schematic view of a rising bubble in its frame of reference.

In a remarkable series of experiments, Savic [8] demonstrated the existence of immobilized surfactant caps on water droplets falling through castor oil. He used aluminum particles to make the internal circulation visible. Fig. 2 shows his photographs of a 1.77-cm water droplet on the left and a 1.21-cm water droplet on the right. It may be seen that the larger droplet contains a vortex that extends throughout the droplet. However, the smaller droplet has a vortex that does not extend to the top of the droplet.

Using asymptotic analysis, Sadhal and Johnson [9] rigorously established the conditions under which an immobilized surfactant cap would form on a bubble or drop in Strokes flow. They also found an exact solution of the Stokes equation for a fluid sphere with an immobilized surfactant cap.

At infinite Reynolds number, potential flow theory may be used to investigate bubble motion. As a consequence of D'Alembert's paradox (see, e.g., Ref. 10), potential flow theory predicts no drag on a steadily rising bubble. However, it provides a result for the force on an accelerating sphere:

$$\mathbf{F} = \frac{1}{2} m_1 \mathbf{a}. \tag{3}$$

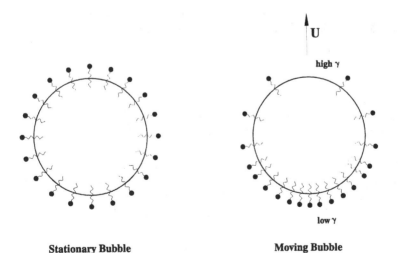

Stationary Bubble   Moving Bubble

FIG. 1 Surfactant-caused retardation of bubble motion.

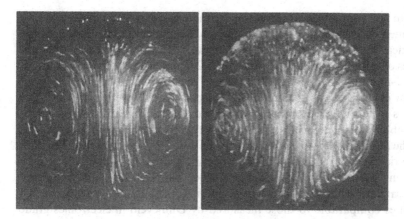

FIG. 2 Water droplets in castor oil by Savic [8]. The photographs show a 1.77- and a 1.21-cm drop.

where, $\mathbf{F}$, m_l, and $\mathbf{a}$ denotes the force on the sphere, the mass of liquid displaced by the sphere, and the acceleration of the sphere, respectively. Thus according to potential flow theory, the acceleration on a massless sphere due to buoyancy should be $2\mathbf{g}$.

When combined with boundary layer analysis, potential flow theory provides estimates for the velocity of a rising bubble. The lowest-order result for the terminal velocity of a freely rising bubble is

$$U_t = -\frac{1}{9}\frac{r^2\Delta\rho g}{\mu}. \tag{4}$$

It is interesting to note that the result in Eq. (4) is the same as the result for a rigid, massless sphere in Stokes flow except for a factor $1/2$. Moore [11] extended the above result to include finite Reynolds number effects. Moore [12] also developed a result for a spheroidal bubble. El Sawi [13] extended Moore's theory to include the effects of fore-aft asymmetry.

Although the above results for the bubble rise velocity at large Reynolds numbers are based on rigorous analyses, one finds discrepancies with experiments. Haberman and Morton [14] found that bubbles smaller than about 5 mm significantly rise more slowly in tap water than in filtered or distilled water. For example, they found that, at room temperature, a 1-mm bubble rose at about 13 cm/sec in tap water. However, a bubble of the same size rose at 19 cm/sec in filtered or distilled water. The latter value is still significantly smaller than the value suggested by Moore's theory as extended by El Sawi (26 cm/sec). This suggests that, although the filtered water used by Haberman and Morton contained less surface-active material than their tap water, it still contained enough surface-active material to significantly slow their bubbles. This idea was recently confirmed by Duineveld [15,16] who measured the velocities of bubbles over the size range 0.72–2.0 mm in electronics grade water. For a 1-mm bubble, he found that the rise velocity was 26 cm/sec.

Sam et al. [17] and Zhang et al. [18] performed experiments on single bubbles in tap water containing various concentrations of frothing agents. The bubbles were released from a capillary tube at sufficiently low frequencies that the bubbles were, to a good approximation, released from rest. The bubbles varied in size from 0.8 to 1.5 mm. They measured the bubble velocity as a function of height above the capillary tube. Although the water contained significant amounts of surface-active materials, the bubbles typically reached velocities comparable to those measured by Duineveld in electronics grade water before slowing down. The terminal velocity was comparable to terminal velocities observed by other investigators in impure water. For very large concentrations, the bubble rise velocity monotonically approached the steady state value for contaminated water. This behavior suggests that, at sufficiently large concentrations, the concentration of surfactant on the bubble surface is limited by sorption kinetics [19].

Both the Stokes flow solutions and the large Reynolds number boundary layer solutions assume that the flow around the bubble is axisymmetric. This assumption fails for some situations at large Reynolds numbers. Saffman [20] and Hartunian and Sears [21] performed experiments on bubbles at large Reynolds numbers. For bubbles in distilled water, they found that bubbles larger than a critical size tend to spiral or zigzag. Hartunian and Sears reported the critical value of the equivalent spherical diameter to be 1.7 mm. Later, Duineveld [15] found that the critical value was 1.9 mm. In tap water, the critical value is smaller. Hartunian and Sears reported the value to be 1.3 mm. Table 1 summarizes experimental results for the onset of oscillations.

B. Thermocapillary Bubble Motion

The above results are for single bubbles that are driven by buoyancy. In low gravity environments, buoyancy may be less important than temperature gradients in producing bubble motion. When a bubble is placed in a temperature gradient, the surface tension on the "hot" end of the bubble is smaller than the surface tension on the "cold" end of the bubble. The resulting surface tension gradient causes bubbles to migrate toward warmer regions. Fig. 3 is a schematic illustration of the phenomenon.

Reviews of thermocapillary migration may be found in Refs. 24 and 25. The book by Subramanian and Balasubramaniam [26] provides discussions of both thermocapillary migration and buoyancy-driven bubble motion.

TABLE 1 Experimental Results for the Onset of Oscillations

Authors	Water purity	T (C°)	d_e (cm)
Rosenberg [22]	Distilled	19.0	0.15
Hartunian and Sears [21]	Distilled	24.5	0.17
Tsuge and Hibino [23]	Distilled	21.0	0.17
Duineveld [16]	Electronics Grade	20.3	0.193
Haberman and Morton [14]	Filtered	19.0	0.14
Haberman and Morton [14]	Tap	21.0	0.13
Hartunian and Sears [21]	Tap	22.5	0.13

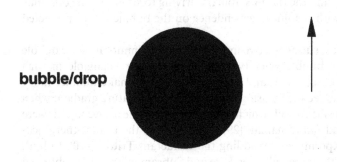

High Temperature

low σ

bubble/drop

high σ

Low Temperature

FIG. 3 Thermocapillary motion of a drop.

In the Stokes flow limit, Young et al. [27] developed an exact solution for the velocity of the bubble and the temperature and velocity fields in the liquids. The velocity of the bubble is given by the following expression:

$$U = \frac{2U_s}{(2 + 3k_\mu)(2 + k_k)},$$ (5)

where k_μ is the ratio of the dynamic viscosity of the drop phase to that of the continuous phase and k_k is the corresponding ratio of thermal conductivities. The quantity U_s is defined by

$$U_s = \frac{|\gamma_T||\mathbf{G}_T|R}{\mu}.$$ (6)

where γ_T is the rate of change of interfacial tension with temperature, and $\mathbf{G}_T$ is the temperature gradient imposed in the continuous phase fluid. For a bubble, it is generally reasonable to assume that k_μ and k_k are both very small compared to unity. Thus the bubble velocity is

$$U = \frac{U_s}{2}.$$ (7)

The analysis leading to the formula in Eq. (5) is based on the assumption that the bubble interface is mobile. Merritt and Subramanian [28] performed measurements of the velocity of air bubbles moving under the combined influence of gravity and the thermocapillary driving force in silicone oil. They found agreement with the linear dependence on the bubble radius predicted by Eq. (5).

If a liquid is polar, surface-active impurities tend to immobilize the bubble interface. Thus the bubble velocity is much smaller or negligible in such situations. Young et al. [27] found that they could not make bubbles move downward in n-hexadecane by using a downward temperature gradient when trace amounts of a silicone oil, which is a surfactant in the above liquid, were present. Barton and Subramanian [29] demonstrated the role of thermocapillarity in their experiments by adding the surfactant Triton X100 to both phases to suppress thermocapillarity. Kim and Subramanian [30,31] obtained asymptotic results for the effects of surfactants on thermocapillary migration for Stokes flow. Nallani and Subramanian [32] found that methanol drops in silicone oil moved at velocities that were smaller than those predicted for an uncontaminated surface. They fit their data to the stagnant surfactant cap model developed by Kim and Subramanian.

Chen and Stebe [33,34] discussed the possibility of remobilizing the interface by increasing the bulk concentration of surfactant to saturate the interface and eliminate concentration gradients.

The Young et al. analysis assumes that both the Reynolds number and the Peclet number of the bubble are small compared to unity. Subramanian [35] used asymptotic analysis to remove the restriction on the Marangoni number for small Reynolds number.

$$U = U_s \frac{1}{2}\left[1 - \frac{301}{7200} Ma^2\right]. \tag{8}$$

Crespo and Manuel [36] and, independently, Balasubramaniam and Chai [37] recognized that the Stokes flow solution for small Marangoni numbers gave aggregate predictions at all Reynolds numbers provided that $Ma \ll 1$ (i.e., that the convective transport of energy is negligible). Balasubramaniam and Chai [37] also showed that, at finite Reynolds numbers, a bubble or drop deforms into an oblate spheroid and obtained an asymptotic result for the shape for small inertial corrections. Their analysis is based on the assumption that both the capillary number and the Weber number are small compared to unity:

$$Ca = \frac{\mu U_s}{\gamma}, \tag{9}$$

$$W = \frac{\rho U_s^2 R}{\gamma}. \tag{10}$$

Haj-Hariri et al. [38] calculated the change to the migration velocity caused by deformation. Balasubramaniam and Subramanian [39] found asymptotic solutions for $Ma \gg 1$ in the limits $Re \gg 1$ and $Re \gg 1$. For $Re \gg 1$, they showed that

$$U = \frac{1}{3} - \frac{1}{8}\log 3 + \frac{1}{\sqrt{Ma}}[0.06845\log(Ma) + 0.6578] \tag{11}$$

For $Re1$, they showed that

$$U = \frac{1}{3} - \frac{1}{8}\log 3. \tag{12}$$

The above analyses ignored the temperature dependence of the bulk properties. Balasubramaniam [40] analyzed the motion of a gas bubble in the limit $Ma \gg \infty$ in a liquid in which the viscosity linearly varies with temperature.

In many situations of possible practical importance, it is not feasible to use asymptotic analysis to obtain solutions of the governing equations. Balasubramaniam and Lavery [41], Ehmann et al. [42], Chen and Lee [43], Nas [44], Treuner et al. [45], Haj-Hariri et al. [93], and Ma et al. [46] reported the results of numerical simulations of thermocapillary migration for bubbles or drops.

C. Bubble Motion in Nonuniform Flows

In many situations, bubbles move through nonuniform flows. The flows may be either laminar or turbulent. For bubbles in sufficiently strong flows, the buoyancy of the bubble may be negligible or have a small effect on the motion of the bubble. In general, such problems are extremely difficult because one must account for bubble deformation and the disturbance created by the bubble and one must make use of empiricism to develop an approximate equation governing the motion of a bubble. However, for sufficiently small bubbles, approximate equations of motion have been developed that account for the deviation of the bubble trajectory from the path of a fluid particle due to buoyancy, fluid inertia, finite size effects, and related phenomena. We will first consider this regime.

1. Small Bubbles

Maxey and Riley [47] derived an equation of motion for a small rigid sphere of radius R in a nonuniform flow. If one considers small bubbles moving in a polar liquid, this equation might be appropriate because surfactants would tend to immobilize the surface of a bubble and make it behave like a rigid sphere. Maxey and Riley assumed that the Reynolds number based on the difference between the sphere velocity and the undisturbed fluid velocity was small compared to unity. In addition, they assumed that the spatial nonuniformity of the undisturbed flow was sufficiently small that the modified drag due to particle rotation and the Saffman [48] lift force could be neglected. Finally, they ignored interactions between particles.

The Maxey–Riley equation takes the following form:

$$m_p \frac{d\mathbf{U}}{dt} = (m_p - m_l)\mathbf{g} + m_l \frac{D\mathbf{U}}{Dt} - \frac{1}{2}m_l \frac{d}{dt}\left[\mathbf{U} - \mathbf{u} - \frac{R^2}{10}\nabla^2\mathbf{u}\right]$$

$$- 6\pi R\mu\left[\mathbf{U} - \mathbf{u} - \frac{R^2}{6}\nabla^2\mathbf{u}\right] \tag{13}$$

$$-6\pi R^2\mu \int_0^t \frac{\frac{d}{dt'}(\mathbf{U} - \mathbf{u} - \frac{R^2}{6}\nabla^2\mathbf{u})}{\left[\pi v(t - t')\right]^{\frac{1}{2}}} dt' \tag{14}$$

where $\mathbf{U}$ is the velocity of the sphere, $\mathbf{u}$ is the velocity of the undisturbed fluid evaluated at the center of the sphere, t is the time, R is the radius of the sphere, m_p is the sphere's mass, m_l is the mass of the displaced fluid, d/dt denotes a time derivative following the sphere, and D/Dt denotes a time derivative using the undisturbed fluid velocity as the convective velocity.

The equation of motion given by Maxey and Riley is valid provided that two Reynolds numbers based on the radius of the sphere are small compared to unity. The two Reynolds numbers are u_0R/v and $R^2u_0/(Lv)$, where u_0 is a velocity that is characteristic of the undisturbed fluid, w_0 is a velocity that is characteristic of the relative motion between the particle and the undisturbed fluid, and L is a characteristic length of the undisturbed flow. These conditions imply that the time required for a significant change in the relative velocity is large compared to the timescale for viscous diffusion, and that viscous diffusion remains the dominant mechanism for the transfer of vorticity away from the sphere.

The terms on the right-hand side of Eq. (14) represent the effects of gravity, pressure gradient, virtual mass, Stokes drag, and history, respectively. If the undisturbed flow is spatially uniform, Eq. (14) reduces to the Basset-Boussineq-Oseen (BBO) equation (see, e.g., Ref. 49). As the sphere's radius goes to zero, the Stokes drag term in Eq. (14) becomes dominant, and the Faxen correction term in the Stokes drag term can be dropped. Another way of viewing this approximation is to assume that the undisturbed flow field, $\mathbf{u}$, varies over a length scale that is much larger than the size of the particle.

Let us consider the specific case of a bubble. In most cases, one can ignore the mass of the bubble, m_p, in comparison with the mass of the displaced liquid, m_l. If one also assumes that the disturbed flow varies slowly enough that the terms involving the Laplacian may be neglected, one can simplify the Maxey–Riley equation as follows:

$$\frac{d\mathbf{U}}{dt} = -2\mathbf{g} + 2\frac{D\mathbf{u}}{Dt} + \frac{d\mathbf{u}}{dt} - \frac{2}{\tau_v}(\mathbf{U} - \mathbf{u}). \tag{15}$$

where τ_v is the viscous relaxation time, which is given by

$$\tau_v = \frac{2}{9}\frac{R^2}{v}. \tag{16}$$

The viscous relaxation time is an estimate of the time needed for a bubble to reach its terminal velocity in a quiescent liquid, $\mathbf{U}_t$:

$$\mathbf{U}_t = -\tau_v\mathbf{g}. \tag{17}$$

More generally, it is a measure of bubble inertia; bubbles with larger values of τ_v have more inertia than bubbles with smaller values of τ_v. If the undisturbed flow is steady, one may further simplify Eq. (15). Maxey [50] performed simulations of bubbles in a steady cellular flow field using Eq. (15) and found that, provided that $|U_t|$ is smaller than the maximum velocity of the cellular flow, bubbles can become trapped in the vortices of the cellular flow.

The Maxey–Riley equation is valid provided that the flow around the particle can be approximated, to lowest order, by unsteady Stokes flow. The Reynolds number, Re, of the flow field in which the particle moves may be very large as long as the particle-scale Reynolds numbers are smaller than unity. In practice, this may be very restrictive. For example, let us consider air bubbles in water. If one estimates the Reynolds number of the bubble from the buoyancy-driven rise velocity of the bubble, one obtains

$$U_t = \frac{2}{9} \frac{R^2 g}{\nu}. \tag{18}$$

The corresponding Reynolds number based on this slip velocity is

$$Re_s = \frac{4}{9} \frac{R^3 g}{\nu^2}. \tag{19}$$

The Reynolds number is equal to unity for a bubble with a diameter equal to 0.12 mm. In more viscous liquids, the Maxey–Riley equation should be valid for larger bubbles provided that the interface may be assumed to be immobilized. For example, in an aqueous sugar solution having a viscosity that is 10 times larger than the viscosity of pure water, one might expect the equation to be valid for 0.55-mm bubbles.

The Maxey–Riley equation does not include inertial effects such as the lift force. Saffman [48] derived an expression for the lift force on a small rigid sphere in a linear shear flow. The leading order lift force on the sphere is caused by an interaction between the slip velocity between the particle and the flow and the shear. Fig. 4 shows a schematic illustration of a particle in a

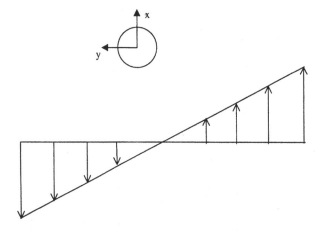

FIG. 4 A schematic illustration of a particle in a vertical flow.

vertical flow. In this case, the slip velocity is caused by the sedimentation of the particle through the fluid.

Let us suppose that the linear shear flow is given by

$$\mathbf{u} = G_u y \mathbf{e}_x, \tag{20}$$

where $\mathbf{u}$ is the fluid velocity and $\mathbf{e}_x$ is a unit vector in the x direction. The Saffman lift force takes the form

$$\mathbf{F}_l = 6.46 \mu R^2 \text{sgn}(G_u) \left(\frac{|G_u|}{\nu} \right)^{1/2} u_s \mathbf{e}_y \tag{21}$$

where $\mathbf{F}_l$ is the lift force, $\mathbf{e}_y$ denotes a unit vector in the y direction and u_s is the slip velocity, $u_x - U_x$. If the slip velocity is negative and G_u is positive, then the lift force points in the negative y direction. A more general way of describing the lift force is that it points in the direction of increasing slip velocity. In the above case, the particle would migrate in the negative y direction. Assuming that the particle possesses sufficient inertia that it retains its x component of velocity, the magnitude of the slip velocity would increase.

With the above assumptions, the equation of motion for small bubbles in a shear flow is obtained from Eq. (14) by neglecting the virtual mass, gravitational, and Bassett history terms as well as the Faxen correction term in the Stokes drag and including the Saffman lift term given in Eq. (21).

$$\frac{d\mathbf{U}}{dt} = -2\mathbf{g} + 2\frac{D\mathbf{u}}{Dt} + \frac{d\mathbf{u}}{dt} - \frac{2}{\tau_v}(\mathbf{U} - \mathbf{u})$$

$$- 2.06 \left(\frac{\nu}{R} \right) \text{sgn}\left(G_u \left(\frac{|G_u|}{\nu} \right)^{1/2} U_s \mathbf{e}_y \right) \tag{22}$$

The Saffman lift force in Eq. (22) does not include a number of effects that are discussed below.

The Saffman lift force is valid provided that the following conditions are satisfied:

$$Re_G \ll 1 \tag{23}$$

$$Re_s \ll 1 \tag{24}$$

$$Re_s \ll Re_G^{1/2}, \tag{25}$$

where $Re_G = 4|G_u|R^2/\nu$ and $Re_s = |u_s|(2R)/\nu$. Asmolov [51] and McLaughlin [52] removed the restriction in Eq. (25). Legendre and Magnaudet [53] performed a similar analysis for a small, spherical bubble with a mobile interface in a linear shear flow.

Saffman's analysis is based on the assumption that the flow in the neighborhood of the sphere is quasi steady. Because the lift force would cause the sphere to migrate across streamlines, the flow field in the sphere's frame of reference would be time dependent. In addition, in many applications, the flow field is inherently time dependent. For these reasons, Miyazaki et al. [54] developed a perturbation formalism that would permit them to compute the lift force on a particle in a time-dependent shear flow. They used their formalism to derive results for the lift force on a particle in an oscillatory flow in the high frequency limit defined by

$$\Omega \gg G_u \qquad (26)$$

where Ω is the angular frequency of the oscillation. Asmolov and McLaughlin [55] used conventional matched asymptotic analysis to obtain results that are valid over a broader range of frequencies.

In turbulent shear flows, the largest shear rates occur in the vicinity of rigid walls. Thus one might expect the lift force to be most important in the vicinity of walls. Saffman's analysis and the other work described above does not account for the presence of a wall. Let us consider the lift on a rigid sphere moving parallel to a flat wall as shown in Fig. 5.

Cox and Hsu [56] and Vasseur and Cox [57,58] presented results for the lift force on a rigid sphere in wall-bounded laminar shear flows. Their analysis assumes that the distance from the closest wall is large compared to the radius of the sphere, but that it is small compared to the smaller of the two characteristic lengths defined by

$$L_s = \frac{v}{u_s} \qquad (27)$$

$$L_G = \left(\frac{v}{|G_u|}\right)^{1/2}. \qquad (28)$$

The physical significance of L_s and L_G is that they are characteristic of the region in which Oseen corrections to Stokes flow are important. By requiring

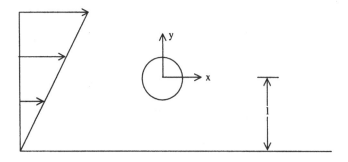

FIG. 5 A rigid sphere translating though a shear flow parallel to a flat wall.

that the distance from the closest wall was small compared to both of these lengths, one justifies the use of regular perturbation methods to compute the leading order effects of inertia. Asmolov [51] and McLaughlin [59] eliminated this restriction and presented results that are valid over a broader range of distances from the closest wall. However, their results are still valid only when the radius of the particle is small compared to the distance from the closest wall. Cherukat and McLaughlin [60] eliminated this restriction and computed results for the lift force that are valid even for very small distances from the closest wall.

Maxey and Riley pointed out that the Magnus force is of order R^3 and, for that reason, it is less important than the Saffman lift force.

2. Larger Bubbles

If one wishes to track larger bubbles through a nonuniform flow, one must rely on empiricism. Intuitively, one might expect the buoyancy and drag forces to be the biggest forces for a large bubble in a turbulent flow field. However, there is some experimental evidence that lift forces can be important near the walls of vertical shear flows (see, e.g., Ref. 61). In addition, Legendre and Magnaude [61] have shown by direct solution that the lift force can be comparable to the drag force for high Reynolds number spherical bubbles in a linear shear flow.

Dandy and Dwyer [62] and Cherukat et al. [63] reported the results of simulations of flows around rigid spheres at finite Reynolds numbers. Mei [64] developed correlations for the lift force on rigid spheres at finite Reynolds numbers. Auton [65] derived an expression for the lift on a bubble in a rotational inviscid flow with a weak shear. Legendre and Magnaudet [61] presented the results of computations of the lift and drag forces acting on a spherical bubble with a clean interface for bubble Reynolds numbers ranging from 0.1 to 500. An interesting result of their calculations is that the drag coefficient is very close to Moore's [11] theoretical value even for fairly large shear rates. For example, if $G_u R / U = 0.2$, the change in the drag coefficient is only 5% even for $Re = 500$ (the worst case). This provides some support for the idea that one can use potential flow results even in a fairly strongly sheared flow.

A limitation of the above results is that they do not account for the presence of walls and it is not clear that this is permissible in situations where the lift force is likely to be important.

Provided that one may approximate the bubble by a sphere, one might use an equation of the following form to track a bubble in a high Reynolds number flow:

$$\frac{dU}{dt} = -2\mathbf{g} + 3\frac{D\mathbf{u}}{Dt} + -\frac{3}{4R} UU C_D + \mathbf{F}_1 \tag{29}$$

where C_D is a suitable drag coefficient and $\mathbf{F}_l$ is the lift force based on the mean shear. Unlike the Maxey–Riley equation, Eq. (29) has not been derived in a rigorous way from the Navier–Stokes equation. It is based on physical intuition. The virtual mass term is based on potential flow theory and, therefore, is based on the assumption that one can ignore vorticity. One might expect Eq. (29) to be reasonable as a rough approximation provided that the bubble is small compared to the length scales of the energy-containing eddies. A difference between Eq. (29) and Eq. (15) is that the derivative of the undisturbed flow is a convective derivative following the liquid rather than the sphere. Auton [66,67] derived this result for a spherical bubble in a nonuniform potential flow. Clift et al. [1] suggest expressions for the drag force. For a spherical bubble with a clean interface, one could use Moore's [11] leading order result:

$$C_D = \frac{64}{Re}. \tag{30}$$

For a bubble with an immobilized interface, one could use the empirical correlation recommended by Clift et al. [1]:

$$C_D = \frac{24}{Re}\left[1 + 0.1935 Re^{0.6305}\right]. \tag{31}$$

The correlation in Eq. (31) is recommended for use in the Reynolds number range $20 < Re < 260$. In practice, the expression for C_D in Eq. (31) is more likely to be relevant than the potential flow result in Eq. (30). Even the minute amounts of surfactant present in tap water are sufficient to immobilize the surface of bubbles small enough to be approximately spherical. The upper Reynolds number range for the correlation in Eq. (31) roughly coincides with the largest Reynolds number for which one might expect contaminated bubbles to be approximately spherical.

III. GOVERNING EQUATIONS

This section will present the governing equations and the boundary conditions for simulating the flow around a bubble.

A. Navier–Stokes Equation

Before discussing the governing equations, it is useful to introduce the Reynolds number, Re, the Weber number, W, and the Morton number, M:

$$Re = \frac{U_t d_e}{\nu}, \tag{32}$$

$$W = \frac{\rho U_t^2 d_e}{\gamma_0},$$ (33)

$$M = \frac{g\mu^4}{\rho\gamma_0^3}.$$ (34)

In the above equations, the symbol d_e denotes the equivalent spherical diameter of the bubble and v, ρ, γ_0, and μ denote the kinematic viscosity, the density, the surface tension, and the dynamic viscosity of the pure liquid, respectively. (Later in this section, we shall consider bubbles in dilute surfactant solutions.)

To analyze the flow around the bubble, the Navier–Stokes equation must be solved. The Navier–Stokes equation in the Cartesian coordinates may be written as follows:

$$\frac{\partial \mathbf{u}}{\partial t} + \mathbf{u} \cdot \nabla \mathbf{u} = -\frac{\nabla p}{\rho} + \mu\nabla^2\mathbf{u} + \mathbf{f}$$ (35)

To solve Eq. (35), the boundary conditions must be specified. In many situations, the boundary conditions at the bubble interface are free-slip or no-slip. For an isothermal uncontaminated interface, the tangential stress in the liquid should equal the tangential stress in the gas:

$$(\mathbf{n} \cdot \tau \cdot \mathbf{s})_{liq} = (\mathbf{n} \cdot \tau \cdot \mathbf{s})_{gas}.$$ (36)

In Eq. (36), $\mathbf{n}$ is a unit vector normal to the interface pointing into the liquid and $\mathbf{s}$ is a unit vector tangential to the interface. Because the viscosity of the gas is negligible, the tangential stress in the gas is zero and Eq. (36) reduces to,

$$(\mathbf{n} \cdot \tau \cdot \mathbf{s})_{liq} = 0.$$ (37)

Equation (37) expresses the free-slip condition.

As discussed in the Sec. II, Haberman and Morton [14] found that small bubbles rise in tap water as if they were rigid spheres. In this case, the boundary conditions in Eqs. (36) and (37) do not apply and one should apply a no-slip boundary condition to the liquid velocity:

$$\mathbf{u} \cdot \mathbf{s} = 0.$$ (38)

This condition should be applied to the immobilized portion of the bubble surface as discussed by McLaughlin [68]. As shown by McLaughlin [68], an immobilized surfactant cap forms on a bubble in steady motion provided that convective transport of the surfactant on the bubble dominates diffusion and that the surfactant is sparingly soluble. More generally, one should compute the surface concentration of surfactant. The transport equation for the surface concentration of surfactant is given by Stone [69] and Stone and Leal

[70]. The bulk concentration of surfactant in the liquid, C_∞, the maximum surface concentration of surfactant, Γ_∞, and the surface tension of clean water, γ_0, may be used to make the equations governing the surfactant dimensionless. The Peclet number in the bulk liquid, Pe, and on the surface of the bubble, Pe_s, take the following forms: $Pe = 2U_t r_e/D$ and $Pe_s = 2U_t r_e/D_s$, where D, D_s, and r_e denote the diffusivity of the surfactant in the liquid, the diffusivity on the surface of the bubble, and the equivalent spherical radius of the bubble, respectively. In dimensionless form, the transport equation for the surface concentration of surfactant, Γ, takes the following form:

$$\frac{\partial \Gamma}{\partial t} + \nabla_s \cdot (\Gamma \mathbf{u}_s) + \Gamma(\nabla_s \cdot \boldsymbol{n})(\mathbf{u} \cdot \boldsymbol{n}) = \frac{1}{Pe_s} \nabla_s^2 \Gamma + \frac{1}{Pe} \frac{\partial C}{\partial n}, \tag{39}$$

where ∇_s is the surface gradient operator and $\mathbf{u}_s = \mathbf{u} - \mathbf{nn} \cdot \mathbf{u}$. Because the flux of surfactant to the interface must equal the net rate of adsorption, it follows that, in dimensionless form,

$$\frac{1}{Pe_s} \frac{\partial C}{\partial n} = K_1 Bi[kC_s(1 - \Gamma) - \Gamma], \tag{40}$$

in which $k = \beta C_\infty/\alpha$ and $Bi = \alpha r_e/U_t$, where α and β are the Langmuir desorption and adsorption rate constants and $K_1 = \Gamma/2r_e C_\infty$ is the dimensionless adsorption length. In Eq. (40), it has been assumed that the sorption kinetics are described by the Langmuir model (see Ref. 71).

The volume concentration of surfactant in the liquid satisfies the following transport equation:

$$\frac{\partial C}{\partial t} + \mathbf{u} \cdot \nabla C = \frac{1}{Pe} \nabla^2 C. \tag{41}$$

When surface diffusion is negligible, $Pe_s 1$, the adsorption and desorption rates are slow (the "insoluble surfactant" limit) and the bubble motion is steady, one obtains stagnant cap behavior for the steady motion. In this regime, Eq. (39) reduces to

$$\nabla_s \cdot (\Gamma \mathbf{u}_s) = 0. \tag{42}$$

The solution is

$$\mathbf{u}_s = 0, \quad \theta < \theta_{cap}, \tag{43}$$

$$\Gamma = 0, \quad \theta > \theta_{cap}, \tag{44}$$

where θ_{cap} is the cap angle. The angles θ_{cap} and θ are measured from the bottom of the bubble. For the clean interface, $\theta > \theta_{cap}$, one applies the free-slip boundary condition, while on the surfactant cap, $\theta < \theta_{cap}$, one applies the no-slip boundary condition.

The normal stress on the surface should balance the surface tension,

$$(\boldsymbol{n} \cdot \tau \cdot \boldsymbol{n}) = \gamma(\nabla_s) \cdot \boldsymbol{n}. \tag{45}$$

The kinematic boundary condition for the interface is

$$\frac{D\mathbf{r}}{Dt} = \mathbf{u},\tag{46}$$

in which $\mathbf{r}$ is a point on the surface. Equation (46) is used to determine the deformation of the bubble.

B. Small Reynolds Numbers

For small Reynolds numbers, one may linearize the Navier–Stokes equation. In the most general situation, one obtains the unsteady Stokes equation:

$$\frac{\partial \mathbf{u}}{\partial t} = -\frac{\nabla p}{\rho} + \mu \nabla^2 \mathbf{u} + \mathbf{f}\tag{47}$$

The unsteady term in Eq. (47) may be neglected if the timescale for the unsteadiness of the flow is large compared to the timescale for the diffusion of vorticity. In this case, one obtains the Stokes equation. One solves either the unsteady or the steady version of the Stokes equation together with the continuity equation:

$$\nabla \cdot \mathbf{u} = 0.\tag{48}$$

The above equations should be solved subject to the usual rigid boundary conditions on solid surfaces. The boundary conditions on the surface of a bubble depend on the presence or absence of surfactant molecules that can immobilize the surface. Stone and Leal [70] reported the results of simulations in which an insoluble surfactant was adsorbed on the surface of the bubble.

C. Large Reynolds Numbers

For large Reynolds numbers, $Re \gg 1$, one may sometimes consider the flow around the bubble to be approximately irrotational. In this case, the velocity may be expressed in terms of a velocity potential:

$$\mathbf{u} = \nabla \phi.\tag{49}$$

As long as the above equation is satisfied, even if the viscosity is finite, the flow is still potential. The Navier–Stokes equation reduces to the irrotational Euler equation,

$$\frac{\partial \mathbf{u}}{\partial t} = -\nabla \Pi.\tag{50}$$

where Π is the pressure head defined as

$$\Pi = p + \frac{1}{2}\rho \mathbf{u}^2.\tag{51}$$

The continuity equation takes the form of Laplace's equation:

$$\nabla^2 \phi = 0. \tag{52}$$

The solution of Eq. (52) will give the velocity potential and Eq. (49) can be used to give the velocity. Once the velocity field is known, the pressure field can be obtained from Bernoulli's law, which follows from Eq. (50).

The usefulness of potential flow theory is somewhat limited because in polar liquids, relatively small concentrations of surfactants can immobilize a sufficient portion of the bubble surface to cause flow separation and create a vortex under the bubble [68].

IV. BOUNDARY INTEGRAL METHODS

Boundary integral methods are particularly useful in situations in which the governing equations are linear. In the context of fluid mechanics problems, this means either the Stokes flow regime or the potential flow regime. Most published simulations involving the application of boundary integral method to problems involving bubbles or drops have been performed in the Stokes flow regime.

Using theoretical results developed by Ladyzhenskaya [72], Youngren and Acrivos [73] formulated the boundary integral equations for a rigid particle in a uniform Stokes flow and obtained solutions for circular cylinders and prolate spheroids. When the cylinder or spheroid had a sufficiently large axis ratio, the drag force was found to be in good agreement with asymptotic results obtained by Cox [74,75] using slender body theory. Youngren and Acrivos [76] used similar methods to obtain numerical results for the shape of an inviscid gas bubble in an axisymmetric extensional flow. Rallison and Acrivos [77] generalized their approach so that they could simulate the shape of a viscous drop in an axisymmetric extensional flow. Finally, Rallison [78] extended the method to nonaxisymmetric shear flows. To treat this three-dimensional problem, he discretized the surface of the drop into triangles.

The basic idea behind the method is to express the velocity at an arbitrary point in a two- or three-dimensional flow in terms of an integral over a surface or several surfaces. To do this, the surface is treated as a collection of point forces and the solution is expressed in terms of an integral of the "Stokeslet" solution. One can then obtain an integral equation that can be solved to obtain the velocity and shape of a bubble or drop. The advantage of the method is that one can obtain the solution of very complicated three-dimensional flow problems by solving linear two-dimensional integral equations. In principle, one can apply the same method to flows at finite Reynolds numbers. This could be carried out by using an iterative procedure in which the nonlinear term is explicitly treated.

For simple problems, one may be able to find asymptotic solutions to the boundary integral equation or at least simplify the numerical solution. For example, Youngren and Acrivos [76] exploited the axisymmetry of their problem to simplify the boundary integral equation by performing the azimuthal integral analytically. However, in general, one must solve the problem numerically. To do this, one discretizes the surface into elements and describes the spatial variations of the variables inside each element with a set of basis functions in a manner similar to the finite element method. This approach is known as the boundary element method. The paper by Rallison [78] provides an early example of this method. The book by Pozrikidis [79] provides an introduction to boundary integral methods including boundary element techniques.

Let us consider a drop, bubble, or particle that is in a Stokes flow. The velocity of the fluid around the object, $\mathbf{u}$, may be written as the sum of the flow that would exist in the absence of the object, $\mathbf{u}_0$, and the disturbance created by the object, $\mathbf{v}$:

$$\mathbf{u} = \mathbf{u}_0 + \mathbf{v}. \tag{53}$$

The flows $\mathbf{u}$ and $\mathbf{u}_0$ are assumed to satisfy the Stokes equation. This implies that $\mathbf{v}$ also satisfies the Stokes equation. For simplicity, an unbounded fluid will be considered for which $\mathbf{v}$ is assumed to vanish at infinite distance from the object. The Stokeslet solution is the disturbance flow created by a point force, $\mathbf{F}$, acting on the fluid:

$$v_i(\mathbf{x}) = J_{ij}(\mathbf{x}, \mathbf{y})F_j/\mu \tag{54}$$

$$p(\mathbf{x}) = p_j(\mathbf{x}, \mathbf{y})F_j/\mu \tag{55}$$

where summation over the repeated indices is understood. In Eqs. (54) and (55), the quantities J_{ij} and p_j are defined as follows:

$$J_{ij}(\mathbf{x}, \mathbf{y}) = -\frac{1}{8\pi}\left[\frac{\delta_{ij}}{r_{xy}} + \frac{(x_i - y_i)(x_j - y_j)}{r_{xy}^3}\right] \tag{56}$$

$$p_{j(x,y)} = \frac{(y_j - x_j)}{4\pi r_{xy}^2}, \tag{57}$$

where $r_{xy} = |\mathbf{x} - \mathbf{y}|$.

The stress tensor associated with the disturbance flow is given by

$$\tau_{ij} = -\delta_{ij}p + \mu\left(\frac{\partial v_i}{\partial x_j} + \frac{\partial v_j}{\partial x_i}\right). \tag{58}$$

One may express the solution for the disturbance flow in terms of the above quantities as follows:

$$v_i = -\int_S J_{ij}(\mathbf{x}, \mathbf{y}) \frac{\tau_j(y)}{\mu} \, dS + \int_S K_{ijk}(\mathbf{x}, \mathbf{y}) n_j u_k(\mathbf{y}) dS, \tag{59}$$

where the integrals are understood to be over the surface of the bubble, drop, or particle and summation over the repeated indices is understood. In Eq. (59), the quantity τ_j is defined by

$$\tau_j(\mathbf{y}) = \tau_{jk} n_k(\mathbf{y}), \tag{60}$$

where $\mathbf{n}(\mathbf{y})$ is a unit normal vector at the point $\mathbf{y}$ on the surface of the object. The quantity K in Eq. (59) is defined as follows:

$$K_{ijk} = -\frac{3}{4\pi} \frac{(x_i - y_i)(x_j - y_j)(x_k - y_k)}{r_{xy}^5}. \tag{61}$$

One can obtain an integral equation for the velocity of the surface of a bubble or drop by combining Eq. (59) with the boundary conditions on the surface of the bubble or drop. For simplicity, let us assume that the viscosity and density of the bubble are negligible and measure the pressure relative to the uniform pressure inside the bubble. In this case, the stress balance on the bubble surface takes the following form:

$$\mathbf{n} \cdot \tau = \gamma(\nabla_\mathbf{s} \cdot \mathbf{n})\mathbf{n} - \mathbf{n}\rho \mathbf{g} \cdot \mathbf{x_s}, \tag{62}$$

where $\mathbf{x_s}$ is a point on the surface of the bubble. In Eq. (62), $\mathbf{g}$ is the acceleration of gravity and $\nabla_\mathbf{s}$ is the surface gradient operator.

The integral equation for the surface velocity takes the following form:

$$\frac{1}{2}\mathbf{u}(\mathbf{x_s}) = \mathbf{U}(\mathbf{x_s}) - \int_S \mathbf{J} \cdot \mathbf{n}\left[\frac{\gamma}{\mu}(\nabla_\mathbf{s} \cdot \mathbf{n}) - \frac{\rho}{\mu}(\mathbf{g} \cdot \mathbf{x_s})\right] ds - \int_S \mathbf{n} \cdot \mathbf{K} \cdot \mathbf{u} ds \tag{63}$$

Once one determines the surface velocity, one can compute the deformation of the surface on each time step by using the kinematic boundary condition.

The pioneering work described above lead to a large number of applications of boundary integral methods to problems in Stokes flow. We shall describe only a few of these applications; the papers in question contain references to many other papers dealing with boundary integral methods applied to Stokes flow problems. One area of interest is the behavior of suspensions of drops or bubbles in viscous liquids. Coalescence and breakup affect the rheological behavior of suspensions and a number of papers in recent years have dealt with these phenomena using boundary integral methods.

Stone and Leal [70] applied the method to a droplet contaminated with an insoluble surfactant in an axisymmetric extensional flow. The effect of the surfactant on the surface tension was modeled by the ideal gas law. Because the flow was axisymmetric, they were able to reduce the surface integral to a line integral following a procedure similar to one employed by Youngren and Acrivos [76]. Because the surfactant distribution affects the surface velocity that, in turn, affects the surfactant distribution, it was necessary to employ an explicit/implicit method to perform the time stepping.

Stone and Leal found that the surfactant tended to concentrate near the poles of the droplet. This tended to reduce the surface tension at the poles, which, in turn, caused more deformation at the poles. They studied the breakup of droplets and determined critical capillary numbers.

Milliken and Leal [80] extended the work of Stone and Leal by considering situations in which the drop becomes highly elongated. They identified conditions under which the ends of the drop become pointed and suggested that this might explain the phenomenon of tip streaming observed by Rumscheidt and Mason [81] and Milliken and Leal [82].

Kennedy et al. [83] used the boundary element method to investigate the rheology of dilute suspensions in a shear flow. Because the suspension was dilute, encounters close enough to cause coalescence or breakup were sufficiently rare to be unimportant.

Manga and Stone [84] carried out experiments and boundary integral simulations of the interactions of bubbles and drops in viscous liquids. Their experiments revealed a mechanism of coalescence between small and large bubbles or drops in which the smaller object is pulled around the larger object and "sucked into" the bottom of the larger object. Their boundary integral simulations of the process appear to be the first three-dimensional boundary integral simulations of buoyancy-driven bubble or drop interactions. They performed simulations of pairs of droplets with viscosities equal to the viscosity of the continuous liquid. Their results are qualitatively in agreement with their experiments. A conclusion of their work was that the rate of coalescence in suspensions of deformable drops or bubbles can be an order of magnitude larger than in suspensions of spherical drops or bubbles.

Loewenberg and Hinch [85] and Zinchenko et al. [86,87] developed new boundary integral techniques that permitted them to simulate the phenomena reported by Manga and Stone for small drop–drop separations. Lowenberg and Hinch developed a "near-singularity" subtraction to improve convergence for drops that are in close proximity. They added artificial "springs" between mesh points to stabilize the grid and facilitate long time simulations. Zinchenko et al. developed a more robust method for computation of the mean curvature as well as a "curvatureless" version of the boundary integral method. They also developed a variational method of global mesh stabilization. With their techniques, they obtained results that clearly demonstrated the "sucking in" mode of bubble coalescence. Results for separations as small as 1% of the drop radius were presented.

V. POTENTIAL FLOW METHODS

Kumaran and Koch [88] studied the interactions between a pair of nondeformable bubbles. They assumed that the flow around bubbles was potential.
The velocity potential was determined using twin spherical expansions, and
the momentum equations were derived by requiring that the net force on each
bubble be zero. The equations were solved in a spherical coordinate system
that was moving with the same speed as one of the bubbles and the velocities
were defined in a fixed frame of reference. As shown in Fig. 6, two bubbles
with radius R_1 and R_2 have velocities U_1 and U_2. The unit vector from the
center of bubble 1 to the center of the bubble 2, e_{a1}, serves as one of the axes.
The two unit vectors perpendicular to this axis are e_{b1} and e_{c1}. The vectors e_{a1}
and e_{a2} are opposite to each other, while the other two pairs of unit vectors are
in the same direction. The distance between the two bubbles is denoted by R_d.

The velocity is expressed in terms of the velocity potential:

$$\mathbf{u} = \nabla\phi, \tag{64}$$

and the velocity potential satisfies the Laplace equation:

$$\nabla^2\phi = 0 \tag{65}$$

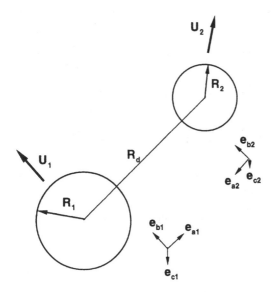

FIG. 6 Coordinate system for simulations of two bubbles.

The liquid pressure is given by the Bernoulli equation. Because the bubble is nondeformable, the component of the fluid velocity normal to surface of the bubble is equal to the velocity of the bubble in that direction:

$$\mathbf{u} \cdot \mathbf{n} = \mathbf{U_i} \cdot \mathbf{n}, \tag{66}$$

where the subscript i denotes bubble i, and $\mathbf{n}$ is the normal vector to the bubble surface. Because the masses of the bubbles are assumed to be zero, the net force on each bubble is zero:

$$-\int_{A_i} pn dA_i + \mathbf{F}_i = 0, \tag{67}$$

where A_i is the surface area of bubble i and $\mathbf{F}_i$ is the sum of the external forces on the bubble. The boundary condition at large distance from the bubble is

$$\mathbf{u} = 0, \quad p = 0, \tag{68}$$

Equations (64)–(67) along with the boundary condition in Eq. (68) are solved to give the accelerations of the bubbles:

$$\frac{dU_{ia}}{dt} = \frac{9R_j^3}{2R_d^4} \left(U_{1b}U_{2b} + U_{1c}U_{2c} - 2U_{ja}^2 \right) + \frac{\mathbf{F}_i \cdot \mathbf{e}_{ia}}{(2/3)\pi\rho R_i^3}, \tag{69}$$

$$\frac{dU_{ik}}{dt} = \frac{9R_j^3}{2R_d^4} U_{ja}(U_{1k} + U_{2k}) + \frac{\mathbf{F}_i \cdot \mathbf{e}_{ik}}{(2/3)\pi\rho R_i^3}, \text{ for } k = b, c \tag{70}$$

The acceleration of the bubble due to the external force, $\mathbf{F_i}$, is equal to the force on the bubble divided by its added mass, which is half the mass of the fluid displaced by it. The external force is the sum of the buoyancy and the drag forces given by

$$\mathbf{F}_i = \frac{\mathbf{U}_{it} - \mathbf{U}_i}{\tau_{vi}}, \tag{71}$$

where $\mathbf{U}_{it}$ is the terminal velocity and τ_{vi} is the viscous relaxation time $[\rho R_i^2/(18\mu)]$. The interaction decays as $1/R_d^4$.

Kumaran and Koch integrated the bubble equations of motion to obtain trajectories for the bubbles. In one set of simulations, the Reynolds number of the smaller bubble was either 100 or 200. Calculations were performed for radius ratios equal to 1.2 and 1.5. The interaction between the bubbles was caused by the approach of the larger bubble from beneath the smaller bubble. In these simulations, collisions did not occur. In a second set of collisions, the Reynolds number of the larger bubble was 200 and the radius ratio was 0.2. In these simulations, repeated collisions occurred between the bubbles. They found that, when the Reynolds number of the bigger bubble was 200, collisions occurred when the radius ratio was smaller than 0.223. When the

Reynolds number of the bigger bubble was 100, collisions occurred when the radius ratio was smaller than 0.365. Kumaran and Koch [89] found that collisions also occurred when the size ratio was between 0.9 and 1.1 and the Reynolds number of one of the bubbles was 100 and between 0.93 and 1.07 when the Reynolds number was 200.

Sangani and Didwania [90] also performed simulations of spherical bubbles using potential flow methods. A difference between their work and that of Koch and Kumaran is that Sangani and Didwania simulated periodic suspensions of bubbles. The periodic box was cubic and it contained as many as 150 bubbles. A second difference is that, in a majority of cases, the bubbles were identical. Bubble Reynolds numbers equal to 100, 500, and 2000 were considered. Bubble–bubble collisions occurred and were treated as elastic. This assumption is based on Kok's [91] experiments. Kok found that when a very small amount of surfactant was present in water, colliding bubbles tended to bounce rather than coalesce. He also analytically calculated the trajectories of the bubbles using a potential flow approximation together with the viscous forces computed with the energy dissipation method. An unusual feature of the results presented by Sangani and Didwania is that the bubbles tended to form planar layers. This behavior does not seem to have been experimentally observed. It was conjectured that vorticity generated by the bubbles and the container wall might play an important role in real suspensions. The bubble distributor at the bottom of the container might also generate shear and mixing.

VI. FINITE DIFFERENCE METHODS

There are many varieties of finite difference methods that have been used in computational fluid mechanics. Typically, the grid lines are parallel to coordinate lines. The coordinate system may be either orthogonal or nonorthogonal. The grid may be either fixed or moving. In a fixed grid simulation, the bubble moves through the grid. Thus at any particular time, the bubble interface does not coincide with the grid. Therefore one must somehow track the interface. There are two general approaches to this problem. One approach is the front tracking method discussed by Unverdi and Tryggvason [92]. The other approach uses the level set technique that involve the use of a "color function." This approach has been discussed by Haj-Hariri and Shi [93].

Moving grid methods have been successfully used to study the motion of rising bubbles because of the advantage that they offer in handling the interface. Ryskin and Leal [94–96] used an adaptive grid technique to develop an orthogonal, curvilinear coordinate system in which one of the coordinate curves coincided with the surface of the bubbles. They applied their methods to the computations of bubbles in steady state motion for Reynolds numbers up to 200 and Weber numbers up to 20. Kang and Leal [97] extended the method developed by Ryskin and Leal to study time-dependent bubble deformation in an extensional flow. McLaughlin [68] used Ryskin and Leal's techniques to simulate the steady state motion of bubbles in water for Reynolds numbers as large as 637, which corresponds to a 2.0-mm bubble.

In what follows, the discretization by finite difference method will be given and a few examples using this method in different grid systems will be presented too.

A. Discretization by Finite Difference Method

By truncating the higher-order terms of the Taylor expansion, one obtains the finite difference formulation. The one-dimensional grid shown in Fig. 7 will be used as an example.

The grid is equally distributed such that $\delta x = x_P - x_W = x_E - x_P$. The dependent variable φ is a function of x, $\varphi = f(x)$. The Taylor expansion for φ_W and φ_E around point P is,

$$\varphi_W = \varphi_P - \delta x\left(\frac{d\varphi}{dx}\right)_P + \frac{1}{2}(\delta x)^2\left(\frac{d^2\varphi}{dx^P}\right)_P - \frac{1}{6}(\delta x)^3\left(\frac{d^3\varphi}{dx^3}\right)_P + O(\delta x^4) \quad (72)$$

$$\varphi_E = \varphi_P - \delta x\left(\frac{d\varphi}{dx}\right)_P + \frac{1}{2}(\delta x)^2\left(\frac{d^2\varphi}{dx^2}\right)_P - \frac{1}{6}(\delta x)^3\left(\frac{d^3\varphi}{dx^3}\right)_P + O(\delta x^4) \quad (73)$$

By adding and subtracting the above two equations, the following two equations are obtained:

$$\left(\frac{d\varphi}{dx}\right)_P = \frac{\varphi_E - \varphi_W}{2\delta x}, \quad (74)$$

$$\left(\frac{d^2\varphi}{dx^2}\right)_P = \frac{\varphi_W + \varphi_E - 2\varphi_P}{(\delta x)^2}. \quad (75)$$

The error in both cases is $O(\delta x^2)$. By substituting such expressions into the differential equation, the finite difference equation is obtained. In general, the

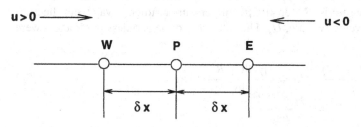

FIG. 7 One-dimensional grid for finite difference method.

dependent variable φ is a function of three spatial coordinates, x,y,z, and time, t. The discretization of $\partial\varphi/\partial t$ is as follows:

$$\left(\frac{\partial\varphi}{\partial t}\right)_P = \frac{\varphi_P^{n+1} - \varphi_P^n}{\Delta t}, \tag{76}$$

in which n is the time step.

When the convection term is present in the equation, the use of central differences may produce numerical instability. One can suppress the convective numerical instability by using the upwind difference technique. In this technique, forward differencing is used when the flow is moving from left to right (negative velocity) and backward differencing is used when the flow is moving from right to left (positive velocity). Referring to Fig. 7 again, if the flow is from point W to point P, $u > 0$, Eq. (72) will take the form

$$\left(\frac{d\varphi}{dx}\right)_P = \frac{\varphi_P - \varphi_W}{\delta x}, \tag{77}$$

otherwise, $u < 0$,

$$\left(\frac{d\varphi}{dx}\right)_P = \frac{\varphi_P - \varphi_E}{\delta x}. \tag{78}$$

In both cases, the error is $O(\delta x)$.

B. Finite Difference Method on a Fixed Grid

When the bubble surface does not deform, it is relatively easy to solve the problem on a fixed grid. Balasubramanian and Lavery [41] used the fixed grid method to study the thermocapillary migration of a spherical bubble under microgravity for large Reynolds and Marangoni numbers. The problem they studied was a stationary bubble surrounded by the liquid with a steady state velocity field. The full Navier–Stokes equation was solved in a three-dimensional spherical (r,θ,ϕ) coordinate system. The origin of the coordinate is at the center of mass of the bubble.

The distance, velocity, pressure, and temperature were made dimensionless by the bubble radius, R, density, ρ, temperature gradient away from the bubble, G_T, and velocity scale, U_s. The following dimensionless parameters were introduced:

$$U_s = \frac{(|\gamma_T||G_T|R)}{\mu} \quad Re = \frac{(|\gamma_T||G_T|R^2)}{\mu\nu} Pr Re \tag{79}$$
$$= Pe = Ma = \frac{(|\gamma_T||G_T|R^2)}{\mu\kappa_t},$$

where κ_t is the thermal diffusivity and Pr is the Prandtl number.

In their study, they considered that the flow was laminar and symmetric in the ϕ direction. The θ direction is measured from the direction of the temperature gradient. Therefore the continuity equation in dimensionless form is

$$\frac{1}{r^2}\frac{\partial}{\partial r}(r^2 u_r) + \frac{1}{r\sin\theta}\frac{\partial}{\partial\theta}(u_\theta\sin\theta) = 0. \tag{80}$$

The r component of Navier–Stokes equation is

$$u_r\frac{\partial u_r}{\partial r} + \frac{u_\theta}{r}\frac{\partial u_r}{\partial\theta} - \frac{u_\theta^2}{r} = -\frac{\partial p}{\partial r} + \frac{1}{Re}\left(\nabla^2 u_r - \frac{2u_r}{r^2} - \frac{2}{r^2}\frac{\partial u_\theta}{\partial\theta} - \frac{2u_\theta}{r^2}\cos\theta\right). \tag{81}$$

The θ component of the Navier–Stokes equation takes the form

$$u_r\frac{\partial u_\theta}{\partial r} + \frac{u_\theta}{r}\frac{\partial u_r}{\partial\theta} + \frac{u_r u_\theta}{r} = -\frac{1}{r}\frac{\partial p}{\partial\theta} + \frac{1}{Re}\left(\nabla^2 u_\theta + \frac{2}{r^2}\frac{\partial u_r}{\partial\theta} - \frac{u_\theta}{r^2}\sin^2\theta\right). \tag{82}$$

The energy equation is

$$-U + u_r\frac{\partial T}{\partial r} + \frac{u_\theta}{r}\frac{\partial T}{\partial\theta} = \frac{1}{Ma}\nabla^2 T, \tag{83}$$

where ∇^2 is the Laplacian and is defined as

$$\nabla^2 = \frac{1}{r^2}\frac{\partial(r^2\partial/\partial r)}{\partial r} + \frac{1}{r^2\sin\theta}\frac{\partial(\sin\theta\,\partial/\partial\theta)}{\partial\theta} \tag{84}$$

The boundary conditions on the surface of the bubble are

$$u_r = 0 \tag{85}$$

$$\frac{(\partial u_\theta/r)}{\partial r} = \frac{\partial T}{\partial\theta} \tag{86}$$

$$\frac{\partial T}{\partial r} = 0. \tag{87}$$

respectively. Because the flow at $r = \infty$ is undisturbed by the presence of the bubble, the boundary conditions are

$$u_r = -u_\infty\cos\theta \tag{88}$$

$$u_\theta = u_\infty\sin\theta \tag{89}$$

$$p = p_\infty \tag{90}$$

$$T = r \cos \theta \left(0 \leq \theta \leq \frac{\pi}{2} \right) \tag{91}$$

$$-\cos \theta \frac{\partial T}{\partial r} + \frac{\sin \theta}{r} \frac{\partial T}{\partial \theta} = -\frac{\partial T}{\partial Z} = -1 \left(\frac{\pi}{2} \leq \theta \leq \pi \right). \tag{92}$$

On the symmetry axis $\theta = 0$ and $\theta = \pi$, the boundary conditions are

$$\frac{\partial u_r}{\partial \theta} = 0, \quad u_\theta = 0, \quad \frac{\partial T}{\partial \theta} = 0, \quad \frac{\partial p}{\partial \theta} = 0. \tag{93}$$

Because the bubble was treated as a void, the net force on the bubble is zero:

$$\int_0^\pi \sin^2 \theta \left(\frac{\partial T}{\partial \theta} \right) + \cos \theta \sin \theta \left[\frac{2}{\sin \theta} \frac{\partial (u_\theta \sin \theta)}{\partial \theta} + Re\, p \right] d\theta \Big|_{r=1} = 0. \tag{94}$$

Because the maximum differences between $r = 5$ and $r = \infty$ for u, v, T, and p were 0.8%, 0.4%, 0.4%, and 0.8% respectively, the computational domain was chosen in the range $1 \leq r \leq 5$ and $0 \leq \theta \leq \pi$. The conditions at infinity were used at $r = 5$. The grid was equally distributed in both the r and θ directions. The temperature and velocity were evaluated at the cell nodes, while the pressures were evaluated at the center of the cell. Equations (81)–(83) were discretized by using the three-point finite difference formula with second-order error. The pressure on the node of the cell was calculated by the interpolation of pressure at the center of the four closest cells. The discretized equation was solved along with the boundary conditions in an iterative procedure.

The solution method involved two iteration loops. The outer loop was a secant method to solve Eq. (94) that was discretized by Simpson's method to give the value of u_∞. The inner loop was Newton's method to compute u, v, T, and p.

The steps in the solution method are outlined below:

1. Assume a value of u_∞, and solve the discretized Eqs. (80)–(93) along with the boundary condition equations to give new values of u, v, T, and p. This is performed by using Newton's method. The iterations of the inner loop are stopped when the relative difference in each of the four variables u, v, T, and p is less than 0.01. At the end of the inner iteration, the value of the integral in Eq. (94) is calculated.

2. A new value of u_∞ is assumed. Again, Eqs. (80)–(93) are solved along with the boundary equation for u, v, T, and p by Newton's method. The starting values for Newton's method are the latest temperature and pressure fields and a rescaled velocity field that is obtained by multiplying the latest velocity field by the ratio $(u_\infty)^{n+1}/(u_\infty)^n$. On each Newton step, the linear system of equations were obtained using a

banded direct solver. Finally, the value of the integral in Eq. (94) is calculated.

3. The velocities $(u_\infty)^{n+1}$ and $(u_\infty)^n$ and the corresponding integrals are used to predict a new value of u_∞ by the secant method.
4. The new value of u_∞ computed in step 3 is used as input in step 2.
5. Steps 2, 3, and 4 are repeated until the relative difference between the two latest value of u_∞ is less than 0.5×10^{-5}.

In their study, the Reynolds numbers ranged from 10^{-7} to 2000 and the Marangoni numbers ranged from 10^{-7} to 1000. They found that the scaled bubble velocity varied by less than one order of magnitude and that the bubble velocity was influenced more by the Marangoni number than by the Reynolds number.

C. Front Tracking Method

Unverdi and Tryggvason [92] used a front-tracking method to study viscous incompressible multiphase flow. The simulation was carried out on a fixed grid using the finite difference method. The bubble interface did not coincide with the grid as shown in Fig. 8. A separate, unstructured grid shown in Fig. 9 that moved through the stationary grid was introduced to explicitly represent the interface. The grid on the surface was restructured as the interface became deformed.

The Navier–Stokes equation for the incompressible, two-phase unsteady flow takes the form

$$\frac{\partial(\rho\mathbf{u})}{\partial t} + \nabla \cdot (\rho\mathbf{u}\mathbf{u}) = -\nabla p + \rho\mathbf{g} + \nabla \cdot (2\mu\mathbf{D}) + \gamma\kappa\mathbf{n}\delta(\mathbf{x} - \mathbf{x}') \qquad (95)$$

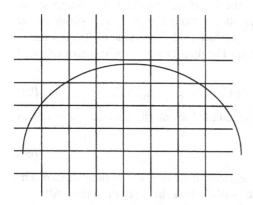

FIG. 8 A schematic of an interface across a structured grid.

FIG. 9 A schematic of an interface across an unstructured grid.

where γ is the surface tension, κ is the curvature, and $\mathbf{D}$ is the rate of deformation tensor. The components of $\mathbf{D}$ are $D_{ij} = 1/2\left(u_{i,j} + u_{j,i}\right) = \partial u_i/\partial x_j + \partial u_j/\partial x_i$. The term $\delta(\mathbf{x}-\mathbf{x}^f)$ is a delta function that is zero everywhere except at the interface, where $\mathbf{x} = \mathbf{x}^f$. The continuity equation is

$$\nabla \cdot \mathbf{u} = 0 \tag{96}$$

The equations for the density and viscosity fields are

$$\frac{D\rho}{Dt} = \frac{\partial \rho}{\partial t} + \mathbf{u} \cdot \nabla \rho = 0, \quad \frac{D\mu}{Dt} = \frac{\partial \mu}{\partial t} + \mathbf{u} \cdot \nabla \mu = 0. \tag{97}$$

Inside each fluid, ρ and μ are constants. Equations (95) and (96) were solved by using a finite difference method on a fixed two- or three-dimensional grid. The spatial terms were discretized by second-order finite differences on a staggered Eulerian grid. The discretization of time was achieved by an explicit Euler method or a second order Adams–Bashforth method. The boundary conditions used in their study were either periodic or full slip in the horizontal directions and rigid, stress-free on the top and bottom.

Instead of directly solving Eq. (97), the density and velocity fields were computed by explicitly tracking the interface. To track the interface, an indicator function $I(\mathbf{x})$ was constructed from the known position of the interface that was defined by the interface grid. The value of $I(\mathbf{x})$ is 1 inside the bubble and 0 outside the bubble. The density and velocity were evaluated on the grid points by

$$\rho(\mathbf{x}) = \rho_0 + (\rho_b - \rho_0)I(\mathbf{x}), \quad \mu(\mathbf{x}) = \mu_0 + (\mu_b - \mu_0)I(\mathbf{x}). \tag{98}$$

The surface tension force was calculated from the surface curvature as follows:

$$\mathbf{F}^{(l)} = \gamma \kappa^{(l)} \mathbf{n}^{(l)} \Delta s^{(l)}, \tag{99}$$

where $\kappa_{(l)}$ is the curvature at the centroid of element (l) on the interface. The grid force field was constructed by distributing the force onto the grid:

$$\mathbf{f}(\mathbf{x}) = \sum_l D\left(\mathbf{x} - \mathbf{x}^{(l)}\right)\mathbf{F}^{(l)}, \tag{100}$$

where $D[\mathbf{x} - \mathbf{x}^{(l)}]$ is the distribution function discussed in detail by Unverdi and Tryggvason [92]. The velocity at the interface point 1 was the sum of contribution from the i points near the point l:

$$\mathbf{u}^{(l)} = \sum_i D\left(\mathbf{x}_i - \mathbf{x}^{(l)}\right)\mathbf{u}_i. \tag{101}$$

The velocity of the surface was determined by

$$\frac{d\mathbf{x}^{(l)}}{dt} = \mathbf{u}^{(l)}. \tag{102}$$

As the surface deforms, to maintain good resolution, the grid points on the interface must be reconstructed and this was performed by three steps, namely, node addition, node deletion, and restructuring. For the curvature calculation, they used different methods for two dimensions and three dimensions. In two dimensions, they fit a local, cubic Hermite polynomial to four points on the interface and found the curvature by differentiation with respect to the arc length. The surface tension force was found by evaluating the curvature at the middle of the element. In three dimensions, they used a method described by Todd and McLeod [98]. The details are given in their paper.

They presented results for buoyancy-driven bubbles in both two and three dimensions. They showed the interaction of two bubbles such as nonaxisymmetric merging of two bubbles and the collision of a large bubble with a smaller one.

D. Level Set Technique

In the front tracking approach, the interface is represented by an unstructured grid. Each point on the grid must be tracked by using the local fluid velocity. If the points on the unstructured grid become too crowded or too sparse, grid reconstruction is necessary. The disadvantage is that for three-dimensional problems, the interface reconstruction will involve extremely costly calculations. Another approach called the level set technique, which involves the use of a "color function," has the advantage that, with local grid refinement, one can avoid the use of an unstructured grid to mark the interface. Haj-Hariri et al. [93] used the above method to study the three-dimensional thermocapillary motion of deformable drops in the presence of convective transport of momentum and energy. The continuity and momentum equation along with the temperature distribution were solved on an equally spaced grid in three coordinate directions. To track the interface without the reconstruction of the grid on the interface, they integrated an adaptive local grid refinement technique into the continuum interface model. This model is an improved version of the volume-of-fluid (VOF) method discussed by Hirt and Nichols [99], wherein the information of surface is retained by advecting a Lagrangian invariant named VOF or "color function." The traditional VOF needs reconstruction of the interface on every time step. Another problem is that the computed surface tends to spread out because the color function "washes away" from the interface due to numerical errors. To overcome the drawback, a level-set function, $s(x)$, was constructed (see Ref. 100). The quantity $s(x)$ is the normal distance from the interface ($s = 0$) and it satisfies the equation

$$|\nabla s| = 1. \tag{103}$$

The function is advected in the interface according to

$$\frac{Ds}{Dt} = \frac{\partial s}{\partial t} + \mathbf{u} \cdot \nabla s = 0. \tag{104}$$

Away from the interface, s is reconstructed to satisfy Eq. (103). The condition in Eq. (103) is imposed by solving the following equation:

$$\frac{\partial s}{\partial \tau_p} = \text{sgn}(s) \left[|\nabla s|^2 - 1 \right]. \tag{105}$$

An iteration method is used the solve the above the equation by marching in the pseudotime, τ_p. The color function Φ is defined as a mollified Heaviside step function based on s. The value of Φ varies from 1 to 0 in moving from the "interior" to "exterior" with a sharp gradient over a thin interface region.

The stress jump, τ_γ, across the interface is given by,

$$\tau_\gamma = [(-\gamma_T)(\mathbf{I} - \mathbf{nn}) \cdot \nabla T_\infty + \gamma \mathbf{n} \nabla \cdot \mathbf{n}], \tag{106}$$

in which $\mathbf{n} = \nabla s$ based on the definition of s. Furthermore, the surface force per unit area is replaced by the volume-distributed counterpart, f, which satisfies

$$\tau_\gamma = \int_{-\infty}^{\infty} \mathbf{f}(s)ds = \int_{\infty}^{\infty} \tau_\gamma \delta(s)ds, \tag{107}$$

where $\delta(s)$ is the Dirac delta function with its singularity on the interface. Because the mollified delta function can be defined as $|\nabla \Phi|$, the body force may be rewritten in the form

$$\mathbf{f} = \tau_\gamma |\nabla \Phi|. \tag{108}$$

The following characteristic scales, radius, R, velocity scale, U_s, timescale, R/U_s, pressure scale, $\mu U_s/R$, and $\mu U_s/R^2$ are used to make the continuity and Navier–Stokes equation dimensionless. The dimensionless density, viscosity, thermal conductivity, and heat capacity distributions are defined as

$$[\rho, \mu, \kappa_t, c] = [\sigma, \lambda, \beta, k]C + (1 - C), \tag{109}$$

where $\sigma = \rho_{in}/\rho_{ex}$, $\lambda = \mu_{in}/\mu_{ex}$, $\lambda = \kappa_{tin}/\kappa_{tex}$, $k = c_{in}/c_{ex}$. The symbol c denotes the specific heat. The subscripts "ex" and "in" denote the values of the physical properties in the external and internal phases, respectively. Therefore the dimensionless continuity and momentum equations take the form

$$\nabla \cdot \mathbf{u} = 0 \tag{110}$$

$$\frac{\partial(\rho\mathbf{u})}{\partial t} + \nabla \cdot (\rho\mathbf{u}\mathbf{u}) = \frac{1}{Re}[-\nabla p + \mathbf{f} + \nabla \cdot \mu(\nabla\mathbf{u} + \nabla\mathbf{u}^+)]. \tag{111}$$

The temperature distribution equation is

$$\rho c\left[\frac{\partial T}{\partial t} + U\right] + \nabla \cdot (\rho c T\mathbf{u}) = \frac{1}{Ma}[\nabla \cdot (\kappa\nabla T) + \mu Br(\nabla\mathbf{u} + \nabla\mathbf{u}^+) \cdot \nabla\mathbf{u}], \tag{112}$$

where Br is the Brinkman number defined as $\mu_{ex}U^2/\kappa_{tex}G_T R$.

The infinite physical domain is truncated into the computational domain, which is increased until the results are not influenced by the size of the domain. The computational domain is formed by equal-sized cubic cells and the governing equations are spatially discretized at the cell centers by using the second-order finite difference approximations. There are two steps in each time step:

$$\frac{\mathbf{u} - \mathbf{u}^n}{\delta t} = -\mathbf{u}^n \cdot \nabla\mathbf{u}^n + \frac{1}{\rho^n Re}\left[\nabla \cdot \mu(\nabla\mathbf{u}^n + \nabla\mathbf{u}^{+n}) + \mathbf{f}_t^n\right]. \tag{113}$$

$$\frac{\mathbf{u}^{n+1} - \mathbf{u}}{\delta t} = \frac{1}{\rho^n Re}\left[-\nabla p^{n+1} + \mathbf{f}_n^n\right], \tag{114}$$

where $\mathbf{f}_t$ and $\mathbf{f}_n$ are the tangential and normal component of $\mathbf{f}$. The solution steps are summarized in what follows. First, the intermediate velocity is determined by Eq. (113). Second, the pressure term on the right side of Eq. (114) is calculated by using the intermediate velocity $\mathbf{u}$ through the following equation:

$$\nabla \cdot \left[\frac{1}{\rho^n}\nabla p^{n+1}\right] = \frac{Re}{\delta t}\nabla \cdot \mathbf{u} + \nabla \cdot \left[\frac{1}{\rho^n}\mathbf{f}_n^n\right]. \tag{115}$$

The above equation is obtained by taking the divergence of Eq. (114) and requiring that the divergence of the new velocity is zero. Third, after calculating the new pressure distribution at $n+1$, Eq. (114) is used to obtain the velocity at $n+1$. Fourth, the temperature distribution is updated by using explicit Euler time-marching in Eq. (112) and, by averaging the velocity within the drop, the migration velocity is determined. Finally, the level-set function in the bulk is calculated using an explicit Euler approximation, and the color function is reconstructed. By using the oct-tree adaptive grid refinement scheme with the level set method, the authors avoided the need for interface reconstruction.

It was found that finite Marangoni number effects, representing the convection of energy, at small Reynolds number had a negligible effect on the bubble shape. However, finite Reynolds number effects caused sufficient deformation to significantly retard the bubble.

E. Finite-Difference Method on Moving Grid

Based on the methods developed by Ryskin and Leal [94–96] and Kang and Leal [97], Liao and McLaughlin [19] presented results for bubble motion in aqueous surfactant solutions. Following Ryskin and Leal, an adaptive grid finite-difference technique was used to simulate axisymmetric bubble motion. As in the problem considered by Kang and Leal [97], the mapping is time dependent because the shape of the bubble changes with time. The mapping is a transformation from a dimensionless cylindrical coordinate system (x,σ,ϕ) to a dimensionless body-fitted coordinate system (ξ,η,ϕ). The infinite region outside the bubble is mapped into a unit square in the ξ–η plane. The surface of the bubble coincides with the coordinate curve $\xi = 1$ and the region at infinite distance from the bubble corresponds to $\xi = 0$. The η coordinate is similar to an angular coordinate. The coordinate curve $\eta = 0$ lies along the portion of the positive x axis outside the bubble, and the coordinate curve $\eta = 1$ lies along the portion of the negative x axis outside the bubble. Fig. 10 shows a typical grid.

As pointed out by Ryskin and Leal, the (ξ,c,ϕ) system is left-handed and this means that pseudovectors such as the vorticity have the opposite sign as in the cylindrical system.

The body-fitted curvilinear coordinate system is obtained by solving the covariant Laplace equations for x and σ considered as functions of ξ and η:

$$\frac{\partial}{\partial \xi}\left(f\frac{\partial x}{\partial \xi}\right) + \frac{\partial}{\partial \eta}\left(\frac{1}{f}\frac{\partial x}{\partial \eta}\right) = 0, \quad \frac{\partial}{\partial \xi}\left(f\frac{\partial \sigma}{\partial \xi}\right) + \frac{\partial}{\partial \eta}\left(\frac{1}{f}\frac{\partial \sigma}{\partial \eta}\right) = 0. \quad (116)$$

On each time step, the shape of the bubble is modified to satisfy the normal stress balance at each point on the bubble surface. The procedures are described in detail in the papers by Ryskin and Leal.

With the exception of the force balance, most of the computations will be performed in a frame of reference moving the same velocity as the bubble. In this frame of reference, the velocity at large distances from the bubble is $-U_b\mathbf{e}_x$, (i.e., the bubble velocity in the laboratory frame of reference is $U_b\mathbf{e}_x$), where $\mathbf{e}_x$ is a unit vector in the x direction. The magnitude of the steady state bubble velocity in pure water is denoted by U_t. The variables of the governing equations will be nondimensionalized with U_t and the equivalent spherical radius of the bubble, r_e. Because U_t is the steady state value of the bubble

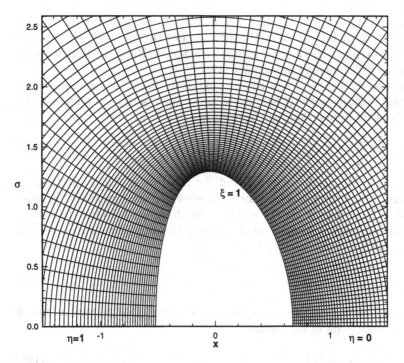

FIG. 10 A typical grid.

speed in pure water, the steady state velocity of the same bubble in contaminated water will be smaller than unity in magnitude.

Because axisymmetric bubble motion will be considered in this paper, the streamfunction-vorticity equations will be used instead of the Navier–Stokes equation. In dimensionless form, the equations for the vorticity, $(0, 0, -\omega)$, and the streamfunction, ψ, are as follows:

$$\frac{Re}{2}\left[\left(\frac{\partial\omega}{\partial t}\right)_{x,\sigma} + \frac{1}{h_\xi h_\eta}\left(\frac{\partial\psi}{\partial\xi}\frac{\partial\omega/\sigma}{\partial\eta} - \frac{\partial\psi}{\partial\eta}\frac{\partial\omega/\sigma}{\partial\xi}\right)\right] = L^2(\omega\sigma) \tag{117}$$

$$L^2\psi + \omega = 0, \tag{118}$$

where

$$L^2 = \frac{1}{h_\xi h_\eta}\left[\frac{\partial}{\partial\xi}\left(\frac{f}{\sigma}\frac{\partial}{\partial\xi}\right) + \frac{\partial}{\partial\eta}\left(\frac{1}{f\sigma}\frac{\partial}{\partial\eta}\right)\right]. \tag{119}$$

In Eqs. (117) and (118), h_ξ and h_η are metric functions, and f is the ratio h_η/h_ξ (the "distortion function"). The subscript (x,σ) in Eq. (117) indicates that the time derivative is to be evaluated at a fixed point. Kang and Leal [97] have discussed the evaluation of such derivatives.

Because the bubble is treated as a void, if the interface is assumed to be contaminated, the velocity field in the liquid must satisfy a condition at the surface of the bubble that may be expressed as follows:

$$\omega = \frac{2}{h_\eta}\frac{\partial u_\xi}{\partial \eta} + 2\kappa_\eta u_\eta - \frac{Re}{h_\eta W}\frac{\partial \gamma}{\partial \eta}, \tag{120}$$

where κ_η is the normal curvature of the interface in the η direction. In addition, the vorticity must vanish for $\eta = 0, 1$ from symmetry. Because $u_\eta = 0$ on the x axis, ψ may be taken to vanish for $\eta = 0$. Unless the motion is steady, the normal component of velocity on the bubble surface is nonzero. On the surface of the bubble, ψ is given by

$$\psi = \int_0^\eta (-u_\xi)\sigma h_\eta d\eta. \tag{121}$$

At each instant of time, the normal stress must balance the surface tension force at each point of the surface:

$$\tau_{\xi\xi} = \frac{4}{W}\gamma(\kappa_\eta + \kappa_\phi), \tag{122}$$

where $\tau_{\xi\xi}$ is the normal stress. In Eq. (122), γ is the dimensionless surface tension made dimensionless by the surface tension of pure water, γ_0.

Kang and Leal [97] give formulas for the normal curvatures, κ_η and κ_ϕ. The condition in Eq. (122) determines the shape of the bubble at each instant of time.

One imposes the tangential stress balance at the surface of the bubble:

$$\tau_{\xi\eta} = -\frac{4}{W h_\eta}\frac{\partial \gamma}{\partial \eta}. \tag{123}$$

The pressure at the interface may be obtained by integrating the Navier–Stokes equation along the bubble surface (see Ref. 97):

$$p = -\left(u_\xi^2 + u_\eta^2\right) + \int_0^\eta \left[2u_\xi \omega h_\eta - \frac{4}{Re}\frac{f}{\sigma}\frac{\partial}{\partial \xi}(\sigma\omega)\right.$$
$$\left. -2\mathbf{e}_\eta \cdot \left(\left(\frac{\partial \mathbf{u}}{\partial t}\right)_{x,\sigma} h_\eta\right)\right] d\eta + B(t). \tag{124}$$

The symbol $\mathbf{e}_\eta$ denotes a unit vector in the η direction. The constant $B(t)$ is determined by the condition that the volume of the bubble is constant.

When a surfactant is present in the water, the convective–diffusion equation for the surface concentration of surfactant, Γ, which is made dimensionless by Γ_∞, has to be solved. Equation (39) is rewritten in the dimensionless form below

$$\frac{\partial \Gamma}{\partial t} + \frac{u_\eta}{h_\eta} \frac{\partial \Gamma}{\partial \eta} + \frac{\Gamma}{h_\eta} \left(\frac{\partial u_\eta}{\partial \eta} + \frac{u_\eta}{h_\eta} \frac{\partial \sigma}{\partial \eta} \right) + \frac{u_\xi}{h_\xi} \left(\frac{1}{h_\eta} \frac{\partial h_\eta}{\partial \eta} + \frac{1}{\sigma} \frac{\partial \sigma}{\partial \xi} \right) \Gamma$$

$$= \frac{2}{Pe_s} \frac{1}{\sigma h_\eta} \frac{\partial}{\partial \eta} \left(\frac{\sigma}{h_\eta} \frac{\partial}{\partial \eta} \right) \Gamma + BikC_s(1 - \Gamma) - \Gamma. \tag{125}$$

The symmetry boundary conditions are

$$\frac{\partial \Gamma}{\partial \eta} = 0, \quad \eta = 0, 1. \tag{126}$$

Equation (125) is coupled to the equations governing the liquid velocity field and the volume concentration of surfactant in the liquid to give the distribution of surface concentration of surfactant. In dimensionless form, the volume concentration equation is as follows:

$$\left(\frac{\partial C}{\partial t} \right)_{x,\sigma} + \frac{u_\xi}{h_\xi} \frac{\partial C}{\partial \xi} + \frac{u_\eta}{h_\eta} \frac{\partial C}{\partial \eta} = \frac{2}{Pe h_\xi h_\eta \sigma} \left[\frac{\partial}{\partial \xi} \left(f\sigma \frac{\partial C}{\partial \xi} \right) + \frac{\partial}{\partial \eta} \left(\sigma/f \frac{\partial C}{\partial \eta} \right) \right], \tag{127}$$

where C is the dimensionless concentration, made dimensionless by the bulk concentration of surfactant, C_∞. One boundary condition is that the concentration at large distance is not affected by the bubble:

$$C = 1, \quad \xi = 0. \tag{128}$$

Because the flux of surfactant to the interface balances the net rate of adsorption, it follows that

$$\mathbf{n} \cdot \nabla C = -\frac{2}{Pe_s h_\xi} \frac{\partial C}{\partial \xi} = K_l B_i [kC_s(1 - \Gamma) - \Gamma], \tag{129}$$

in which $K_l = \Gamma_\infty / 2r_e C_\infty$ is the dimensionless adsorption length.

The Frumkin equation is used to relate the surface tension to the surface concentration of the surfactant:

$$\gamma = 1 + \frac{1}{e} \ln(1 - \Gamma), \tag{130}$$

where $= RT\Gamma_\infty/\gamma_0$. In all cases that were considered by Liao and McLaughlin, $\Gamma1$ so that the Frumkin equation was well approximated by the ideal gas law.

Finally, an overall force balance was used to determine the acceleration of the bubble. Because the bubble is treated as a void, the net force acting on the bubble must vanish:

$$F = 2\pi \int_0^1 \left(-\tau_{\xi\xi}e_{\xi x} + \tau_{\xi\eta}e_{\eta x}\right)\sigma h_\eta \mathrm{d}\eta = 0 \tag{131}$$

where $e_{\xi x}$ and $e_{\eta x}$ are the x components of the unit vectors $\mathbf{e}_\xi$ and $\mathbf{e}_\eta$. For an uncontaminated bubble, $\tau_{\xi\eta} = 0$. In this case, if Eq. (122) were exactly satisfied, Eq. (131) would automatically be satisfied. However, Eq. (122) cannot be satisfied unless the acceleration of the bubble is correct. Thus one must use Eq. (131) with an iterative procedure to determine the acceleration in addition to using Eq. (122) to determine the bubble shape.

The governing equations were advanced in time with a dimensionless time step, Δt. On each time step, it was necessary to determine the following set of dimensionless unknowns: the acceleration, a_b, the velocity, U_b, the streamfunction, ψ, the vorticity, ω, the pressure on the surface of the bubble, p, the coordinate mapping, $x(\xi,\eta)$, $\sigma(\xi,\eta)$, the volume concentration of surfactant, C, and the surface concentration of surfactant, Γ. The covariant Laplace equations for x and σ and the streamfunction-vorticity equations were solved on a rectangular grid in the $\xi–\eta$ plane using the methods described by Ryskin and Leal [94,95]. If the time derivatives in Eqs. (117) and (127) are discretized, all of the governing equations with the exception of Eq. (125) may be written in the following form:

$$0 = f^2 \frac{\partial^2 w}{\partial \xi^2} + \frac{\partial^2 w}{\partial \eta^2} + q_1 \frac{\partial w}{\partial \xi} + q_2 \frac{\partial w}{\partial \eta} + q_3 w + q_4. \tag{132}$$

where w denotes x, σ, ψ, ω, c, or C. The quantities q_i are independent of the dependent variable w and given in detail in the paper by Liao and McLaughlin. The spatial derivatives were discretized using the central difference approximation and the Alternating-Direction Implicit (ADI) method was used to solve the discretized equations. The ADI equations may, following Ryskin and Leal [95], be expressed in the following form:

$$\frac{\tilde{w} - w^n}{\Delta t_a/2} = f^2 \frac{\partial^2 w^n}{\partial \xi^2} + \frac{\partial^2 \tilde{w}}{\partial \eta^2} + q_1^n \frac{w^n}{\partial \xi} + q_2^n \frac{\partial \tilde{w}}{\partial \eta} + \left(\frac{q_3^n - |q_3^n|}{2}\right)$$

$$\times \frac{\tilde{w} + w^n}{2} + \left(\frac{q_3^n + |q_3^n|}{2}\right)w^n + q_4^n, \tag{133}$$

$$\frac{w^{n+1} - \tilde{w}}{\Delta t_a/2} = f^2 \frac{\partial^2 w^{n+1}}{\partial \xi^2} + \frac{\partial^2 \tilde{w}}{\partial \eta^2} + q_1^n \frac{w^{n+1}}{\partial \xi} + q_2^n \frac{\partial \tilde{w}}{\partial \eta} + \left(\frac{q_3^n - |q_3^n|}{2} \right)$$

$$\times \frac{\tilde{w} + w^{n+1}}{2} + \left(\frac{q_3^n + |q_3^n|}{2} \right) w^n + q_4^n. \tag{134}$$

where Δt_a denotes an artificial time step that is chosen to optimize the convergence of the ADI iteration and the superscript n denotes the ADI iteration index and should not be confused with the time step index.

Equation (125) is different from the other equations in that it is one dimensional. It may be put into a form similar to Eq. (132) except that the terms involving derivatives with respect to ξ are absent. It was discretized in η using upwind differencing for the convective term and central differences for the other terms. The resulting tridiagonal matrix equation was directly solved.

They found that a surfactant cap was naturally formed as the bubble is moving through the surfactant solution. When the concentration of surfactant in the liquid is smaller than a critical value, the magnitude of the bubble velocity exhibits a maximum value before decreasing to a steady value as shown in Fig. 11. The maximum value of the magnitude of the velocity can be close to the value for pure water although the ultimate bubble speed may be smaller than this value by a factor of roughly 2.

In the concentration range of their study, the adsorption of surfactant is strongly affected by the sorption rate constant. The adsorption rate constant has a particularly strong effect on the bubble velocity as shown in Fig. 12. This suggests that observations of bubble velocities as a function of time could be used to estimate the adsorption rate constant.

In the concentration range of their study, the rate of adsorption is not purely either diffusion controlled or sorption kinetics controlled. As shown in Fig. 13, the concentration profile is not uniform at the different points on the bubble surface.

Also, as shown in Fig. 14, the concentration of surfactant adjacent to the bubble surface is not equal to the corresponding equilibrium value.

VII. FINITE VOLUME METHOD

The finite volume method has the advantage that it conserves mass, momentum, and energy because the equation is integrated on a control volume. The computational domain is divided into a number of control volumes and one grid point is surrounded by each control volume. The differential equation is integrated over each volume. The divergence theorem is used to express the integral in terms of the values of the dependent variable at points on the

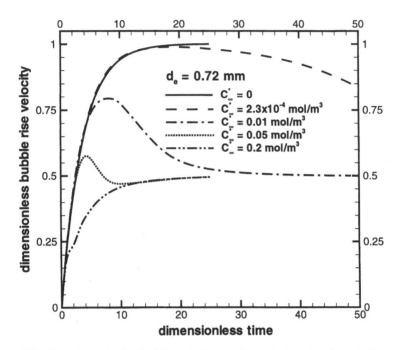

FIG. 11 Dimensionless bubble velocity vs. dimensionless time for a 0.72-mm bubble in aqueous solutions of decanoic acid.

surface of the volume. The discretized equations are then solved either directly or by iteration.

In the following, the formulation of the finite volume method will be given. Then, an example using the finite volume method to simulate bubble motion will be presented.

A. Finite Volume Method Formulation

The formulation will be given for an one-dimensional mass transfer problem. The equation of time-dependent diffusion is

$$\frac{\partial c}{\partial t} = D \frac{\partial^2 c}{\partial x^2} \tag{135}$$

To simplify the problem, the computational domain will be on a Cartesian coordinate, part of which is shown in Fig. 15.

We will focus on the point P, which is between the grid points W in the west and E in the east. The dashed lines are the control volume interface. One can either choose the faces to lie midway between the grid points or choose the

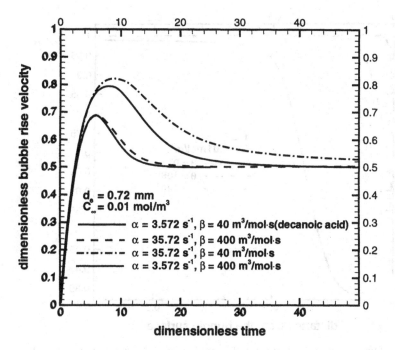

FIG. 12 Dimensionless bubble velocity vs. dimensionless time for a 0.72-mm bubble for various values of α and β, $C^*_\infty = 0.01$ mol m^{-3}.

grid points to lie midway between the faces. The control volume shown in Fig. 15 is $\Delta x*1*1$. Equation (135) is integrated over the control volume assuming D is a constant, which leads to

$$\frac{\partial c}{\partial t}\Big|_p \Delta x = \left(D\frac{\partial c}{\partial x}\right)e - \left(D\frac{\partial c}{\partial x}\right)w, \tag{136}$$

where the term $\partial c/\partial x$ can be discretized as

$$\left(D\frac{\partial c}{\partial x}\right)_e = D\frac{c_E - c_P}{\delta x_e}, \quad \left(D\frac{\partial c}{\partial x}\right)w = D\frac{c_P - c_W}{\delta x_w}. \tag{137}$$

Furthermore, if Eq. (137) is substituted into Eq. (136) and Eq. (136) is integrated from t to $t + \Delta t$,

$$\Delta t\left(c_P^{n+1} - c_P^n\right) = \int_t^{t+\Delta t}\left[D\frac{c_E - c_P}{\delta x_e} - D\frac{c_P - c_W}{\delta x_w}\right]dt \tag{138}$$

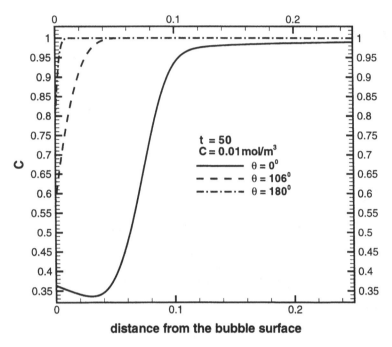

FIG. 13 Dimensionless surfactant concentration as a function of dimensionless distance from the surface of a 0.72-mm bubble at various locations along the bubble surface, $t = 50$, $C^*_\infty = 0.01$ mol m^{-3}.

where the superscripts $n + 1$ and n indicate the new value and the old value. Let us assume that c_P, on the right side of Eq. (138), is a function of time such that,

$$\int_t^{t+\Delta t} c_P dt = \left[\delta c_P^{n+1} + (1 - \delta)c_P^n\right]\Delta t \tag{139}$$

where δ is an interpolation coefficient between 0 and 1. Similar formulas are used for c_E and c_W. If one chooses $\delta = 1$, one obtains the Euler backward method. If $\delta = 1/2$, one obtains the Crank–Nicholson method. Finally, dropping the superscript $n + 1$, Eq. (138) may be written as,

$$a_P c_P = a_E\left[\delta c_E + (1 - \delta)c_E^n\right] + a_w\left[\delta c_W + (1 - \delta)c_W^n\right] + a_P^0$$
$$- (1 - \delta)a_E - (1 - \delta)a_W c_P^n, \tag{140}$$

where

$$a_E = \frac{D}{\delta x_e}, \quad a_W = \frac{D}{\delta x_w}, \quad a_P^n = \frac{\Delta x}{\delta t}, \quad a_P = f a_E + f a_W + a_p^0. \tag{141}$$

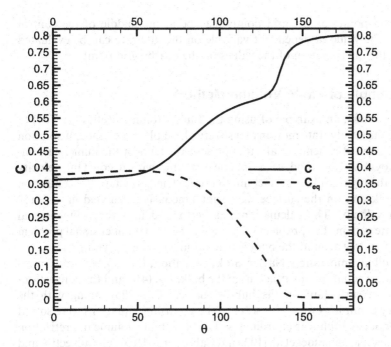

FIG. 14 Dimensionless surfactant concentration adjacent to the bubble surface compared to the corresponding equilibrium values for a 0.72-mm bubble, $t = 50$, $C^*_\infty = 0.01$ mol m^{-3}.

Equation (140) can be solved either using an iteration method such as the Gauss–Siedel method or a direct method such as tridiagonal-matrix algorithm (TDMA). Other methods, such as the Alternating-Direction Implicit (ADI) method combine iteration with the TDMA.

Finally, a few points need to be emphasized. The finite volume method can be used in any orthogonal coordinate grid such as polar coordinates or cylindrical coordinates. The control volume interface can be formed midway

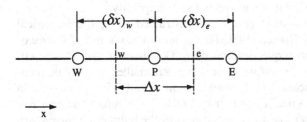

FIG. 15 One-dimensional grid for finite volume method.

of the grid points or the grid points may be in the middle of the control volume. The value of dependent variable on the interface can be evaluated through the interpolation of the values on the nearby grid points.

B. Example of Finite Volume Method

In what follows, an example of using the Finite Volume Method (FVM) to simulate the steady state motion of a spherical bubble in a surfactant solution will be discussed. Cuenot et al. [101] presented results of the numerical simulation by FVM method of a spherical bubble with a constant velocity moving through a dilute surfactant solution. The surfactant concentration in the bulk and on the surface were simultaneously calculated in a steady uniform velocity. The calculation was performed in a three-dimensional coordinate system. The problem was simplified to a two-dimensional problem because the flow around the bubble was assumed to be axisymmetric.

The full incompressible Navier–Stokes equation, Eq. (35), the concentration balance equation of surfactant in the bulk, Eq. (41), and the convection-diffusion of surfactant on the bubble surface, Eq. (125), along with the boundary conditions were also solved. The governing equations were solved in orthogonal curvilinear coordinates. The basic finite volume discretization was based on Magnaudet et al. [102] and Calmet et al. [103]. The advective and diffusive terms were evaluated with a second order centered scheme. Time advancement of Eqs. (35) and (127) were performed through a low-storage, third-order Runge–Kutta scheme, except for the second-order derivatives involved in the diffusion terms that were advanced through a semi-implicit second-order Crank–Nicolson scheme. The time step Δt was split into three intermediate steps, Δt_k, $k = 1,3$, with $\Delta t_1 = (8/15)\Delta t$, $\Delta t_2 = (2/15)\Delta t$, and $\Delta t_3 = (1/3)\Delta t$ [104]. The intermediate velocity and concentration were also calculated. To ensure incompressibility, the Poisson equation was solved for an auxiliary potential ϕ from which the true pressure was deduced and it was solved only once at the end of the time step.

Equation (125) was integrated on the surface to obtain the finite volume discretization formula. This equation was advanced in time using a fully explicit centered scheme for both the advective and the diffusive terms. The adsorption and desorption flux was treated semi-implicitly.

The curvilinear orthogonal grid was constructed using the numerical generator developed by Blanco [105]. This generator made use of the quasi conformal mapping technique developed by Duraiswami and Prosperetti [106]. The concentration boundary layer was much smaller than the momentum boundary layer because the Reynolds number and Schmidt number in this study were on the order of $O(10^5)$ and $O(10^3)$. Therefore the grid was highly refined at the interface and the rear part of the bubble. Because sharp gradients of the surfactant concentration on the surface were expected in the stagnant cap situation, the grid needed to be fine in the tangential direction.

They found that pressure, viscous stress, and vorticity on the interface showed a peak near the surfactant cap region. The bulk concentration of surfactant near the interface was not uniform: There were large variations, especially around the separation point. Although on most of the contaminated part of the bubble, desorption dominated, adsorption was still present on the top part of the bubble.

VIII. LATTICE BOLTZMANN METHODS

The Lattice Boltzmann method (LBM) is an approach to obtaining solutions of fluid mechanics problems. Instead of solving the Navier–Stokes equation or equivalent formulations such as the streamfunction-vorticity equations, one finds solutions of a "mesoscopic" kinetic theory equation. The LBM developed from an earlier method called the lattice gas automaton (LGA). Rothman and Zaleski [107] and Chen and Doolen [108] discuss the LBM.

In a lattice gas, computational particles are allowed to move on a lattice according to certain rules. The lattice may be either two or three dimensional. By considering a hexagonal lattice, Frisch et al. [109] showed that one could obtain solutions of the two-dimensional Navier–Stokes equation by suitable averaging of the lattice gas solutions for appropriate collision rules. Their work was extended to three dimensions by d'Humieres et al [110] who considered a projection of the four-dimensional face-centered-hyper-cubic (FCHC) lattice.

Lattice gas simulations are amenable to parallelization, which, given the simplicity of the method, suggested that such simulations might be useful in performing simulations of turbulent flow on massively parallel computers. However, it was soon discovered that the statistical averaging needed to obtain useful approximations outweighed the other advantages of the method.

Frisch et al. [111] discussed the LBM and its relationship to the LGA. In the LBM, one deals with a single particle distribution function, $N_i(\mathbf{x},t)$, that may be interpreted as the probability of finding a particle with velocity $\mathbf{c}_i$ on lattice site $\mathbf{x}$ at time t. The quantity $N_i(\mathbf{x},t)$ could be obtained from averaging the results of an ensemble of lattice gas simulations. However, McNamara and Zanetti [112] pointed out that one could, instead, solve an evolution equation for the distribution functions and avoid the need of performing ensemble averaging. Higuera and Jiménez [113] simplified the LBM by pointing out that one could linearize the collision operator by assuming that the distributions were close to equilibrium. Higuera and Succi [114] used the LBM to simulate flow around a circular cylinder.

The general form of the lattice Boltzmann equation (LBE) is

$$N_i(\mathbf{x} + \mathbf{e_i}, t + 1) = N_i(\mathbf{x}, t) + \Omega_i(\mathbf{x}, t) \tag{142}$$

In Eq. (142), Ω_i is the collision term and all quantities have been made dimensionless in terms of lattice units. In lattice units, the unit of length is the lattice spacing and the unit of velocity is either the speed of the particles or the magnitude of a component of the particle velocity along one of the coordinate axes. The dimensionless particle speed, c, is 1 for the hexagonal (2-D) lattice and $\sqrt{2}$ for the FCHC (4-D) lattice. The number of velocities, b, is 6 for the hexagonal lattice and 24 for the FCHC lattice.

For simplicity, most of the following discussion will be based on the FCHC lattice. However, a number of other lattices have been used for LBM simulations. For example, Chen et al. [115] showed that a 15 velocity could be used with the LBM Sankaranarayanan et al. [116] used this lattice in LBM simulations of bubble motion. He et al. [117] used the lattice in simulations of the Rayleigh–Taylor instability. He and Luo [118] showed that the LBM could be derived through a finite difference quadrature of the Bhatnagar–Gross–Krook (BGK) [119] equation of kinetic theory. The discretization involves the exact evaluation of low-order moments of the Maxwellian distribution by use of Gauss–Hermite quadrature. They point out that a large number of discretizations are possible—including variable mesh sizes. Shan and He [120] generalized the analysis presented by He and Luo.

The collision operator must satisfy conservation of mass. This condition may be expressed as follows:

$$\sum_i \Omega_i(\mathbf{x}, t) = 0. \tag{143}$$

In the absence of nonlocal interactions (i.e., interactions between particles on different lattice sites), one also requires that momentum be conserved by the collisions at each lattice site at each time step:

$$\sum_i \mathbf{c_i} \Omega_i(\mathbf{x}, t) = \mathbf{f}. \tag{144}$$

In Eq. (144), $\mathbf{f}$ is a dimensionless body force in the Navier–Stokes equation. In one approach to two phase flow simulations, one introduces a nonlocal interaction between the particles for which Eq. (144) is not satisfied; this approach will be discussed later in this section.

The LBM was established for two-dimensional hexagonal lattices by Frisch et al. [109] and by d'Humières et al. [110] for the four-dimensional FCHC lattice. Projections of the FCHC lattice may be used for either three-dimensional or two-dimensional simulations. There are 24 velocities on the four-dimensional FCHC lattice. They may be expressed as permutations of the following four velocities:

$$\mathbf{c}_i = (\pm 1, \pm 1, 0, 0).\tag{145}$$

Fig. 16 shows the projection of the FCHC lattice into three dimensions and Fig. 17 shows the projection of the FCHC lattice into two dimensions.

When projected onto a three-dimensional lattice, four velocities project onto each direction that is parallel to a cube edge, while there is a one-to-one relationship between the directions parallel to face diagonals in three dimensions and four dimensions. To account for the above degeneracy, let us introduce the weighting factor, m_i, which is equal to unity in four dimensions. In three dimensions, m_i has the following values:

$$m_i = 2, \quad \mathbf{c_i} = (\pm 1, 0, 0), (0, \pm 1, 0), (0, 0, \pm 1)\tag{146}$$

$$m_i = 1, \quad \mathbf{c_i} = (\pm 1, \pm 1, 0), (\pm 1, 0, \pm 1), (0, \pm 1, \pm 1).\tag{147}$$

It is easily verified that

$$\Sigma_i m_i = 24.\tag{148}$$

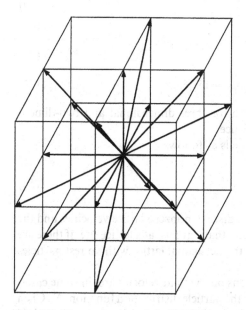

FIG. 16 The projection of the FCHC lattice into three dimensions.

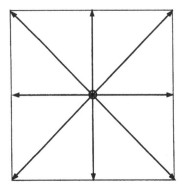

FIG. 17   The projection of the FCHC lattice into two dimensions.

In two dimensions, $m_i = 4$ along cube edges, while $m_i = 1$ along face diagonals. In the two-dimensional case, four of the FCHC velocities project to a "rest" direction for which the velocity vanishes.

If one uses the notation $c_{i\alpha}$ to denote the component of $\mathbf{c_i}$ along the α direction, in a Cartesian coordinate system aligned with its axes parallel to the edges of the cubic lattice, one can derive the following identities:

$$\Sigma_i m_i c_{i\alpha} = 0 \tag{149}$$

$$\Sigma_i m_i c_{i\alpha} c_{i\beta} = 12\delta_{\alpha\beta} \tag{150}$$

$$\Sigma_i m_i c_{i\alpha} c_{i\beta} c_{i\gamma} = 0 \tag{151}$$

$$\Sigma_i m_i c_{i\alpha} c_{i\beta} c_{i\gamma} c_{i\delta} = 4\delta_{\alpha\beta}\delta_{\gamma\delta} + 4\delta_{\alpha\gamma}\delta_{\beta\delta} + 4\delta_{\alpha\delta}\delta_{\beta\gamma}. \tag{152}$$

The above identities are valid both for three-dimensional and two-dimensional projections of the FCHC lattice.

The speed of sound on the lattice is as follows:

$$c_s^2 = \frac{b_m c^2}{bD}, \tag{153}$$

where b is the sum of the number of distinct nonzero lattice velocities and the number of particles with zero velocity that can exist at a lattice site. If there are no rest particles, $b = b_m$. For both the hexagonal lattice with no rest particles and the FCHC lattice, $c_s = 1/\sqrt{2}$.

In the LBM, boundary conditions on the fluid velocity are, in some cases, imposed by applying conditions to the particle distribution function, N_i. Chen and Doolen [43] discuss boundary conditions. If one wishes to impose a rigid boundary condition on a motionless flat surface, one imposes a "bounce-back" condition on the particle distribution function as illustrated in Fig. 18.

Ziegler [121] showed that if the rigid boundary was located midway between the nearest lattice sites, the bounce-back scheme would produce second-order accuracy. Therefore the physical boundary is assumed to lie midway between the closest lattice points in the flows and the closest boundary point (i.e., a point that lies inside the solid surface).

On stress-free surfaces, one can use a symmetry boundary condition. Fig. 19 shows the symmetry boundary condition.

The condition states that the particle distribution functions with equal and opposite normal components of velocity on opposite sides of the symmetry surface are equal.

Finally, Fig. 20 shows the outflow (no variation) boundary condition.

Eggels [122] used a force field method proposed by Goldstein et al. [123] to impose boundary conditions on curved and moving surfaces in a lattice Boltzmann simulation of a turbulent stirred tank. Derksen and Van den Akker [124] used the same technique in higher-resolution studies of a stirred tank and found good agreement with experimental results for the mean flow and turbulent statistics. The idea underlying the use of the force field is to add an artificial body force to the Navier–Stokes equation and choose the force so that points inside the solid objects move with the correct velocity. The technique provides a relatively simple way of handling complex geometries and moving objects such as the impeller blades in the stirred tank. Fig. 21 shows a side view of the instantaneous flow field in a stirred tank obtained with the above approach.

There are at least two different general approaches to the LBE that are currently in use. The method described by Eggels and Somers [125] is the more complicated of the two, but appears to have the advantage that it can be used for larger Reynolds numbers. The other approach is called the BGK method.

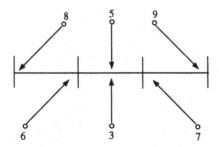

FIG. 18 Bounce back boundary condition.

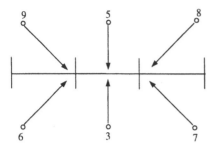

FIG. 19 Symmetry boundary condition.

It is based on the simplified kinetic equation for dilute gases suggested by Bhatnagar et al. [119]. Both approaches will first be described in the context of single-phase flow. In principle, one could also treat a bubble as a void and use an artificial force field as described by Derksen and Van den Akker [124] to impose the boundary conditions at the surface of the bubble. This would permit one to use the single phase LBM to simulate bubble motion in a liquid.

A. Eggels–Somers Method

Eggels and Somers [125] obtained an expression for the collision operator in Eq. (142) by using an asymptotic expression for N_i that is valid provided that the lattice gas is close to equilibrium and assuming that the magnitude of the fluid velocity, **u**, is small compared to unity. The asymptotic result was obtained by Frisch et al. [111] by using a multiple timescale analysis together

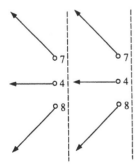

FIG. 20 Outflow boundary condition.

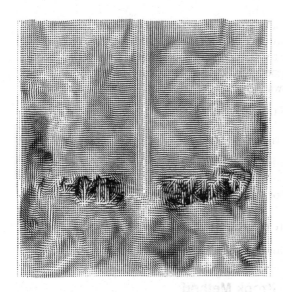

FIG. 21 The instantaneous flow in stirred tank.

with an expansion of N_i around the equilibrium solution. The result, correct to terms of order u^3 and $u\Delta u$, is as follows:

$$
N_i = \frac{m_i\rho}{24}\left(1 + 2\mathbf{c_i}\cdot\mathbf{u} + 3\left[\mathbf{c_i}\mathbf{c_i} : \mathbf{uu} - \frac{1}{2}tr(\mathbf{uu})\right]\right.
$$
$$
\left. -6v\left[(\mathbf{c_i}\cdot\nabla)(\mathbf{c_i}\cdot\mathbf{u}) - \frac{1}{2}\nabla\cdot\mathbf{u}\right]\right).
$$

(154)

The weighting factor m_i is included in Eq. (154) to account for the degeneracy in directions for the projections of the FCHC lattice that were discussed above. Eggels and Somers also include a subgrid-scale stress in their formulation that can be used for Large Eddy Simulation (LES). However, for simplicity, this stress will not be included in the present discussion.

When the asymptotic expression in Eq. (154) is substituted in Eq. (142), one obtains an expression for the collision operator:

$$
\Omega_i = \frac{m_i\rho}{12}\left[(\mathbf{c_i}\cdot\nabla)(\mathbf{c_i}\cdot\mathbf{u}) - \frac{1}{2}\nabla\cdot\mathbf{u}\right] + \frac{m_i}{12}\mathbf{c_i}\cdot\mathbf{f}.
$$

(155)

It can be verified that the expression for the collision operator in Eq. (155) satisfies the conservation laws in Eqs. (143) and (144).

To solve the LBE, Eggels and Somers factorized the asymptotic expression for N_i in Eq. (154) into a product of a matrix, E, that depends only on the velocity c_i and a column vector that depends only on the flow variables $\mathbf{u}$, ρ, and $\mathbf{f}$. The inverse of E can be analytically calculated. The solution strategy can be summarized as follows:

1. The particle move to their new lattice sites:

$$N_i^I(\mathbf{x}, t) = N_i(\mathbf{x} - \mathbf{c_i}, t - 1). \tag{156}$$

2. Using the factorization described above, the flow variables are computed.
3. The collision operator is applied to the flow variables obtained in step 2.
4. The flow variables are transformed into a new value of N_i.

The boundary conditions can be applied in steps 1 and 4.

B. Bhatnagar–Gross–Krook Method

Qian [126], Chen et al. [127], Qian et al. [128], and Chen et al. [129] discussed the use of the BGK version of the LBE. The BGK version takes the following form:

$$N_i(\mathbf{x} + \mathbf{c_i}, t + 1) = N_i(\mathbf{x}, t) - \frac{(N_i(\mathbf{x}, t) - N_i^{eq}(\mathbf{x}, t))}{\Theta}, \tag{157}$$

where N_i^{eq} is the equilibrium particle distribution function and Θ is a dimensionless relaxation time. The dimensionless kinematic viscosity is related to Θ as follows:

$$\nu = \frac{c^2}{D + 2}(\Theta - 0.5). \tag{158}$$

For the FCHC model, $c_2 = 2$ and $D = 4$. Thus

$$\nu = \frac{(\Theta - 0.5)}{3}. \tag{159}$$

The BGK formulation of the LBE is explicit in the particle distribution function. For that reason, it is plausible that it would be numerically unstable for small values of ν. On the other hand, it is significantly easier to program than the Eggels–Somers method.

C. Bubble Motion

Sankaranarayanan et al. [116] reported LBM results for the motion of buoyancy-driven bubbles in a quiescent liquid. Periodic boundary conditions were imposed in all three directions. Thus the motion of a periodic array of bubbles was simulated. In all cases, the Reynolds numbers were small enough that the flow was laminar. The Morton numbers of the liquids varied from 10^{-6} to 10^{-4}. Numerical stability placed the lower limit on the Morton number.

Sankaranarayanan et al. used the BGK method with two sets of particle distributions, N_i^α, where $\alpha = 1,2$. The gas phase corresponds to $\alpha = 1$ and the liquid phase corresponds to $\alpha = 2$. A nonlocal interparticle force was used to make the two phases immiscible and to create surface tension at the interface. The LBE for each phase takes the form

$$N_i^\alpha(\mathbf{x} + \mathbf{c_i}, t + 1) = N_i^\alpha(\mathbf{x}, t) - \frac{\left(N_i^\alpha(\mathbf{x}, t) - N_i^{\alpha, eq}(\mathbf{x}, t)\right)}{\Theta}. \tag{160}$$

The quantity $N_i^{\alpha, eq}$ is the second-order Hermite expansion of the Maxwellian distribution as discussed by Shan and He [120]:

$$N_i^{\alpha, eq} = w_i \rho^\alpha \left[1 + \mathbf{c_i} \cdot (\mathbf{u} + \Theta \mathbf{f}^\alpha) + \frac{1}{2} (\mathbf{c_i} \cdot (\mathbf{u} + \Theta \mathbf{f}^\alpha))^2 \right. \tag{161}$$
$$\left. - \frac{1}{2} (\mathbf{u} + \Theta \mathbf{f}^\alpha) \cdot (\mathbf{u} + \Theta \mathbf{f}^\alpha) \right]$$

where $\mathbf{f}$ is the sum of all forces per unit mass acting on the component α at a given lattice point. The intermolecular part of $\mathbf{f}$ is absorbed into the pressure term and results in a nonideal gas equation of state. The symbol w_i denotes quadrature weights that are described by Shan and He [120].

Shan and He [130,131] described a method for performing LBM simulations of flows with multiple phases and components. A central idea of their method is the use of an interparticle potential:

$$V(\mathbf{x}, \mathbf{x}') = G_{\alpha\bar{\alpha}}(\mathbf{x}, \mathbf{x}')\psi^\alpha(\mathbf{x})\psi^{\bar{\alpha}}(\mathbf{x}'). \tag{162}$$

In Eq. (162), $G_{\alpha\bar{\alpha}}(\mathbf{x}, \mathbf{x}')$ is a Green's function. The quantity ψ^α is an "effective mass." It is a function of the number density at a given lattice:

$$\psi^\alpha(\mathbf{x}) = \psi^\alpha(n^\alpha(\mathbf{x})), \tag{163}$$

where the number, $n^\alpha(\mathbf{x})$, is given by

$$n^\alpha(\mathbf{x}) = \sum_i N_i^a(\mathbf{x}). \tag{164}$$

In homogeneous systems, the Green's function depends only on the distance between lattice points:

$$G_{\alpha\bar{\alpha}}(\mathbf{x}, \mathbf{x}') = G_{\alpha\bar{\alpha}}(|\mathbf{x}, \mathbf{x}'|) \tag{165}$$

Shan and He considered only nearest-neighbor interactions:

$$G_{\alpha\bar{\alpha}}(|\mathbf{x},\mathbf{x}'|) = 0, \quad |\mathbf{x},\mathbf{x}'| > c \tag{166}$$

$$G_{\alpha\bar{\alpha}}(|\mathbf{x},\mathbf{x}'|) = \mathscr{G}_{\alpha\bar{\alpha}}, \quad |\mathbf{x},\mathbf{x}'| = c, \tag{167}$$

where c is the distance from a lattice site to its nearest neighbors and $\mathscr{G}$ is a constant. In what follows, we will restrict attention to systems in which a single component is present and discuss conditions under which a phase transition can occur.

One property of the nearest-neighbor interactions is that, although momentum is conserved globally, it is not conserved on individual lattice sites. The force due to the nearest neighbor interactions that acts on a lattice site is given by

$$\mathbf{f}(\mathbf{x}, t) = -\mathscr{G}\psi(\mathbf{x}, t) \sum_i \psi(\mathbf{x} + \mathbf{c}_i, t)\mathbf{c}_i, \tag{168}$$

where b is the number of links connecting a site to its nearest neighbors. This result underlies the fact that the quantity $\mathbf{u} + \Theta\mathbf{f}^{\alpha}$ appears in the argument of the equilibrium particle distribution function in Eq. (161).

Shan and Chen [130] show that the pressure in a single component system is given by

$$p = \frac{c^2}{D}\left[(1 - d_0)n + \frac{b}{2}\mathscr{G}\psi^2(n)\right], \tag{169}$$

where D is the number of spatial dimensions. The first term on the right-hand side of Eq. (169) represents ideal gas behavior. The second term represents nonideal behavior. If p is not a monotonic function of the density, ρ, phase transitions will occur. Shan and Chen illustrated this phenomenon for the following choice:

$$\psi = n_0[1 - \exp(-n/n_0)] \tag{170}$$

Shan and Chen [131] also derived an expression for the dimensionless surface tension, γ, in single component two phase systems:

$$\gamma = \frac{c^4 b\mathscr{G}}{2D(D + 2)} \int_{n_g}^{n_l} \psi'^2(y(n))^{1/2} \mathrm{d}n, \tag{171}$$

where $y(n) = (\mathrm{d}n/\mathrm{d}z)^2$ and z is the dimensionless distance measured from the interface.

In their analysis of two-component bubbly systems, Sankaranarayanan et al. [116] set the function $\mathscr{G}$, which means that component 1 (the gas component) obeyed the ideal gas law. The function ψ^2 was chosen to produce a nonideal equation of state so that component 2 could exist as a liquid phase.

With their approach, they can simulate bubbly flows with density ratios as large as 100.

IX. SPECTRAL METHODS

Spectral simulation methods have been used since the late 1960s. The books by Gottlieb and Orszag [132], Canuto et al. [133], and Boyd [134] discuss spectral methods. Typically, one expands the velocity field and other dependent variables in Fourier series or other functions such as Chebyshev polynomials that are solutions of second order Sturm–Liouville problems.

An attractive feature of spectral methods is the fact that they exhibit exponential convergence when the spatial grid spacing is sufficiently small. In addition, one can make use of fast Fourier transform (FFT) methods. However, there are also a number of disadvantages. First, the traditional spectral methods are limited to very simple geometries. Karniadakis and Henderson [135,136] have alleviated this problem by developing "spectral element" methods in which spectral methods are combined with finite element methods.

A second difficulty in using spectral methods is the evaluation of nonlinear terms. If one evaluates such terms in "spectral space" (i.e., by working with the spectral coefficients), one must evaluate convolution sums. One can avoid this problem by using FFTs to transform the velocity and its spatial derivatives to "physical space" and evaluating the products on the grid points. One then transforms back to spectral space for the remaining calculations. This technique introduces aliasing errors [137] because when one forms a product in physical space, one introduces functions that lie outside the basis functions. The "two-thirds rule" is a commonly accepted way of reducing aliasing errors [138] in spectral simulations. Although this procedure works well in practice, it introduces additional cost in CPU time and storage, which reduces the competitive advantage of the spectral method relative to the finite difference or other methods.

Spectral methods have been most heavily used in simulations of turbulence. Orszag and Patterson [139] reported the first simulation of turbulent flow. They used spectral methods to simulate homogeneous turbulence in a periodic cubic box. Kim et al. [138] published the first results for fully developed turbulent channel flow. The Reynolds number, based on the hydraulic diameter and the bulk velocity, was roughly 13,660. Their results were generally in good agreement with experimental results obtained for wall-bounded turbulent shear flows. This approach has been widely used to track particles or drops in recent years as will be discussed in the next section.

Yuan and Prosperetti [140] used a hybrid spectral-finite difference method to simulate the motion of two identical buoyancy-driven bubbles that were

vertically aligned. Because the flow was axisymmetric, they used a vector po-
tential-vorticity formulation. (In principle, they could also have used a
streamfunction-vorticity method, but the spectral representation of the vector
potential is simpler.) The simulations were performed in a bispherical
coordinate system, ξ, η, where ξ is similar to a radial coordinate and η is
similar to an angular coordinate.

Two types of simulations were performed. In one type of simulation, the
velocities of the bubbles were prescribed to equal and the development of the
flow field around the bubbles was studied. In the other type, the bubbles were
assigned initial velocities and the velocities of the bubbles and the surrounding
flow field were studied as a function of time. It was found that the bubbles
attained a steady state value of the vertical separation that depended on the
Reynolds number.

X. BUBBLE TRACKING IN TURBULENCE

In many industrial applications such as bubble columns and stirred tank reactors, it is of interest to know the local concentration of bubbles. In general,
this is an extremely difficult problem because the bubbles modify the flow and
one must compute shape and velocity of the bubbles simultaneously with the
motion of the liquid phase. However, for dilute flows, one may be able to
obtain some progress with the so-called one-way coupling approximation. In
this approach, one ignores the effect of the bubbles on the motion of the
liquid. This is reasonable provided that the gas volume fraction is very small
and if the bubbles are smaller than the energy-containing eddies so that the
turbulence created by the bubbles is unimportant. One then integrates an
approximate equation of motion for the bubble.

McLaughlin [141] used the one-way coupling approach to simulate the
motion of aerosol particles in a turbulent channel flow. The particle Reynolds
number was smaller than unity for most of the particles although some
particles occasionally attained Reynolds numbers larger than unity as a result
of interaction with unusually strong eddies.

Small bubbles in polar liquids can be treated as rigid spheres because of the
effects of surfactants. Provided that the bubble Reynolds number is $O(1)$ or
smaller, one can use the Maxey–Riley equation, which is given in Eq. (14).
However, this may be a very restrictive assumption. For example, based on its
rise velocity, a 120-μm bubble in water has a Reynolds number roughly equal
to unity. It reasonable to assume that bubbles as large as 1 mm are spherical in
water. Using Ryskin and Leal's [94,95,96] finite difference methods,
McLaughlin [68] found that the axis ratio of a freely rising 1-mm bubble in
pure water (i.e., a mobile interface) was 1.12. However, the axis ratio of a fully
contaminated 1-mm bubble was 1.01. In the latter case, the Reynolds number
based on the rise velocity and the equivalent spherical diameter was 110. This

is sufficiently large that one could reasonably use an equation of motion based on boundary layer analysis. For the clean interface, one can use Moore's [11,12] results for the drag coefficient. For the contaminated bubble, one can use the drag correlation for a rigid sphere suggested by Clift et al. [1].

The added mass and pressure gradient terms in Eq. (14) has the same form as that predicted by potential flow theory (see, e.g., Ref. 142). Therefore these terms do not require modification. Finally, Odar [143] reported results for the memory term for rigid spheres at large Reynolds numbers. Thus one can develop an approximate equation of bubbles that should be reasonable for bubbles that are not highly deformed and which have Reynolds numbers that are $O(10^2)$. Such an approach could, for example, be used to track small bubbles in turbulent channels or stirred tanks.

In general, the position of a bubble will not coincide with a grid or lattice point. Therefore one must use an interpolation scheme to determine the liquid velocity and its spatial derivatives at the location of the bubble. Yeung and Pope [144], Balachandar and Maxey [145], and Kontomaris et al. [146] have discussed a variety of such interpolation schemes.

The simulation of the liquid phase turbulence may be performed by a variety of methods. For simple geometries such as channels or periodic boxes, spectral methods may be used [139,138]. For more complex geometries, finite difference [147,148], spectral element [135], or lattice Boltzmann methods [124] may be used.

XI. SUMMARY

In this chapter, we have discussed several approaches to the numerical simulation of bubbles as well as some of the results that have been obtained with these simulations. We first wish to emphasize that we have not included either all of the techniques that are currently being used or even all of the results that have been obtained with these methods. For example, we have not discussed finite element methods and we have not attempted to discuss modeling approaches or commercial software that is available. We have omitted entire areas such as research on bubbles in non-Newtonian liquids, the behavior of foams, and boiling.

We have chosen to emphasize fundamental approaches in which one directly solves the governing equations. We have also mainly discussed simulations in which one deals with single bubbles or relatively small numbers of bubbles. In doing this, we have tried to emphasize the areas that we think are the most promising in terms of obtaining fundamental understanding of phenomena involving bubbles.

In our opinion, there is no simple answer to the question "What is the best numerical method for simulating bubbles?." The choice of method seems to depend on the type of problem and the background of the person who will be performing the work. For simulations of bubbles in low Reynolds number flows, the boundary integral method is very powerful. Recent developments in the technique open up the possibility of simulating bubble coalescence and other strong interactions. On the other hand, people have published finite difference solutions of low Reynolds number problems in the past decade. The finite difference method is, perhaps, simpler in some respects and it also has the advantage that it is straightforward to use it over a broad range of Reynolds numbers. The VOF technique enables one to perform simulations of highly deformed bubbles.

For simulations of bubbles in complex high Reynolds number flows, the lattice Boltzmann method appears to be promising. The technique has already been used in LES of single phase flow in stirred tanks with very encouraging results. Recent developments appear to open the way for simulations of two phase flow in such highly complex geometries in the near future. The lattice Boltzmann method has the additional advantage of being relatively simple to program. Also, the ease with which one can parallelize lattice Boltzmann programs is important. The primary disadvantage is that the technique is still in the process of being developed and refined at the time of writing.

In conclusion, we hope that this chapter will provide a useful introduction to students and colleagues who wish to begin a research program that involves some aspect of numerical simulations of bubbles.

XII. NOTATION

a	bubble acceleration, m sec^{-1}
A	surface area of bubble, m^2
b	total number of moving directions, dimensionless
b_m	number of moving directions, dimensionless
Bi	Biot number for the surfactant, dimensionless
Br	Brinkman number, dimensionless
$B(t)$	constant in expression for pressure, dimensionless
$\mathbf{c_i}$	i_{th} velocity vector, dimensionless
c_s	speed of sound, dimensionless
C	volume concentration of surfactant in liquid, dimensionless
C_s	value of C at the interface, dimensionless
C^∞	volume concentration of surfactant in liquid, mol m^{-3}
Ca	Capillary number, dimensionless
C_D	drag coefficient, dimensionless
d_e	equivalent spherical diameter of a bubble, m
D_g	bulk diffusivity of solute, m^2 sec^{-1}
D	bulk diffusivity of surfactant, m^2 sec^{-1}
D_s	surface diffusivity of surfactant, m^2 sec^{-1}

$\mathbf{D}$	deformation rate tensor, $\sec^{-1}$
$\mathcal{D}$	lattice dimension, dimensionless
$\mathbf{e}$	unit vector for coordinates
f	distortion function, dimensionless
$\mathbf{f}$	body force kg $m^{-2} \sec^{-2}$
$\mathbf{F}$	force kg m $\sec^{-2}$
$\mathbf{F_d}$	drag force kg m $\sec^{-2}$
$\mathbf{F_l}$	lift force kg m $\sec^{-2}$
g	gravitational acceleration, m $\sec^{-2}$
$\mathbf{G_T}$	temperature gradient, dimensionless
$\mathbf{G}_u$	velocity gradient, dimensionless
$\}_{\alpha\alpha'}$	constant in Green's function, dimensionless
$G_{\alpha\alpha'}$	Green's function, dimensionless
h_ξ	metric scale factor for ξ, dimensionless
h_η	metric scale factor for η, dimensionless
$\mathbf{I(x)}$	indicator function
k	dimensionless ratio of adsorption and desorption constants of surfactant
k_k	ratio of thermal conductivity of bubble and continuous phase, dimensionless
k_μ	ratio of viscosity of bubble and continuous phase, dimensionless
k_ρ	ratio of density of bubble and continuous phase, dimensionless
K_{ijk}	tensor in expression for disturbance flow
K_l	dimensionless adsorption length
L	characteristic length, m
L_s	characteristic length based on slip velocity, m
L_G	characteristic length based on velocity gradient, m
m_i	weight factor, dimensionless
m_l	the mass of displaced liquid, kg
m_p	the mass of sphere, kg
M	Morton number, dimensionless
Ma	Marangoni number, dimensionless
n	unit vector normal to the surface pointing into the liquid, dimensionless
N_i	particle distribution function, dimensionless
p	pressure, Pa
p_∞	pressure at infinity, Pa
Pe	liquid phase Peclet number for surfactant, dimensionless
r_e	equivalent spherical radius of a bubble, m
r_{xy}	distance between $\mathbf{x}$ and $\mathbf{y}$, m
$\mathbf{r}$	a point on the surface
R	sphere radius, m

R_{gas}	gas constant, $\text{m}^3 \text{ Pa mol}^{-1} \text{ K}^{-1}$
R_{d}	the distance between two bubbles
Re	bubble Reynolds number, dimensionless
Re_{s}	slip velocity Reynolds number, dimensionless
Re_{G}	Reynolds number based on velocity gradient, dimensionless
R_i	the radius of bubble i, m
$s(x)$	level set function, m
s	unit vector tangential to the surface, dimensionless
t	time, sec
t_{p}	pseudotime
t'	integration variable
T	temperature, K or dimensionless
u_{c}	characteristic velocity m sec^{-1}
u_{s}	slip velocity m sec^{-1}
u_r	r component of the velocity, dimensionless
u_θ	θ component of the velocity, dimensionless
u_∞	velocity far away from the bubble, dimensionless
$\mathbf{u}$	liquid velocity, m sec^{-1}
$\mathbf{u}_{\text{s}}$	velocity on the bubble surface, dimensionless
$\mathbf{u}_0$	undisturbed velocity, m sec^{-1}
U	bubble rise velocity, m sec^{-1}
U_{s}	velocity scale for thermocapillary migration m sec^{-1}
U_{t}	tangential bubble terminal velocity m sec^{-1}
$\mathbf{v}$	disturbance velocity created by bubble, m sec^{-1}
$V_{\mathbf{xx'}}$	interparticle potential, dimensionless
W	Weber number, dimensionless
$\mathbf{x}_{\text{s}}$	a point on the surface of bubble
$\mathbf{z}$	the axis of spherical coordinate

Greek Letters

α	desorption rate constant, sec^{-1}
β	adsorption rate constant, $\text{m}^3 \text{ sec}^{-1} \text{ mol}^{-1}$
β_h	relaxation parameter for normal stress balance, dimensionless
β_ω	relaxation parameter for vorticity b.c., dimensionless
γ	surface tension, kg sec^{-2} or dimensionless
γ_0	surface tension of clean interface, kg sec^{-2}
γ_T	the rate of change of interfacial tension with temperature, kg sec^{-2} k^{-1}
Γ	surface concentration of surfactant, mol m^{-2} or dimensionless
Γ_∞	value of Γ at close-packing, mol m^{-2}

Δt time step, dimensionless
Δt_a artificial time step, dimensionless
ϵ factor involving close-packing surfactant concentration and surface tension, dimensionless
η body fitted orthogonal coordinate, dimensionless
θ angle measured from the bottom of the bubble
θ_{cap} surfactant cap angle
Θ relaxation time, dimensionless
κ normal curvature, dimensionless
κ_T thermal diffusivity, dimensionless
μ dynamic viscosity of liquid, kg m^{-1} sec^{-1}
ξ body fitted orthogonal coordinate, dimensionless
Π pressure head, dimensionless
ρ liquid density, kg m^{-3}
σ radial cylindrical coordinate, dimensionless
τ stress tensor, Pa
τ_{ij} stress tensor for disturbance flow, kg m^{-1} sec^{-2}
τ_p pseudotime, dimensionless
τ_v viscous relaxation time, sec
τ_γ force per unit area, kg m^{-1} sec^{-2}
ϕ velocity potential, m^2 sec^{-1}
ψ stream function, dimensionless
ψ^α effective mass function, dimensionless
ω vorticity, dimensionless
ω_p vorticity of singular part of stream function, dimensionless
Ω angular frequency, sec^{-1}
Ω_I collision operator, dimensionless

Subscripts

l an element on the bubble surface
0 initial value
∞ bulk value
g property for the solute
s property near the surface
im   immobile interface

REFERENCES

1. Clift, R.; Grace, J.R.; Weber, M.E. *Bubbles, Drops, and Particles*; Academic Press: New York, 1978.

2. Hadamard, J.S. C. R. Acad. Sci. 1911, *152*, 173.
3. Rybczynski, W. Bull. Int. Acad. Pol. Sci. Lett., Cl. Sci. Math. Nat., Ser. A 1911, *40*.
4. Bond, W.N. Philos. Mag. Ser. 1927, *4*, 889.
5. Bond, W.N.; Newton, D.A. Philos. Mag. 1928, *5*, 794.
6. Frumkin, A.N.; Levich, V.G. Zhur. Fiz. Khim. It. 1947, *21*, 1183.
7. Levich, V.G. *Physicochemical Hydrodynamics*; Prentice-Hall: Englewood Cliffs, New York, 1962.
8. P. Savic. Natl. Res. Counc. Can. Rep. No. MT-22, 1953.
9. Sadhal, S.S.; Johnson, R.E. J. Fluid Mech. 1983, *126*, 237.
10. Lamb, H. *Hydrodynamics*; Cambridge University Press: Cambridge, 1932.
11. Moore, D.W. J. Fluid Mech. 1963, *16*, 113.
12. Moore, D.W. J. Fluid Mech. 1965, *23*, 749.
13. El Sawi, M. J. Fluid Mech. 1972, *62*, 163.
14. Haberman, W.L.; Morton, R.K. Proc. Am. Soc. Civ. Eng. 1954, *387*, 227.
15. Duineveld, P.C. Ph.D. Dissertation, University of Twente, 1994.
16. Duineveld, P.C. J. Fluid Mech. 1995, *292*, 325.
17. Sam, A.; Gomez, C.O.; Finch, J.A. Int. J. Miner. Process 1996, *47*, 177.
18. Zhang, Y.; Gomez, C.O.; Finch, J.A. Column '96 1996, *63*.
19. Liao, Y.; McLaughlin, J.B. submitted to journal, 1999.
20. Saffman, P.G. J. Fluid Mech. 1956, *1*, 249.
21. Hartunian, R.A.; Sears, W.R. J. Fluid Mech. 1957, *3*, 27.
22. Rosenberg, B. David Taylor Model Basin, Rep. No.727, 1950.
23. Tsuge, H.; Hibino, S.I. J. Chem. Eng. Jpn. 1977, *10*, 66.
24. Subramanian, R.S. In *Transport Processes in Bubbles, Drops, and Particles*; Chhabra, R.P., De Kee, D., Eds.; Hemisphere: New York, 1992; 1–41.
25. Subramanian, R.S.; Balasubramaniam, R.; Wozniak, G. Fluid mechanics of bubbles and drops. In *Physics of Fluids in Microgravity*; Monti, R. Ed.; Gordon and Breach: Paris, 2000, Chap 6.
26. Subramanian, R.S.; Balasubramaniam, R. *The Motion of Bubbles and Drops in Reduced Gravity*; Cambridge University Press: Cambridge, England, 2000.
27. Young, N.O.; Goldstein, J.S.; Block, M.J. J. Fluid Mech. 1959, *6*, 350.
28. Merritt, R.S.; Subramanian, R.S. J. Colloid Interface Sci. 1988, *125*, 333.
29. Barton, K.D.; Subramanian, R.S. J. Colloid Interface Sci. 1989, *133*, 211.
30. Kim, H.S.; Subramanian, R.S. J. Colloid Interface Sci. 1989, *127*, 417–428.
31. Kim, H.S.; Subramanian, R.S. J. Colloid Interface Sci. 1989, *130*, 112.
32. Nallani, M.; Subramanian, R.S. J. Colloid Interface Sci. 1993, *157*, 24.
33. Chen, J.; Stebe, K.J. J. Colloid Interface Sci. 1996, *178*, 144.
34. Chen, J.; Stebe, K.J. J. Fluid Mech. 1997, *340*, 35.
35. Subramanian, R.S. AIChE J. 1981, *27*, 646.
36. Crespo, A.; Manuel, F. Proc. 4th European Symposium on Materials Sciences under Microgravity, 45–49, ESA SP-11, 1983.
37. Balasubramaniam, R.; Chai, A. J. Colloid Interface Sci. 1987, *119*, 531.
38. Haj-Hariri, H.; Nadim, A.; Borhan, A. J. Colloid Interface Sci. 1990, *140*, 277.
39. Balasubramaniam, R.; Subramanian, R.S. Int. J. Multiph. Flow 1996, *22*, 593.

40. Balasubramaniam, R. Int. J. Multiph. Flow 1998, *24*, 679.
41. Balasubramaniam, R.; Lavery, J.E. Numer. Heat Transfer A 1989, *16*, 175.
42. Ehmann, M.; Wozniak, G.; Siekmann, J. Z. Angew. Math. Mech. 1989, *8*, 347.
43. Chen, J.C.; Lee, Y.T. AIAA J. 1992, *30*, 993.
44. Nas, S. Ph.D. thesis in Aerospace Engineering, University of Michigan, 1995.
45. Treuner, M.; Galindo, V.; Gerbeth, G.; Langbein, D.; Rath, H.J. J. Colloid Interface Sci. 1996, *179*, 114.
46. Ma, X.; Balasubramaniam, R.; Subrmanian, R.S. Numer. Heat Transfer, A 1999, *35*, 291.
47. Maxey, M.R.; Riley, J.J. Phys. Fluids 1983, *26*, 883.
48. Saffman, P.G. J. Fluid Mech. 1965, *22*, 385.Corrigendum 1968, *31*, 624..
49. Landau, L.D.; Lifshitz, E.M. *Fluid Mechanics*; Pergamon: Oxford, 1975.
50. Maxey, M.R. Phys. Fluids 1987, *30*, 1915.
51. Asmolov, E.S. Fluid Dyn. 1990, *25*, 886.
52. McLaughlin, J.B. J. Fluid Mech. 1991, *224*, 261.
53. Legendre, D.; Magnaudet, J. Phys. Fluids 1997, *9*, 3572.
54. Miyazaki, K.; Bedeaux, D.; Bonet Avalos, J. J. Fluid Mech. 1995, *296*, 373.
55. Asmolov, E.S.; McLaughlin, J.B. Int. J. Multiph. Flow 1999, *25*, 739.
56. Cox, R.G.; Hsu, S.K. Int. J. Multiph. Flow 1977, *3*, 201.
57. Vasseur, P.; Cox, R.G. J. Fluid Mech. 1976, *78* (Part 2), 385.
58. Vasseur, P.; Cox, R.G. J. Fluid Mech. 1977, *80* (Part 3), 561.
59. McLaughlin, J.B. J. Fluid Mech. 1993, *246*, 249.
60. Cherukat, P.; McLaughlin, J.B. J. Fluid Mech. 1994, *265*, 1–18.
61. Legendre, D.; Magdaunet, J. J. Fluid Mech. 1998, *368*, 81.
62. Dandy, D.S.; Dwyer, H.A. J. Fluid Mech. 1989, *216*, 381.
63. Cherukat, P.; McLaughlin, J.B.; Dandy, D.S. Int. J. Multiph. Flow 1999, *25*, 15.
64. Mei, R. Int. J. Multiph. Flow 1992, *18*, 145.
65. Auton, T.R. J. Fluid Mech. 1987, *183*, 199.
66. Auton, T.R. Ph.D. thesis, Cambridge University, 1984.
67. Auton, T.R.; Hunt, J.C.R.; Prud'homme, M. J. Fluid Mech. 1988, *197*, 241.
68. McLaughlin, J.B. J. Colloid Interface Sci. 1996, *184*, 613.
69. Stone, H.A. Phys. Fluids. A 1989, *2* (1), 111.
70. Stone, H.A.; Leal, L.G. J. Fluid Mech. 1990, *220*, 161.
71. Dukhin, S.S. *Dynamics of Adsorption at Liquid Interface*; Elsevier: Amsterdam, 1995.
72. Ladyzhenskaya, O.A. *The mathematical theory of viscous incompressible flow*; Gordon and Breach: New York, 1963.
73. Youngren, G.K.; Acrivos, A. J. Fluid Mech. 1975, *69*, 377.
74. Cox, R.G. J. Fluid Mech. 1970, *44*, 791.
75. Cox, R.G. J. Fluid Mech. 1971, *45*, 625.
76. Youngren, G.K.; Acrivos, A. J. Fluid Mech. 1976, *76*, 433.
77. Rallison, J.M.; Acrivos, A. J. Fluid Mech. 1978, *89*, 191.
78. Rallison, J.M. J. Fluid Mech. 1981, *109*, 465.
79. Pozrikidis, C. *Boundary Integral and Singularity Methods for Linearized Viscous Flow*; Cambridge University Press: Cambridge, U.K., 1992.

80. Milliken, W.J.; Leal, L.G. Phys. Fluids. A 1993, 5, 69.
81. Rumscheidt, F.D.; Mason, S.G. J. Colloid Sci. 1961, 16, 238.
82. Milliken, W.J.; Leal, L.G. J. Non-Newton. Fluid Mech. 1991, 40, 355.
83. Kennedy, M.R.; Pozrikidis, C.; Skalak, R. Comput. Fluids 1994, 23, 251.
84. Manga, M.; Stone, H.A. J. Fluid Mech. 1995, 300, 231.
85. Loewenberg, M.; Hinch, E.J. J. Fluid Mech. 1996, 321, 395.
86. Zinchenko, A.Z.; Rother, M.A.; Davis, R.H. Phys. Fluids 1997, 9, 1493.
87. Zinchenko, A.Z.; Rother, M.A.; Davis, R.H. J. Fluid Mech. 1999, 391, 249.
88. Kumaran, V.; Koch, D.L. Phys. Fluids A 1993, 5(5), 1123.
89. Kumaran, V.; Koch, D.L. Phys. Fluids A 1993, 5(5), 1135.
90. Sangani, A.S.; Didwania, A.K. J. Fluid Mech. 1992, 250, 307.
91. Kok, J.B.W. Ph.D. thesis, University of Twente, 1989.
92. Unverdi, S.O.; Tryggvason, G. J. Comput. Phys. 1991, 100, 25.
93. Haj-Hariri, H.; Shi, Q.; Borhan, A. Phys. Fluids 1997, 9, 845.
94. Ryskin, G.; Leal, L.G. J. Comput. Phys. 1983, 50, 71.
95. Ryskin, G.; Leal, L.G. J. Fluid Mech. 1984a, 148, 1.
96. Ryskin, G.; Leal, L.G. J. Fluid Mech. 1984b, 148, 19.
97. Kang, I.S.; Leal, L.G. Phys. Fluids 1987, 30, 1929.
98. Todd, P.H.; Mcleod, R.J.Y. Comput.-Aided Design 1986, 18, 33.
99. Hirt, C.W.; Nichols, B.D. J. Comput. Phys. 1981, 39, 201.
100. Osher, S.; Sethian, J.A. J. Comput. Phys. 1988, 79, 12.
101. Cuenot, B.; Magnaudet, J.; Spennato, B. J. Fluid Mech. 1997, 339, 25.
102. Magnaudet, J.; Rivero, M.; Fabre, J. J. Fluid Mech. 1995, 284, 97.
103. Calmet, I.; Magnaudet, J.; Rivero, M.; Fabre, J. Phys. Fluids 1997, 9, 435.
104. Rai, M.M.; Moin, P. J. Comput. Phys. 1991, 96, 15.
105. Blanco, A. Ph.D. thesis, Inst. Natl. Polytech. Toulouse, France, 1992.
106. Duraiswami, R.; Prosperetti, A. J. Comput. Phys. 1992, 98, 254.
107. Rothman, D.H.; Zaleski, S. Lattice-Gas Cellular Automata; Cambridge University Press: Cambridge, UK, 1997.
108. Chen, S.; Doolen, G.D. Annu. Rev. Fluid Mech. 1998, 30, 329.
109. Frisch, U.; Hasslacher, B.; Pomeau, Y. Phys. Rev. Lett. 1986, 56, 1505.
110. d'Humières, D.; Lallemand, P.; Frisch, U. Europhys. Lett. 1986, 2, 291.
111. Frisch, U.; d'Humères, D.; Hasslacher, B.; Lallemand, P.; Pomeau, Y.; Rivet, J.-P. Complex Syst. 1987, 1, 649.
112. McNamara, G.R.; Zanetti, G. Phys. Rev. Lett. 1988, 61, 2332.
113. Higuera, F.J.; Jiménez, J. Europhys. Lett. 1989, 9, 663.
114. Higuera, F.J.; Succi, S. Europhys. Lett. 1989, 8, 517.
115. Chen, S.; Wang, Z.; Shan, X.; Doolen, G.D. J. Stat. Phys. 1992, 68, 379.
116. Sankaranarayanan, K.; Shan, X.; Kevrekidis, I.G.; Sundaresan, S. Chem. Eng. Sci. 1999, 54, 4817.
117. He, X.; Zhang, R.; Chen, S.; Doolen, G.D. Phys. Fluids 1999, 11, 1143.
118. He, X.; Luo, L.-S. Phys. Rev. E 1997, 55, R6333.
119. Bhatnagar, P.L.; Gross, E.P.; Krook, M. Phys. Rev. 1954, 94, 511.
120. Shan, X.; He, X. Phys. Rev. Lett 1998, 80, 65.
121. Ziegler, D.P. J. Stat. Phys. 1993, 71, 1171.

122. Eggels, J.G.M. Int. J. Heat Fluid Flow 1996, *17*, 307.
123. Goldstein, D.; Handler, R.; Sirovich, L. J. Comput. Phys. 1993, *105*, 354.
124. Derksen, J.; Van den Akker, H.E.A. AIChE J. 1999, *45*, 209.
125. Eggels, J.G.M.; Somers, J.A. Int. J. Heat Fluid Flow 1995, *16*, 357.
126. Qian, Y.H. Ph.D. Thesis, Université Pierre et Marie Curie, Paris, 1990.
127. Chen, S.; Chen, H.D.; Martinez, D.; Mattheus, W. Phys. Rev. Lett. 1991, *67*, 3776.
128. Qian, Y.H.; d'Humières, D.; Lallemand, P. Europhys. Lett. 1992, *17*, 479.
129. Chen, H.; Chen, S.; Matthaeus, W.H. Phys. Rev. A 1992, *45*, 5339.
130. Shan, X.; He, H. Phys. Rev., E 1993, *47*, 1815.
131. Shan, X.; He, H. Phys. Rev., E 1994, *49*, 2941.
132. Gottlieb, D.; Orszag, S.A. *Numerical Analysis of Spectral Methods: Theory and Applications. CBMS-NSF Regional Conference Series on Applied Mathematics, No. 26*, 1977.
133. Canuto, C.; Hussaini, M.Y.; Quarteroni, A.; Zang, T.A. *Spectral Methods in Fluid Dynamics*; Springer-Verlag: Berlin, 1988.
134. Boyd, J.P. *Chebyshev & Fourier spectral methods, Lecture Notes in Engineering*; Springer-Verlag: Berlin, 1989.
135. Karniadakis, G.E. Appl. Numer. Math. 1989, *6*, 85.
136. Henderson, R.; Karniadakis, G.E. J. Sci. Comput. 1991, *6*, 79.
137. Orszag, S.A. J. Atmos. Sci. 1971, *28*, 1074.
138. Kim, J.; Moin, P.; Moser, R. J. Fluid Mech. 1987, *177*, 133.
139. Orszag, S.A.; Patterson, G.S. Phys. Rev. Lett. 1972, *28*, 76.
140. Yuan, H.; Prosperetti, A. J. Fluid Mech. 1994, *278*, 325.
141. McLaughlin, J.B. Phys. Fluids, A 1989, *1*, 1211.
142. Crowe, C.; Sommerfeld, M.; Tsuji, Y. *Multiphase Flow with Droplets and Particles*; CRC Press: Boca Raton, 1998.
143. Odar, F. J. Fluid Mech. 1996, *25* (Part 3), 591.
144. Yeung, P.K.; Pope, S.B. J. Comput. Phys. 1988, *79*, 373.
145. Balachandar, S.; Maxey, M.R. J. Comput. Phys. 1989, *83*, 96.
146. Kontomaris, K.; Hanratty, T.J.; McLaughlin, J.B. J. Comput. Phys. 1992, *103*, 231.
147. Eggels, J.G.M.; Unger, F.; Weiss, M.H.; Westerweel, J.; Adrian, R.J.; Friedrich, R.; Nieuwstadt, F.T.M. J. Fluid Mech. 1994, *268*, 175.
148. Salvetti, M.V.; Banerjee, S. Phys. Fluids. 1995, *7* (11), 307.

5

Role of Capillary Driven Flow in Composite Manufacturing

SURESH G. ADVANI University of Delaware, Newark, Delaware, U.S.A.

ZUZANA DIMITROVOVA Technical University of Lisbon, Lisbon, Portugal

I. INTRODUCTION

Capillary-driven flows or wicking of fluids into a porous medium have been investigated since the early nineteenth century. Lucas [1] and Washburn [2] were among the first to understand and characterize the spontaneous flow as a result of capillary action in a porous medium. Their model is still one of the most adequate ones and is widely used. Capillary action in porous media plays a key role in a variety of fields such as dyeing, paper and textiles manufacturing, soil science, and petroleum industry [3–6]. Recently, the importance of such flows is being investigated in composite manufacturing applications. This chapter will review and summarize the state of our understanding of capillary effects in polymer fiber reinforced composites processing and manufacturing.

In the last three decades, fiber-reinforced composite materials have become an important class of engineering materials as they allow flexibility in the design of the component, a possibility of tailoring their properties to the industrial requirements exists, and because of the development of efficient manufacturing processes. The two constituents materials, resin (also known as the matrix) and fibers, are brought together to make a fiber-reinforced composite. Usually, the goal of the fibers is to provide the stiffness and strength. The objective of the matrix is to bind the fibers together and protect them from adverse environmental effects and maintain the component shape, surface appearance, and overall durability. However, as the fibers carry most of the structural loads, they dictate the macroscopic stiffness and strength.

Hence more fibers translate into stiffer and stronger composite. Fig. 1 shows some examples of fiber preforms. To obtain these high-fiber volume fractions, the fibers have to be aligned like matches in a matchbox [usually called as fiber tows or bundles that may contain from 2000 to 48,000 fiber strands (fibrils) as seen in Fig. 1b]. These fiber tows can then be arranged in a pattern by knitting, weaving, or braiding them. This is called a fiber preform. Random fiber preform can also be formed from fiber strands entangled together like a plate of spaghetti (Fig. 1a). However, the volume fraction of random preform in a composite is much less than the other preforms.

The empty spaces between the fibers have to be filled with the resin as any unfilled region known as a void or dry spot can be detrimental to the mechanical performance of the composite. The interface within a composite is volume occupied by the fibers in a composite divided by the fiber diameter and multiplied by 4, approximately. As the diameter of these fibers is of the order of a few microns, the interfacial area can be very large. For example, for a part of the size of a desktop with 50% fibers, the interfacial area would be about the area of a football field [7]. To have a good interface to transfer loads to fibers and carry transverse loads, the resin must come in intimate contact with the fibers. This contact usually takes place if the resin can "wet" the fibers in a natural way by spreading over the fiber surface.

The thermodynamics of wetting dictate that a drop of fluid will spread on a solid only if the surface energy of the solid is greater than the combined

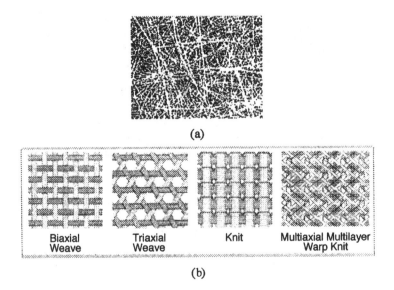

FIG. 1 (a) Random preform; (b) woven preforms.

surface energies of the liquid and the liquid–solid interface. Usually, low-energy fluids will wet high-energy solids. In composites, the polymers have surface energies of the order of 30–40 dyn/cm^2 and the glass fibers of the order of 500 dyn/cm^2, thus allowing for easy wetting. The surface free energy of glass fibers varies depending on whether the glass is bare or sized. Sizings and finishes are a mixture of ingredients (organic or aqueous based) applied on the fiber surface at the time of fiber manufacture to protect the surface and aid in handling. Bare glass fibers have a very high surface free energy value that promotes wetting. However, the glass fiber surface is very prone to corrosion. If the glass fibers are bare for even a microsecond of exposure to water, the strength of the fiber drops by a factor of 5. To prevent this loss of strength, glass can be coated with sizings or finishes to protect the surface. The coatings are often much more organic than glasslike, so they tend to reduce surface free energy. Wetting or spreading generally is not as favorable with these sizings as it is with bare fibers. Carbon and Kevlar fibers have usually an order of magnitude less surface energy than glass, making wetting less natural. Carbon fibers are primarily made of graphitic fibrils. These fibril surfaces are treated via electrolytic oxidation thus increasing the surface energy and promoting wetting [8,9].

The thermodynamics of wetting will not tell us about the rate at which the resin spreads over the fiber surfaces and impregnate the empty spaces between all the fibers to form the interface. This phenomenon is governed by resin flow. To completely wet a fiber tow, the resin must advance along and through the bundle. The bulk flow is usually controlled by the pressure gradient in the fluid with the final degree of wetting accomplished by capillarity.

There are many types of composite manufacturing processes. We can basically divide them into two categories. One in which the resin and the fibers are premixed and stored in the solid form called "prepregs." By applying heat and pressure, these prepregs can be formed into the shape of the component. Autoclave molding, fiber tow placement, and filament winding are some of the manufacturing processes that belong to this group. The second group of manufacturing method places the preform into a mold or over a tool, and injects the resin into it to fill all the empty spaces between the fibers and the fiber tows. They are collectively known as liquid composite molding (LCM). Resin transfer molding (RTM) and structural reaction injection molding (SRIM) are some of the examples of these fabrication processes [10].

The polymer resins used are of two types: thermoplastics and thermosets. Thermoplastic resins are very viscous, usually about 10^6 to 10^{12} more viscous than water. Hence it is very difficult to drive the resin with the help of capillary pressure into the empty spaces between the fibers. Consequently, external pressure is usually used to promote resin flow. Capillary action may help but is not the dominant mechanism of wetting. Thermoset resins are only about 10^2 to 10^4 times more viscous than water and capillary action can play a role in resin flow to cover the fibers and the spaces between them. We will discuss mainly the composite processes that use thermoset resins.

One way to manufacture "prepregs" is to drag fiber tows or preforms through a resin bath. Here the resin infiltrates inside the fiber tows because of the pressure gradient created as the fibers move and also because of the capillary action along the fibers. In manufacturing processes such as RTM and SRIM, the resin is injected into a closed mold in which the compacted fiber preform is placed. The resin flows because of the external pressure difference imposed at the injection locations. The resistance to the flow is offered by the fiber preforms. The problem is usually modeled as flow through porous media, where the resistance offered by the preform is characterized by the permeability of the preform. In most of the fiber preforms, the distance between the fibers in the fiber tows is about 1–2 orders of magnitude smaller than the distance between the fiber tows as shown in Fig. 2. Hence there are two scales of permeabilities. The resin will usually move very quickly in between the fiber tows mainly because of the higher permeability in those regions in response to the pressure force. However, inside a fiber tow, the resin can rapidly move because of capillary action as the pressure forces may be insufficient to drive the fluid because of the low permeability. The permeability is at least 2 to 4 orders of magnitude lower across the fibers inside a tow; thus the movement of resin inside a tow is largely driven by capillary action and can be modeled as wicking of fluids into a porous media. Wicking of resin

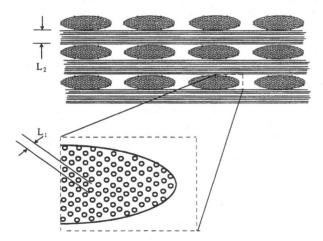

FIG. 2 Dual-scale architecture of a bidirectional preform: L_1 = characteristic length of microscopic domain; L_2 = characteristic length of mesoscopic domain; $L_1 \ll L_2$.

has been studied along the fibers or across the fibers inside a tow. Both these phenomena are important in forming a good interface between the resin and the fibers and also in elimination of the microvoids that may get trapped inside the fiber tows.

In this chapter, we will discuss the phenomena and the theory used to describe the role of capillary action in the impregnation process. The approaches on how these concepts have been applied to liquid molding processes will be presented and the issues of when it is important will be discussed with the help of examples. Finally, our outlook on how one can refine the theory to accurately capture the physics will be discussed.

II. FLOW THROUGH POROUS MEDIA

Composites that are fabricated by liquid composite molding (LCM) rely on infusion of the thermoset resin into the stationary network of fiber preforms draped over a tool surface under a pressure gradient. The resin usually undergoes a chemical reaction after the mold is filled. This process is known as curing, wherein the resin viscosity rapidly increases and the resin eventually solidifies. Hence it is important to cover all the empty spaces between the fibers before the resin viscosity increases and the resin solidifies. Thus it is important to understand and model the resin impregnation process into the fibrous preform. Practice has converged on modeling this as flow through porous media, where the porous medium is created by the fiber network [11–16]. Basic literature on flow through porous media is covered in books such as Refs. 17–19.

In reality, all nonrandom preform create a fibrous network that is a dual-scale porous medium because of the two scales of pore sizes as shown in Fig. 2. Hence the permeability in between the fiber tows is usually higher than the permeability within the fiber tows. Because of this complexity, the infusion process plays a vital role in the successful manufacturing of the part. The filling process is controlled by the inlet pressure or the inlet flow rate. High injection pressure or flow rate can cause the resin to quickly infiltrate the empty spaces between the fiber tows because of their high permeability, creating microvoids inside the fiber tows. Low injection pressure or flow rate allows sufficient time for the resin to impregnate the fiber tows but might cause the resin to cure and solidify before all the empty pores are filled. Low pressure gradient could also cause the capillary action to become dominant along the fiber tow directions, causing voids in between the tows.

Physically, the resin movement is promoted by pressure gradient and capillary action and is resisted by viscous forces. In this context, it is useful to separate the pressure in the hydrodynamic part (corresponding to the externally applied contribution) and the capillary part [20]. Gravitational force plays a minor role in the LCM processes and is commonly neglected. The role of capillary forces is also ignored as the hydrodynamic pressure is usually more dominant at the scale of the distance between the fiber tows, which we will refer to as the mesoscale.* In the last two decades, the focus has been to predict the movement of the resin flow front on the macroscale; thus the porous media have been treated as single-scale media. Under isothermal conditions, the governing equations in the saturated region for steady state flow of incompressible Newtonian fluid through porous medium (without inertia or surface tension effects) are expressed as:

$$\nabla \cdot v^D = 0, \tag{1}$$

$$v^D = -\frac{\mathbf{K}}{\mu} \cdot \nabla P, \tag{2}$$

where v^D is the average velocity vector and P is the macroscopic (global) pressure; $\mathbf{K}$ stands for the permeability tensor of the fiber preform that may be determined by analytical [21–24], experimental [16,25–34], or numerical methods [19,35–39]. However, originally, it was an empirical relation proposed by Darcy [40]; equations (1) and (2) were analytically verified by homogenization techniques, namely by asymptotic expansion methods [35,41] and by local averaging methods [19]. Regarding the asymptotic expansion methods, Eq. (1) is the macroscopic equation of the homogenization theory. When related to incompressible flow, it is known as the continuity equation. Equation (2) expresses Darcy's law in the region encompassing the impregnated (saturated) fiber preform. By substituting Eq. (2) in Eq. (1), one can solve for the Laplacian equation for pressure once the boundary conditions are specified. For LCM process, they are:

at the resin front: $\partial f / \partial t + (v^D \cdot \nabla f) / \phi = 0,$ \hfill (3a)

$P = 0,$ \hfill (3b)

at the mold walls: $\partial P / \partial n = 0,$ \hfill (3c)

at the injection gates: $v^D = v_0(t)$ or $P = P_0(t),$ \hfill (3d)

where ϕ is the porosity and the function $f[\mathbf{x}(t),t] = 0$ describes the location of the moving front position. Equation (3a) is known as the kinematic, while Eq. (3b) as static free boundary condition; they should be simultaneously applied at the boundary of the saturated region. However, the global analysis of the mold filling phase can be assumed as a quasi steady state process and the free

---

* This term is introduced according to other fields where some homogenization techniques are used to keep the macrolevel or macroscale for the full medium scale and the microscale for the lowest necessary scale needed for the description.

boundary conditions may not be simultaneously applied, but by explicit time integration approach. At each time step, first, the Laplacian equation with boundary conditions [Eqs. (3b)–(3d)] is solved for the pressure field; then, Eq. (2) is used to calculate the global velocities. A sufficiently small time step is chosen; the additional resin mass enters through the gates and the total resin mass is redistributed. Because of incompressibility, solely global velocities at the flow front can be used and the new front is calculated exploiting Eq. (3a). At this point, either the control volume approach [11,14,15,42,43] or other approaches that use nonconforming finite elements for pressure to improve the numerical stability of the continuity equation [44–46] are invoked.

Regarding the other boundary condition [Eq. (3c)], no flow out of the mold walls is imposed by $\partial P/\partial n = 0$, but exact satisfaction of the no-slip condition at the mold walls $\mathbf{v}^D = 0$ cannot be ensured by $\partial P/\partial n = 0$. The error is negligible only when $\mathbf{K}/\mu$ is small as pointed out by van der Westhuizen and du Plessis [47], where an approach permitting application of the no-slip condition at the macrolevel is presented. The impossibility of the no-slip condition implementation is related to the effect of the boundary layer of the homogenized region. Anyway, because race tracking is present in the boundary layer, this discrepancy is dealt differently than by forcing the no-slip condition [33, 48–51].

The specified approach has been implemented in many numerical simulations of the impregnation of resin into the fibrous preforms such as LIMS or LCMFlot [11–13,15,44,45,52,53]. These simulations have proven to be useful in understanding the flow of resin inside anisotropic and heterogeneous* porous media and in design and control of such manufacturing processes to avoid large areas of dry spots. Dry spots are fiber preform regions where the resin is not able to displace the air because of lack of a vent location at that position or because of the premature gelling of the resin before the injection process is complete as shown in Fig. 3.

The above approach does not address the dual-scale nature of the preforms†; hence the prediction and the presence of microvoids cannot be captured by this method. Microvoids are small air pockets that get entrapped within the fiber tows or in between the fiber tows because of the dual-scale nature of the fiber preform and the importance of the capillary action in the fiber tow regions (see Fig. 4). The next section will introduce the physics of the

---

*Implying that permeability tensor is anisotropic and varies with spatial coordinate.

†Previous approaches that account for the dual porosity either estimate or calculate the permeability with all single fibrils introduced, either directly [54,55] or in form of a porous medium [38,56,57]. This can yield correct results only in the saturated regions.

FIG. 3 Dry spot formation in a window mold as a result of the absence of vents where the flow fronts meet.

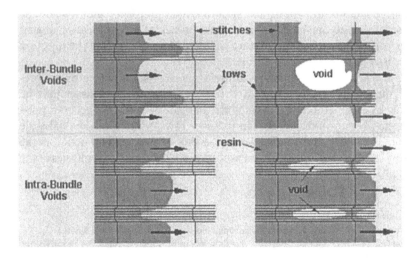

FIG. 4 Intertow and intratow void formation.

capillary flow and highlight its importance in the composites manufacturing process.

III. THE PHYSICS OF CAPILLARY DRIVEN FLOW

Externally applied pressure gradient during the filling phase in LCM processes is responsible for the macroscopic resin flow. Comparison of experimental and numerical simulation* results for fibrous media consisting of random preforms show excellent agreement implying that capillary forces can be neglected. For woven or stitched preforms with dual-length scales, macroscopic flow fronts do reasonably match when visually observed. However, as evident from Fig. 5, the saturation near the flow front is not instantaneous and partially saturated region of finite depth near the flow front is visible. This occurs because the permeability of the fiber tows is usually couple of magnitudes lower than the permeability of the region in between the tows. Secondly, the capillary forces at the microscale become at least as significant as the external pressure gradient in deciding the movement of the resin. Generally, filling of the fiber tows is delayed because the flow must overcome very low permeability of the tows as shown in Fig. 6. On the other hand, the capillary effects that are stronger inside the fiber bundle than in between the fiber bundles facilitate flow. To accurately understand capillary contribution, we will address the simplest case first.

In the simplest case, i.e., under the assumptions mentioned in the previous section and at the microlevel, we describe the resin flow inside the resin domain with Stokes flow. Thus the governing equations are:

$$\nabla \cdot \mathbf{v} = 0 \quad \text{and} \quad \nabla p = \mu \Delta \mathbf{v}, \tag{4}$$

where $\mathbf{v}$ and p denote the local velocity vector and pressure, respectively, and μ is the coefficient of the resin viscosity. As usual, $\nabla = \{\partial/\partial x_1, \partial/\partial x_2, \partial/\partial x_3\}$, $\mathbf{x}$ is the spatial variable, and $\Delta = \nabla \cdot \nabla$. Fibers are assumed impermeable and rigid, with perfectly fixed locations. Because the air influence and surface tension effects are usually neglected, the boundary conditions reduce to [58]:

$$\text{at the resin front}: \partial f/\partial t + \mathbf{v} \cdot \nabla f = 0, \tag{5a}$$

$$\sigma \cdot \mathbf{n} = 0 \tag{5b}$$

$$\text{at the fiber boundary}: \mathbf{v} = 0. \tag{5c}$$

---

*For standard approaches as specified in the previous section.

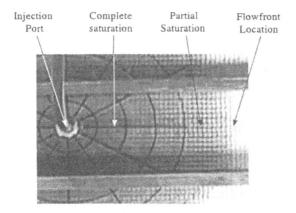

Injection   Complete   Partial   Flowfront
Port   saturation Saturation   Location

FIG. 5 Partial saturation near the flow front region.

Here the function $f[\mathbf{x}(t),t] = 0$ describes the position of the moving front, $\mathbf{n}$ is the unit normal vector to the corresponding surface, and σ is the fluid stress tensor.

To evaluate the importance of the capillary forces, the boundary conditions on the free boundary must be introduced with the surface tension effect, then the kinematic condition (5a) is retained, while the static condition at the resin front (neglecting the air influence) can be written as [58]:

$$\tau^{v} \cdot \mathbf{n} = 0, \tag{6a}$$

$$p = -\gamma \left(\frac{1}{R_1} + \frac{1}{R_2} \right) = -2\gamma H, \tag{6b}$$

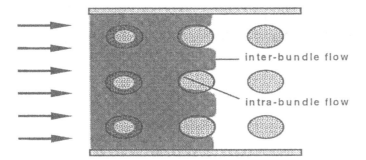

inter-bundle flow

intra-bundle flow

FIG. 6 Schematic of delayed impregnation of fiber tows.

where τ^v is viscous stress, R_1 and R_2 are the radii of the free surface,* H is the mean curvature, and γ is the resin surface tension. Here, as usual, the viscous normal stresses were neglected as compared to pressure values.

Equation (6b) is based on the concept of the interfacial (Hemholtz) free energy described in Refs. 17, 19, 58, and 59, and introduced first by Gibbs. It states that "if a liquid is in contact with another liquid (when both are immiscible) or gas, there is a free interfacial energy between the two substances." The free energy (W_{LA}, e.g., for liquid–air interface) manifests itself as interfacial tension,† γ_{LA}. The phenomenon of interfacial tension results from the fact that liquid has a certain freedom in the molecule movement permitting to reach an equilibrium interfacial shape, but on the other hand liquid cannot freely expand like a gas, because of its molecular attractive forces. For the surface molecules, because half of their neighbors are missing, they are subject to the inward attraction force, giving rise to the phenomenon of the interfacial tension at the interface. Thus the interface has a tendency to minimize its surface area. The interface has generally finite thickness, but for common purposes it is possible to assume that this thickness is in fact zero (Young's concept) and that there is a phase discontinuity at the interface [19]. The concept of the interfacial tension is mathematically equivalent to the thermodynamic concept of the interfacial free energy, defined as the work required to form "more surface," i.e., to bring more molecules from the interior to the interface. Then, the thermodynamical definition of interfacial (liquid–air) tension can be expressed as:

$$\gamma_{LA} = \frac{\partial W_{LA}}{\partial A_{LA}}, \tag{7}$$

where A_{LA} is the interfacial area and the process is valid under constant pressure and temperature. The units for γ_{LA} are [force/length], which is, of course, equivalent to [energy/area]. Under strict definition, the term surface tension, γ_L, is used when the air is replaced by vacuum. However, surface tension is often used (for the sake of simplicity) for the interfacial tension.

When the interface is curved and at equilibrium, interfacial tension, γ_{LA}, is equivalent to a pressure drop Δp called the capillary pressure, acting across the interface. Functional dependence can be derived from the equivalence of

---

* The radii of curvature are related to perpendicular planes but they are not necessarily the main radii. Anyway, it is well known that the sum ($2H = 1/R_1 + 1/R_2$) is independent of the choice of the perpendicular planes.

† The concept of the interfacial tension is based on classical Newtonian mechanics and was separately introduced by Laplace, Young, and Gauss in the early nineteenth century.

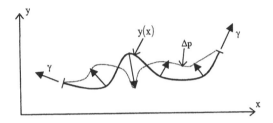

FIG. 7 Two-dimensional justification of the Young–Laplace equation.

the work, which is needed to increase the liquid–air interfacial area [59]. Thus p_c can be expressed as

$$p_c = \Delta p = p_A - p_L = \gamma_{LA}\left(\frac{1}{R_1} + \frac{1}{R_2}\right) = 2H\gamma_{LA}. \tag{8}$$

Equation (8) can also be derived differently, e.g., as presented in Ref. 19, where gravitational effect is directly included in terms of the Bond number (defined later on in this section). Equation (8) was first introduced in 1805 by Young and Laplace (hence Young–Laplace equation) and is considered as the basic equation of capillarity. Equation (8) is equivalent to Eq. (6b) with the influence of air accounted for in Eq. (8).

It can be easily verified that on a cross-sectional cut made across any surface, the capillary pressure is equivalent to the surface tension applied in the cut, i.e., instead of surface tension, a pressure drop is applied. Let us take a fictitious cross-sectional cut, for the sake of simplicity in two dimensions (Fig. 7). Then, when the following curvature equations,

$$\frac{1}{R_1} = \frac{y''}{(1 + y'^2)} \text{ and } \frac{1}{R_2} = 0 \tag{9}$$

are introduced into Eq. (8), integration along the surface will yield equivalent load conditions. Thus Eq. (8) calculates the pressure drop that should be applied* but it cannot define the equilibrium interfacial shape.

The concept of the surface tension of solids is described in Refs. 17, 19, and 59, but it is generally the same as for liquids. Interfacial solid–liquid tension is

---

* It is useful to remark that the radii of curvature (if both are of the same sign) always lie on that side of the interface having the greater pressure. Thus the capillary pressure (or the surface tension) has always tendency to "smooth up" the liquid–air interface.

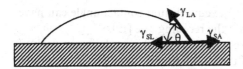

FIG. 8 Interfacial surface tensions between solid, liquid, and air, allowing determination of the contact angle.

defined in terms of the work, W_{SL}, which must be done to separate the unit area of the two substances [17]:

$$W_{SL} = \gamma_S + \gamma_L - \gamma_{SL}. \tag{10}$$

When two fluids meet a solid, then the shape of the interface between the two fluids is such that a contact angle with a solid is formed depending on the properties of the three substances. Let us again assume liquid–air interface meeting the solid. The contact angle θ is given by the following equation, presented in Refs. 9, 17, 19, 58, and 59:

$$\cos \theta = \frac{\gamma_{SA} - \gamma_{SL}}{\gamma_{LA}}. \tag{11}$$

Equation (11) is called Young's equation and can be demonstrated as mechanical equilibrium condition, as shown in Fig. 8. When $\cos \theta$ is positive, i.e., when $\theta \in (0, \pi/2)$, the liquid wets the solid. Obviously, when $\gamma_{SA} - \gamma_{SL} > \gamma_{LA}$, no contact angle can be defined by Eq. (11) and the liquid spreads over the solid. The driving force of this motion can be characterized by the spreading coefficient, $S_{L,SA}$, defined in Ref. 9:

$$S_{L,SA} = \gamma_{SA} - \gamma_{SL} - \gamma_{LA}, \tag{12}$$

$\gamma_{SA} - \gamma_{SL}$ is sometimes called the adhesion tension [60] and $\cos \theta$ the relative adhesion coefficient [61]. In fact, in Fig. 8, the equilibrium (static) contact angle is shown. It is different from the contact angle during a liquid motion* (apparent dynamic or moving contact angle) and moreover a phenomenon of (wetting) hysteresis is present; that is, the advancing contact angle is always higher than the receding contact angle. Surface tension exhibits the phenomenon of hysteresis as well; however, contact angle hysteresis is sometimes explained only by the solid surface roughness and heterogeneity. Detailed

---

* When a liquid moves along the solid against another liquid, then the former is called the wetting liquid and the latter the nonwetting liquid, and this "wetting" is not related to the contact angle.

discussion can be found in Ref. 62. The equilibrium contact angle can have any value between the advancing and the receding one [19].

For the sake of completeness, we introduce the thermodynamic work of adhesion W_a, which is a measure of the interaction between phases. It can be expressed as shown below [8]:

$$W_a = \gamma_{LA} + \gamma_{SA} - \gamma_{SL} = \gamma_{LA}(1 + \cos\theta). \tag{13}$$

For LCM processes, a high value for work of adhesion is desirable.

Interfacial tension between two fluid phases is a definite and accurately measurable property depending on the properties of both phases. Also, the contact angle, depending now on the properties of the three phases, is an accurately measurable property. Experimental approaches are described, e.g., in Refs. 8, 60, and 63 and in Ref. 62, where especially detailed discussion of the Wilhelmy technique is presented. Theories such as harmonic mean theory, geometric mean theory, and acid–base theory (reviewed, e.g., in Refs. 8, 20, and 64) allow calculation of the solid surface energy (because it is difficult to directly measure) from the contact angle measurements with selected test liquids with known surface tension values. These theories require introduction of polar and dispersive components of the surface free energy.

If a certain volume of liquid is poured into an open container and allowed to stabilize, the liquid–air interfacial shape is not known in advance. However, the shape is formed in a way that equilibrium is reached between all unbalanced forces, i.e., the corresponding mechanical energy functional reaches a minimum [61]. Even if the container is a semi-infinite cylindrical tube with a reasonably shaped base, the problem of existence of a unique solution is not trivial. If only surface tension effects are considered and all other contributions are neglected, then there are only two terms in the functional. One corresponds to the interfacial area (positive term) and the other to the contact angle contribution (negative when cos θ is positive), justifying the possibility of the curved surfaces* formation. Further examining the functional suggests that the constant mean curvature surfaces are formed [61,65]. Then the problem can have a nonunique solution,[†] as it is simply demonstrated in Fig. 9 [61]. In the case of a narrow circular tube, the two terms in the functional are dominant and a spherical cap can approximate the interfacial surface. Because of the significant curvature, the capillary pressure is substantial and narrow tubes are called capillaries.

When a capillary tube is vertically inserted into an open container filled with a liquid, and a gravitational field is present, then the liquid density,

---

*Obviously, in a container, a straight surface always has the minimum area.

[†] If the gravity influence was omitted.

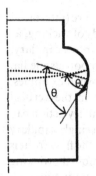

FIG. 9 Nonunique solution of the interfacial surface when gravitational force is neglected.

interfacial tension (fluid–air), and the contact angle determine the capillary rise or depression, as shown in Fig. 10. The pressure drop in Eq. (8) is now opposed by the gravitational forces. In practical applications, more than the exact interfacial shape, the height of the liquid column is important. Anyway, the integral of the pressure drop over the interface can be expressed by the surface integral along the contact, as already mentioned in this section. The capillary rise or depression, h, in circular capillary of radius, r, is expressed as:

$$h = \frac{2\gamma_{LA}\cos\theta}{rg\Delta\rho}, \tag{14}$$

where $\Delta\rho$ denotes the difference in density between the liquid and air phase and g is the acceleration due to gravity.

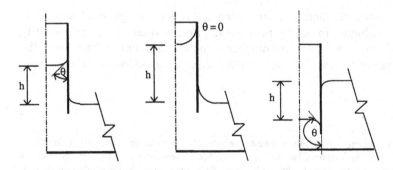

FIG. 10 Capillary rise and depression.

Spontaneous liquid flow originated purely by capillary pressure (interfacial tension effects) is called wicking flow. There are different kinds of wicking, as specified in Ref. 60. The simplest case is the spontaneous flow into a capillary. The practical importance of this is having wicking flows into porous media. Microlevel flows related to LCM processes are one of many such examples. Direction of the wicking flow is not restricted. When it is in the vertical direction, it is opposed by gravitational effects. However, capillary action can be so strong that gravity can be neglected. Wicking flow into a single capillary is different than flow into a porous medium. In the latter case, the flow front is not macroscopically sharp and if the fluid source is removed, wicking continues until an equilibrium state is reached. This is in contradiction with the former case, where the motion immediately stops [66].

Some facts were omitted in the discussion of this section. First of all, interfacial tension (generally) monotonically decreases with increasing temperature; thus the correct values must be assumed. However, temperature variations along the surface can cause additional motions along the surface, known as Marangoni convection under zero gravity conditions. These motions can also be invoked by variation in relative concentrations of the two phases or by variations in electric potential [65]. Local temperature variation can be related to the wetting process itself, where the heat is liberated because of the changes in the interfacial free energies, as a result of the chemical reaction occurring between the resin and fiber, etc. [17].

The important dimensionless parameters, which indicate the relative importance of gravity, viscosity, and capillarity are Bond and capillary numbers* [19,63]. They are usually introduced for isotropic media; however, in our case, anisotropy is expected. Thus it is useful to make the capillary number directional. Therefore the Bond and capillary numbers are

$$N_b = \frac{\Delta\rho g a^2}{\gamma}, \ N_{c,x} = \frac{v_x \mu}{\gamma}, \tag{15}$$

where a is some characteristic dimension and v_x is some typical velocity of the related problem. In LCM processes, gravitational forces are usually neglected; thus only capillary number come into account to "measure" the importance of the capillary vs. viscous forces in a specified direction.

---

*The Bond and capillary numbers definition might differ in different sources according to the problem under consideration, but the basic concept (relative importance of gravity vs. viscosity and capillarity vs. viscosity) is always the same. Sometimes, the ratio for capillary number is inverted.

A. Application to Liquid Composite Molding Processes

The theory presented in this section can be applied to LCM processes in many ways.

1. Direct Application of Thermodynamic Characteristics Related to Proper Wetting, Adhesion, and Local Viscosity Changes in the Interfacial Region

As mentioned in the beginning of this section, filling of the fiber tows is usually delayed. To reduce this delay, proper resin wetting must be achieved; thus the surface free energy of the fiber should be high while the surface free energy of the resin matrix should be low [9], which is true in case of polymer resin and glass or carbon fibers.

Even if the filling is successful, improper wetting (high contact angle) can result in poor fiber–matrix adhesion [9]. Critical values of the liquid surface energy can be defined. Then, liquids with the surface energy higher than the critical one may leave microvoids at the fiber–matrix interface. Surface energy of resin can be influenced by adding some soluble agents or by changing the processing temperature; however, it was concluded in Ref. 67 that the surface tension changes because of the temperature variation between 20°C and 50°C were negligible. Surface energy of fiber can be affected by surface treatments, such as sizing and finishing. Sizings and finishes are mixtures of ingredients applied at the time of fiber manufacture. The fiber–matrix interface is not instantaneously created; sufficient time must be allowed for its creation to obtain optimal adhesion. Soluble sizing influences the mechanical properties of the final composite generally in favorable direction. However, related to the void formation, it was noticed in Ref. 67 that removing of sizing significantly decreased void formation.

In addition to its influence on adhesion, one has to be aware of possible local viscosity changes because of the interfacial region formation, which would result in time-dependent reduction in resistance to infusion of resin within the fiber matrix. Fiber–resin interactions, interface formation, sizing and finishing, and their influence on wicking and mechanical properties of the final composite and many other aspects are studied, mainly experimentally, in Ref. 8. Valuable conclusions related to the work of adhesion influence on the final properties are in Ref. 68.

2. Microlevel Void Formation and Transport

The motion of a bubble in a tube, presented in Ref. 19, can be directly utilized for void transport in LCM processes, namely such study allows one to understand how to drive the bubbles out of the filled area. The bubble motions are a strong function of the Bond and capillary numbers; a is the characteristic dimension that can be equal to the tube radius and v can either be the velocity of the bubble front or some other typical velocity in the problem. Lundstrom [69] concluded that the bubble motion requires not only the pressure gradient to be sufficiently high and the liquid surface tension to be sufficiently low, but also the cross-sectional area of the channels must be sufficiently uniform. He experimentally verified theoretical predictions and some discrepancies for very low capillary numbers were explained by Marangoni effects. Additional experimental work in this area is presented in Ref. 70 where the theory is extended to the macrolevel.

3.   Analytical Tow Filling Study

Interesting analytical tow filling study is presented in Ref. 71. It is assumed that the resin radially reaches the tow. The curved surfaces are then immediately formed (Fig. 11). When the capillary number is small, the surfaces can be approximated by the constant curvature shapes—cylindrical surface in this case. Then, it is assumed that additional motion is solely directed by capillary action. If the fibrils are not close enough when the resin front reaches the straight shape, then no more capillary pressure is present and the motion stops (Fig. 12a). It is possible to analytically express critical fibril distance. We extended the results from Ref. 71 to the staggered arrangement. Continuous resin motion in this case is shown in Fig. 12b. In Fig. 12c, the dimensionless maximum free distance between fibrils ensuring the continuous resin motion is plotted together with the volume fraction against the contact angle for both packing arrangements. The staggered packing arrangement seems to be more favorable, but it is necessary to point out that in this case, the initial resin direction is important. It is obvious from Fig. 12b that if we rotate the initial direction by 45°, the motion would not be continuous (in-line packing arrangement). Nevertheless, if hydrodynamic pressure is present, the resin front can further move and capillary action becomes dominant again when it touches other fibrils.

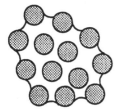

FIG. 11   Situation when the resin radially touches the tow.

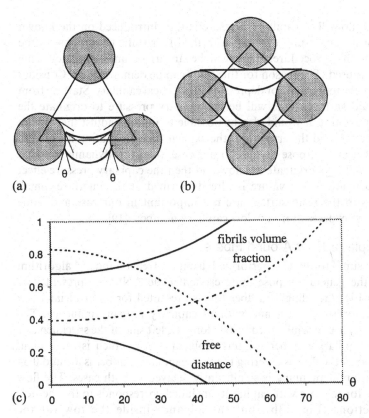

FIG. 12 (a) Resin motion stops because of high contact angle and large distance between fibrils in hexagonal packaging. (b) Continuous resin motion for $\theta = 0$ and suitable distance between fibrils in staggered packaging. (c) Maximum free distance between fibrils (dimensionless with respect to the fiber radius) ensuring continuous wetting and the corresponding fibrils volume fraction in hexagonal (full line) and staggered (dashed line) packaging as a function of the contact angle.

It is also mentioned in Ref. 71 that very high capillary pressure on the tows can significantly change the single fibrils positions and actually close some spaces between them. This fact will generate random or irregular positions other than hexagonal, in-line or staggered arrangement.

B. Numerical Simulations of Transient Flows at the Microlevel

In numerical simulation, it is possible to model the local transient flows as quasi steady state processes and use explicit time integration. Then, free boundary conditions [Eqs. (5a) and (6a,b)] can be used in a similar way to the macroscopic flow. The capillary pressure is thus introduced on the known surface (known curvature) together with the other static conditions. In the first step of this procedure, there will be an error, unless implicit time integration is used. The reason for this error can be demonstrated. Consider the example of capillary rise or depression in a vertical capillary. Starting from the flat liquid surface, there will be no capillary pressure to originate the motion. To avoid it, it would also be possible to initially adjust the contact angle. This error and the error from the explicit time integration is not very important for our purpose because in our case, the hydrodynamic pressure gradient will always originate the flow and then the capillary pressure effect will add to it after a curvature is already formed at the interface. Small inaccuracies in the resin surface are not important in our case and some oscillation are expected for very low capillary numbers [19].

1. Example of Flow Around Fibers

Numerical simulation was performed using a free boundary algorithm exploiting the general-purpose finite element code ANSYS,* presented in Ref. 55. In Fig. 13, filling of a fiber tow is presented for an artificially low number of fibrils [48] with and without capillary pressure influence. The inflow velocity is uniformly introduced along the left side of the specimen as 1 m/sec and capillary number for this problem is 3.4, which is quite high. However, when the flow is entering the tow, capillary number is around 0.34 at this location. Tow filling is significantly delayed in both cases. The flow front shape for the case with capillarity included corresponds to the theoretical predictions (Fig. 13b), but the advance inside the tow (at the corresponding time) is in both cases approximately the same. This happens because capillarity is not very strong in this case and as it is clearly seen in Fig. 13b, the resin front in between fibrils becomes straight before the other fibril is reached. Therefore hydrodynamic pressure is needed to move the resin further. Capillarity cannot be dominant because distances between fibrils are too large and the contact angle is too high. For this case, the volume fraction of fibrils inside the tow is 0.48 and the dimensionless free distance between fibrils is 0.6, which is more than the distance specified in Fig. 12.

A better understanding of the phenomena is offered with the following example by separately studying flow across single fibril, again with inflow velocity uniformly distributed along the left side of the specimen. Capillary number variation was chosen as 3.4, 0.34, 0.034, and it was modeled as inlet velocity variation. As the capillary number decreases, capillarity becomes stronger and the interfacial shape might approach the constant curvature shape directed by the contact angle (see Fig. 14). Simultaneously, the front

---

*For Stokes flow, analogy with incompressible elasticity is used [72].

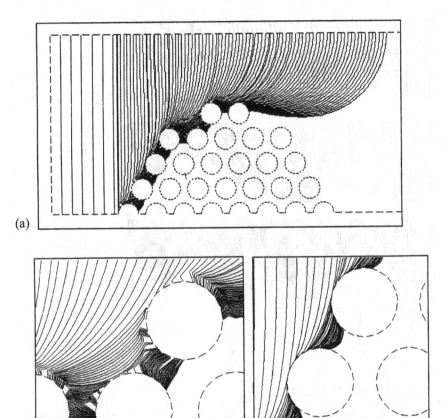

FIG. 13 (a) Numerical simulation results of the fiber tow filling with capillary pressure included, $N_{c,x} = 3.4$, $v_x = 1$ m/s, $\gamma = 0.067$ N/m, $\mu = 0.226$ Pa·sec, $\cos\theta = 0.45$, $r_{tow} = 0.3$ mm, $r_{fibril} = 0.03$ mm, $t = 0.57 \times 10^{-3}$ sec. (b) Typical details from the numerical simulation of the fiber tow filling with (left, $t = 0.57 \times 10^{-3}$ sec) and without (right, $t = 0.37 \times 10^{-3}$ sec) capillary pressure included; different progression front shape is visible [resin and tow parameters are the same as for (a); however, close to the tow, $N_{c,x} \approx 0.34$].

becomes more sensitive (also numerically) to the capillary pressure applied there. In case (c), the front would not be really smooth and thus the resin progression would oscillate, if the parameter controlling the maximum advance would have been kept the same. In case (a), flow front reaches the second fibril before the space between the fibrils is filled; therefore a void is formed and our results correspond to the vacuum assistance. In case (c), flow front looks as resembling the constant curvature shape specified in Fig. 14d,

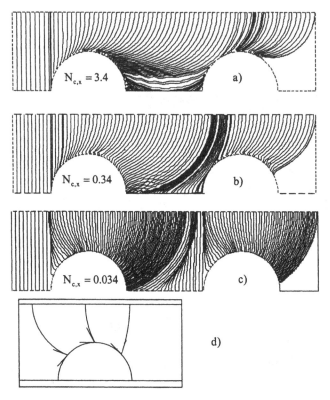

FIG. 14 Flow front progression along single fibril (r_{fibril} = 0.025 mm); resin parameters are γ = 0.067 N/m, μ = 0.226 Pa·sec, cos θ = 0.45, and decrease in capillary number is modeled as decrease in inlet velocity: v_x = 1 m/sec, v_x = 0.1 m/sec, v_x = 0.01 m/sec for cases (a), (b), and (c), respectively. Case (d) shows the theoretical front progression with constant curvature surface.

more time steps were necessary to fill the specimen and in Fig. 14c only each fifth front is plotted. However, the progression is helped by the hydrodynamic pressure, which is still quite strong. This can be proved by further decreasing of the capillary number to 0.0034 (see Fig. 15). Now when the front reaches the fibril, the upper part moves back to adjust the constant curvature shape determined by the contact angle, because the hydrodynamic pressure is insignificant. But the flow front shape cannot be kept like that and cannot be solely directed by capillarity because the fibrils are too far and the contact angle is too high. Therefore constant curvature surface directed by the contact angle is not developed. Contact angle is kept only locally close to the fibril (it is assumed that once the fibril surface is wetted, it keeps this stage) and the bulk front is delayed. In this case, oscillation of the front is stronger;

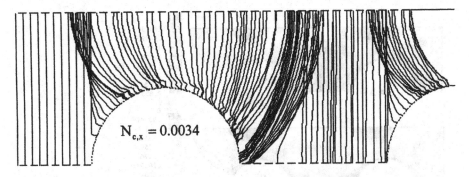

FIG. 15 Flow front progression along single fibril (r_{fibril} = 0.025 mm); resin parameters are γ = 0.067 N/m, μ = 0.226 Pa·sec, $\cos\theta$ = 0.45, and inlet velocity is v_x = 0.001 m/sec.

especially close to the fiber surface thus only each fifth to fiftieth front is plotted in Fig. 15, for the sake of clarity depending on the analysis progress. Analysis proceeds very slowly and practically stops for the reasons specified above; however, theoretically finite number of time steps is necessary to completely fill the specimen. For this analysis, the fibrils volume fraction is 0.2 and the free distance is 2, which is extremely high and justifies the conclusion that in this analysis, resin front progression cannot be directed by capillarity even if the capillary number is low.

The surface tension effects are negligible for the case (a), but simulation without them would again cause oscillation. Flow pattern is shown in Fig. 16 for reduced front advance. Surface tension smoothes the interfacial shape in any case and thus it should be always included, sometimes only to improve the numerical performance.

It was seen that with variation of the capillary number, flow front progression is very different. However, once all the space is filled, the velocity distribution (Fig. 17) or the pressure gradient corresponds to the steady state

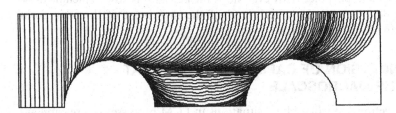

FIG. 16 Flow front progression for the case from Fig. 14 or 15, but without capillary pressure influence.

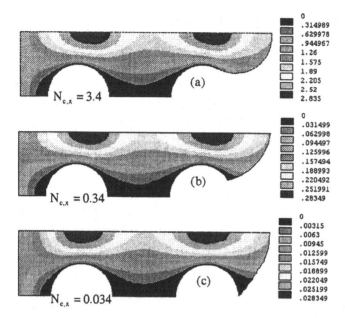

FIG. 17 Final v_x distribution for cases (a)–(c) from Fig. 14.

saturated solution. Consequently, capillarity cannot influence the saturated permeability. Steady state solution is easily obtainable by application of the periodicity boundary conditions, viscosity enters only as a parameter of the linear analysis and the surface tension does not appear at all. In Fig. 17, the velocity distribution for cases (a)–(c) is not exactly the same, as the final front is different; but after the front would pass several fibrils, the distribution would stabilize. Naturally, velocity values in Fig. 17 are different by a factor of 10, as the capillary number variation was made.

In summary, any microlevel transient flow study should include surface tension effects. As we will see in the following sections, at the macro and meso level, this is not always necessary. The capillary number will tell us about the level of such importance. However, now, the key issue is how to include these microscale effects at the macro level of the movement of the resin front infiltrating the fiber preform.

IV. INCLUSION OF CAPILLARY EFFECTS AT THE MACROSCALE

In this section, we will address situations in LCM processes where capillary effects (wicking flows) might be important and put forth approaches on how to model and account for them, when conducting macro or meso level analysis. From the previous section, we can safely assume that capillary effects are nonnegligible when capillary number [Eq. (15)] is small, allowing the resin front that is formed to approximately have a constant mean curvature surface if spaces in between the fibers to be filled are narrow. In addition, when the contact angle θ is small, this mean curvature is significant.* Capillary effects must also be included in the analysis when curvature at the interface develops for other reasons. Possible scenarios in LCM processes in which capillary action may play not an insignificant role could be:

In fiber tows, the resin flow rates are very low because of the low permeability of fiber tows. Moreover, the spaces in between individual fibers are narrow and low contact angle condition is always satisfied because this is characteristics for LCM processes. Thus these conditions boost the importance of capillary action.

Nonuniform filling of the mold can create voids and dry spots and analysis of their motion or volume changes should include influence of capillary pressure because of the significant curvature of the resin–air (or vacuum) interface.

Hence proper modeling of the capillary effects permit understanding of the microvoids and dry spot formation, transport, and volume changes, and allows for the correct flow front progression in regions where fiber tows are present.

In macrolevel analysis, capillary number can be modified while retaining Darcy's law[†] to describe the bulk impregnation of the resin into the fiber preforms. It is useful to point out that the anisotropic nature of the fiber preforms will now appear in the permeability tensor and we can expect some directionality of the capillary pressure influence, which is not necessary in almost all other research areas, where capillary pressure is used. Thus

$$N_{c,x} = \frac{K_{xx}}{\gamma} \left| \frac{\partial P}{\partial x} \right|. \tag{16}$$

However, as pointed in Ref. 19, permeability tensor cannot uniquely substitute characterization of the internal geometry of the porous medium; thus capillary number cannot be the only parameter to "measure" the importance of the capillary vs. viscous action in a specified direction.

Because of the relatively high externally applied pressure, the regions of the preform with higher permeability, e.g., the spaces between the fiber tows, are filled first (viscous action is dominant and the capillary number is high). While inside the tows, the filling is usually delayed. Thus at the mesolevel, it is possible to distinguish the generally smooth primary flow front (viscous

---

*Will depend on the internal geometry of the porous medium.
†We present definition for principal directions of the anisotropy.

action) between the tows and wicking flow front (combined viscous and capillary action or solely capillary action) inside the fiber tows [20].

The phenomenon is different for flow along and across the fiber tows. In flow across the fiber tows, only a thin strip along the fiber tow circumference is filled when the primary front has passed over the tow (Fig. 6). The air is compressed inside the tow until it is balanced with the surrounding resin pressure. Then, the capillary action becomes the only factor that can drive the resin inside the tow. When the air pressure becomes higher than the surrounding pressure, the air can escape from the tow in the form of microvoids. Because the resin surrounds the tow from all sides, the wicking flow for this case is called the *immersional* wicking [60]. Such situation can arise not only in radial impregnation of the fiber tows, but also in filling of some dry spots. Related to the quality of the final composite, this is the most elusive type of dry spot because it can be eliminated only by the capillary action. Dry spot changes of this type have been studied analytically, numerically as well as experimentally by Han [73]. For flows along fiber tows, the proper term for wicking flow is *penetration*, i.e., liquid source faces the porous medium from one side only [60].

In summary, proper modeling of the situation where immersional wicking might occur cannot neglect capillarity and is important irrespective of the fact whether vacuum assistance is used or not. Regarding the capillary number, permeability across the tows (or radial permeability) is much lower (at least four times) than along the tows, while capillary pressure is higher (around twice*) when resin passes along the tows. Hence the capillary number is lower for flows across the fiber tows. Nevertheless, this is just an academic case. If the resin is not forced to flow only across the fiber tows, it will naturally choose the easier (higher permeability) direction, i.e., along the fiber tows in this case.

A. Approaches

Three approaches are generally adopted to determine the governing equations for flow through regions with fiber tows:

1. Assumption of two regions with very different permeability and application of Darcy's law in both regions [28,74–77]. Radial Darcy's law is used to determine a sink term corresponding to the tow filling for the perpendicular flow in [78,79]; in Ref. 36, both fronts are simultaneously tracked.

---

*This directly yields from the Young–Laplace equation [Eq. (8)] and it is related to the fact that we can expect spherical interfaces when resin passes along the fibrils and cylindrical interfaces for the flow across the fibrils and the radius of these interfaces is approximately the same. This will be discussed in more detail in the next section.

2. Assumption of Stokes flow to describe the resin flow between the tows and Brinkman's equation to describe the flow inside the tows (in this case, it is useful to use the Lattice–Boltzman method as a numerical approach [80]).

3. Assumption of Stokes flow between the tows, Darcy's flow inside the tows and suitable slip condition on the boundary of the "porous" part. Many theoretical studies can be found about such appropriate slip boundary condition [33,81–84].

Nevertheless, the studies mentioned in this paragraph are concerned with transient (flow front progression) problems, but no capillary contribution is introduced in them.

Experimental results regarding flow across the fiber tows are presented in Refs. 28 and 79. Parnas et al. [28] present experimental results for flow along the fiber tows. In their work, the formation of voids was not related to the dual-scale nature of the porous media, but to defects on the fiber tow surface. Other experimental results are also presented in some of the references from the previous paragraph. Usually, satisfactory agreement of the theoretical predictions with the experimental results is concluded by the authors, suggesting that capillarity might be neglected.

Other experiments concerning with the formation of voids are presented in Ref. 9. The void formation problem for flows along the tows was also studied in Ref. 20. Several liquids with different properties were injected into unidirectional stitched fiberglass mat. The authors manage to relate the percentage aerial void area for different liquids to a single parameter, the modified capillary number (with the contact angle included), introduced as:

$$N_c = \frac{\nu\mu}{\gamma\cos\theta} \tag{17}$$

that serves as the "measure" better than the capillary number. Nevertheless, some discrepancies were explained by the fact that the surface energy of the mold surface is not included in the modified capillary number. Visualization clearly showed the void formation as a consequence of the flow front nonuniformity and curvatures.

Flow through woven fabrics was experimentally treated in Ref. 85. Such situation is far more complex. At low flow rates, mainly macrovoids (spherical) between the tows were formed (usually stacked close to the stitches) with good wetting of the tows. At high flow rates, microvoids (cylindrical in form) were trapped inside the tows. Results were again fit to the modified capillary number. It was experimentally verified [20,85] that sometimes the wicking flow front was delayed but sometimes it was advanced with respect to the primary front. Obviously, the advance of the wicking front cannot be explained without invoking capillarity. It is extremely important to address the question, under which circumstances the capillary action can be neglected and when should one account for it in modeling the LCM processes. The modified capillary number will play a crucial role in such decisions. Numerical simulations have been conducted to explore the role of the capillary pressure. First, numerical simulation accounting for voids and dry spots formation with capillarity included was developed by Han [73]. A generally applicable sink term with capillarity included is developed in Ref. 86, and it is included in the LIMS code; the new name for the simulation code is SLIMS [77].

So far in this section, we discussed the cases where the capillarity might be important. The key challenge is how to determine the macroscopic (averaged, homogenized) capillary pressure, which is not related to the macroscopic front curvature but to the average of the local capillary pressures at the microlevel. Hence it is not possible to incorporate capillary effects in the macrolevel (or mesolevel) analysis by direct application of the Young–Laplace equation [Eq. (8)] because any macroscopic front or surface is not very sharp in reality. This fact is not related to the capillary effects, but to the homogenization techniques used. Each macroscopic location, in fact, corresponds to the representative volume* at the microlevel and thus a parameter called saturation having values in closed interval (0,1) and corresponding to the ratio of the filled void volume to the total void volume (in this representative volume) is meaningful. In standard approaches, it was assumed that the macroscopic flow front is sharp and partly saturated control volumes or nodal factors of values in the open interval (0,1) were only consequences of the full domain discretization. Numerical simulations of the LCM filling phase have already started to account for the nonsharpness of the flow front.

If the flow front is not sharp, an intermediate (partly saturated or transition) region, where the saturation is in the open interval (0,1), is formed during the filling phase. Proper description of this region has to be determined from the microlevel problem.

B. Approaches to Modify Liquid Composite Molding Models to Include Capillary Effects

Possible modifications or "extensions" to the governing equations of LCM models can be made in three ways:

1. Introduction of the relative permeability into the Darcy's law and the time derivative of the saturation into the continuity equation [87].
2. Addition of the influence of the (averaged) capillary pressure, which cannot be carried out as before in the form of a boundary condition,

---

*For instance, in periodic media, such representative volume can be taken as the basic cell, which is defined as the (smallest possible) medium part, which can generate the full medium by its periodic repetition.

but the capillary pressure will act in the full intermediate region and vary with the degree of saturation.

3. Resurrection of the two-phase flow theory [17–19,88].

Thus depending on the type of complexity that needs to be modeled, the corresponding method can be chosen. These can also be thought of as increasing level of sophistication in the theory and are not mutually exclusive. The modifications mentioned above will be cast into equations and presented in the next section.

Regarding (1) and (2), better intermediate region model is obtained, however, if no voids will be formed, the final results in the saturated area will be the same as for standard approaches. We note in this context that neither relative permeability nor capillary pressure can influence the saturated permeability as also demonstrated in the previous section. Nevertheless, either by (1) or by standard approaches, macroscopic dry spots can be captured and then the air pressure can be added inside the dry spot, varying according to the dry spot volume. This feature is included in the available simulation codes as LIMS or LCMFlot. It is not the exact solution but in many cases it is sufficient. Approach (2) is necessary for proper filling of the tows and dry spots, but usually in the macroscopic analysis can be replaced by the sink term.

The complexity of (2) and (3) is necessary for microvoids formation and mainly for the analysis at the mesoscale. If fiber tows would be modeled as porous regions, inclusion of capillary effects would involve determination of the homogenized (averaged) analog of the capillary pressure. It was proposed in Ref. 20 that the flow at this level should have three zones: zone I, where only capillary pressure is operative; zone II, where both (capillary and hydrodynamic) are important; and zone III (saturated zone), where only hydrodynamic pressure applies as demonstrated in Fig. 18.

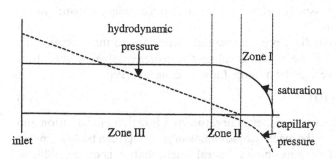

FIG. 18 Schematic displaying of the three zones of the flow at the mesolevel. (From Ref. 20.)

Averaged capillary pressure has been the subject of active research in many other fields such as oil reservoir industry [88], soil science, [4], transportation of pollutants in the environment [5], paper manufacturing [60], textiles industry, water source industry, and other areas of chemical or sanitary engineering. Very few studies relate it to composites manufacturing [89,90]. Unfortunately, other research fields deal with macroscopically isotropic porous media with different internal geometry than the architecture of fiber preforms. In addition, the liquid flow in other research areas is rarely viscous. Thus experimental data from these research studies are not directly relevant for composites manufacturing. However, the overall theoretical approaches relating the capillary pressure to the degree of saturation can be adopted.

V. CAPILLARY PRESSURE VS. SATURATION

Generally, the capillary pressure is a monotonically decreasing function of the only variable, which is saturation. The graph of this function is sometimes called the capillary or the saturation curve. Naturally, saturation curves are specific to specific porous media and the impregnating liquids. Regarding the hysteresis in interfacial tension, saturation curves must also exhibit the effect of hysteresis. As pointed out in the previous section, in composites manufacturing, only the region that is partially saturated can be studied, because it can be assumed that once the preform is fully saturated, it continues to remain saturated. This partially saturated region exists only near the flow front and can also apply in the region where the resin-starved regions such as dry spots and voids are undergoing redistribution of resin as a result of wicking effects. Methods of determination of the saturation curve in composites manufacturing have so far not been systematically studied. Some experimental results and theoretical discussions are presented in Refs. 73 and 91. In Ref. 73, the theoretical predictions from Ref. 87 are generalized. As mentioned in the previous section, experimental results from other research areas cannot be directly applied to composites. On the other hand, few theoretical studies that have addressed this issue could be adopted to relate capillary pressure to the degree of saturation.

First of all, the capillary pressure should correspond to the average of its microscopic values while liquid is filling the representative volume, thus generally we can average the Young–Laplace equation [19]:

$$P_c(s) = 2\gamma_{LA}\langle H \rangle(s), \tag{18}$$

where $\langle H \rangle(s)$ is the averaged mean curvature as a function of saturation and we used capital P_c for the macroscopic analog of p_c. $\langle H \rangle(s)$ can be different for different wicking directions and its general determination is quite a difficult problem.

The simplest approaches estimate the capillary pressure in porous media by adapting the Young–Laplace equation [Eq. (8)], assuming more or less uniform flow front with no voids behind the front [92]. Thus there is no dependence of P_c on the saturation. Either the relation proposed by Carman, or the notion of the hydraulic radius, or the direct use of Eq. (8) leads to [17,63,93]:

$$P_c = \frac{S\gamma_{LA}\cos\theta}{\phi}, \; P_c = \frac{\gamma_{LA}\cos\theta}{m}, \; P_c = \frac{4\gamma_{LA}\cos\theta}{D_e}, \tag{19}$$

S is the ratio of the surface area of the medium to its pore volume and D_e stands for equivalent diameter of the pores. The hydraulic (mean) radius m is defined as the ratio of the average pore cross-sectional area to the average wet perimeter, in line with the concept of the equivalent loads (as explained in Section III). All the geometrical parameters from Eq. (19) can be estimated for particulars of the porous media. For example, in the case of aligned fibers, hydraulic radius and equivalent diameter can be expressed by:

$$m = \frac{d_f\phi}{\alpha(1-\phi)} \text{ and } D_e = \frac{4d_f}{F} \cdot \frac{\phi}{1-\phi}, \tag{20}$$

where d_f is the single fiber diameter. In the above relation, $\alpha = 2$ corresponds to the flow across the fibrils, while $\alpha = 4$ to the flow along the fibrils. The latter expression is presented in Ref. 93 and F is the shape factor. For flow along the tows, the interface can be approximated by spherical cup, thus $F = 4$; for flow across the tows, the approximation is cylindrical, so $F = 2$ (compare with the Young–Laplace Equation [8]).* Obviously, both expressions in Eq. (20) yield the same capillary pressure.† On the other hand, the first relation from Eq. (19) is suitable only for isotropic media. For aligned fibers, we cannot obtain different values for different flow directions; the parameter obtained will always correspond to the flow along the fibers. Regarding woven fabrics, the capillary pressure as a function of porosity is presented in Ref. 93.

Some models were developed to account for the dependence on saturation. Among them, capillary models should be mentioned. We slightly modify the model presented in Ref. 17. It could be assumed that the porous medium is composed of straight circular capillaries of diameter δ and that a spherical cap

---

* Thus it is proposed that the capillary pressure should be twice as high for flows along the fibrils than across the fibrils.

† We would point out that when liquid flows along fibrils, the contacting surface coincides with the flow direction. When flowing across fibrils, the contacting surface is curved; consequently, $\cos\theta$ in Eq. (19) should be replaced by $\beta\cos\theta$, where $\beta = \pi/4$, which further decreases the capillary pressure.

FIG. 19   Relation between the diameter and the radius of the curvature for spherical and cylindrical approximations.

can approximate the flow front curvature. Then, the capillary pressure in a single capillary is given by (Fig. 19):

$$p_c = \frac{2\gamma_{LA}}{R} \text{ with } R = \frac{\delta}{2\cos\theta} \tag{21a}$$

yielding

$$p_c = \frac{4\gamma_{LA}\cos\theta}{\delta}. \tag{21b}$$

If liquid displaces the air and if it wets the solid, then the smaller capillaries will be filled first and the intermediate saturation can be expressed as [$\alpha(\delta)$ is the probability distribution function of the capillary with diameter δ]:

$$s = 1 - \int_{\delta=4\gamma_{LA}\cos\theta/p_c}^{\infty} \alpha(\delta)d\delta, \tag{21c}$$

giving the desired relation between the capillary pressure and saturation. This is obviously a very rough estimate and could be improved if more geometrical characteristics about the porous medium would be known.

   Another attempt was made in 1941 by Leverett, who, using a semiempirical approach, proposed the following expression for the capillary pressure, applicable to isotropic porous media

$$P_c(s) = J(s)\gamma_{LA}\sqrt{\frac{\phi}{K}}. \tag{22}$$

In Eq. (22), K and ϕ represent the permeability and porosity, respectively. No particular forms were given for the function $J(s)$, only certain shapes for a selected group of porous materials.

   Generally, in liquid–air system, the capillary curve resembles the shape given in Fig. 20. Usually, the pressure values are well defined only in interval $[s_r,1] \subset [0,1]$, where s_r is the residual saturation,* because in the exsorption

---

* There is a hysteresis in this value as well; sometimes an initial saturation s_i is used for the other value because it is in fact the initial value for the absorption (imbibition) process.

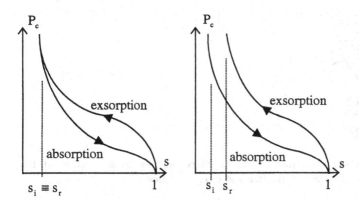

FIG. 20 Typical liquid–air capillary pressure curve without (left) and with (right) hysteresis in the residual saturation.

(drainage) process, there usually remains some liquid, which cannot be expelled. In two-liquid saturation curve, the saturation 1 is also unreachable, but this is not important for our purpose. In LCM processes, the resin is able to completely displace the air. From the shape of the curve in Fig. 20, it is obvious why the dry spots, which can be eliminated only by the wicking flow, are so fatal. The reason is that the capillary pressure rapidly decreases with the saturation thus the wicking action becomes weaker and eventually ceases.

Saturation curves suitable for LCM processes have not been studied systematically so far. Apparently the only theoretical attempt given in Ref. 87 for flow across the fibrils was generalized in Ref. 73 for the case of arbitrary contact angle of the resin–air–fiber system, $\theta \in [0, \pi]$. Following Han [73], let a medium corresponding to the highest dense cubic (in-line) packaging of circular straight fibers be given. If the wicking flow takes place in the plane across the fibers and in radial direction, and if the interface curvature is approximated by a circular cylinder, the saturations in terms of the geometrical angles α and β is given by (see Fig. 21):

$$s = \frac{\sin\alpha - \alpha + \xi\sin\alpha\sin\beta - \xi^2\beta + \xi^2\sin\beta\cos\beta}{1 - \pi/4}, \tag{23a}$$

where $\xi = R/r$, $1/R$ being the resin surface curvature and r the fiber radius. Two more equations, to substitute for the angles α and β, are needed. They are:

$$\xi\sin\beta = 2\sin^2(\alpha/2) \quad \text{and} \quad \alpha + \beta = \pi/2 - \theta. \tag{23b}$$

Because the contact angle keeps its value fixed, the interface curvature varies with saturation. Then, Eq. (23a) defines the relation between the saturation

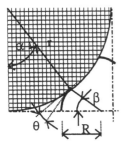

FIG. 21 Analytical derivation of the capillary pressure as a function of the saturation for the flow across the fibers in radial direction.

and the interfacial curvature, which by adopting Eq. (8) yields the dependence of the capillary pressure on the saturation. Generalization of Eq. (23a) to more general fiber geometries is difficult; thus experimental results cannot be completely substituted by theoretical predictions.

A general constitutive relation proposed by Han [73] for LCM processes is expressed as:

$$P_c(s) = \frac{a}{(s-b)^d},$$ (24)

where a, b, and d are constants; a is related to the interfacial tension, but generally all the constants depend on the particular geometry and should be experimentally determined. In Ref. 73, experimentally determined constants for bidirectional and random fiber mat are suggested.

Because surface tension and contact angle are two of the main parameters of the capillary pressure expression, capillary pressure curves are very sensitive to the state of the internal surface of the porous medium and apparently the experimental results under the same conditions can vary.

When it is not necessary to introduce the capillary pressure into the analysis and wicking can be modeled by some estimate or sink term, then it is sufficient to determine the rate of wicking. Various attempts have been made in this area so far and the theoretical predictions were usually experimentally verified by measuring the wicking rise against the gravitational forces. These approaches are: (1) based on the model of the porous medium as a bundle of capillary tubes (of uniform radius r) and exploiting Poiseuille law, at first presented in Refs. 1 and 2 and extended by Refs. 60, 62, and 63; (2) based on the Darcy's law [63,77,91]; and (3) based on the diffusion law [66]. Regarding (3), it should be pointed out that in these cases, sometimes discrete theories such as percolation theory are more appropriate than the continuous theories [19]. Generally, for LCM processes (1) and (2) are more suitable.

It was shown in Ref. 63 that Washburn's theory coincides with the approach exploiting the Darcy's law, if simply an appropriate relation is substituted for the permeability of the bundle of capillaries medium. Kozeny equation is suitable for this case.

Let us focus on one-dimensional wicking. Both approaches can be simultaneously reviewed. Flow is assumed to be laminar and inertia effects are neglected; hence flow laws for the two approaches can be respectively expressed as:

$$v = \frac{dh}{dt} = \frac{r^2}{8\mu} \cdot \frac{dp}{dh} \text{ and } v^D = \phi v = \frac{dh}{dt} = -\frac{K}{\mu} \cdot \frac{dP}{dh}. \tag{25a}$$

It is supposed that the only driving pressure is the capillary pressure and, moreover, because at the actual wicking position, h_a, the flowing velocity is spatially constant (because of the incompressibility condition), the capillary pressure gradient is also constant along the wetted length. In addition, capillary pressure at the wicking front is constant throughout the process; hence the pressure gradient and the wicking velocity vary along with time. The capillary pressure can be estimated from Eq. (19) as:

$$p_c \text{ or } P_c = \frac{\gamma_{LA}\cos\theta}{m} \text{ and } \frac{dp}{dh} = \frac{p_c}{h_a} \text{ or } \frac{dP}{dh} = \frac{P_c}{h_a}. \tag{25b}$$

If none of the other parameters from Eq. (25a) depend on the actual wicking position, it can be easily integrated, yielding

$$h(t) = \sqrt{t} \cdot \sqrt{\frac{r^2}{4\mu}p_c} = A\sqrt{t} \text{ and } h(t) = \sqrt{t} \cdot \sqrt{\frac{2K}{\mu}P_c} = A\sqrt{t}, \tag{25c}$$

where A is called the Washburn slope. Hydraulic radius for the Washburn medium is $r/2$ and thus the well-known relation for A can be written as:

$$A = \sqrt{\frac{r\gamma_{LA}\cos\theta}{2\mu}} \Rightarrow \frac{1}{\tau}\sqrt{\frac{r\gamma_{LA}\cos\theta}{2\mu}}. \tag{25d}$$

In the last expression, tortuosity $\tau > 1$ was added as a generalized factor to allow for a correction factor if the capillaries are not straight. Then, τh should be used instead of h [19]. Obviously, the other approach exploiting the permeability tensor offers better opportunity to account for the internal geometry of the porous medium because many theoretical predictions are already available for K, or permeability can be measured or calculated. However, the hydraulic radius must be estimated to properly account for the capillary pressure. Nevertheless, expressions valid for wicking along and across the fibers were already given in the beginning of this section [Eq. (20)]. Other estimates of the capillary pressure, exploiting equation based on the free energy changes, is presented in Ref. 91, where all these approaches are compared with each other and with experimental results for flow across the fibrils.

The dependence $h = A\sqrt{t}$ is called Washburn's law and A must be determined for a particular porous medium. The law was experimentally confirmed and is widely used. Discrepancies appear in the initial and final stages; thus the term Washburn regime is used for the region where the "linearity" is maintained (see Fig. 22). Reasons for the nonlinearities are summarized in Ref. 60 and they are not important for the composites manufacturing. In Ref. 62, modifications accounting for two- and three-dimensional wicking are suggested and verified. Other generalizations regarding the nonuniformity of the capillary radius are presented in Ref. 63. Because wicking rate depends on the interfacial tensions, it can be influenced by sizing and finishing that are applied to the fibers as discussed in Section III.

Another important fact is that microvoids will convect with the flow. To have an idea of maximum void size, which can be driven by the resin, let us assume an ellipsoidal shaped void as shown in Fig. 23 with the flow in the x direction. Following Ref. 73, the resin flow is approximately expressed as

$$v_x^D = -\frac{K_{xx}}{\mu} \cdot \frac{dP}{dx}. \tag{26a}$$

The void can only be driven by the flow, if the air pressure gradient inside it is equal or lower (it is negative) than the one corresponding to the resin flow. At the interfacial points 1 and 2, the capillary pressure is present and acts on both sides inside the void. Thus the pressure gradient is only caused by the

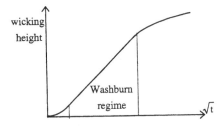

FIG. 22 Schematic plot of the Washburn's law.

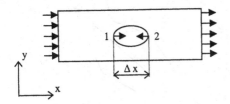

FIG. 23 Determination of the maximum void size, which can be driven by the resin flow.

hysteresis in the capillary curve. Naturally, at point 1, the wetting* value, P_{cw}, and at point 2 the nonwetting value, P_{cn}, is used. Thus:

$$\frac{v_x^D m}{K_{xx}} \leq \frac{P_{cn} - P_{cw}}{\Delta x} \Rightarrow \Delta x \leq \frac{K_{xx}(P_{cn} - P_{cw})}{\mu v_x^D}. \tag{26b}$$

A. Outlook on Modeling and Simulation of Partially Saturated Region

Let us examine the three possible ways to modify or extend the governing equations accounting for the proper modeling of the intermediate (partly saturated) region, which were given at the end of Section IV. To implement extension (1), one must eliminate the capillary pressure in Eq. (27a). Here we present how one can directly extend (2) with the help of an example case study.

Darcy's law and the continuity equation (which must now account for the continuous change of the saturation with time) take the form [87]:

$$\mathbf{v}^D = \frac{-k}{\mu} \mathbf{K} \cdot \nabla(P - P_c), \tag{27a}$$

$$\phi \frac{\partial s}{\partial t} = -\nabla \cdot \mathbf{v}^D. \tag{27b}$$

In Eq. (27a), k denotes the relative permeability, having values in closed interval $(0,1)$ and being a function of the degree of saturation. More about the relative permeability will be discussed later. It is useful to point out that if $s = 1$, then $P_c = 0$ and Eqs. (27a) and (27b) (in the fully saturated region) coincide with Eqs. (1) and (2). In numerical simulations, it is necessary to introduce an initial saturation, s_i, in the unfilled part, because the capillary pressure for zero saturation tends to infinity. Modifications of Darcy's law according to Eq.

---

*In general, wetting phase is the resin because it displaces the air; however, here, at the point 1, resin displaces the air but at the point 2, resin is displaced by the air and thus becomes the nonwetting phase. These terms will be specified later on in this section.

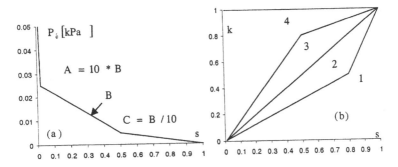

FIG. 24   (a) Capillary pressure variation. (b) Relative permeability variation.

(27a) are rarely used in the context of LCM. We can only mention Refs. 73, 93, and 94. Up to now, relative permeability has not been measured or studied for composites manufacturing; however, the term "unsaturated permeability" has been introduced and experimentally verified.

The influence of the relative permeability and capillary pressure distribution on the length of the intermediate region and on the pressure and saturation distribution in this region can be numerically studied in simple problems using the general-purpose finite element code ABAQUS. Moreover, in ABAQUS, the compressibility of the porous medium itself could be included into the analysis. However, more complicated problems related to LCM processes,* as, e.g., sharp permeability variations, would cause convergence difficulties because only backward time integration (unsuitable for the filling processes) is allowed in ABAQUS. The ABAQUS analysis is intended for soils industry (coupled pore fluid flow and stress analysis), where wicking is strong and the backward time integration gives the best results.

1.   Example Case Study

Let us present simple one-dimensional example of a porous medium 30-mm long being filled by constant inflow velocity 0.2 mm/sec. Parameters of the porous medium are $\phi = 0.5$, $s_i = 0.012$ (thus the theoretical filling time is 74.1 sec), $K = 5.37 \times 10^{-3}$ mm^2 corresponding, e.g., to the perpendicular flow across in-line packed cylindrical fibers of diameter 1.35 mm with distances between their centers being 1.7 mm. The resin viscosity is $\mu = 0.226$ Pa·sec and consequently the maximum pressure, when the specimen is fully filled, is 0.25 kPa. Three different capillary pressure variations were chosen according to Fig. 24a (case B is plotted and cases A and C are schematically explained) and

---

*ABAQUS was firstly used in LCM context by Wocke and van der Westhuizen [95].

four relative permeability distributions (cases 1–4) according to Fig. 24b were selected. Thus case A was chosen as the case where the maximum hydrodynamic and the minimum (is negative) capillary pressure at initial saturation are the same. Combining all the possibilities from Fig. 24, 12 different situations were obtained; however, cases C1 and C2 did not converge; from cases C3–C4 only C3 and from B cases B1 and B4 were examined further.

Mainly the influence of the relative permeability distribution and capillary pressure variation on the behavior of the intermediate region was studied. Relative permeability influence is significant for high capillary effect (case A), while it is almost negligible for low capillarity (cases B and C). On the other hand, the stronger the capillarity, the larger the intermediate region. These conclusions are well seen in Fig. 25, where for $t = 40$ sec, the pressure (Fig. 25a) and saturation (Fig. 25b) distribution along the specimen is shown for the seven cases. Obviously, pressure gradient in the fully saturated part is the same for all cases (it is proportional to the inlet velocity), and on the other hand, integral of saturation distribution is also the same for all cases (it is proportional to the total resin volume injected until that time).

According to the length of the intermediate region, the cases can be arranged in descending order as follows: A4, A3, A2, A1, B4, B1, and C3. This order is intuitively obvious and the primary influence is the capillary pressure distribution. Case A4 exhibits the longest intermediate region; it even exceeds the specimen. It is seen that at time 40 sec, the end of the specimen is already partly saturated, while the fully saturated region is very short. Passing from A4 to A1, the intermediate region is getting shorter, the saturation at the end of the specimen becomes lower, and for case A1 it just started to increase. Cases A1–A4 offer the comparison of the relative permeability influence for the same strong capillarity. Then, it is seen that for lower capillary effect (case B), the variation as a result of the relative permeability is not so significant (compare B4 and B1). The end of the intermediate region for B4 is again beyond the specimen, but now the saturation decreases much faster than for A4. When the capillary action is not very strong, the relative permeability variation does not cause big differences in the intermediate region behavior and the flow front becomes sharper. Cases C exhibit very short intermediate region, hence the convergence difficulties. In this case, one can ignore the capillary effects anyway. It should be pointed out that because of the different saturation distribution in the selected cases, the inlet pressure at the same time is very different, increasing according to the order specified above (Fig. 25a).

It is also interesting to see how the inlet pressure varies during the filling (Fig. 25c). In the saturated region, the functions are convex with different curvatures (arranged from bottom to top in the same order as before); however, the final value (when the specimen is completely filled) must be the same

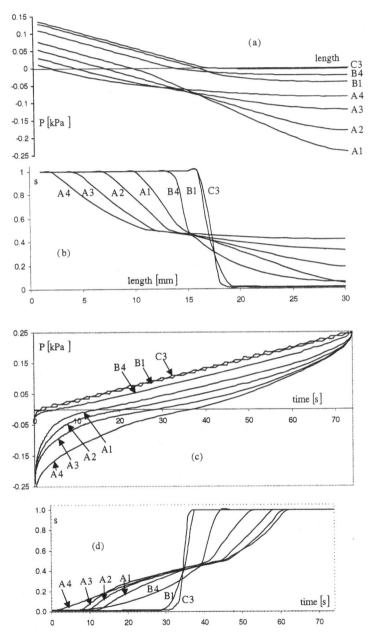

FIG. 25 (a) Pressure distribution along the specimen at $t = 40$ sec. (b) Corresponding saturation distribution. (c) Pressure variation at the inlet. (d) Saturation variation at the middle point.

for all cases. Cases with low capillarity (B and C) are approaching the linear distribution typical for the standard approaches. We point out that the initial capillary pressure is different for cases A–C. Oscillations in the almost linear distributions are caused by numerical difficulties. Again, in the same order (from bottom to top), variations of saturation along the time can be plotted for the center of the specimen (Fig. 25d). It is seen that for cases A4 and B4, the saturation starts to increase almost immediately; however, B1 and C3 exhibit typical curves as from standard approaches, which ignore capillary effects.

This type of case study allows one to identify cases for which the intermediate region is significant and situations in which the standard approaches would be sufficient.

It is seen that the saturation distribution in the intermediate region (Fig. 24b) is rather convex than concave as proposed in Ref. 20 (compare with Fig. 18).

2. Extension (3)

To account for the air influence and the voids formation, two-phase immiscible fluid flow through porous media needs to be modeled [approach (3)]. The physical principles for such flows are outlined in Refs. 17 and 19. The governing equations for the continuum model (the discrete model is inappropriate for LCM processes) can be found in Refs. 17, 18, and 88. Usually, the advancing (displacing) phase is denoted as wetting phase and the displaced phase as the nonwetting one. In LCM processes, the wetting phase is the resin and the nonwetting phase is the air. This "wetting" has nothing to do with the contact angle; however, in our case, the resin is also wetting in the sense of covering the fibers. We are going to use the subscripts "w" and "n" for the wetting and nonwetting phase, respectively. The source or sink terms, dispersion/diffusion, and the gravitational forces are neglected. In addition, the porous medium itself is assumed rigid and again the filling process is taken to be isothermal. Under these conditions, the governing equations become

$$\mathbf{v}_w^D = -\frac{k_w}{\mu_w}\mathbf{K}\cdot\nabla P_w, \qquad \mathbf{v}_n^D = -\frac{k_n}{\mu_n}\mathbf{K}\cdot\nabla P_n,$$

$$\phi\frac{\partial(\rho_w s_w)}{\partial t} = -\nabla\cdot(\rho_w\mathbf{v}_w^D), \quad \phi\frac{\partial(\rho_n s_n)}{\partial t} = -\nabla\cdot(\rho_n\mathbf{v}_n^D), \tag{28}$$

$$\rho_w = \rho_w(P_w), \qquad \rho_n = \rho_n(P_n),$$

$$s_w + s_n = 1, \qquad P_n - P_w = P_c(s_w),$$

Here we retained the compressibility even for the resin. In Eq. (28), $\mathbf{K}$ represents the absolute (saturated) permeability, while k_w and k_n are the relative permeabilities of the wetting and nonwetting phases, respectively. They generally depend only on the saturation of the wetting phase, s_w, and have values in the closed interval (0,1). Sometimes notation $\mathbf{K}_w = k_w\mathbf{K}$ and $\mathbf{K}_n = k_n\mathbf{K}$ is used and $\mathbf{K}_w$, $\mathbf{K}_n$ are called effective permeabilities. For the sake of completeness, it can be mentioned that the terms $\mathbf{K}_w/\mu_w$, $\mathbf{K}_n/\mu_n$ are also referred to as mobilities of the respective phases.

The extension of Darcy's law in Eq. (28) to two-fluid systems was first suggested in 1936 by Muskat et al.; however, the capillary pressure term was first introduced by Leverett in 1941. The governing equations [Eq. (28)] were initially only a heuristic procedure suggested by the analogy with the single-phase flow without a solid theoretical background. Then, they were verified, e.g., by the local averaging methods (even in a more general form); the Whitaker approach is summarized in Ref. 19. Equation (28) makes sense only if the relative permeabilities can be defined in such a way that they are velocity and pressure independent. This fact was previously experimentally verified, but it was recently pointed out by Kaviany [19] that the dependence on the saturation is not sufficient and that for instance Marangoni effect must be negligible in this case. For LCM processes, Eq. (28) is sufficient. More about the physical aspects of the two-phase flows can be found in Refs. 17 and 19.

To use Eq. (28) to predict the flow behavior of the resin, the dependence of the relative permeabilities and of the capillary pressure on the saturation of the wetting phase must be known. The capillary pressure dependence was already discussed in this section. The first serious theoretical treatments of the relative permeability theory, however, irrespective of the LCM processes, can be found in Ref. 17. Typical curves, e.g., from oil industry are presented in Fig. 26. Numerical studies of the relative permeability in oil or soil mechanics are presented in Refs. 5 and 96. The simplest approximation would be the linear dependence on the saturation, yielding that the sum of the two relative permeabilities would be unity. It was experimentally confirmed that this sum is

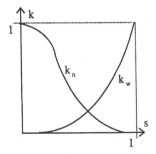

FIG. 26 Typical relative permeability curves.

less than unity for intermediate saturations; thus the relative permeability of the wetting phase is a nonlinear function of the saturation. The determination of relative permeabilities is even more complicated than determination of the capillary pressure curve because both (wetting and nonwetting) curves must be simultaneously studied and we do not have a simple starting point as Young–Laplace equation for the capillary pressure. Usually, exponential expressions are suggested with the exponent from unity to 7 [19]. Relative permeabilities are specific for the three phases; thus internal porous geometry and interfacial parameters strongly influence the curves. As a consequence, the relative permeability exhibits the phenomenon of hysteresis. Theoretical models for the determination of the relative permeabilities are given in Refs. 17 and 19.

There are some differences in general aspects of relative permeabilities for the fiber preforms, as noticed by Ref. 73. For instance, according to Fig. 26, the relative permeability of the wetting liquid at low saturations is almost zero, while in fiber preforms, it exhibits experimentally a finite value. Expressions for relative permeability from semiempirical approaches are as follows:

$$k_w = a(s - s_i)^n \text{ or } k_w = \frac{\int_0^s P_c^{-n} ds}{\int_0^1 P_c^{-n} ds}, \tag{29}$$

where a is a constant and n is a suitable integer. In numerical simulations of wicking flows presented in Ref. 73, the former relation was used, as it is similar to Eq. (24) used by them as well. They substituted $n = 1$ because for higher relative permeability, the capillary effects play a minor role and thus for the filling of dry spots by immersional wicking, this is the worst-case scenario.

Eqs. (28) are very difficult to solve, even numerically. The set can be simplified to the Buckely–Leverett formulation used in LCM processes by Pillai and Advani [77,86,91,92] and by Chui et al. [98]. Such formulation first of all neglects the capillary pressure and moreover both phases (air as well) has to be taken as incompressible; thus the Buckely–Leverett formulation is not suitable for our purpose.

Nowadays, numerical simulations of the LCM filling phase can already account for the wicking flow [73], for the voids and dry spots formation, and changes [73] (LCMFlot, LIMS) for the sink term (filling of the fiber tows) with capillary action included (LIMS or with the new name SLIMS) [77,86]. Thus the problem which is still not satisfactorily resolved is the correct determination of saturation curves and relative permeabilities for particular preform geometries.

VI. CONCLUDING REMARKS

In conclusion, one can definitely account for most of the macro level influences by ignoring the capillary effects in composites processing. However, if one wants to address the issues of interphase strengthening, dry spot wicking, and microvoid formation, one must include the capillary effects. We have presented how the existing approaches of capillary effects in porous media have been adopted from other research areas to address the needs in composite manufacturing processes. With the help of case studies at the micro and macro level, we have also suggested possible three levels of sophistication that could be added to the existing flow models in LCM processes to capture some of the physics of wicking and capillary action. However, as the level of sophistication increases, so does the number of required parameters that need to be characterized.

REFERENCES

1. Lucas, R. Veber das zeitgesetz des kapillaren aufstiegs von flussigkeiten. Kolloid-Z. 1918, 23, 15–22.
2. Washburn, E.W. The dynamics of capillary flow. Phys. Rev. 1921, 17, 273–283.
3. Krueger, J.J.; Hodgson, K.T. The relationship of single fiber wettability to sizing performance. TAPPI J. 1995, 78, 154.
4. Wu, T.H. *Soil Mechanics*; Allyn and Bacon, Inc.: Boston, 1966.
5. Keller, A.A. *Single and Multiphase Flow and Transport in Fractured Porous Media, PhD dissertation*; The Stanford University: California, USA, 1996.
6. Parker, J.C.; Lenhard, R.J. A model for hysteretic constitutive relations governing multiphase flow, 1. Saturation–pressure relations. Water Resour. Res. 1987, 23, 2187–2196.
7. Gutowski, T.G.P. Ed.; (1997). *Advanced Composites Manufacturing*; John Wiley & Sons Inc.: New York, 1997.
8. Larson, B.K. *Wetting, Spreading and Interface Formation in a Liquid Composite Molding Environment, PhD dissertation*; The Michigan State University: Michigan, USA, 1993.
9. ISL (Intelligent Systems Lab). *Tutorial on Polymer Composite Molding*; The Michigan State University: Michigan, USA, 1997 (http://islnotes.cps.msu.edu/ trp/rtm/inj_wet.html).
10. Advani, S.G.; Bruschke, M.V.; Parnas, R.S. Resin transfer molding. *Flow and Rheology in Polymeric Composites Manufacturing*. Advani, S.G., Ed.; Elsevier Publishers: Amsterdam, 1994; 465–516.
11. Bruschke, M.V.; Advani, S.G. A finite element/control volume approach to mold filling in anisotropic porous media. Polym. Compos. 1990, 11, 398–405.
12. Bruschke, M.V.; Advani, S.G. RTM: Filling simulation of complex three-dimensional shell-like structures. SAMPE Q. 1991, 22, 2–11.
13. Bruschke, M.V.; Advani, S.G. A numerical approach to model non-isothermal,

viscous flow with free surfaces through fibrous media. Int. J. Numer. Methods Fluids 1994, *19*, 575–603.

14. Lee, L.J.; Young, W.B.; Lin, R.J. Mold filling and cure modeling of RTM and SRIM processes. Compos. Struct. 1994, *27*, 109–120.

15. Liu, B.; Bickerton, S.; Advani, S.G. Modeling and simulation of resin transfer molding (RTM)—Gate control, venting, and dry spot prediction. Composites A 1996, *27A*, 135–141.

16. Gauvin, R.; Trochu, F.; Lemenn, Y.; Diallo, L. Permeability measurement and flow simulation through fiber reinforcement. Polym. Compos. 1996, *17*, 34–42.

17. Scheidegger, A.E. *The Physics of Flow Through Porous Media*; The Macmillan Company: New York, 1957.

18. Bear, J. *Dynamics of Fluids in Porous Media*; American Elsevier Publishing Company, Inc: New York, 1972.

19. Kaviany, M. *Principles of Heat Transfer in Porous Media* (2nd edition). Springer-Verlag, Inc.: New York, 1995.

20. Patel, N.; Rohatgi, V.; Lee, L.J. Microscale flow behavior and void formation mechanism during impregnation through a unidirectional stitched fiberglass mat. Polym. Eng. Sci. 1995, *35*, 837–851.

21. Gebart, B.R. Permeability of unidirectional reinforcements for RTM. J. Compos. Mater. 1992, *26*, 1100–1133.

22. Bruschke, M.V.; Advani, S.G. Flow of generalized Newtonian fluids across a periodic array of cylinders. J. Rheol. 1993, *37*, 479–498.

23. van der Westhuizen, J.; du Plessis, J.P. An attempt to quantify fibre bed permeability utilizing the phase average Navier–Stokes equation. Composites A 1996, *27A*, 263–269.

24. Simacek, P.; Advani, S.G. Permeability model for a woven fabric. Polym. Compos. 1996, *17*, 887–899.

25. Parnas, R.S.; Salem, A.J. A comparison of the unidirectional and radial in-plane flow of fluids through woven composite reinforcements. Polym. Compos. 1993, *14*, 383–394.

26. Wang, T.G.; Wu, C.; Lee, L.J. In-plane permeability measurement and analysis in liquid composite molding. Polym. Compos. 1994, *15*, 278–288.

27. Wu, C.; Wang, T.G.; Lee, L.J. Trans-plane fluid permeability measurement and its application in liquid composite molding. Polym. Compos. 1994, *15*, 289–298.

28. Parnas, R.S.; Salem, A.J.; Sadiq, T.A.K.; Wang, H.P.; Advani, S.G. The interaction between micro- and macroscopic flow in RTM preforms. Compos. Struct. 1994, *27*, 93–107.

29. Parnas, R.S.; Luce, T.; Advani, S.G.; Howard, G. Permeability character-ization, Part 1: A proposed standard reference material. Polym. Compos. 1995, *16*, 430–446.

30. Carter, E.J.; Fell, A.W.; Griffin, P.R.; Summerscales, J. Data validation procedure for the automated determination of the two-dimensional perme-ability tensor of a fabric reinforcement. Composites A 1996, *27A*, 255–261.

31. Mogavero, J.; Advani, S.G. Experimental investigation of flow through multi-layered preforms. Polym. Compos. 1997, *18*, 649–655.

32. Bickerton, S.; Advani, S.G. Experimental investigation and flow visualization of resin transfer molding process in a non-planar geometry. Compos. Sci. Technol. 1997, *57*, 23–33.

33. Bickerton, S.; Advani, S.G.; Mohan, R.V.; Shires, D.R. Experimental analysis and numerical modeling of flow channel effects in resin transfer molding. Polym. Compos. 1999, *20*, 12–23.

34. Ballata, B.; Walsh, S.; Advani, S.G. Measurement of the transverse permeability of fiber preforms. J. Reinf. Plast. Compos. 1999, *18*, 1450–1464.

35. Sanchez-Palencia, E. Fluid flow in porous media. In: *Non-Homogeneous Media and Vibration Theory*; Ehlers, J. Kippenhahn, R., Zittartz, J., Eds.; Lecture Notes in Physics. Springer-Verlag: Berlin, 1980; Vol. 127, 129–157.

36. Chang, W. *Modeling and Numerical Analysis of Composite Manufacturing Processes, PhD dissertation*; The University of Michigan: Michigan, USA, 1993.

37. Calado, V.M.A.; Advani, S.G. Effective permeability of multi-layer preforms in resin transfer molding. Compos. Sci. Technol. 1996, *56*, 519–531.

38. Ranganathan, S.; Phelan, F.R.; Advani, S.G. A generalized model for the transverse fluid permeability in unidirectional fibrous media. Polym. Compos. 1996, *17*, 222–230.

39. Dimitrovova, Z.; Faria, L. Finite element modeling of the resin transfer molding process based on homogenization techniques. Comput. Struct. 2000, *76*, 379–397.

40. Darcy, H. *Les Fontaines Publiques de la Ville de Dijon*; Dalmont: Paris, 1856.

41. Ene, H.I.; Polisevski, D. *Thermal Flow in Porous Media*; D. Reidel Publishing Company: Dordrecht, 1987.

42. Fracchia, C.A.; Castro, J.; Tucker, C.L. A finite element/control volume simulation of resin transfer mold filling, Proceedings of the American Society for Composites. 4th Technical Conference, Novel Processing Techniques-II, 1990; 157–166.

43. Cloete, T.J.; van der Westhuizen, J. An alternative approach to volume flux prediction during FEM modeling of RTM. Proceedings of the 1st South African Conference on Applied Mechanics, July; Midrand: South-Africa, 1996; 53–60.

44. Trochu, F.; Gauvin, R.; Zhang, Z. Simulation of mold filling in resin transfer molding by non-conforming finite elements. In *Computer Aided Design in Composite Material Technology III*; Advani, S.G. Blain, W.R., de Wilde, W.P., Gillespie, J.W., Griffin, O.H., Eds.; Computational Mechanics Publications, Southampton Boston and Elsevier Applied Science, 1992; 109–120.

45. Trochu, F.; Gauvin, R.; Gao, D.M. Numerical analysis of the resin transfer molding process by the finite element method. Adv. Polym. Technol. 1993, *12*, 329–342.

46. Gallez, X.E.; Advani, S.G. Resin infusion process simulation (RIPS): A fast method for three dimensional geometries, Proceedings of the 55th Annual Technical Conference, ANTEC, Part 2 (of 3) April–May, Toronto, Canada, Society of Plastics Engineers Brookfield, CT, USA, 1997; 2454–2458.

47. van der Westhuizen, J.; du Plessis, J.P. Numerical solution of the phase average Navier–Stokes equation using the finite element method. R&D J. 1996, *12*, 45–53.

48. Gupte, S.; Advani, S.G. Flow near the permeable boundary of aligned fiber preform. Polym. Compos. 1997, *18*, 114–124.

49. Gupte, S.; Advani, S.G. Non-Darcy flow near the permeable boundary of a porous medium: An experimental investigation using LDA. Exp. Fluids 1997, *22*, 408–422.

50. Hammami, A.; Gauvin, R.; Trochu, F. Modeling the edge effect in liquid composite molding. Composites A 1998, *29A*, 603–609.

51. Hammami, A.; Gauvin, R.; Trochu, F.; Touret, O.; Ferland, P. Analysis of the edge effect on flow patterns in liquid composite molding. Appl. Compos. Mater. 1998, *5*, 161–173.

52. Simacek, P.; Sozer, E.M.; Advani, S.G. *LIMS User's Manual, Version 4.0*; The University of Delaware: Delaware, USA, 1998 (http://barbucha.ccm.udel.edu/lims42/).

53. L3P (Liquid Process Performance Prediction), Inc., LCMFlot Software Web Site, http://www.l3p.qc.ca/lcmflot/, 1999.

54. Papathanasiou, T.D. A structure oriented micromechanical model for viscous flow through square arrays of fibre clusters. Compos. Sci. Technol 1996, *56*, 1055–1069.

55. Dimitrovova, Z.; Faria, L. Finite element modeling of the RTM process. Proceedings of the 5th International Conference on Flow Processes in Composite Materials, July, Plymouth, UK, 1999; 125–135.

56. Phelan, F.R.; Wise, G. Analysis of transverse flow in aligned fibrous porous media. Composites A 1996, *27A*, 25–33.

57. Spaid, M.A.A.; Phelan, F.R. Lattice Boltzman method for modeling microscale flow in fibrous porous media. Phys. Fluids 1997, *9*, 2468–2474.

58. White, F.M. *Fluid Mechanics*; 3rd Ed.; McGraw-Hill, Inc.: New York, 1994.

59. Adamson, A.W. *Physical Chemistry of Surfaces*; 5th Ed.; John Wiley & Sons, Inc: New York, 1982.

60. Hodgson, K.T. *The Role of Single Fiber Wettability in Wicking in Fiber Networks, PhD dissertation*; The University of Washington: Washington, USA, 1986.

61. Concus, P.; Finn, R. Capillary surfaces in microgravity. In: *Low-Gravity Fluid Dynamics and Transport Phenomena*; Koster, J.N. Sani, R.L., Eds.; Progress in Astronautics and Aeronautics. American Institute of Aeronautics and Astronautics, Reston, VA, USA, 1990; Vol. 130, 183–205.

62. Hedvat, S. *Capillary Flow Studies on Fibrous Substrates, PhD dissertation*; The Princeton University: New Jersey, USA, 1980.

63. Kim, J. *Studies on Liquid Flow in Fibrous Assembles, PhD dissertation*; The Princeton University: New Jersey, USA, 1987.

64. Greco, P.D. *A Study into the Effect of Wicking due to Capillary Action During the Curing of Epoxy Resins, Master thesis*; The University of Massachusetts Lowell: Massachusetts, USA, 1997.

65. Legros, J.C.; Dupont, O.; Queeckers, P.; Van Vaerenbergh, S.; Schwabe, D. Thermohydrodynamic instabilities and capillary flows. In: *Low-Gravity Fluid*

Dynamics and Transport Phenomena; Koster, J.N. Sani, R.L., Eds.; Progress in Astronautics and Aeronautics. American Institute of Aeronautics and Astronautics, Reston, VA, 1990; Vol. 130, 207–239.

66. Nguyen, H.V.; Durso, D.F. Absorption of water by fiber webs: an illustration of diffusion transport. TAPPI J. 1983, *66*, 76–79.

67. Lundstrom, T.S.; Gebart, B.R. Influence from process parameters on void formation in resin transfer molding. Polym. Compos. 1994, *15*, 25–33.

68. Lee, M.C.H. Effects of polymer-filler adhesion on the properties of polychloroprene elastomers filled with surface-treated fillers. J. Appl. Polym. Sci. 1987, *33*, 2479–2492.

69. Lundstrom, T.S. Bubble transport through constricted capillary tubes with application to resin transfer molding. Polym. Compos. 1996, *17*, 770–779.

70. Lundstrom, T.S. Void collapse in resin transfer molding. Composites A 1997, *28A*, 201–214.

71. Potter, K. *Resin Transfer Molding*; Chapman & Hall: London, 1997.

72. Hughes, T.J.R. *The Finite Element Method. Linear Static and Dynamic Finite Element Analysis*; Prentice-Hall International, Inc, Englewood Cliffs, NJ, 1987.

73. Han, K. *Analysis of Dry Spot Formation and Changes in Liquid Composite Molding, PhD dissertation*; The Ohio State University: Ohio, USA, 1994.

74. Chan, A.W.; Morgan, R.J. Modeling preform impregnation and void formation in resin transfer molding of unidirectional composites. SAMPE Q. 1992, *23*, 48–52.

75. Chan, A.W.; Morgan, R.J. Tow impregnation during resin transfer molding of bi-directional nonwoven fabrics. Polym. Compos. 1993, *14*, 335–340.

76. Binetruy, C.; Hilaire, B.; Pabiot, J. The interactions between flows occurring inside and outside fabric tows during RTM. Compos. Sci. Technol. 1997, *57*, 587–596.

77. Pillai, K.M.; Advani, S.G. A model for unsaturated flow in woven fiber preforms during mold filling in resin transfer molding. J. Compos. Mater. 1998, *32*, 1753–1783.

78. Parnas, R.S.; Phelan, F.R. The effect of heterogeneous porous media on mold filling in resin transfer molding. SAMPE Q. 1991, *22*, 53–60.

79. Sadiq, T.A.K.; Advani, S.G.; Parnas, R.S. Experimental investigation of transverse flow through aligned cylinders. Int. J. Multiph. Flow 1995, *21*, 755–774.

80. Spaid, M.A.A.; Phelan, F.R. Modelling void formation dynamics in fibrous porous media with the Lattice Boltzman method. Composites A 1998, *29A*, 749–755.

81. Beavers, G.S.; Joseph, D.D. Boundary conditions at a naturally permeable wall. J. Fluid Mech. 1967, *30*, 197–207.

82. Saffman, P.G. On the boundary condition at the surface of a porous medium. Stud. Appl. Math. 1971, *50*, 93–101.

83. Ross, S.M. Theoretical model of the boundary condition at a fluid–porous interface. AIChE J. 1983, *29*, 840–846.

84. Parnas, R.S.; Cohen, Y. Coupled parallel flows of power-law fluids in a channel and a bounding porous medium. Chem. Eng. Commun. 1987, *53*, 3–22.

85. Patel, N.; Lee, L.J. Effects of fiber mat architecture on void formation and removal in liquid composite molding. Polym. Compos. 1995, *16*, 386–399.
86. Pillai, K.M.; Advani, S.G. A numerical simulation of unsaturated flow in woven fiber preforms during resin transfer molding process. Polym. Compos. 1998, *19*, 71–80.
87. Collins, R.E. *Flow of Fluids Through Porous Media*; Reinhold Publishing Corporation: New York, 1961.
88. Peaceman, D.W. *Fundamentals of Numerical Reservoir Simulation*; Elsevier Scientific Publishing Company: New York, 1977.
89. Dave, R. A unified approach to modeling resin during composite processing. J. Compos. Mater. 1990, *24*, 22–41.
90. Skartis, L.; Khomami, B.; Kardos, J.K. The effect of capillary pressure on the impregnation of fibrous media. SAMPE J. 1992, *28*, 19.
91. Pillai, K.M.; Advani, S.G. Wicking across a fiber-bank. J. Colloid Interface Sci. 1996, *183*, 100–110.
92. Pillai, K.M.; Advani, S.G. Numerical and analytical study to estimate the effect of two length scales upon the permeability of a fibrous porous medium. Transp. Porous Media 1995, *21*, 1–17.
93. Ahn, K.J.; Seferis, J.C.; Berg, J.C. Simultaneous measurements of permeability and capillary pressure of thermosetting matrices in woven fabric reinforcements. Polym. Compos. 1991, *12*, 146–152.
94. Williams, J.G.; Morris, C.E.M.; Ennis, B.C. Liquid flow through aligned fiber beds. Polym. Eng. Sci. 1974, *14*, 413–419.
95. Wocke, C.; van der Westhuizen, J. RTM: Mould fill simulation using ABAQUS Standard. Proceedings of the SA Finite Element Conference, FEMSA95, 1995; 741–755.
96. Honarpour, M.; Koederitz, L.; Harvey, A.H. *Relative Permeability of Petroleum Reservoirs*; CRC Press, Inc., 1986.
97. Chui, W.K.; Glimm, J.; Tangerman, F.M.; Jardine, A.P.; Madsen, J.S.; Donnellan, T.M.; Leek, R. Process modeling in resin transfer molding as a method to enhance product quality. SIAM Rev. 1997, *39*, 714–727.

6

Surfactant Solution Behavior in Quartz Capillaries

N. V. CHURAEV Russian Academy of Sciences, Moscow, Russia

Surfactants are widely used to control wetting, capillary penetration, and evaporation. Adsorption of surfactants accelerates mass transfer processes, such as impregnation of hydrophobic porous bodies by aqueous solutions, cleaning of greasy oiled surfaces, and crude oil recovery. Surface modification of adsorbents, membranes, and catalysts by surfactants is often used to control their properties.

Processes of liquid–gas and liquid–liquid displacement are of practical importance in chemical technology. They have been intensively investigated both experimentally and theoretically using model porous bodies and networks, channels with variable cross section, or periodically shaped capillaries. However, in these cases it is difficult to separate the geometrical effects from physicochemical ones associated with adsorption, dynamic interface tension, and dynamic contact angles of moving menisci. For investigation of the latter effects, which determine the rates of spontaneous and forced displacement, it is preferable to use cylindrical capillaries with molecularly smooth surface having radii on the order of micrometers to avoid the gravity forces.

I. MATERIALS AND METHODS

Thin capillaries were prepared using a method of high-speed stretch of a melted quartz tube 0.5 mm in inner diameter. The rotating tube with thick walls was heated locally using acetylene–oxygen burners [1]. Quartz capillaries were drawn from high-purity quartz tubes (more than 99.99% SiO_2) previously treated with 20% fluoric acid to remove the surface layers of tubes enriched by ionic admixtures. After that the tubes were washed with triply distilled water up to neutral reaction. The freshly drawn, up to 5 m long, thin quartz capillary was cut into pieces 10–15 mm in length, which are stored with sealed ends.

The earlier developed experimental device [2] is shown in Fig. 1. A capillary with length L of about 15 cm is glued with epoxy resin onto the walls of two high-pressure chambers. The sealed ends of the capillary were broken just before experiments. The capillary ends are placed into the vessels containing liquids under investigation. The left vessel may be shifted back or forth (using a sylphon) to bring the capillary end in contact with a solution. The vessels contain two nonmiscible liquids or a liquid and gas. In the latter case, the left vessel is empty or shifted back. The cell containing the capillary is covered with transparent glass plates for observing microscopically the position of the meniscus and measuring the rate of its motion. A pressure difference in the chambers, $\Delta P = P_1 - P_2$, is measured using a mercury manometer. The nitrogen pressures P_1 and P_2 in the chambers could be maintained and regulated separately. The experiments were performed in a thermostatted room.

Capillaries with radii r from 1.75 to 20 μm have been used. Because of high capillary pressure, $P_c = 2\gamma_{12} \cos \theta/r$, the latter may be measured very precisely. Here γ_{12} is the interface tension and θ is the contact angle. Capillary radii were assessed preliminarily using an optical microscope (from the side view in immersion) and after that determined more precisely by the Poiseuille equation measuring the flow rate v of a liquid through the capillary

$$v = r^2 \Delta P/8\eta L, \tag{1}$$

where ΔP is the pressure drop applied to the capillary ends and η is the bulk viscosity. Liquids that completely wet the capillary surface ($\theta = 0$) and having well-known viscosity must be used in this case.

The second way to determine the r values consists in measuring the rate of meniscus movement in a capillary partly filled with liquid under an applied

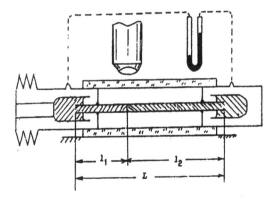

FIG. 1 Schematic representation of the experimental device.

pressure difference. According to the Washburn equation [3], the rate of meniscus movement in this case is equal

$$v = r^2(\Delta P + P_c)/8\eta l, \tag{2}$$

where l is the length of the liquid column and P_c is the capillary pressure of a meniscus.

In this case, the left part of the capillary (Fig. 1) contains nitrogen. When a liquid that completely wets the capillary is used, the capillary pressure $P_c = 2\gamma/r$, where γ is the surface tension, remains constant during meniscus motion. Depending on pressure difference, the direction of meniscus movement could be changed from an advancing ($v > 0$) at $\Delta P < P_c$ to a receding one ($v > 0$) at $\Delta P > P_c$. The rate of the meniscus movement v was determined measuring time τ of its travel between two marks on the microscope scale of a horizontal comparator IZA-2. The position of the meniscus was determined with an accuracy of 1 μm. The device shown in Fig. 1 was placed on a movable plate of the comparator. The rates v of the meniscus back and forth motion were measured within a small part of the capillary, the length of which ($\Delta l < 1$ mm) is much smaller than the column length l. In this case, the l value in Eq. (2) may be considered as constant, and therefore the rate of meniscus displacement v, according to this equation, depends linearly on ΔP.

The results obtained are shown, as an example, in Fig. 2. The linear dependencies $v(\Delta P)$ intersect the pressure axis at a capillary pressure value P_c. In the case of complete wetting, this allows one to determine the capillary radius from the Laplace equation $r = 2\gamma/P_c$, where γ is the surface tension of the liquid. As follows from Fig. 2, the used quartz capillary was completely wetted by triply distilled water, and no hysteresis was observed. The data shown in Fig. 2 relate to two different lengths of water column in the capillary, $l = 9.28$ cm (curve 1) and $l = 5.28$ cm (curve 2). The radii calculated from the capillary pressure values P_c are equal to $r = 5.69$ and 5.63 μm, respectively, assuming $\gamma = 72.58$ mN/m for water at room temperature, 19°C.

Radii of these two capillaries may be determined also from the slope of the graphs, $dv/d(\Delta P) = r^2/8\eta l$ using the viscosity of water $\eta = 1.03$ cPs at the same temperature. Calculated values are equal to 5.63 and 5.66 μm, respectively. The last method gives the r values averaged over the length l. The results are close to the determined values using the first method. Some difference between the calculated r values may be caused by very small, but finite conicity of the capillaries, ranged usually from 10^{-6} to 10^{-7}. The accuracy of capillary radii determination is better than 1–2%.

Hydrophobic quartz capillaries with methylated surface were prepared using two methods. In the first case, the capillaries with open ends were stored at 250°C in an evacuated chamber in an atmosphere of trimethylchlorosilane vapor for 12 h. The second method consisted in pumping 5% trimethylchlor-

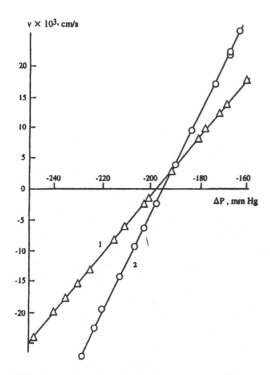

FIG. 2 Calibration of a thin quartz capillary by measuring rates of meniscus motion v in dependence on applied pressure difference ΔP. The length of water column in the capillary was equal to $l = 9.28$ cm (curve 1) and 5.28 cm (curve 2); $t = 19°$C.

osilane solution in benzene through a capillary during several hours and consecutive washing of the capillary with pure benzene and drying. Advancing contact angles of water in hydrophobed capillaries are in the range from 90° to 110° and the receding angles from 70° to 85°.

A. Immiscible Liquid–Liquid Displacement in the Absence of Surfactants

Spontaneous and forced displacement of oil–water systems was studied previously in the absence of surfactants [4–6]. Toluene, tetradecane, and dibutyl phthalate were used as oils. The main results have been summarized in a review paper [2].

At low rates of spontaneous displacement ($v < 5 \times 10^{-3}$ cm/s) dynamic advancing contact angles are close to their static values. In the case of forced displacement, at flow rates from 10^{-2} to 10^{-1} cm/s, thick nonequilibrium wetting films are formed after retreating meniscus. Their thickness, according to Derjaguin's theory [7], depends linearly on the parameter $(v\eta/\gamma^{12})^{2/3}$. Capillary counterpressure of the displaced liquid reaches in this case the maximum value $P_c = 2\gamma_{12}/r$ because $\theta_R \to 0$. At still higher flow rates, $v > 0.1$ cm/s, coalescence of dynamic films took place. This resulted in emulsification of the displaced liquid near the meniscus zone, which decrease the efficiency of the displacement.

In hydrophobed capillaries, advancing contact angles of the above-mentioned nonpolar liquids during a spontaneous displacement are smaller and also coincide with static values. The kinetics of forced displacement is quite different. This is caused by complete wetting of the capillary surface by nonpolar liquids and its nonwetting by water. As a result, in the case of forced displacement of water dynamic wetting films are not formed. Advancing contact angles of tetradecane, $\theta_A = 77°$, and of dibutyl phthalate, $\theta_A = 40°$ to $53°$, are constant in the range of flow rates up to 10^{-2} cm/s.

When nonpolar liquids are displaced by water, dynamic wetting films are formed that results in complete wetting of the capillary wall by a retreating meniscus. For viscous dibutyl phthalate $\theta_R = 0$. Thinner wetting films form in the case of less viscous tetradecane, which results in $\theta_R = 55°$. The amount of nondisplaced oil in hydrophobic capillaries decreases at very low rates of displacement by water when much thinner (equilibrium) wetting films of oils were formed.

II. NONIONIC SURFACTANT SOLUTION IN HYDROPHILIC CAPILLARIES

A. Aqueous Solution–Air Systems

At first aqueous solutions of a nonionic surfactant, oxyethylated isononyl-phenol (EO_{10}) with molecular mass $M = 666$ and critical micelle concentration (CMC) $= 0.01\%$, were used [8]. The experiments were performed at solution concentration $C_0 = 0.025\%$, much higher than CMC. The solution was prepared using triply distilled water with electrical conductivity of 10^{-6} Ω^{-1} cm^{-1} and pH 6.5. Surface tension of bulk solution equals $\gamma = 31$ mN/m at $t = 20°C$.

Two quartz capillaries of different radii were used in these experiments. Capillary radii were determined using water as a reference liquid on the basis of Eq. (2) at the column length $l = 7.75$ cm. For the first capillary, the intersection point of a linear $v(\Delta P)$ dependence with the pressure axis gives $P_c = 2\gamma/r = 218$ Torr, which results in $r = 5$ μm using the surface tension of water $\gamma = 72.7$ mN/m at room temperature (20°C). Nearly the same value, $r = 5.1$ μm, was determined from the slope of the linear graph $v(\Delta P)$. The latter value characterizes the mean radius averaged over the length l. For the second capillary, the values calculated from capillary pressure are $r = 1.76$ μm at $l =$

10.8 cm and 1.74 μm at $l = 3.8$ cm. Mean capillary radius determined from the slope of the $v(\Delta P)$ graphs is equal to $r = 1.75$ μm. The agreement between measured r values is also quite good.

Fig. 3 shows the results of spontaneous suction of the solution into the capillary having radius $r = 5$ μm, and of subsequent displacement of the solution under external pressure drop ΔP. The suction–displacement cycles are repeated 10 times. The distance of solution penetration, x, was nearly the same and is equal to about 9 cm. Curves 1 and 1′ correspond to experiments with pure water, and curves 2 and 2′ to the first cycle with EO_{10} solution. In both cases the dependencies of l^2 values on time τ are linear, in agreement with Eq. (2), when the latter was integrated over the l values after substitution $v = dl/d\tau$ and assuming $P_c = $ constant:

$$l^2 = r^2\tau(\Delta P + P_c)/4\eta \tag{3}$$

Values of wetting tension, $\gamma \cos \theta_A = rP_c/2 = 55$ mN/m, calculated from the slope of the graphs $l^2(\tau)$ for capillary suction, are the same for both pure

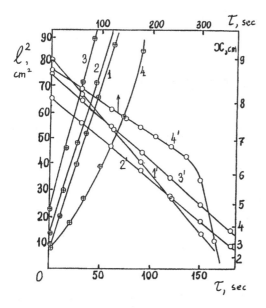

FIG. 3 Time dependence of the length l of a liquid column in the quartz capillary ($r = 5$ μm) obtained for water (curves 1 and 1′) and for nonionic EO_{10} surfactant solution (curves 2 and 2′ in the first cycle; 3 and 3′ in the third cycle; 4 and 4′ in the tenth cycle). The cycle consists in a spontaneous suction (curves 1 to 4) and subsequent forced displacement at 271, 255, 254, and 226 mmHg, respectively (curves 1′ to 4′).

water (curve 1) and the first cycle of solution suction (curve 2). This means that surfactant was completely adsorbed on capillary walls and the surface tension of the meniscus does not differ from water surface tension. Advancing contact angle of both water and solution is equal to $\theta_A = 40°$. In the course of displacement, the wetting tension of the retreating menisci of water (curve 1') and of the solution (curve 2') equals $\gamma \cos \theta_R = 72$ mN/m. Therefore, in the first cycle with solution the surface tension of the meniscus is not changed as compared with water, and $\theta_R = 0$.

In the course of consecutive cycles the situation changes gradually (curves 3 and 3'), and at the 10th cycle (curves 4 and 4') the kinetic changes and the $l^2 (\tau)$ dependencies become nonlinear. This leads to the conclusion that in thin capillaries, where the ratio of surface area to unit volume (proportional to $1/r$) is high, to obtain reproducible results the capillary surface must be preliminarily set in equilibrium with solution, thus avoiding uncontrollable loss of surfactant during capillary suction. In this regard, in all the following experiments capillaries were equilibrated with each solution under investigation by pumping the solution through the capillary 1–3 h before the experiments. The thinner the capillary, the longer must be the time of formation of an equilibrium adsorption layer on capillary walls. The necessary time of equilibration was estimated on the basis of adsorption capacity of quartz surface, and controlled experimentally [8].

Fig. 4 shows the results of measurements of flow rates v of the EO_{10} solution in dependence on an applied pressure drop ΔP in the first of the capillaries used ($r = 5\ \mu m$, $L = 15$ cm) when the capillary was filled up to the length $l = 10$ cm. Open points characterize the advancing ($v > 0$) and the receding ($v < 0$) motion of the meniscus (left scale of the v values). The flow rates are measured during a back-and-forth motion of the meniscus within a short portion of the capillary ($\Delta l < 1$ mm), when the length l of the column in Eq. (2) was considered to be constant. In the course of back motion, a wetting film is formed after retreating meniscus. As a result, during the subsequent advancing motion the meniscus meets the capillary surface covered with a thin equilibrium wetting film. Using the smearing-off method [9], the film thickness was assessed to be 3 to 5 nm. The method consists in measurement of a change in the length of a small column of the solution after shifting of the column on a defined distance inside the capillary. The film thickness was evaluated from the volume of the smeared-off liquid. The measurements were performed in a separate quartz capillary having the same radius.

Black points in Fig. 4 relate to the region of much lower flow rates (10^{-5} to 10^{-6} cm/s) shown on the right scale of v values. They correspond to open points situated on the pressure axis. This means that the meniscus is always movable, but the rate of motion decreases sharply in some range of pressure drops. The capillary pressures are in this case calculated by means of Eq. (2)

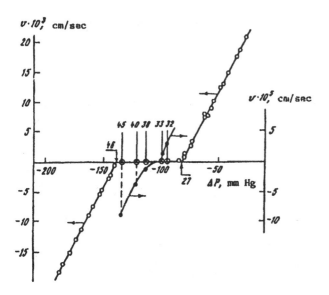

FIG. 4   Rates of advancing ($v > 0$) and receding ($v < 0$) motion of the meniscus of the EO_{10} solution ($C_0 = 0.025$ wt%) vs. pressure drop ΔP in a single quartz capillary, $r = 5$ μm. Length of water column $l = 10$ cm.

using measured low values of v (right scale in Fig. 4) and applied pressure drops ΔP. Calculated values of wetting tension $\gamma_m \cos \theta$ are indicated by figures near each point. The wetting tension changes in the region of very low flow rates gradually from 32 to 45 mN/m. Here γ_m is the dynamic surface tension of the meniscus.

The intersection point of the linear part of the graph for $v > 0$ with pressure axis gives, for advancing meniscus, the wetting tension $\gamma_m \cos \theta = 27.3$ mN/m, somewhat lower than the surface tension of bulk solution $\gamma_0 = 31$ mN/m. This means that the meniscus forms a dynamic advancing contact angle $\theta_A = 28°$. It is remarkable that the angle is not flow rate dependent, at least in the region of v from 1×10^{-3} to 2×10^{-2} cm/s.

Let us now consider the results obtained for retreating meniscus (Fig. 4, region of $v < 0$), when wetting tension $\gamma_m \cos \theta_R = 46$ mN/m. Assuming that in this case complete wetting takes place, the surface tension of retreating meniscus becomes equal to $\gamma_m = 46$ mN/m. This value corresponds to the solution concentration near the meniscus that is 10 times lower than in bulk solution. This is caused by transfer of surfactant molecules from the receding meniscus to the wetting film formed behind it. Similar effect of transfer of surfactant molecules from a receding meniscus to solid surface was observed earlier [10,11].

Fig. 5a gives a schematic representation of the processes that occur near a retreating meniscus of surfactant solution resulting in formation of a dynamic surface tension γ_m. In the steady state, the flux q_f of surfactant molecules from the meniscus onto the wetting film interface is compensated by the diffusion flux q_d of surfactant molecules to the meniscus surface from the solution inside the capillary. The flux q_f is assumed to be proportional to the flow rate v and the surface concentration of surfactant molecules Γ_f on the wetting film interface:

$$q_f = 2\pi r v \Gamma_f \ \text{(mol/s)}. \tag{4}$$

The adsorption Γ_f is regulated, in turn, by the surface diffusion flux q_s along the surface of a transition zone between meniscus and film

$$q_s = 2\pi r D_s (\Gamma_m - \Gamma_f)/\Delta_s, \tag{5}$$

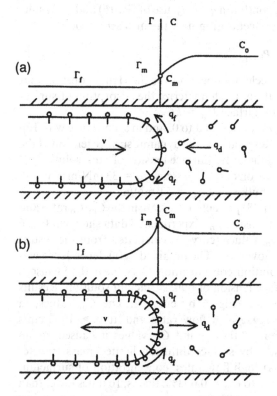

FIG. 5 Schematic representation of the processes of mass transfer of surfactant molecules in the near-to-meniscus zone: (a) retreating meniscus; (b) advancing meniscus.

where Γ_m is the adsorption on the meniscus, D_s is the coefficient of surface diffusion, and Δ_s is the corresponding diffusion length.

In the steady state $q_s = q_f$, and this results in an expression for the flux q_f:

$$q_f = 2\pi r v \Gamma_m / (1 + Pe_s), \tag{6}$$

where $Pe_s = v\Delta_s/D_s$ is the surface Peclet number.

The outgoing flux q_f from the meniscus is compensated for by the incoming diffusion flux to the meniscus surface

$$q_d = 2\pi r^2 D(C_0 - C_m)/\Delta, \tag{7}$$

where $2\pi r^2$ is the area of the spherical meniscus, C_0 and C_m are concentrations of surfactant solution inside the capillary and near the meniscus, respectively. Here D is the coefficient of diffusion in bulk solution, and Δ is the corresponding diffusion length.

Because in the steady-state condition $q_f = q_d$, use of Eqs (6) and (7) results in an expression for solution concentration near the meniscus zone:

$$C_m/C_o = \{1 + [kPe/r(1 + Pe_s)]\}^{-1}, \tag{8}$$

where $Pe = v\Delta/D$ is the bulk Peclet number and k is the Henry constant that was used, as a first approximation, to characterize the adsorption of surfactant molecules on the meniscus surface $\Gamma_m = kC_m$.

At very low flow rates, the Pe values tend to 0, and in accordance with Eq. (8) the ratio C_m/C_0 tends to 1. Respectively, the dynamic surface tension of the meniscus γ_m approaches the value of the surface tension of bulk solution γ_o. Fig. 4 shows that the surface tension of the meniscus, $\gamma_m = 33$ mN/m at $v \approx 0$, does not differ much from the bulk value, $\gamma_o = 31$ mN/m.

When flow rate increases, Eq. (8) predicts a decrease in the C_m/C_o ratio and an increase of dynamic surface tension γ_m. Experimental data shown in Fig. 6 confirm this prediction: the γ_m values (curve 1) calculated from the results shown in Fig. 4 increase with flow rates. The obtained $\gamma_m(v)$ dependence was used for calculation of the solution concentration $C_m(v)$ near the meniscus using the isotherm of the surface tension of bulk surfactant solution. Results of the calculations are shown in Fig. 6 by curve 2. The C_m values, in accordance with Eq. (8), decrease with flow rates and at $v > 10^{-4}$ cm/s become constant. Stabilization of the γ_m and C_m values is caused, as an analysis of Eq. (8) has shown, by surface diffusion limited mass transfer between meniscus and film. At high flow rates, concentration of surfactant solution near the meniscus drops to $C_m = 0.003\%$, that is, 10 times lower than bulk solution concentration $C_o = 0.025\%$.

From Eq. (8), it follows that at high flow rates the ratio C_m/C_o ceases to change and becomes a constant value

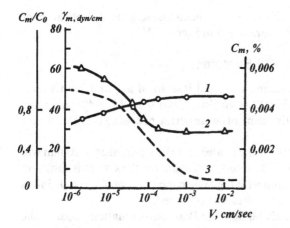

FIG. 6 Dependencies of dynamic surface tension of the retreating meniscus γ_m (curve 1) and corresponding concentration of EO_{10} surfactant solution near to the meniscus C_m (curve 2) on flow rate v in a quartz capillary, $r = 5\ \mu m$. By dotted line (curve 3) the results of calculation of the ratio C_m/C_0 using Eq. (8) are shown.

$$C_m/C_0 = [1 + (k\Delta D_s/r\Delta_s D)]^{-1} \tag{9}$$

As follows from this equation, the decrease in C_m/C_0 values is more pronounced in thin capillaries and at large k values that characterize adsorption for surfactant molecules on the solution–air interface.

Dotted curve 3 in Fig. 6 shows the results of calculations on the basis of Eq. (8) of the ratio C_m/C_0 in dependence on flow rate v. The results are in qualitative agreement with experimental data. When plotting the graphs, it was assumed that $\Delta_s = 5 \times 10^{-4}$ cm, $\Delta = 5 \times 10^{-3}$ cm, $D = 10^{-6}$ cm^2/s, and $D_s = 5 \times 10^{-7}$ cm^2/s. Henry's constant $k = 10^{-3}$ cm corresponds to the area per molecule, 0.02 nm^2. A quantitative difference between calculated and experimental data is caused not only by not well known values of the parameters entering Eq. (8), but rather by the simplified approach used when deriving this equation.

In a similar manner, the effect of reverse mass exchange in the case of advancing meniscus when the latter meets a wetting film (Fig. 5b) may be considered. The derived equation, because of adopted reversibility of the mass exchange, differs from Eq. (9) only by the sign of the term in parentheses:

$$C_m/C_0 = [1 - (k\Delta D_s/r\Delta_s D)]^{-1} \tag{10}$$

In this case, the concentration of surfactant solution near the meniscus, C_m, is higher than C_0 because the flux of surfactant molecules is directed from the wetting film interface to the advancing meniscus. Results of application of Eq. (10) will be discussed in more detail in Section IV.

B. Aqueous Solution–Oil Systems

In these experiments tetradecane was used instead of air, and both water-soluble (EO_{10}) and oil-soluble (EO_4 with molecular mass $M = 402$) nonionic surfactants were used [12]. Experiments were performed using the same device shown in Fig. 1.

First, the results obtained for the (water + EO_{10})–tetradecane system will be discussed. The isotherm of the interface tension for the system is shown in Fig. 7 by curve 1. EO_{10} solution with the same concentration $C_o = 0.025\%$ as in the case of the solution–air system was prepared.

For two viscous immiscible liquids, the Washburn equation acquires the following form:

$$v = r^2(\Delta P + P_c)/8(\eta_2 l_2 + \eta_1 l_1) = r^2(\Delta P + P_c)/8\eta_1 l_{1e} \tag{11}$$

where l_1 and η_1 are the length and viscosity of the replacing liquid, and l_2 and η_2 correspond to the liquid to be replaced. Here l_{1e} is the effective length

$$l_{1e} = l_1[(\eta_2 l_2/\eta_1 l_1) + 1]. \tag{12}$$

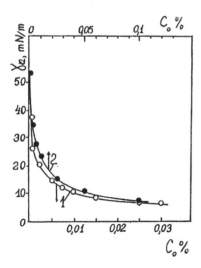

FIG. 7 Isotherms of interface tension of aqueous EO_{10} solutions in contact with tetradecane (curve 1) and of the EO_4 surfactant solution in tetradecane in contact with water (curve 2).

In this case Eq. (11) may be rewritten in another form:

$$vl_{1e} = r^2(\Delta P + P_c)/8\eta_1 = C, \tag{13}$$

from which follows that when capillary pressure is constant, the vl_{1e} values depend linearly on pressure drop ΔP and on the parameter C. When plotting graphs $vl_{1e}(\Delta P)$, the capillary radius r may be determined from the slope of the linear graph. Intersection of the graph with the pressure axis allows one to determine the capillary pressure of the meniscus

$$P_c = 2\gamma_{12}\cos\theta/r, \tag{14}$$

where γ_{12} is the interface tension and θ is the contact angle.

Eq. (13) was used when flow rates v are measured during reversible motion of a meniscus within a small portion of the capillary ($\Delta l < 1$ mm). In this case, the lengths l_1 and l_2 are considered to be practically constant.

The second method of treatment of experimental results is based on integration of Eq. (11) from l_{10} at $\tau = 0$ to l_1 at a given time τ after expressing the flow rate as $v = dl_1/d\tau$:

$$vl_{1e}\tau = C\tau = (l_1 - l_{10})\{(\eta_1/\eta_2)[L - (l_1 + l_{10})/2] + (l_1 + l_{10})/2\}$$

$$= B(\tau) \tag{15}$$

At $C = $ constant, the B values depend linearly on the time of displacement τ. This allows determination of the parameter C in Eq. (13) from the slope of the linear graph $B(\tau)$. For this purpose, a change in the position of the meniscus l_1 during time at a constant pressure drop ΔP was measured. Results of the measurements may be transformed into a $vl_{1e}(\Delta P)$ function using the relation $vl_{1e} = B/\tau$ and corresponding ΔP values.

The $B(\tau)$ graphs allow determination of the capillary pressure and wetting tension independently:

$$P_c = (8\eta_1/r^2)(B/\tau) - \Delta P \tag{16}$$

$$\gamma_{12}\cos\theta = P_c r/2 = (4\eta_1/r)(B/\tau) - (r\Delta P/2) \tag{17}$$

Fig. 8 shows the results obtained for a simpler case of water–tetradecane system without surfactant. In this case, the condition of constant interface tension γ_{12} (in contrast to surfactant solutions) is strictly fulfilled. Some changes in capillary pressure may be caused only by changes in contact angles. The measurements are performed in a quartz capillary $r = 4.45$ μm at 20 °C.

Fig. 8 illustrates both methods of treatment of the experimental results. On the left (Fig. 8a), the $B(\tau)$ dependencies obtained at three constant magnitudes of the pressure drop ΔP are shown. All the graphs are linear, corresponding to constant values of contact angle, at least in the range of flow rates $v < 10^{-2}$ cm/s. Graph 1 corresponds to spontaneous displacement of tetradecane by water (at $\Delta P = 0$) and graphs 2 and 3 to a forced displacement of water by

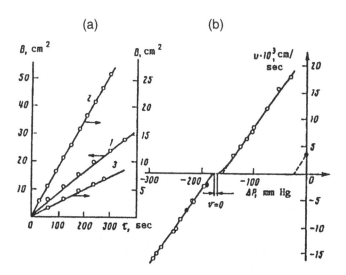

FIG. 8 The results obtained for the water/tetradecane system in a quartz capillary, $r = 4.45$ μm, at $t = 20°C$. (a) $B(\tau)$ dependencies obtained at $\Delta P = 0$ (curve 1), -233 (curve 2), and -189 mmHg (curve 3). (b) Dependence of flow rate v on pressure drop ΔP.

tetradecane at $\Delta P = -233$ and -189 mmHg, respectively. From Eq. (17) it follows that advancing contact angle in the case of spontaneous displacement is equal to $\theta_A = 82°$. Similar values were obtained earlier [2,4]. In the case of forced displacement, values of wetting tension of retreating water menisci γ_{12} $\cos \theta_R$ are equal to 51.5 and 51.2 mN/m, which gives $\theta_R = 15°$ and $13°$ at ΔP $= -189$ and -233 mmHg, respectively. When calculating contact angles, the interface tension $\gamma_{12} = 53$ mN/m taken from Fig. 7 (at $C_s = 0$) was used. The contact angles were calculated using Eq. (11) and bulk viscosity values for water (0.993 cPs) and for tetradecane (2.3 cPs).

Fig. 8b shows the experimentally obtained dependencies $v(\Delta P)$ when the meniscus was shifted within a small portion of the capillary back and forth at various values of the pressure drop ΔP. In contrast to the $B(\tau)$ graphs shown in Fig. 8a, when meniscus motion was not limited to a small portion of the capillary, the l_{1e} and l_1 values were practically constant. Positive values of v correspond to advancing motion of water, when $l_{1e} = 23.5$ cm and $l_1 = 3.5$ cm. Negative values of v correspond to receding motion of water, when $l_{1e} = 10.13$ cm and $l_1 = 8.61$ cm. From the slope of the graphs (the same in the cases $v < 0$ and $v > 0$), the viscosity of water and of tetradecane in thin capillary was calculated using Eq. (11). Calculated viscosities, 0.994 and 2.35 cPs, are close to bulk values. This shows evidence for the applicability of the Washburn equation in the form of Eq. (11) for viscous flow of immiscible liquids in thin capillaries.

Advancing contact angle $\theta_A = 27°$ was determined from the intersection point of the linear graph at $v > 0$ with the pressure axis. Receding contact angle determined in the same way for $v < 0$ is equal to $\theta_R = 9°$, somewhat smaller than that determined using $B(\tau)$ graphs. The difference may be related to different conditions of meniscus motion, namely, with higher wetting film thickness in the case when the meniscus passes many times over the same small portion of the capillary.

Three black points on the graphs $v(\Delta P)$ in Fig. 8b show the results obtained from three $B(\tau)$ graphs in Fig. 8a. In the case of receding motion of water ($v < 0$), the position of the black points coincide with the $v(\Delta P)$ graphs. However, for advancing motion ($v > 0$), the position of the black point is shifted to lower pressure values. The advancing contact angle is equal to $\theta_A = 82°$, similar with the above considered case of spontaneous displacement of tetradecane by water. This is associated with different conditions in the zone near the meniscus. In the case of spontaneous displacement, water meniscus contacts tetradecane directly. In contrast, during reversible meniscus motion inside a small portion of the capillary advancing meniscus meets thin wetting film of water that is retained on capillary walls under tetradecane after preceding retreating motion of water. Contact with water film substantially reduces the θ_A value.

Let us now consider the results of similar experiments when water contains nonionic surfactant EO_{10} at a concentration $C_o = 0.025\%$, much higher than the CMC. Equilibrium interface tension decreases in this case to $\gamma_{12} = 7$ mN/m (see Fig. 7). The results shown in Fig. 9 do not differ much from that obtained for the same system without surfactant (Fig. 8) because they were obtained for a capillary that was not equilibrated previously with surfactant solution. As a result, concentration near the meniscus was 100 times lower than C_o due to adsorption of surfactant on the capillary surface. Advancing contact angle $\theta_A = 80°$, calculated using the $B(\tau)$ graph (two black points in Fig. 9b), is close to the value calculated from Fig. 8b in the absence of surfactant.

The results obtained for the same ($r = 4.45$ μm), but equilibrated, capillary are shown in Fig. 10. The bulk solution was prepumped through the capillary during 1 h at a pressure drop of about 500 mmHg. Fig. 10a shows $B(\tau)$ dependencies, and Fig. 10b the $vl_{1e}(\Delta P)$ data, when the meniscus was shifted back and forth within a small portion of the capillary.

However, in this case the $B(\tau)$ dependencies could be obtained only when the surfactant solution displaces tetradecane [these data are shown by two black points on the $vl_{1e}(\Delta P)$ graph in Fig. 10b]. When tetradecane displaces the surfactant solution ($v < 0$), intense emulsification takes place just at low

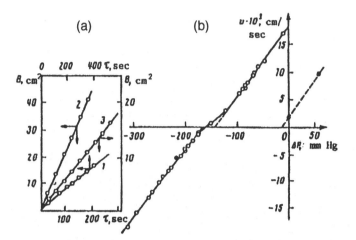

FIG. 9 The results obtained for (water + EO$_{10}$)/tetradecane system in a quartz capillary, r = 4.45 μm, at t = 20°C. Concentration of aqueous solution C_o = 0.025 wt%. (a) $B(\tau)$ dependencies determined at ΔP = 0 (curve 1), 60 (curve 2), and −214 mmHg (curve 3). (b) Dependence of flow rate v on pressure drop ΔP.

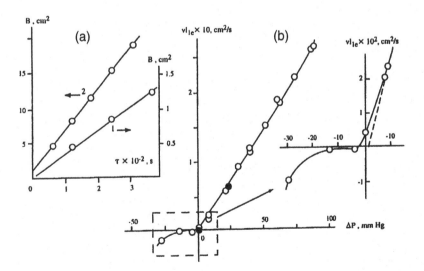

FIG. 10 The results obtained for the (water + EO$_{10}$)/tetradecane system in a quartz capillary, r = 4.45 μm, preequilibrated with 0.025 wt% EO$_{10}$ solution at t = 20°C. (a) $B(\tau)$ dependencies determined at ΔP = 0 (curve 1) and 23.5 mmHg (curve 2). (b) Dependence of the vl_{1e} values on pressure drop ΔP. The region of small flow rate is shown in a larger scale in the insertion.

flow rates of about 10^{-3} cm/s. In the absence of surfactants, emulsification was observed only at higher (by two orders of magnitude) flow rates [4,5]. In the region $v > 0$ (Fig. 10b), when the surfactant solution is displaced, tetradecane emulsion is not formed. The $vl_{1e}(\Delta P)$ graph is linear in the region of $v > 10^{-3}$ cm/s. Continuation of the linear part of the graph is shown in the insertion by dotted lines. A deviation from linearity is observed at $vl_{1e} < 0.02$ cm^2/s. The rates of spontaneous displacement at $\Delta P = 0$ are very small, 2×10^{-4} cm/s, but positive which corresponds to the dynamic advancing angle $\theta_A = 87°$. The value was calculated from the wetting tension $\gamma_{12} \cos \theta = 0.34$ mN/m, assuming $\gamma_{12} = 7$ mN/m as in the case of bulk solution. The static advancing angle calculated using P_c value at $v = 0$ is smaller, $\theta_{AS} = 84°$.

At higher flow rates, the intersection of the continuation of the linear part of the graph with the pressure axis (shown by dotted lines in the insertion) gives a negative wetting tension equal to -0.36 mN/m, which results in dynamic advancing angle of the solution $\theta_A = 93°$. This means that at high flow rates, tetradecane wet better, as compared with the solution, the capillary surface covered with adsorbed layer of surfactant molecules. Deviation of interface tension from the adopted bulk value may influence the magnitude of the contact angle, but not the sign of the wetting tension. For instance, when γ_{12} of the moving meniscus is assumed to be 20 mN/m due to some loss of surfactant as a result of adsorption on tetradecane film interface, the calculated advancing angle decreases only slightly to 91°.

In the next experiments, nonionic surfactant EO$_4$ was added to the tetradecane phase and water was free from surfactant. Concentration of EO$_4$ in tetradecane was equal to 0.025%, which corresponds to the interface tension with water, $\gamma_{12} = 15$ mN/m (Fig. 7, curve 2). Quartz capillary, $r = 3.2$ μm, was equilibrated with EO$_4$ solution in tetradecane by pumping the solution through the capillary during 3 h at a pressure drop of 600 mmHg.

The results obtained are shown in Fig. 11. The $B(\tau)$ functions shown in Fig. 11a are linear at all values of the pressure drop, from $+145$ to -190 mmHg. It should be noted that the sign of ΔP values depends on absolute values of P_1 and P_2 (Fig. 1). In the case of advancing motion on water meniscus ΔP values are positive, and in the case of receding motion are negative.

The position of black points shown on the $v(\Delta P)$ graph (Fig. 11b) were determined from the $B(\tau)$ dependencies. Coincidence of the points with $v(\Delta P)$ data shows that both methods of studying flow rates gave the same results. This also shows that surfactant added to the tetradecane phase does not influence the advancing contact angle of water, which is equal to $\theta_A = 105°$. The value was calculated from the wetting tension $\gamma_{12} \cos \theta = -3.8$ mN/m determined from the intersection point of the linear part of the graph (at $v > 0$) with the pressure axis ($P_c = -17$ mmHg) using the bulk value of interface tension. In the region of small flow rates, when the graph deviates from linear dependence, the θ_A value decreases, tending to 90° at $v = 0$. At nearly the

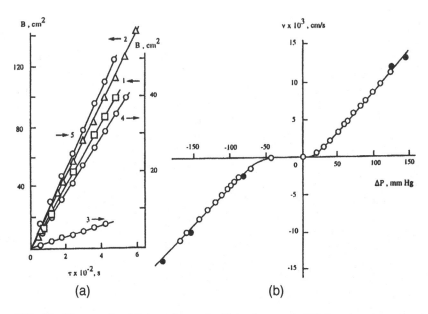

FIG. 11 The results obtained for water/(tetradecane + EO$_4$) system in a quartz capillary, $r = 3.2$ μm, preequilibrated with EO$_4$ solution in tetradecane, $t = 20°$C. (a) $B(\tau)$ dependencies determined at $\Delta P = +123$ (curve 1), $+145$ (curve 2), -69 (curve 3), -152 (curve 4), and -190 mmHg (curve 4). (b) Dependence of flow rate v on pressure drop ΔP.

same wettability of capillary surface by water and EO$_4$ solution in tetradecane, when the contact angle is close to 90°, receding motion of water starts at $\Delta P = -45$ mmHg that corresponding to $\gamma_{12} \cos \theta = 10.1$ mN/m and $\theta_{RS} = 48°$. Dynamic receding contact angle determined from the intersection point of the linear part of the graph at $v < 0$ with the pressure axis is lower, $\theta_R = 31°$.

As seen from Fig. 11b, the slopes of the graphs at $v > 0$ and $v < 0$ are the same. Calculated from the slopes, using Eq. (13), viscosity values of water, 0.97 cPs, and of tetradecane, 2.31 cPs, are close to the known bulk data.

Therefore, these experiments have shown that the measured flow rates are in agreement with the Washburn equation (11), which describes the viscous flow of immiscible liquids. The presence of surfactants influences only the values of wetting tension and contact angles.

C. Influence of Surfactants on Evaporation from Quartz Capillaries

Surfactants are often used to suppress or depress evaporation from a free liquid surface [13]. Surfactant monolayers formed on the water surface provide a barrier that molecules must overcome to transfer into the vapor from liquid phase. This is a kinetic effect that results in a decrease in vapor pressure Δp over an evaporating water surface covered with a surfactant adsorption layer. In the state of equilibrium, the vapor pressure is equal to that of saturated vapor over the surfactant solution. At low concentration of surfactant molecules in the bulk solution this pressure is practically the same as for pure water, p_s.

According to the Hertz–Knudsen theory, the drop of vapor pressure depends on the rate of evaporation v (cm/s):

$$\Delta p = (v\rho/\alpha_0)(2\pi RT/\mu)^{1/2} \tag{18}$$

where ρ is the density, μ is the molecular mass of water, and α_0 is the coefficient of condensation. Its value, $\alpha_0 \sim \exp(-\varepsilon/kT)$, depends on the height ε of the potential barrier. For pure water under normal conditions, $\alpha_0 = 0.04$–0.05, which corresponds to ε of about $3kT$. This is close to the mean energy of the intermolecular hydrogen bonds in water.

The presence of surfactant adsorption layers on the water surface sharply reduces, by several orders, the coefficient of condensation. Therefore, for the same conditions of evaporation as in the case of pure water, the vapor pressure drop Δp becomes significant, and the evaporation is controlled by kinetic rather than by diffusion factors. The rate of evaporation v becomes dependent primarily from the coefficient α_0. For example, $\alpha_0 = 3.5 \times 10^{-5}$ for a saturated monolayer of insoluble cetyl alcohol [14].

To study the effect of surfactants, rates of evaporation of a nonionic surfactant solution from single quartz capillaries with radii from 10 to 20 μm were measured [15]. The results obtained for evaporation of pure water (curve 1) and 0.25% solution of syntamide-5 (curve 2) from capillaries of equal radii, $r = 8.2$ μm, are shown in Fig. 12. Here L is the distance of evaporating meniscus from the open capillary end and t is the time. The capillaries filled with water and with surfactant solutions, respectively, were placed in an evacuated chamber [16] near each other. Curves 1 and 2 refer to evaporation in vacuum ($p/p_s = 0$) at $K = 5 \times 10^{-4}$ cm^2/s, where K is the coefficient characterizing the rate of evaporation. The coefficient $K = L^2/4t$ depends on external conditions of evaporation and is proportional to the difference between vapor pressure over meniscus, p_m, and in surrounding media, p_0. At first, the curves 1 and 2 practically coincide, but later on evaporation from the capillary filled with the surfactant solution gradually slows down. This can be explained by the concentration of surfactant molecules near the evaporating meniscus surface.

At less than half the value of the coefficient of evaporation ($K = 2 \times 10^{-4}$ cm^2/s), when higher relative vapor pressure $p/p_s = 0.54$ was maintained in the chamber, there was no marked difference in the rate of evaporation of pure water and aqueous surfactant solutions (curve 3). Probably, in this case

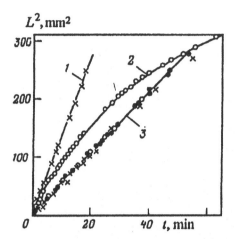

FIG. 12 Kinetics of evaporation $L^2(t)$ of water (curve 1) and syntamide-5 solution, $C_o = 0.25$ wt%, (curve 2) from a quartz capillary, $r = 8.2$ μm, at $p/p_s = 0$. The coinciding results of evaporation of water and syntamide-5 solutions ($C_o = 0.05$; 0.1 and 0.25 wt%) from a capillary, $r = 7.4$ μm, at $p/p_s = 0.54$ are shown by curve 3.

surfactant molecules effectively escaped from the meniscus into the bulk of the capillary by diffusion. Similar results were also obtained for lower rates of evaporation, at $p/p_s = 0.915$ and 0.95, when K values are on the order of 10^{-5}. No effect of syntamide-5 (at $C_s = 0.05$ and 0.25%) on the rate of evaporation from the capillaries into dry air, when K values were much lower ($K = 3.3 \times 10^{-6}$ cm^2/s), has been revealed.

Consequently, it must be concluded that the surfactant studied impede water evaporation only at a sufficiently high rate of evaporation, when K values are higher than 10^{-3} cm^2/s.

For further experiments, cetyl alcohol (CA), a surfactant insoluble in water whose adsorption on water surface significantly decreases the coefficient of condensation [14], was used. However, the problem arose of how to apply the surfactant on the meniscus surface in a narrow capillary. First, CA was adsorbed onto the inner capillary surface from its solution in ethanol. The capillary was filled with 0.5–1% solution, which was then displaced, retaining on the capillary surface a thin wetting film. After that the capillary was dried, filled with water, and sealed. Observation of evaporation was initiated after the capillary was broken inside the region filled with water, so that at the beginning there was no CA on the meniscus.

The result of such experiments are shown in Fig. 13. At first the rates of evaporation from pure capillary (curve 1) and from those pretreated with CA

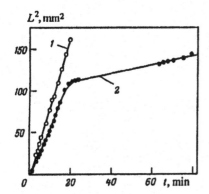

FIG. 13 Kinetics of evaporation $L^2(t)$ of water from native quartz capillary (curve 1) and from the capillary pretreated with CA solution in ethanol (curve 2) at $p/p_s = 0.48$, $r = 7.6 \ \mu m$.

(curve 2) practically coincide. Then, as the meniscus becomes covered by CA molecules due to their diffusion from the adsorbed layer of CA on the capillary surface, the rate of evaporation decreases. As was shown in Ref. 14, a sharp decrease of the coefficient of condensation occurs after a complete monolayer is formed. This corresponds to a sharp decrease in the rate of evaporation after 20 min (curve 2, Fig. 13).

In the next experiment, CA was deposited on the capillary surface locally. For this purpose, only a part of the capillary was filled with CA solution in ethanol and the boundary was marked on the external capillary surface. Ethanol was then evaporated from the capillary, the empty side of which was previously sealed. Thus, only a part of the capillary surface was pretreated with CA. Next, the capillary was filled with pure water and evaporation was observed from the nontreated part of the capillary. As seen from Fig. 14, the rate of evaporation from the pretreated capillaries (curves 1–3) was equal to that from the control capillary not treated with CA (curve 4), unless the meniscus approached the region covered with CA. At this moment the evaporation rate sharply slowed due to transfer of CA molecules from the capillary surface onto the meniscus. Fig. 14 shows that the evaporation decreases sharply when the distance from the entrance of the capillary to the meniscus was $L = z$, where z is the coordinate of the boundary of the pretreated part.

In the first case (Fig. 13), the CA concentration on the capillary surface was low. Therefore, a monolayer of CA was formed only when the meniscus has traveled a rather long distance. In the second case (Fig. 14), due to evaporation of ethanol much more CA was deposited locally on the capillary surface. Thus, the saturation of the monolayer on the meniscus surface was achieved

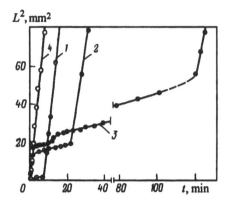

FIG. 14 Kinetics of water evaporation $L^2(t)$ from quartz capillaries, $r = 9\,\mu m$, at p/p_s = 0.54 with locally deposited adsorption CA layers (curves 1–3) and from a nontreated capillary (curve 4). The distance z between capillary entrance and the position of deposited CA layer was equal to 0.8 (curve 1), 4.23 (curve 2), and 6,14 mm (curve 3).

quickly, practically at the moment when meniscus contacts the deposited CA layer on the capillary surface.

It is interesting to note that after the meniscus has passed the region of the capillary surface covered with CA, the initial rate of evaporation reestablishes corresponding to the rate of evaporation of pure water (Fig. 14). This means that the meniscus displaces the CA molecules due to their reverse diffusion onto the clean capillary surface. A sharp transition to the initial rate of evaporation is evidence of rapid disruption of the compact CA monolayer.

Effective inhibition of evaporation is associated with a sharp decrease of the coefficient of condensation α_o by the formation of a monolayer of CA molecules. On the basis of Eq. (18), the values of α_0 might be found using measured rates of evaporation and $(p_s - p_o)$ values, where p_o is the vapor pressure in the surrounding medium. The diffusion resistance to the vapor flux inside the capillary between the meniscus and the entrance may, in this case, be neglected. For different regions of the graphs presented in Fig. 14, values of α_0 range from 1.8×10^{-4} to 2.7×10^{-4}. In another experiment with a capillary of the radius 8.7 μm ($z = 4$ mm, $p/p_s = 0.54$) almost the same value, $\alpha_0 = 3 \times 10^{-4}$, was calculated.

The values of α_0 obtained from observations of evaporation from the capillaries are two orders of magnitude lower than those for pure water (0.03–0.04), but approximately by one order of magnitude higher than those obtained for CA earlier in experiments with evaporating droplets [14].

Probably, the conditions under which the surfactant adsorption layer forms on the meniscus contacting the capillary walls are different from the conditions in the case of an evaporating droplet.

III. NONIONIC SURFACTANT SOLUTIONS IN HYDROPHOBIC CAPILLARIES

Methylation of quartz surface was used for preparation of hydrophobic capillaries. The methods are based on a reaction of OH groups on quartz surface with trimethylchlorosilane molecules from vapor phase or from a solution in benzene (see Section I).

Penetration of water into hydrophobic capillaries is retarded or becomes impossible when advancing contact angle exceeds 90°. Wetting of a hydrophobic surface may be improved by adsorption of surfactant molecules, which are directed with polar groups towards the water phase. Thus, an increase in surfactant concentration leads to gradual decrease of contact angles accelerating spontaneous imbibition.

However, adsorption of surfactant molecules occurs also on the meniscus surface that decreases the interface tension γ and therefore the capillary pressure $P_c = 2\gamma \cos \theta_A / r$, which is the driving force of imbibition. Considering the kinetics of impregnation, one needs to take into account the influence of surfactants on both contact angles and surface tension.

A. Acceleration of Impregnation of Hydrophobic Capillaries by Addition of Surfactants

Investigations of rates and mechanism of impregnation of hydrophobed quartz capillaries with nonionic surfactant solutions have been performed in Refs. 17–21. It was shown that addition of the surfactant to accelerate the imbibition makes sense only if the resulting decrease in the contact angle prevails over the counteraction of a decreased surface tension.

1. Theory

Let us consider the spontaneous imbibition of a surfactant solution into a hydrophobic capillary taking into account the possibility of the eventual passing of some surfactant molecules through the meniscus interface to the nonwetted portion of the capillary (Fig. 15). $C(x,t)$ and $\alpha(x,t)$ are the local surfactant concentrations (in g/cm^3) of the solution inside the capillary and on the capillary surface (in adsorbed state), respectively. A constant surfactant concentration, $C(0,t) = C_o =$ constant, is maintained at the capillary inlet. First, the case $C_o < C_c$, where C_c is the CMC, will be considered.

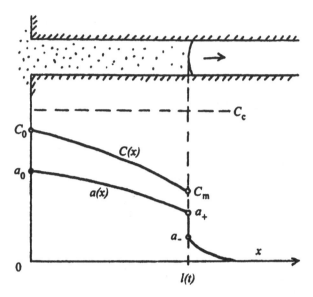

FIG. 15 Distribution of surfactant solution concentration inside a hydrophobic capillary, $C(x)$, and in adsorbed layer on capillary surface, $a(x)$, during imbibition of the solution at $C_o < C_i = \text{CMC}$.

The transport of surfactant inside the solution-filled part of the capillary may be described by the following equation:

$$d(a + C)/dt = D_1 \, d^2C/dx^2 + D_{1s} \, d^2a/dx^2 - v \, dC/dx, \tag{19}$$

where D_1 and D_{1d} are diffusion coefficients of surfactant molecules in the bulk phase and over the wetted part of the capillary surface; $v = dl/dt$ is the rate of the meniscus displacement.

Since the rate of imbibition of surfactant solutions into a hydrophobed capillary is rather slow, the kinetics of adsorption may be ignored and it is possible to include into Eq. (19) an equilibrium adsorption isotherm $a = f(C)$. As a first approximation, the linear form of the isotherm will be used here:

$$a(x, t) = (2/r)G_1C(x, t) = F_1C(x, t), \tag{20}$$

where G_1 is Henry's constant contained in the adsorption equation $\Gamma = G_1C$ (g/cm^2). In the case of a cylindrical capillary, $a = 2\Gamma/r$.

Substituting Eq. (20) into Eq. (19) and introducing the effective diffusion coefficient $D = (D_1 + F_1 D_{s1})/(1 + F_1)$ for the region $0 \leq x \leq l(t)$, we obtain:

$$dC/dt = d^2C/dx^2 - [(dl/dt)/(l + F_1)]dC/dx. \tag{21}$$

For the empty part of the capillary, $l(t) \leq x \leq \infty$, a corresponding equation for surfactant transport takes the form:

$$da/dt = D_{s2}d^2a/dx^2, \tag{22}$$

where D_{s2} is the coefficient of surface diffusion of surfactant molecules over the unwetted capillary surface.

Eqs. (21) and (22) may be solved under the following boundary and initial conditions:

$$C(0, t) = C_0 = \text{constant}; \quad a(\infty, t) = 0, \tag{23}$$

$$l(0) = 0; \quad a(x, 0) = 0, \tag{24}$$

When the rates of penetration are controlled by the viscosity of a liquid, the time dependence of the length l of the imbibed liquid column is described by following equation:

$$l(t) = (r\gamma \cos \theta_A / 2\eta)^{1/2} = K\sqrt{t} \tag{25}$$

where γ is the surface tension and η is the viscosity of the liquid.

In this case, the goal of the theory consists in determination of the solution concentration at the meniscus $C_m < C_0$, where C_0 is the surfactant concentration at the capillary inlet. The respective values of $\gamma(C_m)$ and $\theta_A(C_m)$ may be determined from corresponding experimental dependencies obtained for a given system.

Equation (19) under the conditions (23) and (24) has a solution only if C_m = constant, which corresponds to $l = K\sqrt{t}$, the law of spontaneous imbibition. The following mass-balance condition on the moving meniscus surface will be used in this case:

$$[D_{s2} \, da/dx]_{l_+} - [D_1 \, dC/dx + D_{s1} \, da/dx]_{l_-} = [a_- - a_+]dl/dt, \tag{26}$$

where l_+ and l_- represent positions on the opposite sides of the meniscus: on the liquid phase side (l_-) and on the nonwetted one (l_+). The corresponding values of a_- and a_+ characterize the jumpwise change in adsorbed amount (Fig. 15).

When C_m = constant, the value of a_- is also constant because they are interrelated by the adsorption isotherm (20): $a_- = F_1C_m$. Assuming that the imbibition goes rather slowly to preserve the equilibrium between C_m and a_+, an adsorption equation similar to Eq. (20) is here used:

$$a_+ = (2/r)G_2C_m = F_2C_m, \tag{27}$$

where G_2 is the corresponding Henry's constant.

When C_m = constant, the solution of Eqs. (21) and (22) can be found in a parametric form [20]: $C = C(\xi)$ and $a = a(\xi)$, where $\xi = x/\sqrt{t}$. After some transformation we get:

$$(-\xi/2)dC/d\xi = D\, d^2C/d\xi^2 - [K/2(1 + F_1)]dC/d\xi; \qquad (28)$$

$$(-\xi/2)da/d\xi = D_{s2}\, d^2a/d\xi^2. \qquad (29)$$

Solving Eqs. (28) and (29), we arrive at a transcendental equation that contains two unknown variables: K and C_m. To determine K and C_m, Eq. (25) should be added, which can be rewritten in the form:

$$K = (r\varphi_m/2\eta)^{1/2} \qquad (30)$$

where $\varphi_m = \gamma_m(C_m) \cos \theta_A(C_m)$ is the wetting tension that depends on the surfactant concentration C_m at the meniscus, which regulates its surface tension and forms contact angles.

Fig. 16 shows the results of calculations of the ratio C_m/C_0 vs. capillary radii r for different values of K characterizing the rate of penetration [21]. The following values of parameters entered in Eqs. (26)–(29) were used: $G_1 = 10^{-4}$

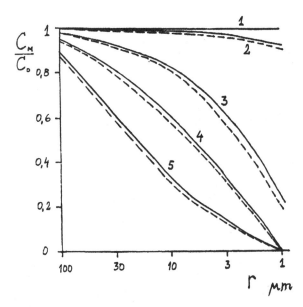

FIG. 16 Relative decrease of surfactant concentration near the meniscus, C_m/C_0, in dependence on capillary radii (r) and rates of penetration (K). The adopted values of K are equal: 10^{-4} (curve 1), 10^{-3} (curve 2), 5×10^{-3} (curve 3), 10^{-2} (curve 4), and 2×10^{-2} cm/s$^{1/2}$ (curve 5).

cm, $G_2 = 10^{-5}$ cm, $D_1 = 10^{-5}$ cm^2/s, $D_{s2} = 10^{-6}$ cm^2/s (solid curves) and 10^{-5} cm^2/s (dashed curves).

At a very low rate of penetration (curves 1 and 2), the concentration C_m does not differ appreciably from C_0 due to the small Peclet number ($Pe = vl/D$) and predominance of diffusion. At very high values of K, the concentration close to the meniscus tends to zero. This also takes place in the case of forced penetration under the action of an external pressure [19], when $Pe >> 1$. In this case, surfactant molecules cannot reach the meniscus and influence the contact angle. The effect of capillary radius is similar: the smaller the values of r, the more pronounced is the influence of surfactant adsorption. An increase in D_{s2} values leads to some decrease in C_m due to the enhanced diffusion along the nonwetted capillary surface.

It should be noted that because the values of K and r are interrelated the results show in Fig. 16 should be considered at r = constant. In this case, K values depend only on γ_m and $\cos \theta_A$.

$$K = (r\gamma_m \cos \theta_A/2\eta)^{1/2} \tag{31}$$

The case of high concentration, $C_0 > C_c$, when adsorption is accompanied by micelle destruction, should be considered separately (Fig. 17). Here the wetting tension $\varphi_0 = \gamma_o \cos \theta_A$ is constant and the rate of penetration is controlled by the K value that is equal to:

$$K_0 = (r\varphi_0/2\eta)^{1/2} = \text{constant} \tag{32}$$

Unlike the situation discussed above, the equation of the meniscus displacement is known: $l = K_0 \sqrt{t}$. In this case, the surfactant adsorption on the capillary surface is accompanied by continuously decreasing concentration C_m (near the meniscus): from C_0 at the start of imbibition ($t = 0$) to $C_m = C_c$. The solid curve in Fig. 17a represents the distribution of concentration $C(x)$ at time $t > 0$, when $C_0 > C_m > C_c$.

When C_m becomes lower than C_c (Fig. 17b, solid curve), the condition K_0 = constant is no longer valid. The values of K go down below K_0, and the rate of penetration decreases. As soon as the condition $C_m = C_c$ is attained at the meniscus, a further decrease in concentration ($C_m < C_c$) will cause separation of the micelle front from the meniscus surface (Fig. 17b). The stage of imbibition, occurring when $C_m < C_c$, has been described above.

2. Experimental Results

The results of measurement of the meniscus displacement l with time t are shown in Fig. 18. The changes in the penetration length were measured using the comparator IZA-2 with an accuracy of ± 1 μm. All the experiments were performed with aqueous solutions of a nonionic surfactant syntamide-5 (CMC = 0.1%) of various concentrations.

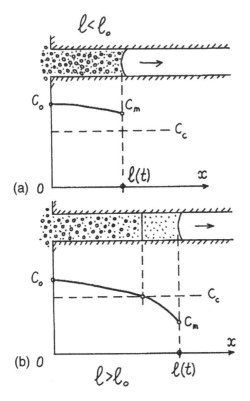

FIG. 17 Distribution of surfactant concentration in the course of imbibition inside a hydrophobic capillary, $C(x)$, at $C_o > C_c$ in the cases $l < l_o$ (a) and $l > l_o$ (b).

Let us consider first the case of high concentration of the surfactant solution, $C_0 > C_c$ (curves 3–5). A sharp decrease in the penetration rate at $l = l_0$ (Fig. 18) corresponds to the transition from $C_m > C_c$ to $C_m < C_c$, where C_m is the surfactant concentration at the meniscus. The length l_0 may be estimated from the mass-balance condition on the meniscus surface [18,20,21]:

$$a_k \, dl/dt = -(dU_M/dt)r/2. \tag{33}$$

The left-hand side of this equation characterizes the rate of surfactant adsorption on the wetted surface (g/cm^2 s), and the right-hand side characterizes the rate of micelle disintegration per unit surface area. It is assumed that at $C = C_c$ the limiting adsorption $a_k = F_1 C_c = $ constant is achieved. The difference $U_M = C_M - C_c$ is equal to the micelle concentration and C_M is the concentration of micellar solution. The value $dV/dS = r/2$ represents the ratio

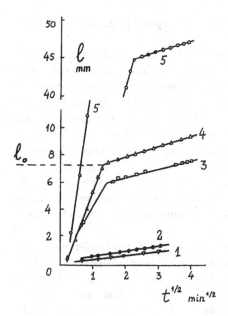

FIG. 18 Dependencies of the length of penetration, l, on the time for syntamide-5 solution of various concentration C_o = 0.05 (1); 0.1 (2); 0.4 (3); 0.5 (4) and 1 wt% (5). Hydrophobed quartz capillary, r = 16 μm, θ_A = 101°, θ_R = 97° against water.

of an elementary volume of the solution close to the meniscus, $\pi r^2 dl$, to its surface area, $2\pi r dl$.

From Eq. (33) solved at the initial condition, $t = 0$, $U_M(0) = C_0 - C_c$, follows:

$$U_M = (C_0 - C_c) - (4C_c G_1 K_0 \sqrt{t})/r^2, \tag{34}$$

Values of U_M decrease as all micelles near the meniscus disintegrate, i.e., as the concentration C_M reduces to C_c, when U_M becomes equal to 0. This allows one to find from Eq. (33) the time, $t = t_0$, corresponding to complete micelle destruction, $U_M = 0$. After that, the t_0 value is used to calculate the length l_0 of the rapid imbibition:

$$l_0 = K_0 \sqrt{t_0} = [(C_0/C_c) - 1]r^2/4G_1, \tag{35}$$

According to this equation, values of l_0 decrease with C_0 in agreement with experimental results (Fig. 18). By calculating $\varphi_0 = \gamma_m \cos \theta_A$ values from Eq. (32) and using the known values of $\gamma_m = \gamma_m$ (CMC) = 30 mN/m, the advancing contact angle can be determined. The slope of the linear part of the graphs for the first stage of imbibition ($l < l_0$) gives the values of $\cos \theta_A$ equal to 0.02–0.04 that corresponds to $\theta_A \cong 88°$. Corresponding K_0 values for $C_0 \geq$ 1% are on the order of 10^{-1} cm/s$^{1/2}$.

The values of K for the second stage of imbibition (when $l > l_0$) increase with C_0 due to an increase in absolute value of C_m. However, an increase in K values (from 3×10^{-3} to 1.7×10^{-2} cm/s$^{1/2}$) is limited because the rate of meniscus motion influences the C_m values in the opposite direction (Fig. 16).

Using Eq. (30), values of wetting tension $\varphi = \gamma_m \cos \theta_A$ were calculated from the experimental data. At low values of $K \cong 10^{-3}$ cm/s$^{1/2}$ (Fig. 18, curves 1 and 2) and $r \cong 10$ μm, surfactant concentrations at the meniscus C_m do not much differ from C_0 values (Fig. 16). In this case $\gamma_m = \gamma_o$, and therefore it becomes possible to calculate advancing contact angle $\cos \theta_A$ from $\varphi = \gamma_o \cos \theta_A$ using the known dependence $\gamma_o = \gamma_o(C_0)$ for the bulk solution. The values of $\cos \theta_A$ obtained are on the order of 10^{-7}, which corresponds to θ_A values near 90°.

Values of advancing θ_A, receding θ_R, and static θ_0 contact angles measured for sessile droplets on a methylated quartz plate in dependence on surfactant concentration [22] are shown in Fig. 19. To determine θ_A and θ_R, the drop was disposed on a polished surface of a flat cut of a thick-walled quartz capillary over its orifice. For solution concentration $C_0 < 0.1\%$, advancing angles are higher than 90° and, therefore, the capillary suction is formally impossible. However, as is shown in Fig. 18, a slow penetration of low concentrated solutions takes place. Similar results were obtained in other experiments with

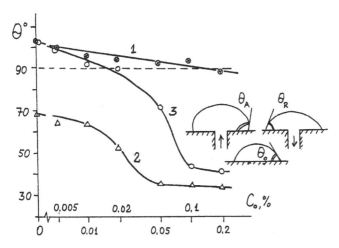

FIG. 19 Advancing θ_A (curve 1), receding θ_R (curve 2), and static θ_o (curve 3) contact angles formed on a flat methylated quartz surface vs. concentration C_o of syntamide-5 solution.

hydrophobic capillaries [23,24] and also with hydrophobed quartz powders [25]. It was supposed that the rate of imbibition is controlled here not by the capillary forces but by a diffusion mechanism.

B. Diffusion Mechanism of Penetration of Surfactant Solutions

Let us first discuss the results obtained in the capillary rise experiments. Fig. 20 shows the obtained time dependencies of the meniscus position, H, for syntamide-5 surfactant solution in vertically oriented capillaries. Two upper curves, 1 and 2, show in a larger scale the results obtained during the first hour of observation. The lower curves 1 and 2 represent the results obtained during the complete time of the experiment, up to 5 days. The linear dependence of H values on $\sqrt{t}$ corresponds to the constant value of $K = H/\sqrt{t} = 2.36 \times 10^{-3}$ cm/s$^{1/2}$ (curve 1) and $K = 1.3 \times 10^{-3}$ cm/s$^{1/2}$ (curve 2). These values are close to those calculated from curve 2 (Fig. 18) for the same bulk solution concentration of syntamide-5, $C_0 = 0.1\%$. This gives grounds for supposing that the same mechanism of penetration was operating in both cases.

If the known mechanism of imbibition, governed by the capillary forces, is assumed here, the values of K obtained may be used to calculate the wetting tension $\varphi = \gamma_m \cos \theta_A = 2\eta K^2/r$. Calculations give the values of $\varphi = 10^{-4}$ mN/m and $\cos \theta_A = 3.4 \times 10^{-6}$ (curve 1) and 2.2×10^{-5} mN/m and 7.3×10^{-7} (curve 2), respectively. The calculated values of advancing angles are

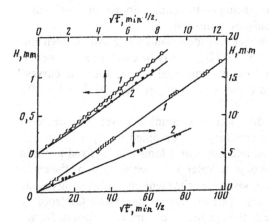

FIG. 20 Height of the rise H of syntamide-5 solution, $C_o = 0.1$ wt%, in dependence on time t in two vertically oriented quartz capillaries: $r = 11$ μm, $\theta_A = 92°$, $\theta_R = 80°$ (curves 1) and $r = 15$ μm, $\theta_A = 102°$, $\theta_R = 86°$ (curves 2); 25°C.

only slightly lower than 90°, as in experiments with horizontally oriented capillaries.

In the framework of the mechanism of capillary forces, the rise of a liquid cannot exceed a limiting value $H_{max} = 2\gamma \cos\theta_A/\rho gr$, where ρ is the density of the liquid and g is the gravity acceleration. Substituting the above calculated values of wetting tension φ into this formula gives: $H_{max} = 2 \times 10^{-3}$ mm (for curve 1) and 3×10^{-4} mm (for curve 2). However, the meniscus rises up to several centimeters and continues to move further (Fig. 20).

The only reasonable explanation of this effect is that the meniscus moves following the front of surface diffusion of surfactant molecules hydrophilizing the dry capillary surface. The meniscus is held at a given height H by the receding contact angle formed to equilibrate the hydrostatic pressure, ρgH.

The coefficient of surface diffusion, D_s, may be estimated from the equation of surface diffusion

$$d\Gamma/dt = D_s(d^2\Gamma/dx^2), \tag{36}$$

where Γ is the adsorption and x is the distance from the capillary inlet. Under the boundary conditions

$$\Gamma(0) = \Gamma_0 = \text{constant}, \quad \Gamma(\infty) = 0, \tag{37}$$

the solution of Eq. (36) takes the form:

$$\Gamma* = \Gamma_0[1 - \text{erf}(K/2D_s^{1/2})]. \tag{38}$$

Here the adsorption $\Gamma*$ corresponds to the position of the moving meniscus (Fig. 21). When $\Gamma \geq \Gamma*$, the advancing contact angle approaches 90°, which makes it possible to set the meniscus in motion.

Supposing that in the case under consideration $\Gamma* \ll \Gamma_0$ the known Einstein equation $H \cong (2D_s t)^{1/2}$ or $K \cong (2D_s)^{1/2}$ may be used, an assessment of coefficients of surface diffusion from experimental values of K from curves 1 and 2 (Fig. 20) give reasonable values of D_s ranging from 10^{-6} to 3×10^{-6} cm²/s.

Direct evidence of surface diffusion of syntamide-5 over the nonwetted part of a hydrophobed quartz capillary was obtained in experiments [24], when the meniscus was stopped and held for a long time at fixed position. After 25 days the contact angles of pure water were measured along the empty capillary part close to the former position of the meniscus. Advancing contact angles θ_A measured in the 7-mm zone are lower than those far from the meniscus. The estimated value of the coefficient of surface diffusion, D_s, was on the order of 10^{-7} cm²/s.

The surfactant diffusion affecting the rate of an advancing front may be a reason for the unstable movement of the wetting line [26,27]. Surface diffusion

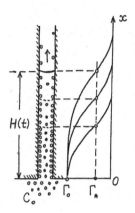

FIG. 21 Schematic representation of a diffusion mechanism of penetration of a surfactant solution into a hydrophobic capillary with advancing contact angles θ_A more than $90°$.

of surfactants influences also the time it takes for droplets to form an equilibrium contact angle on a mica surface [28].

Not only surfactant molecules may diffuse in advance of the wetting front. Spreading of pure liquids by surface diffusion of molecules from a microdoplet over a solid surface was comprehensively studied using microellipsometric measurements [29–34]. It has been observed that on the top of the first monolayer, a second and subsequent layers form, and the corresponding coefficients of surface diffusion were calculated. For liquid polydimethylsiloxan (PDMS) on a hydrophobed silicone wafer, coefficients of surface diffusion in the first monolayer grow with decreasing molecular mass M of the PDMS from $D_s = 4 \times 10^{-8}$ cm^2/s for $M = 28{,}400$ to 7×10^{-7} cm^2/s for $M = 6700$. Correlation between the D_s values and bulk viscosity of the liquid PDMS have been established.

C. Forced Imbibition

Influence of external pressure on imbibition of surfactant solution has been studied using a horizontally oriented hydrophobed capillary ($r = 6$ μm, $\theta_A = 96°$) [1]. The capillary was first washed three times with $C_0 = 0.1\%$ solution of syntamide-5 up to the length $l = 3.7$ cm (Fig. 22, point c). Thus, this part of the capillary might be supposed to be in equilibrium with the surfactant solution.

Then, at the same concentration C_0, a spontaneous imbibition of surfactant solution into the capillary was examined. The linear a–b part of the graph in Fig. 22 corresponds to imbibition at a constant value of advancing contact

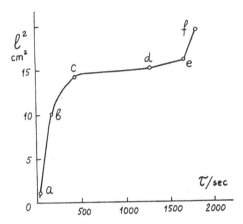

FIG. 22 Rates of spontaneous imbibition of syntamide-5 solution (C_o = 0.1 wt%) into horizontally oriented hydrophobed quartz capillary, r = 6 μm, θ_A = 96° against water. The part a–c of the capillary was previously equilibrated with the solution. The regions c–d and d–e–f relate to nonequilibrated parts of the capillary. Rates of forced imbibition were measured at pressure differences 1.3 N/m² (part d–e) and 2.6 N/m² (part e–f).

angle, θ_A = 86°, considerably lower than θ_A = 96° for pure water. Near point c, the spontaneous imbibition is retarded, because the complete surfactant adsorption in the b–c section of the capillary has not been reached. The next part of the graph, c–d, corresponds to extremely slow movement of the meniscus. The kinetics, as discussed above, is determined here by the surface diffusion of surfactant molecules in advance of the meniscus. The slope of this part of the graph gives the value of K equal to 6.7×10^{-4} cm/s$^{1/2}$ and cos θ_A = 5×10^{-7}, as in the cases considered above.

Subsequently, an external gas pressure was applied to the capillary inlet equal to 1.3 N/m² (Fig. 22, part d–e of the graph) and 2.6 N/m² (part e–f). The calculated values of θ_A for these regions are equal to 94° and 96°, respectively, i.e., are practically the same as for bulk water.

The latter results show that in the case of forced imbibition an addition of surfactant has little or no effect on K values as compared with water. At high Peclet number, surfactant molecules cannot approach the meniscus zone because of their adsorption on capillary walls. This explains, for instance, why an addition of surfactants alone cannot enhance the recovery of oils, especially when a surfactant solution is forced by the pressure gradient.

Spontaneous imbibition of microemulsions into hydrophobed capillaries occurs at much higher rates ($K \cong$ 0.1–1 cm/s$^{1/2}$). Under these conditions complete wetting of the capillary surface is realized [23].

The results obtained lead to the conclusion that three mechanisms of penetration of nonionic surfactant solutions into hydrophobic capillaries are possible. The first takes place at a high concentration of surfactant, $C_0 > C_c$. Spontaneous imbibition advances at a high rate but is limited to some finite length l_0, which depends on C_0, r, and G_1. The second mechanism is realized when $C_0 < C_c$. The rate of penetration in this case is much lower, being controlled by the reduced concentration, $C_m = $ constant, near the meniscus. At still lower concentration of bulk solution C_0, the diffusion mechanism of penetration takes place in thin capillaries. The rate of penetration is determined here by the surface diffusion of surfactant molecules in advance of the meniscus.

Penetration of surfactant solutions into hydrophobic capillaries under an external pressure is not effective because the surfactant molecules, being adsorbed, cannot approach the advancing meniscus at high Peclet numbers.

D. Trisiloxane Surfactants in Hydrophobed Quartz Capillaries

Aqueous solutions of trisiloxane surfactants were recognized to spread very rapidly over hydrophobic surfaces. The trisiloxane D-8 surfactant (Dow Corning Corp.) with a molecular structure $M(D'E_8OH)M$, where $M = (CH_3)_3SiO$, $E = CH_3O(OCH_2CH_2)_8$ and $D' = Si (CH_3)C_3H_6$, was used as an effective superspreader [35,36]. By the presence of hydrophobic methyl groups that are shorter, wider, and occupy larger volume than, for instance, $C_{12}H_{25}$ groups, the surface tension of the methyl-saturated surface of water is about 20 mN/m, much lower than that for other surfactants [35].

In the range of concentrations $C_o \geq 0.007$ wt% the D-8 aqueous solution contain vesicles. The rate of droplet spreading over hydrophobic surface grows with concentration and reach a maximum at $C_o = 0.16$ wt% [36].

Experiments with solution–air and solution–silicon oil systems were performed using the same device (Fig. 1) and methylated quartz capillaries with radii from 5.25 to 3.6 μm. Advancing contact angles against water range from 101° to 105° and receding angles from 78° to 71°. The D-8 solution with concentration $C_o = 0.16$ wt% was sonicated before experiments to destroy aggregates of vesicles. As in the above considered cases, the meniscus was shifted back and forth within a small portion of the capillary under a pressure drop ΔP. Surface tension of the bulk solution is equal to $\gamma_o = 21$ mN/m [37].

Fig. 23 shows the obtained $v(\Delta P)$ dependencies for the solution–air system in methylated quartz capillary, $r = 5.25$ μm. Positive values of flow rates, $v > 0$, characterize the advancing motion of the meniscus and negative ones relate to its receding motion. The results are shown by black points for the capillary nonequilibrated previously with the D-8 solution. Intersection of the contin-

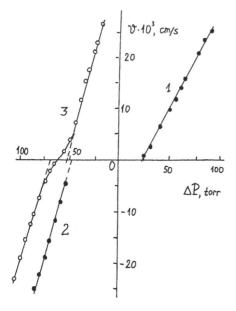

FIG. 23 Dependencies of flow rate v on pressure difference ΔP for the D-8 solution/air system in a hydrophobed quartz capillary, $r = 5.25\ \mu m$, before (curves 1 and 2) and after equilibration with surfactant solution (curve 2).

uation of the linear part of the graph l with the pressure axis gives the dynamic wetting tension $\gamma_m \cos \theta_A = 8.75$ mN/m. Because the capillary was not equilibrated, a loss of surfactant due to adsorption was possible and meniscus surface tension γ_m may be higher than γ_o. Therefore, the dynamic advancing contact angles lie in the region from $115°$, assuming $\gamma_m = \gamma_o$, to $\theta_A = 87°$ and assuming that the meniscus surface due to adsorption is free from surfactant and $\gamma_m = 72$ mN/m as for water.

In the case of retreating meniscus, intersection of the graph 2 with pressure axis gives the wetting tension value equal to $\gamma_m \cos \theta_R = 17.5$ mN/m. Because surface tension cannot be lower than γ_o, the dynamic receding contact angles lie in the region between $\theta_R = 34°$ (when $\gamma_m = \gamma_o$ was used) and $76°$, assuming that $\gamma_m = 72$ mN/m.

From the slope of the linear graphs 1 and 2, the viscosity η of the solution was calculated using the Washburn equation [Eq. (2)]. The viscosity value $\eta = 0.96$ cPs calculated from graph 2 coincides .with known values for bulk solution. However, the slope of graph 1 is higher and corresponds formally to $\eta = 1.7$ cPs. This discrepancy may be caused by additional viscous resistance near the meniscus zone. Transfer of surfactant molecules from meniscus to adsorbing solid wall results in formation of gradients of surface tension along the meniscus profile, which causes Marangoni flux directed toward the walls. Arising circulation of the liquid results in supplementary viscous loss.

When a capillary was previously equilibrated with the solution, the $v(\Delta P)$ dependence obtained is shown in Fig. 23 by curve 3. The slopes of the linear parts of the graph are the same and correspond to the viscosity of the bulk solution. The points of intersection of the extensions of the linear parts with the pressure axis give wetting tensions equal to $\gamma_A \cos \theta_A = 18$ mN/m and $\gamma_R \cos \theta_R = 24$ mN/m. The intersection of curve 3 with the pressure axis gives the static value (at $v = 0$) of surface tension $\gamma = 21$ mN/m, close to the bulk surface tension of the solution γ_o. Because capillary surface was equilibrated with the solution, it may be supposed that the surface tension of the advancing meniscus remains constant and equal to γ_o. This results in dynamic advancing contact angle $\theta_A = 23°$. Dynamic surface tension of the retreating meniscus γ_R exceeds the equilibrium value due to transfer of surfactant molecules to the forming wetting film (Fig. 5a). Assuming that $\theta_R \geq 0$, we arrive at $\gamma_R \geq 24$ mN/m.

Pretreatment of a hydrophobed capillary with the D-8 solution made the surface hydrophilic, and this secured spontaneous suction of the solution. Nontreated capillary remains hydrophobic, and advancing contact angles differ not much from contact angles for water. However, diffusion mechanism of very slow penetration discussed in Section III.B may also take place in this case.

Fig. 24 shows the results obtained for the D-8 solution–silicon oil system in a methylated quartz capillary, $r = 3.6$ μm. Interface tension of silicon oil (molecular mass 2400, viscosity $\eta_2 = 21$ cPs) in contact with bulk D-8 solution is equal to $\gamma_o = 2.5$ mN/m [37]. Three methylated quartz capillaries were prefilled with the silicon oil. The $v(\Delta P)$ dependencies are obtained in the same way as for the solution–air system. Positive values of flow rates, $v > 0$, correspond to displacement of silicon oil by D-8 solution. The inverse case, when D-8 solution was displaced by oil, corresponds to negative v values. As seen from Fig. 24, to displace oil one needs to overcome the dynamic capillary counterpressure of oil, $P_c = 2\gamma_{12} \cos \theta_R/r$. The intersection point of the linear part of the graph at $v > 0$, shown by an arrow, gives $\gamma_{12} \cos \theta_R = 10$ mN/m.

For two other capillaries ($r = 5.2$ and 5.5 μm) close values, 10 and 12.4 mN/m, were obtained. Because of adsorption of surfactant on oil-free hydrophobic surface, the interface tension becomes higher than γ_o. Therefore, the dynamic retreating angle of silicon oil may range from $\theta_R = 0$ at $\gamma_{12} \approx 10$ mN/m to $\theta_R = 75°$, when the meniscus surface becomes free from surfactant and γ_{12} is equal to 41 mN/m, as in the bulk water–oil interface.

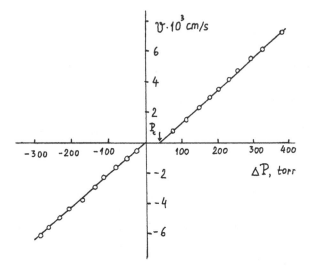

FIG. 24 Dependencies of flow rate v on pressure difference ΔP in a hydrophobed quartz capillary, $r = 3.6\,\mu m$, when D-8 solution displaced silicon oil ($v > 0$) and when silicon oil displaced surfactant solution ($v < 0$).

The capillary counterpressure, which is proportional to the wetting tension $\gamma_{12}\cos\theta_R \approx 10$ mN/m, is about three times lower than when silicon oil was displaced by pure water. The value of wetting tension measured in hydrophobed capillaries equals in this case 30 mN/m. Therefore, use of the D-8 trisiloxane surfactant may facilitate displacement of oils from hydrophobic pores. Besides, when water displaces silicon oil a wetting film of the oil remains on the hydrophobed capillary surface. Because the thickness of the remaining dynamic film grows according to Derjaguin [7] with flow rate, the low rates of displacement are preferable in these cases.

In the case of trisiloxane surfactant, complete removal of oil films from hydrophobed surface becomes possible. This was shown in special experiments with methylated capillaries covered with relatively thick wetting films of silicon oil. The wetting film was formed by smearing off a small oil column (2 to 3 mm long) over the capillary surface. In two experiments the formed films are 16 and 18 nm thick and 5.5 cm and 4.6 cm long, respectively. When a column of the D-8 surfactant solution is brought in contact with the edge of the wetting film, the latter converts into an oil column in front of the moving meniscus of surfactant solution.

The rate of meniscus motion follows diffusion kinetics (see Section III.B) with coefficients of surface diffusion 10^{-4} to 10^{-5} cm^2/s. To the end, the volume of the collected oil column becomes equal to the volume of the pre-formed wetting film. Therefore, the hydrophobic surface of the capillary becomes free from oil. The mechanism of diffusion penetration of surfactants into hydrophobic capillaries was discussed in Section III.

Diffusion of surfactant molecules results also in a directly observed detachment of oil droplets and films from flat hydrophobic surfaces contacting the bulk D-8 solution. Similar effects are known also for other ionic surfactants [38,39].

When silicon oil displaces the D-8 solution ($v < 0$), the dynamic advancing contact angle of oil is close to 90°. As seen from Fig. 24, wetting tension $\gamma_{12} \cos \theta_A$ is very low, near zero. In this case, bulk interface tension $\gamma_o = 2.5$ mN/m is maintained on the meniscus due to desorption of surfactant molecules from hydrophobic walls in front of the advancing meniscus.

IV. CATIONIC SURFACTANT SOLUTIONS IN HYDROPHILIC CAPILLARIES

A. Aqueous Solution–Air Systems

As an example, Fig. 25 shows the dependencies of flow rate v on pressure difference ΔP obtained for a cationic surfactant in a 10^{-4} M KCl background electrolyte. In the experiments, aqueous solutions of cetyltrimethylammonium bromide (CTAB; Merck), of various concentrations C_o from 10^{-6} to 5×10^{-4} M were used [40,41]. The solutions were prepared using water tridistillate with electrical conductivity 10^{-6} Ω^{-1} cm^{-1} and pH 6.5. The capillaries were equilibrated with each CTAB solution under investigation by pumping the solution during 2 to 3 h before experiments.

The rate of meniscus movement in quartz capillaries was measured, as in the above-considered cases, within a small portion of the capillary. In this case the length l of the solution column remains practically constant. The direction of motion was changed by turns from advancing ($v > 0$) to receding, when $v < 0$.

Linearity of the $v(\Delta P)$ graphs in both $v > 0$ and $v < 0$ regions shows that dynamic capillary pressures are independent of flow rate values, at least at $v < 10^{-2}$ cm/s. Dynamic capillary pressures were obtained from the intersection of the extension of linear parts of the graphs with the pressure axis. After that, using capillary radii, values of wetting tension $\gamma_m \cos \theta$ were calculated. For receding menisci these values in the region of CTAB concentration, from 10^{-6} to 5×10^{-5} M (curves 1 to 3), are equal to 72 mN/m, that is, to the surface tension of water in contact with air. Surface tension of the used bulk CTAB solutions was slightly lower: 69 mN/m for $C_o = 10^{-5}$ M and 65 mN/m for 5×10^{-4} M CTAB. Some loss of surfactant molecules on receding menisci is

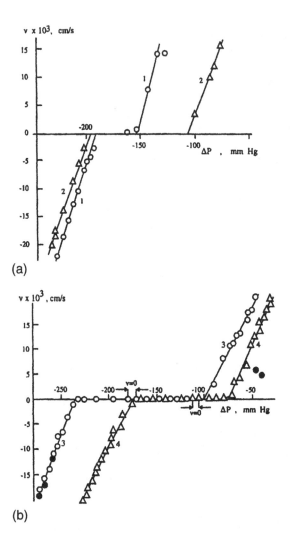

FIG. 25 Rate of advancing ($v > 0$) and receding ($v < 0$) motion of menisci of CTAB solution in thin quartz capillaries in dependence on pressure difference ΔP. (a) $C_o = 10^{-6}$ M, $l = 9.28$ cm, $r = 5.64\,\mu m$, $t = 21°C$ (curve 1); $C_o = 10^{-5}$ M, $l = 9.21$ cm, $r = 5.63\,\mu m$, $t = 18°C$ (curve 2). (b) $C_o = 5 \times 10^{-5}$ M, $l = 9.35$ cm, $r = 4.64\,\mu m$, $t = 26°C$ (curve 3); $C_o = 5 \times 10^{-4}$ M, $l = 9.35$ cm, $r = 4.64\,\mu m$, $t = 24°C$ (curve 4).

caused by adsorption of CTAB on forming wetting film interface (Fig. 5b). Because dynamic surface tension cannot be higher than 72 mN/m, dynamic receding contact angles are equal nearly to zero. Formation of wetting films after a retreating meniscus was observed also earlier [4,6,8,42,43].

At higher CTAB concentration, 5×10^{-4} M (curve 4, Fig. 25), the wetting tension is equal to 53 mN/m, also larger than the surface tension of bulk solution, $\gamma_o = 40$ mN/m. Using the isotherm of surface tension of CTAB solutions, $\gamma_o(C_o)$, the concentration of CTAB solution near the meniscus was estimated. This concentration, $C_m = 2.5 \times 10^{-4}$ M, is two times lower than the concentration of the bulk solution.

Advancing dynamic contact angles are calculated assuming that CTAB concentration near the meniscus differs not much from bulk surface tension γ_o. The results of calculation of dynamic advancing contact angles are shown in Fig. 26 by curve 1. As seen from the graph, the dynamic θ_A values pass through a maximum at CTAB concentration C_o between 10^{-5} and 10^{-4} M. Water tridistillate in the same capillary completely wets the capillary wall. At CTAB concentration near CMC, 10^{-3} M, complete wetting takes place again. Therefore, at $C_o = 0$ and 10^{-3} M, both θ_A and θ_R are equal to 0. Correspondingly, no hysteresis was observed in these cases, and linear $v(\Delta P)$ dependencies intersect the pressure axis at $\Delta P = 0$.

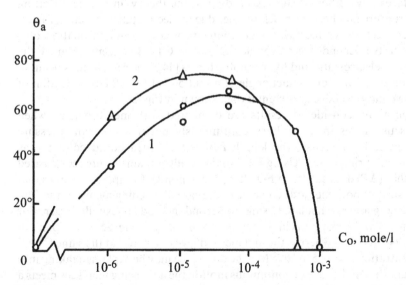

FIG. 26 Dependencies of dynamic advancing contact angles θ_A on CTAB concentration C_o obtained on the basis of the $v(\Delta P)$ data (curve 1) and using air bubble method (curve 2).

Dynamic advancing contact angles were determined for the same solutions using an air bubble (0.3 to 0.6 mm long) trapped in a capillary filled with CTAB solution. The bubble is formed in the middle of the capillary, the ends of which are placed into vessels filled with the same solution. In this case, the rate of motion under applied pressure difference ΔP does not depend on the direction of the bubble motion. The intersection of the obtained $v(\Delta P)$ dependencies measured for various concentrations of CTAB solution gives in this case only the difference between two tensions of wetting, $\gamma_R \cos \theta_R - \gamma_A \cos \theta_A$, which characterizes the hysteresis. The advancing angles are calculated using γ_R values obtained above and assuming $\gamma_A = \gamma_0$. The results of calculations are shown in Fig. 26 by curve 2. The maximum values of advancing contact angles correspond to the same region of CTAB concentration at about 10^{-5} M.

The obtained dependencies of contact angles on the concentration of CTAB solution, shown in Fig. 26, may be explained by the theory of surface forces [44,45]. In the case of low CTAB concentration both wetting film interfaces (film–air and film–quartz) possess negative electrical potentials, which results in repulsive forces. Contact angles are small because they contact relatively thick wetting film. The same situation takes place also at high CTAB concentration, when both interfaces acquire positive potentials. In the intermediate region of concentration electrical potentials of the interfaces have different sign. According to the theory, in this case wetting films rupture and become much thinner due to electrostatic attraction forces, and contact angles increase. A computer program allows calculation of a quantitative dependence of contact angles on CTAB concentration by the theory of electrostatic and hydrophobic forces [44]. The calculations result in a theoretical value of contact angle of about $50°$ at CTAB concentration of 10^{-5} M, close to the experimental data shown in Fig. 26.

The above-considered results are compared with measurements when meniscus moves in a capillary continuously under a constant pressure difference. In this case, the length l of the imbibed liquid column was measured during time τ. Using Eq. (3), the results obtained were recalculated into the $v(\Delta P)$ data shown in Fig. 25 by black points. Comparison shows that the results of both methods coincide in the case of receding meniscus motion. Advancing angles calculated using the second method are usually larger. For instance, two black points in the region $v > 0$ (curve 3, Fig. 25) correspond to $\theta_A = 74°$ and $76°$, whereas the first method gives $\theta_A = 62°$ at the same CTAB concentration, $C_0 = 5 \times 10^{-5}$ M. The difference may be explained taking into account that in the case of continuous motion the advancing meniscus meets a "dry" capillary surface. In contrast, when the meniscus moves back and forth within a small portion of the capillary, the advancing meniscus meets a wetting film remaining after the receding motion of the meniscus.

Regions of practically immobile states of a meniscus are shown in Fig. 25 by arrows on the pressure axis for solution concentration $C_o = 5 \times 10^{-5}$ (curves 3) and 5×10^{-4} M (curves 4). This makes it possible to assess static values of contact angles. Because of small hysteresis (the regions shown by arrows are short) the mean value of static contact angle is equal to $40°$ for $C_o = 5 \times 10^{-4}$ M and $36°$ at 5×10^{-5} M. The calculated values are close to those measured using captive bubbles [45] and differential ellipsometry method [46] on quartz surface for the same solutions.

B. Aqueous Solution–Oil Systems

In these experiments n-tetradecane and silicon oil ($M = 2400$; Rhone Poulenc) were used as oil phases. Viscosity of the silicon oil, $\eta_2 = 21.9$ cPs, and surface tension in contact with air, $\gamma_o = 20.6$ mN/m, were measured using a thin calibrated quartz capillary ($r = 6.14$ µm) and the same device (Fig. 1). The data obtained are close to those known for the bulk oil. Similar results were obtained for tetradecane. The silicon oil and tetradecane completely wet hydrophilic quartz capillaries; no hysteresis was observed.

When air is substituted by oils, the capillary shown in Fig. 1 was partly filled with CTAB solution (length l_1) and partly with oil (length l_2). The sum of the lengths l_1 and l_2 gives the complete length L of the capillary. In this case, an oil was placed into the left chamber and surfactant solution into the right one. The pressure difference was assumed here to be $\Delta P = P_2 - P_1$.

The rate of displacement v follows in this case the Washburn equation [Eq. (10)], where η_1 and η_2 are the viscosity of a CTAB solution and of an oil, respectively.

Rates of meniscus back-and-forth movements are measured within a small portion of the capillary, $\Delta l < 1$ mm, much smaller than either l_1 or l_2. According to Eq. (10), the rate of displacement must linearly depend on the pressure drop ΔP when the dynamic capillary pressure P_c is constant. In this case, the lengths l_1 and l_2 and viscosity values η_1 and η_2 influence only the slope of the linear $v(\Delta P)$ graphs.

In Fig. 27 are shown, as an example, some of the dependencies $v(\Delta P)$ obtained for CTAB solution–silicon oil systems. Curve 1 represents the results obtained for water–silicon oil system in the absence of surfactant. The background electrolyte concentration was 10^{-4} M KCl. Curves 2 and 3 represent the results obtained for CTAB concentration $C_o = 10^{-6}$ and 10^{-5} M, respectively. Positive values of v correspond to displacement of silicon oil by aqueous solutions and negative ones to the inverse process. In accordance with Eq. (10), the $v(\Delta P)$ dependencies are linear, cutting, however, different dynamic capillary pressures for advancing and receding motion of the menisci. The difference between these values characterizes the hysteresis,

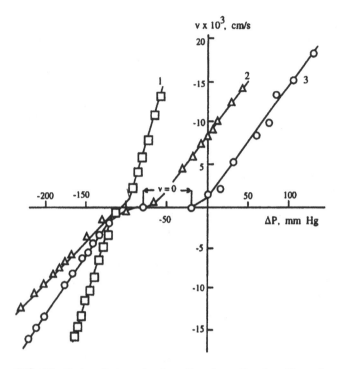

FIG. 27 Rates of advancing ($v > 0$) and receding ($v < 0$) meniscus motion for silicon oil/CTAB solution system on surfactant concentration: $C_o = 0$ (curve 1), 10^{-6} M (curve 2), and 10^{-5} M (curve 3); $r = 5.64$ μm, $t = 20°C$.

which grows with CTAB concentration. The region of static hysteresis is shown by an arrow. Static advancing contact angle is 75°, and the receding angle is near 0°. The dynamic hysteresis was larger: dynamic contact angle is 85°, and the receding angle remains zero. However, in the latter case dynamic receding interface tension was also changed. As in the above-considered cases, values of wetting tension, $\gamma_R \cos \theta_R$, are always larger than the interface tension of the bulk system, γ_o. For CTAB concentration $C_o = 10^{-6}$ M (curve 2), dynamic wetting tension is 43.6 mN/m, whereas $\gamma_o = 38$ mN/m. At higher concentration $C_o = 10^{-5}$ M (curve 3), when $\gamma_o = 30$ mN/m, dynamic wetting tension is 42.2 mN/m. This points to a decreased CTAB concentration near the meniscus.

Values of dynamic advancing contact angles θ_A were calculated assuming $\gamma_A = \gamma_o$. The results of calculation are shown in Fig. 28 by curve 1. The contact angle value reached a maximum in the region of CTAB concentration near 10^{-5} M. As was concluded in Refs. 44 and 45, this is the result of the

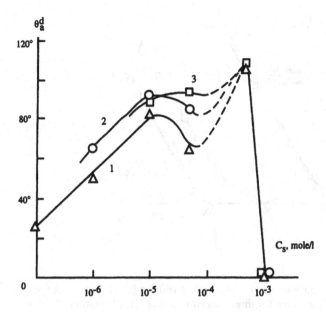

FIG. 28 Dynamic advancing angles of CTAB solution–silicon oil (curves 1 and 2) and CTAB solution–tetradecane (curve 3) systems vs. CTAB solution concentration. Contact angles θ_A were calculated on the basis of the $v(\Delta P)$ data (curves 1 and 2) and using droplet method (curve 3).

action of hydrophobic attraction forces in the thin aqueous interlayer between quartz surface and oil. In the region of concentration near 10^{-5} M, the silicon oil interface remains still hydrophobic, whereas the quartz surface becomes already hydrophobic. This corresponds to the highest manifestation of hydrophobic attraction forces between film interfaces.

At $C_o > 5 \times 10^{-5}$ M, when interface tension decreases to 13 mN/m, emulsification begins, which hinders the measurements of flow rates in thin capillaries. This region is shown in Fig. 28 by dotted curves.

Similar data were obtained using another method, when flow rates v of a small oil column (with length $l_0 \cong 0.5$ mm) in a capillary filled with CTAB solution were measured. In this case, both vessels shown in Fig. 1 contain CTAB solution of the same concentration. Fig. 29 shows part of the results obtained in this way. In this case, the $v(\Delta P)$ data do not depend on the direction of the column motion. Because of this, the branches of the graphs ($v < 0$ and $v > 0$) lie symmetrically with respect to the origin of coordinates. As seen from the graph, hysteresis of capillary pressures in this case decreases with CTAB concentration. At a concentration near CMC, $C_o = 10^{-3}$ M,

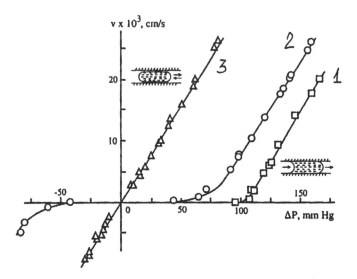

FIG. 29 Rate of motion v of silicon oil droplets with length l_o in dependence on pressure difference ΔP measured for three concentration of CTAB solution: $C_o = 5 \times 10^{-5}$ (curve 1, $l_o = 0.45$ mm); 5×10^{-4} (curve 2, $l_o = 0.48$ mm), and 10^{-3} M (curve 3, $l_o = 0.47$ mm).

(curve 3) complete wetting takes place, and the oil column moves along the capillary without hysteresis.

It is obvious that only the difference between dynamic tensions of wetting, $\delta = \gamma_R \cos \theta_R - \gamma_A \cos \theta_A$, can be measured in this case. However, assuming $\gamma_A \cong \gamma_o$ and using the γ_R values obtained by the above-considered first method, it becomes possible to estimate the dynamic advancing contact angle θ_A from a simplified expression, $\delta = \gamma_R - \gamma_o \cos \theta_A$. The results of calculation, shown by curve 2 in Fig. 28, are close to those obtained using the first method (curve 1). Because of the short length l_o of the column, viscous resistance of the solution in the capillaries was only a few percent higher than that for surfactant solution without an oil column.

Nearly the same results were obtained for CTAB solution–tetradecane system (curve 3, Fig. 28). However, emulsification begins to show up here at somewhat lower CTAB concentration. The process of emulsification is induced by formation after a retreating meniscus of a dynamic wetting film whose thickness is proportional, according to Derjaguin's theory [7], to the parameter $r(v\eta/\gamma_o)^{2/3}$. Decrease of interface tension with CTAB concentration promotes formation of thicker dynamic wetting films. Capillary instability of such films gives rise to the process of undulation and subsequent coalescence

[48]. This results in formation of both oil-in-water and water-in-oil emulsions depending on the direction of meniscus motion.

When tetradecane displaces aqueous solution, the receding contact angle decreases from its static value, $\theta_s = 76°$, to a dynamic one, $\theta_R = 35°$. At this point, an emulsion of tetradecane in aqueous solution ($C_o = 5 \times 10^{-4}$ M CTAB) starts to form (Fig. 30, region of flow rates $v < 0$). When the meniscus moves in the inverse direction, static receding contact angle θ_s is equal to $36°$, and dynamic angle tends to 0. In this case, emulsification of aqueous solution in tetradecane takes place (Fig. 30, region $v > 0$). The same effect was observed at $C_o = 10^{-3}$ M. At CTAB concentrations higher than 5×10^{-4} M, emulsification processes are facilitated due to decreasing interface tension, up to 7 mN/m. In the case of tetradecane, emulsification starts easier than for the more viscous silicon oil when the process of undulation and coalescence proceeds much slowly.

Now, let us consider the effects of dynamic contact angles and dynamic interface tensions in more detail. For this purpose, the $v(\Delta P)$ dependencies were obtained at much lower pressure differences and correspondingly at much lower flow rates, $v < 10^{-4}$ cm/s. The results obtained are shown in Fig. 31 [41].

Solid curves 1 in Fig. 31 relate to the results of measurements at $v < 10^{-4}$ cm/s (left scale). Dotted lines 2 and 3 give the flow rates v measured using the first of the above-considered methods at much higher flow rates (right scale).

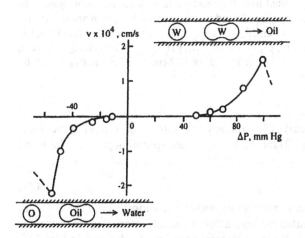

FIG. 30 Processes of emulsification illustrated using obtained $v(\Delta P)$ dependencies for tetradecane–CTAB solution system at $C_o = 5 \times 10^{-4}$ M in a quartz capillary, $r = 4.64$ μm, at $t = 22°C$.

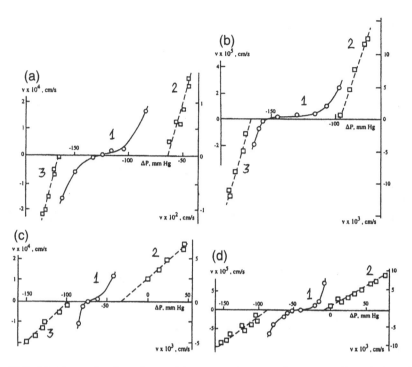

FIG. 31 Dependencies of v on ΔP measured at very low rates of menisci motion (curves 1) for CTAB solution–air (a, b) and CTAB solution–silicon oil (c, d) systems, $r = 6.56$ μm, $t = 20°C$. By dotted lines the results obtained for the same systems at higher flow rates for advancing (lines 2) and receding (lines 3) motion are shown. (a) $C_0 = 5 \times 10^{-4}$ M, pH 3.4, $l = 9.9$ cm, $\theta_{AS} = 0$, $\theta_A = 49°$. (b) $C_0 = 10^{-5}$ M, pH 3.3, $l = 10$ cm, $\theta_{AS} = 42°$, $\theta_A = 53°$. (c) $C_0 = 5 \times 10^{-4}$ M, pH 3.4, $l_1 = 10.3$ cm, $l_2 = 3.4$ cm, $\theta_{AS} = \theta_A \approx 0$. (d) $C_0 = 5 \times 10^{-5}$ M, pH 9.5, $l_1 = 11.4$ cm, $l_2 = 2.9$ cm, $\theta_{AS} = 21°$, $\theta_A = 77°$.

For every point of the curves 1, capillary pressure P_c of a moving meniscus was calculated using Eq. (2) on the basis of corresponding pairs of the v and ΔP values:

$$P_c = -\Delta P + \left(8\eta l v/r^2\right) \tag{39}$$

The P_c values were recalculated into wetting tensions: $\gamma \cos \theta = P_c r/2$. In Fig. 32 are shown calculated dependencies of the ratio $\gamma \cos \theta/\gamma_0$ on flow rates v (curves 1–4), where γ_0 is the bulk interface tension. Each curve in Fig. 32 is bounded by two vertical dotted lines, which give the limiting values of $\gamma_A \cos \theta_A/\gamma_0$ and $\gamma_R \cos \theta_R/\gamma_0$ obtained at high flow rates, $v > 10^{-4}$ cm/s, from curves 2 and 3 in Fig. 31, respectively.

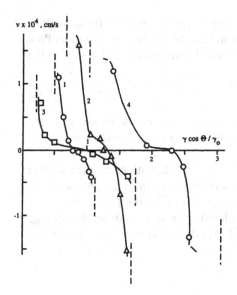

FIG. 32 Calculated from Fig. 31: values of the ratio $\gamma \cos \theta/\gamma_0$ in dependence on flow rates v for CTAB solution/air (curves 1 and 2) and CTAB solution/silicon oil (curves 3 and 4) systems: (1) $C_0 = 10^{-4}$ M, pH 9.5, $\theta_{AS} = 40$, $\theta_A = 66°$; (2) $C_0 = 5 \times 10^{-4}$ M, pH 3.4, $\theta_{AS} = 0$, $\theta_A = 49°$; (3) $C_0 = 5 \times 10^{-5}$ M, pH 9.5, $\theta_{AS} = 21°$, $\theta_A = 77°$; (4). $C_0 = 5 \times 10^{-4}$ M, pH 3.4, $\theta_{AS} = \theta_A \approx 0$.

In the case of low flow rates (curves 1–4 in Fig. 32) the $\gamma \cos \theta/\gamma_0$ values change gradually from $\gamma_A \cos \theta_A/\gamma_0$ at $v > 0$ to $\gamma_R \cos \theta_R$ at $v < 0$. The intersections of curves 1–4 with the pressure axis give the static values of $\gamma \cos \theta_s/\gamma_0$ determined at $v \approx 0$. Assuming that in this case $\gamma \approx \gamma_0$, static contact angles θ_s become equal to 40° (curve 1), 0° (curve 2), 21° (curve 3), and again 0° (curve 4). The zero values of the static contact angles correspond to high concentration of CTAB solution, $C_0 = 5 \times 10^{-4}$ M, in agreement with data shown in Fig. 26.

In the case of slow receding motion, as was shown above, $\cos \theta_R \approx 1$, and therefore $\gamma_R \cos \theta_R/\gamma_0 \approx \gamma_R/\gamma_0$. In the case of slow advancing motion ($v > 0$), both γ_R and θ_R depend on v. However, because flow rates are small, receding contact angles are influenced not only by v, but rather by dynamic interface tension. In the region of advancing motion of menisci, it is not possible to separate the effect of the γ_R and of the θ_R in the value of the product $\gamma_R \cos \theta_R$. The data shown in Fig. 32 demonstrate the changes in dynamic tensions of wetting in the region of small flow rates, $v < 10^{-4}$ cm/s. At higher flow rates (dotted lines in Figs. 31 and 32) both wetting tension and interface tension cease to depend on flow rates.

The dependencies shown in Fig. 32 may be explained in the framework of the approach considered above in Section II.A. In the case of very small flow rates, when $Pe_s \ll 1$, Eqs. (9) and (10) may be simplified. This results in the following expression

$$C_m/C_o = [1 \pm (kv/D)]^{-1}, \tag{40}$$

where C_m is the concentration of CTAB solution in the region near the meniscus. The plus sign corresponds to advancing motion and minus sign to the receding one.

In order to compare the derived expression (40) with experimental data shown in Fig. 32, the C_m/C_o values were recalculated into γ_m/γ_o values using isotherms of surface or interface tensions $\gamma(C)$ for CTAB solutions in contact with air or silicon oil. Fig. 33 shows, calculated in this way, the dependence of γ_m/γ_o on v for a particular case of 5×10^{-4} M CTAB–air system ($\gamma_o = 43$ mN/m),

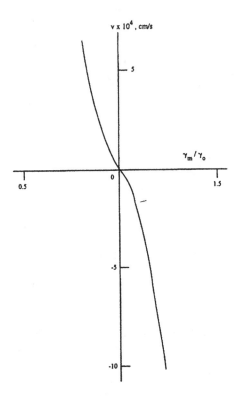

FIG. 33 Theoretical dependence of the ratio γ_m/γ_o on flow rates v calculated using Eq. (40).

assuming $k = 10^{-3}$ cm and $D = 10^{-6}$ cm^2/s. Here γ_m is the surface tension of the meniscus at $C = C_m$.

Comparison of the results of calculations (Fig. 33) with experimental data shown in Fig. 32, demonstrates qualitative agreement between both. The quantitative differences may be associated with the fact that the theory starts from the condition $\gamma = \gamma_0$ at $v = 0$. However, in experiments the conditions $v = 0$ corresponds to $\gamma \cos \theta < \gamma_0$ because $\cos \theta > 1$. In some cases, experimental values of $\gamma \cos \theta$ at $v = 0$ are higher than 1. This suggests that in the case under consideration the interface tension, even near to $v \approx 0$, has not yet relaxed to the equilibrium value γ_0.

Therefore, we can conclude that the proposed theoretical approach is in qualitative agreement with the experimental data, also in the case of low flow rates. The details of the mechanism of exchange of surfactant molecules between moving meniscus and capillary walls need further refinement and understanding.

V. CAPILLARY ELECTROKINETICS WITH SURFACTANT SOLUTIONS

A. Method and Principles

Thin quartz capillaries represent an excellent model for measuring their surface charge in contact with surfactant solutions. Testing of electrosurface properties with thin single capillaries [49–51] affords more reliable information than application of porous diaphragms or electrophoretic measurements. This is primarily because of the absence of polarization and end effects, which are difficult to account for, and also due to molecular smoothness of a juvenile surface and simple geometry of strictly cylindrical quartz capillaries.

A schematic representation of the device is shown in Fig. 34. The quartz capillary l is glued using epoxy resin into the cover of a high-pressure chamber 4. The broken end of the capillary is put into a polyethylene vessel 2 containing an aqueous solution under investigation. The upper end of the capillary is inserted into an orifice in the bottom of a Teflon vessel 3 containing the same solution. Hydrophobization of the outer surface of the capillary prevents outflow of the solution from the vessel.

The apparatus enables a pressure difference ΔP up to 100 atm at the capillary ends to be maintained. This makes it possible to perform electrokinetic measurements over a wide range of flow rates v, up to 1 m/s, while preserving laminar flow. The solution flows through the capillary from the lower to the upper vessel under the nitrogen pressure ΔP. A pipeline connects the compressed nitrogen in a bottle with the stainless steel high-pressure chamber 4 via reductor and a gas filter. The apparatus is coated on the outside

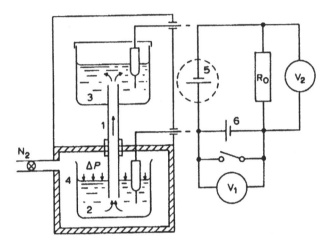

FIG. 34 Schematic representation of an electrokinetic cell with a thin single capillary.

with thermal insulation, and is thermostatted by pumping water through water jackets. The upper vessel in protected against the influence of external fields by a brass screen. The lower chamber is earthed.

The capillary with flowing solution acts as a current source 5 incorporated into electric circuit using two Ag/AgCl electrodes. Another voltage-controlling current source 6 is also included into the circuit. This enables the complete electrical characteristic of a solution in the capillary, i.e., the dependence of the potential difference E across the capillary ends on the current I to be obtained.

Fig. 35 shows, as an example, the dependencies $E(I)$ obtained for 10^{-4} M KCl solution in a thin quartz capillary ($r = 9.1$ μm) at three pressure differences $\Delta P = 10$, 20, and 30 atm. The intersections with the E axis give values of the streaming potential ΔE, and the intersections with the I axis give values of the steaming current, I_o. For determination of the ΔE and I_o values for a given solution, computer treatment of experimental points was done.

Use of high-pressure differences results in high and, consequently, easy-to-measure ΔE values, up to several hundreds of volts. The slope of the $E(I)$ graph is equal, according to Ohm's law, to the electrical resistance $R_s = dE/dI$ of a solution in the capillary. From the data shown in Fig. 35, it follows that the resistance R_s does not depend on the flow rate and on the action of the current source 6 included into the circuit. The accuracy of the resistance determination is usually about 1%.

In the range of ionic strengths of electrolyte solution $i > 10^{-4}$ M, the diffuse electrical layers in capillaries with radii $r > 1$ μm are not overlapped.

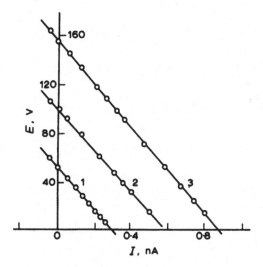

FIG. 35 Dependencies between drop of electrical potential E and current I when 10^{-4} M KCl solution flows through a quartz capillary, $r = 9.1$ μm, under various pressure differences $\Delta P = 10$ (curve 1), 20 (curve 2), and 30 atm (curve 3).

Therefore, the electrokinetic potential ψ can be calculated using a simple Helmholtz–Smoluchowsky equation:

$$\Delta E/\Delta P = e\psi/4\pi\eta K, \tag{41}$$

where e is the electron charge, η is the viscosity of a solution, and $K = L/\pi r^2 R_s$ is the electrical conductivity of the solution in a capillary with length L. In the case of bare quartz surface the potential ψ may be considered as Stern potential ψ_1. In the presence of adsorption layers of polymers or surfactants, the ψ values relate to the position of a slipping plane, and $\psi = \zeta$.

The geometrical characteristics of a capillary were determined measuring the resistance R_o of the capillary filled with 0.1 M KCl. In this case, surface conductivity may be neglected, and capillary radius r may be determined measuring resistance of the solution R_s and using bulk values K_o of the solution conductivity: $r = (L/\pi K_o R_s)^{1/2}$.

Fig. 36 shows the results of measurements of the streaming potential ΔE in a wide range of pressure differences ΔP. The open and filled points correspond to increasing and, subsequently, decreasing of the pressure differences, respectively. The linear form of the dependencies shows that the electrokinetic potential ψ_1 remains still constant, in spite of large values of a shear stress, up to 500 Pa acting on the capillary surface. This provides evidence for the ab-

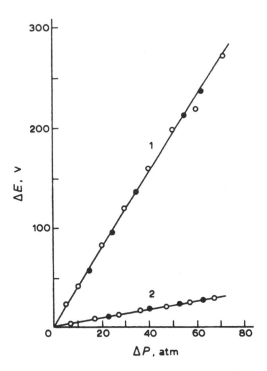

FIG. 36 Values of the streaming potential ΔE measured over a wide range of pressure differences ΔP in a quartz capillary (r = 9.2 μm, L = 6 cm) for KCl solution concentration 10^{-4} M (curve 1) and 10^{-3} M (curve 2).

sence of stagnant water layers near a quartz surface, and measured values of ψ_1 may be really considered as Stern potentials.

Measured values of surface potentials enable calculation of corresponding surface charge for symmetrical electrolytes

$$\sigma = (2\varepsilon iRT/\pi)^{1/2}\sinh(z\psi_1 F/2RT) \tag{42}$$

where ε is the dielectric constant, i is the ionic strength of the solution, z is the valence of ions, R is the gas constant, T is the temperature, and F is the Faraday constant.

Unfortunately, the small surface area of a single capillary (about 0.03 cm^2) makes measurements of adsorption of surfactant on a capillary wall by usual methods impossible. However, in the case of ionic surfactants, use of capillary electrokinetics allows measuring of the surface charge in dependence of surfactant concentration. The dependence, as will be shown below, may sometimes be used for determination of adsorption isotherms of ions.

B. Cationic Surfactant Solutions

Fig. 37a shows the results of measurements of Stern potentials ψ_1 in dependence on CTAB concentration C_o. Calculated using Eq. (42), the corresponding dependence for surface charges, $\sigma = \sigma(C_o)$, is shown in Fig. 37b [46]. Measurements were performed at pH 6.5 in the background electrolyte, 5×10^{-4} M KCl. Experimental data are shown in Fig. 37 by points, whereas solid lines are plotted using Langmuir isotherm, which was recommended to describe adsorption data for weakly charged surfaces [52]:

$$\theta = (c/c_o)\exp(-\Phi/kT)[1 + (c/c_o)\exp(-\Phi/kT)]^{-1} \tag{43}$$

where θ is the degree of coverage of active surface sites by ions, Φ is the energy of adsorption, c is the equilibrium concentration of ions and c_o is the limiting

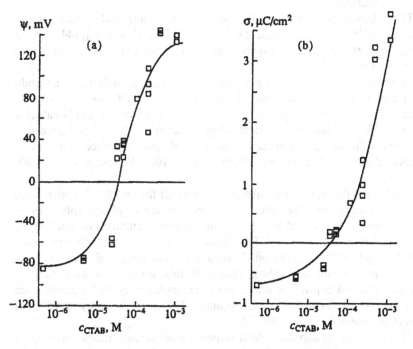

FIG. 37 Dependencies of the electrokinetic potential ψ_1 (a) and of surface charge σ (b) of quartz capillaries on concentration of CTAB solution at pH 6.5.

concentration of solvent molecules (for water $c_o = 55.51$ M); k is the Boltzman constant.

As was shown in Refs. 53–55, the monolayer Langmuir-type model of CTA^+ ion adsorption on silica may be used for describing the experimental data. The model presumes independent simultaneous adsorption of ions on two types of surface centers, charged and noncharged. In this case N_1 cations are attached to N_c charged centers, and N_2 cations are adsorbed on N_n neutral centers. The ratio $m = N_n/N_c$ characterizes the initial state of the surface. The number of charged sites N_c may be determined measuring the initial surface charge $\sigma_i = eN_c$ in the absence of surfactant at $C_o = 0$.

Consideration of the rates of processes of adsorption and desorption of ions in the state of a dynamic equilibrium in the framework of the Langmuir approach results in the following equation for monolayer adsorption:

$$\sigma/\sigma_i = (1 - mK_1K_2C^2)/[1 + K_1C(1 + K_2C], \tag{44}$$

where $C = C_o/c_o$, $K_1 = \exp(-\Phi_1/kT)$ and $K_2 = \exp(-\Phi_2/kT)$. Here Φ_1 and Φ_2 are potentials of specific adsorption of cations on charged and noncharged sites, respectively.

The values of potentials of adsorption of CTA^+ ions and of parameter m were determined comparing the data shown in Fig. 37 with Eq. (44) using a computer procedure that ensures the best fit of the experimental and theoretical data.

Measurement of the Stern potentials ψ_1 in quartz capillaries with radii from 5 to 10 μm were performed in the range of CTAB concentrations C_o from 10^{-6} to 10^{-3} M at various pH, from 3 to 9.5. By changing pH values it was possible to obtain concentration dependencies $\psi_1(C_o)$ and $\sigma(C_o)$ starting from the different initial surface charge σ_i of quartz surface at $C_o = 0$. Measured values of Stern potentials vary from -105 mV at pH 9.5 to -10 mV at pH 3.

Analysis of the results obtained has shown that the best fit of experimental data with theory gives the following values of potential of adsorption $\Phi_1 = -15\,kT$, $\Phi_2 = -11\,kT$, and $m = 8$. This means that the number of noncharged sites that are able to adsorb CTA^+ ions is much larger than charged ones. However, adsorption energy on charged centers is much higher due to electrostatic mechanism of binding. Owing to this, adsorption starts first on charged sites. Adsorption on noncharged centers becomes predominant when the main part of charged sites is neutralized. This cause the charge reversal of quartz surface at $C_o = 3 \times 10^{-5}$ M.

Therefore, using capillary electrokinetics one can in principle investigate the adsorption of cationic surfactant on a silica surface and explain the results obtained on the basis of Langmuir-type isotherms of adsorption.

C. Poly(Ethylene Oxide) Solutions

Water-soluble polymers that adsorb due to hydrogen bonds on the quartz surface may be considered in the Gibbs sense as surfactants. As differentiated from ionic surfactants, capillary electrokinetics allows in this case investigation not of the adsorption state but the state of adsorbed polymer layers and its influence on surface charge of quartz capillaries. The experiments were performed with aqueous poly(ethylene oxide) (PEO) solutions of various concentration in a background KCl solution. As an example, the results obtained for PEO with molecular mass $M = 6 \times 10^5$, $M_w/M_n = 1.1$ (Union Carbide) will be discussed.

Measuring electrokinetic potentials before and after polymer adsorption, that is, the Stern potential of the bare quartz surface ψ_1 and ζ potential, which reflect a shift in the position of slipping plane, it becomes in principle possible to assess the hydrodynamic thickness δ of an adsorbed polymer layer. Assuming that presence of polymer does not change significantly the exponential distribution of local potential values $\psi(x)$ in the electrical double layer, the hydrodynamic thickness may be calculated from the Gouy equation

$$\delta = (1/\kappa) \ln[\tanh(\psi_1 F/4RT)/\tanh(\zeta F/4RT)] \qquad (45)$$

where $1/\kappa$ is the Debye radius and F is the Faraday constant.

However, correct determination of the δ value is possible when the surface potential of quartz preserves its initial magnitude ψ_1 and is not influenced by formation of hydrogen bonds between quartz surface and polymer molecules. It was shown in Refs. 47, 56 and 57 that potential of quartz surface under an adsorbed polymer layer, ψ_{10}, is lower than Stern potential ψ_1 of a bare quartz surface.

In Fig. 38 are shown results of measurements of ζ potentials of quartz capillaries, the surface of which was covered with adsorbed layers of PEO, in dependence on KCl concentration. Equilibrium adsorption layers are formed by pumping PEO solutions of various concentrations $C_p = 0.1$ (curve 1), 10^{-2} to 10^{-3} (curve 2), and 10^{-4} (curve 3) g/dm^3. By curve 4 are shown the results of measurements of Stern potentials ψ_1 for bare quartz capillaries in KCl solutions before adsorption of PEO.

At low concentration of PEO, $C_p = 10^{-4}$ g/dm^3, molecules of PEO are adsorbed in a flat conformation and therefore the slipping plane is shifted on very small distance of about 0.3 nm [56]. This allows, as a first approximation, consideration of the measured electrokinetic potential $\zeta = -72$ mV at $C = 10^{-4}$ M KCl (curve 3, Fig. 38) as ψ_{10} potential because polymer in this case forms an adsorbed monolayer.

Now, using obtained values of quartz surface potential under adsorption layer of PEO, $\psi = -72$ mV and $\zeta = -32$ mV, for saturated adsorption layer

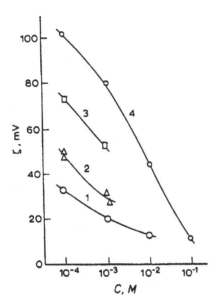

FIG. 38 Electrokinetic potentials of quartz capillaries ($r = 7.2\,\mu m$) equilibrated with PEO solutions of various concentration C_p in dependence on KCl concentration C. $C_p = 10^{-1}$ (curve 1), 10^{-2} to 10^{-3} (curve 2), and 10^{-4} g/dm^3 (curve 3). The ψ_1 potentials measured for bare quartz capillary are shown by curve 4.

of PEO with concentration 10^{-1} g/dm^3 at the same KCl concentration (curve 1, Fig. 38), Eq. (45) gives the hydrodynamic thickness of the layer $\delta = 20$ nm. In the case of higher electrolyte concentration, 10^{-3} M KCl, similar calculations result in $\delta = 10$ nm. Calculated values of the hydrodynamic thickness are in agreement with independent measurements [56,58,59].

After establishment of an equilibrium adsorption layer of PEO at small pressure differences (2.5 to 3 atm), the influence of flow rate of the same solution was studied. For this purpose, measurements were performed at increasing pressure differences ΔP, up to 60 atm.

Fig. 39 shows the dependence of hydrodynamic thickness δ of an adsorbed saturated PEO layer on the tangential shear stress $\tau = r\,\Delta P/2L$ acting on the peripheric part of the layer, where r is the capillary radius and L is the length of the capillary. The obtained dependence is reversible. This points to elastic or viscoelastic deformation of adsorbed layers. Thin quartz capillaries represent a simple model system that may be effectively used for investigation of both electrosurface phenomena and mechanical properties of adsorbed layers of surface-active polymers.

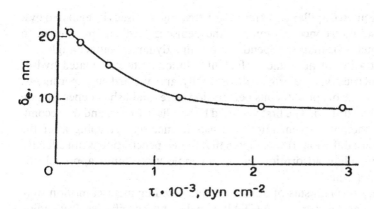

FIG. 39 Dependence of the hydrodynamic thickness δ of a saturated adsorption PEO layer on shear stress τ, when 10^{-1} g/dm^3 PEO solution was pumped through the same quartz capillary ($r = 7.2$ μm). Concentration of the background KCl electrolyte $C = 10^{-4}$ M.

VI. CONCLUSIONS

1. Using thin, single, quartz capillaries with radii of about 5 to 10 μm, advancing and receding motions of surfactant solution menisci contacting gas or an immiscible liquid were investigated. Nonionic surfactant solutions EO_{10} and EO_4 with concentration 0.025 wt% much higher than CMC and cationic surfactant CTAB with concentration from 10^{-6} M to CMC were used.

2. The first method consisted in measuring the rates of meniscus movement under pressure difference applied to the capillary ends. The second method was used for measuring the rates of advancing and receding motion when the meniscus is shifted back and forth within a small portion of the capillary. In this case the length of a liquid column in the capillary remains practically constant. In both cases, measured flow rates follow the Washburn equation, which describes viscous flow of two immiscible fluids. The presence of surfactants influences only the values of wetting tension and contact angles.

3. Flow of surfactant solutions is drastically influenced by adsorption of surfactant on capillary surface. This results in a change of surface or interface tension of moving menisci. In this connection, investigation of physicochemical mechanisms of dynamic tension and dynamic contact angles was possible only when the solution is previously equilibrated with the capillary.

4. Even in equilibrated capillaries, capillary pressure of moving menisci is influenced by transfer of surfactant molecules between meniscus and wetting films formed on the surface. The effects were discussed on the basis of a proposed theoretical approach that takes into account diffusive transport of sur-

factant molecules. Capillary pressure of a retreating meniscus is regulated by a decrease in solution concentration near the meniscus. Capillary pressure of an advancing meniscus mainly depends on forming dynamic contact angles.

5. Penetration of nonionic surfactant solutions into nontreated hydrophobic capillaries was described theoretically and verified in experiments. Two mechanisms of spontaneous penetration were established depending on solution concentration. The first is caused by capillary forces, and the second, at lower surfactant concentration, consists in meniscus creeping after the front of surface diffusion. It was shown that forced penetration is not effective because forestalling adsorption surfactant molecules cannot approach the advancing meniscus.

6. Similar mechanisms of advancing and receding menisci motion were observed for cationic surfactant CTAB in hydrophilic capillaries. Depending on solution concentration, dynamic advancing angles pass through a maximum and tend to zero when approaching CMC. This secures acceleration of capillary suction and oil displacement. However, substantial decrease in this case of interface tension may result in formation near the meniscus zone of direct or inverse emulsions, depending on the direction of meniscus motion.

REFERENCES

1. Sobolev, V.D. A Method of Manufacturing of Quartz Capillaries, Licence of USSR No. 833588. Bulletin of Invention 1981, No. 20.
2. Zorin, Z.M.; Churaev, N.V. Adv. Colloid Interface Sci. 1992, *40*, 85.
3. Washburn, E.W. Phys. Rev. 1921, *17*, 273.
4. Abbasov, M.; Zorin, Z.M.; Iskandarjan, G.A.; Churaev, N.V. Kolloid. Zh. USSR 1989, *51*(N 4), 627.
5. Abbasov, M.; Zorin, Z.M.; Churaev, N.V. Kolloid. Zh. USSR 1989, *51*(N4), 634.
6. Ershov, A.P.; Zorin, Z.M.; Churaev, N.V. Kolloid. Zh. USSR 1990, *52*(N2), 250.
7. Derjaguin, B.V. C.R. Acad. Sci. USSR 1943, *39*, 13; Acta Phys.-Chim. 1945, *20*, 137.
8. Ershov, A.P.; Zorin, Z.M.; Svitova, T.F.; Churaev, N.V. Kolloid. Zh. USSR 1993, *55*(N 3), 39.
9. Derjaguin, B.V.; Zhelezny, B.V.; Tkachev, A.P. Dokl. Akad. Nauk USSR 1972, *206*, 1146.
10. Hopf, W.; Geigel, Th. Colloid Polym. Sci. 1987, *265*, 1075.
11. Tchalowska, S.; Herder, P.; Pugh, R.; Stenius, P.; Eriksson, J.C. Langmuir 1990, *20*, 349.
12. Ershov, A.P.; Zorin, Z.M.; Svitova, T.F.; Churaev, N.V. Kolloid. Zh. RAS 1993, *55*(N 4), 45.
13. La Mer, V.K.; Healy, T.W. Science 1965, *148*, 36.
14. Derjaguin, B.V.; Fedosejev, V.A.; Rozenzveig, L.A. Dokl. Akad. Nauk USSR 1966, *167*, 617.

15. Ershova, I.G.; Zorin, Z.M.; Churaev, N.V. Kolloid. Zh. USSR 1976, *38*(N 5), 973.

16. Churaev, N.V.; Ershova, I.G. Kolloid. Zh. USSR 1971, *33*(N 6), 913.

17. Berezkin, V.V.; Zorin, Z.M.; Frolova, N.V.; Churaev, N.V. Kolloid. Zh. USSR 1975, *37*(N 6), 1040.

18. Zolotarjev, P.P.; Starov, V.M.; Churaev, N.V. Kolloid. Zh. USSR 1976, *38*(N 5), 895.

19. Berezkin, V.V.; Ershov, A.P.; Esipova, N.E.; Zorin, Z.M.; Churaev, N.V. Kolloid. Zh. USSR 1979, *41*(N 5), 849.

20. Churaev, N.V.; Martynov, G.A.; Starov, V.M.; Zorin, Z.M. J. Colloid Polym. Sci. 1981, *259*, 747.

21. Churaev, N.V.; Zorin, Z.M. Colloids Surf. A 1995, *100*, 131.

22. Zorin, Z.M.; Iskandarjan, G.A.; Churaev, N.V. Kolloid. Zh. USSR 1978, *40*(N 4), 671.

23. Esipova, N.E.; Zorin, Z.M.; Trapeznikov, A.A.; Shchegolev, G.G. Kolloid. Zh. USSR 1984, *46*(N 6), 1098.

24. Berezkin, V.V.; Zorin, Z.M.; Iskandarjan, G.A.; Churaev, N.V. In *Physicochemistry of Surfactants (in Russian)*; Vneshtorgisdat: Moscow, 1978; Vol. 2. Part 2, 329 pp.

25. Majevskaja, L.N.; Zorin, Z.M.; Churaev, N.V. Kolloid. Zh. USSR 1977, *39*(N 6), 1081.

26. Cohen Stuart, M.A.; Cazabat, A.M. Prog. Colloid Polym. Sci. 1987, *74*, 64.

27. Princen, H.M.; Cazabat, A.M.; Cohen Stuart, M.A.; Heslot, F.; Nicolet, S. J. Colloid Interface Sci. 1988, *126*, 84.

28. Mc Guiggan, P.M.; Pashley, R.M. Colloids Surf. A 1987, *27*, 277.

29. Cazabat, A.M.; Heslot, F.; Fraysse, N.; Carles, P.; Levinson, P. Prog. Colloid Polym. Sci. 1990, *82*, 82.

30. Cazabat, A.M.; Heslot, F.; Fraysse, N. Prog. Colloid. Polym. Sci. 1990, *83*, 52.

31. Cazabat, A.M. Adv.Colloid Interface Sci. 1991, *34*, 73.

32. Cazabat, A.M.; Fraysse, N.; Heslot, F. Colloids Surf. A 1991, *53*, 1.

33. Cazabat, A.M. Adv.Colloid Interface Sci. 1992, *42*, 65.

34. Fraysse, N. Doctoral thesis, Universite de Paris V1, 1991.

35. Hill, R.M. Curr. Opin. Colloid Interface Sci. 1998, *3*, 247.

36. Zhu, S.; Miller, W.G.; Scriven, L.E.; Davis, H.T. Colloids Surf. A 1994, *90*, 63.

37. Svitova, T.F. Personal communication.

38. Mahe, M.; Vignes-Adler, M.; Rousseau, A.; Jacquin, G.G.; Adler, P.M. J. Colloid Interface Sci. 1988, *126*, 314.

39. Kao, R.L.; Wasan, D.T.; Nikolov, A.D.; Edwards, D.A. Colloids Surf. A 1989, *34*, 389.

40. Churaev, N.V.; Ershov, A.P.; Esipova, N.E.; Iskandarjan, G.A.; Sobolev, V.D.; Svitova, T.F.; Zakhrova, M.A.; Zorin, Z.M.; Poirier, J.E. Colloids Surf. A 1994, *91*, 97.

41. Churaev, N.V.; Ershov, A.P.; Zorin, Z.M. J. Colloid Interface Sci. 1996, *177*, 589.

42. Churaev, N.V. Adv.Colloid Interface Sci. 1995, *58*, 87.

43. Churaev, N.V.; Sobolev, V.D. Adv.Colloid Interface Sci. 1995, *61*, 1.
44. Brown, C.E.; Jones, N.J.; Neustadter, E.L. J. Colloid Interface Sci. 1980, *76*, 582.
45. Mumley, T.E.; Radke, C.J.; Williams, M.C. J. Colloid Interface Sci. 1986, *109*, 398.
46. Sergeeva, I.P.; Sobolev, V.D.; Madjarova, E.A.; Churaev, N.V. Kolloid. Zh. RAS 1995, *57*(N 6), 849.
47. Zorin, Z.M.; Churaev, N.V.; Esipova, N.E.; Sergeeva, I.P.; Sobolev, V.D.; Gasanov, E.K. J. Colloid Interface Sci. 1992, *152*, 170.
48. Berdinskaja, N.V.; Kussakov, M.M. Kolloid. Zh. USSR 1972, *34*(N 5), 755; 1973, *35*(N 1), 137.
49. Sergeeva, I.P.; Sobolev, V.D.; Churaev, N.V.; Derjaguin, B.V. J. Colloid Interface Sci. 1981, *84*, 451.
50. Sergeeva, I.P.; Sobolev, V.D.; Churaev, N.V. *Surface Forces and Boundary Layers of Liquids (in Russian)*; Nauka: Moscow, 1983; 102 pp.
51. Churaev, N.V.; Sergeeva, I.P.; Sobolev, V.D.; Ulberg, D.E. J. Colloid Interface Sci. 1992, *151*, 490.
52. Lyklema, J. In *Adsorption from Solution at the Solid/Liquid Interface*; Parfitt, G.D., Rochester, C.H., Eds.; Academic Press, 1983.
53. Muller, V.M.; Sergeeva, I.P.; Churaev, N.V. Kolloid. Zh. RAS 1995, *57*(N 3), 368.
54. Sergeeva, I.P.; Muller, V.M.; Zakharova, M.A.; Sobolev, V.D.; Churaev, N.V. Kolloid. Zh. RAS 1995, *57*(N 3), 400.
55. Sergeeva, I.P.; Sobolev, V.D.; Churaev, N.V.; Jacobasch, H.J.; Weidenhammer, P.; Schmitt, F.J. Kolloid. Zh. RAS 1998, *60*(N 5), 650.
56. Churaev, N.V.; Nikologorskaja, E.A. Colloids Surf. A 1991, *71*, 59.
57. Eremenko, B.V.; Sergienko, Z.A.; Poljakova, V.V. Kolloid. Zh. USSR 1980, *42*(N 6), 1064.
58. Kawaguchi, M.; Mikura, M.; Takahashi, A. Macromolecules 1984, *17*, 2063.
59. Van der Beek, G.P.; Cohen Stuart, M.A. J. Phys. France 1988, *49*, 1449.

7

Contact Angle and Surface Tension Measurement

KENJI KATOH   Osaka City University, Osaka, Japan

I. INTRODUCTION

The wetting phenomenon is an important issue in various technological processes. In some fields, liquids are desired to spread over solid surfaces, e.g., lubrication oils on metallic surfaces or paint on paper. On the other hand, it is necessary for hydrophobic coatings to repel water such as Teflon film on frying pans. The behavior of bubbles on solid surfaces immersed in liquid often has important effects on the performance of industrial apparatus dealing boiling or condensation. In these problems regarding wetting, it is known that the behavior of a drop or bubble on a solid surface is dependent on the three interfacial tensions between solid, gas, and liquid phases, as shown in Fig. 1. The tangential force balance between these interfacial tensions on the three-phase contact line leads to the following well-known Young's equation [1]:

$$\sigma_{SV} - \sigma_{SL} = \sigma_{LV} \cos \alpha_Y. \tag{1}$$

σ_{SV}, σ_{SL}, and σ_{LV} indicate solid–vapor, solid–liquid, and liquid–vapor interfacial tensions, respectively. The environmental atmosphere is assumed to be filled with saturated vapor of liquid. When a drop is exposed to air, however, σ_{LV} usually does not change because a thin layer of saturated vapor may be formed around the drop [2]. In the right-hand side of Eq. (1), α_Y is the angle between the solid surface and the liquid–vapor interface measured from the inside of the liquid phase and is called the contact angle. When the difference between the two interfacial tensions on the left-hand side of Eq. (1) is large enough to make α_Y on the right-hand side small, the solid is favorably wetted by the liquid. As the drop size becomes sufficiently small and the curvature of the solid–gas–liquid contact line becomes quite large, we should add a term representing the effect of line tension to the above equation [3]. In this

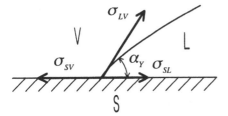

FIG. 1   Force balance on solid–liquid–gas interface.

case, such a term, i.e., line tension times the curvature of the three-phase contact line, appears in the right-hand side of Eq. (1), and the contact angle changes from α_Y to balance with the left-hand side. The line tension might play an important role in the wetting phenomenon in some problems, although this is not yet understood clearly. In this section, we do not touch on the effect of line tension and we use Eq. (1) exclusively. Readers interested in the role of line tension are referred to references [3].

We can regard the interfacial tension represented in Eq. (1) as the free energy per unit area of the interface [4]. This may be readily understood by imagining a soap film surrounded by a square wire with one movable side. The liquid–vapor interfacial tension is equal to the work necessary to spread the film surface by a unit area, opposing the tension acting on the movable side [5]. Hence we can regard Eq. (1) as the relation between three kinds of energies, i.e., we can calculate the reversible energy change ($\sigma_{SV} - \sigma_{SL}$) using σ_{LV} cos α_Y without knowledge of σ_{SV} or σ_{SL} themselves when a unit area of the solid surface is wetted by liquid. It is usually not easy to directly measure the interfacial tensions on the left-hand side of Eq. (1), and the wetting behavior is often discussed using σ_{LV} and α_Y.

It is noted that Eq. (1) holds for an ideally smooth and homogeneous solid surface. We cannot write the same relation as Eq. (1) for the contact angle macroscopically observed on a practical solid surface with inhomogeneity such as roughness or heterogeneity due to adsorption. As shown in Fig. 2, different contact angles usually appear, depending on the direction of liquid movement, i.e., the advancing contact angle θ_A is observed when the liquid wets the solid surface, and the receding contact angle θ_R is observed as the solid surface becomes dry upon the retreat of the liquid. This is called the "hysteresis phenomenon of the contact angle." When we put a liquid drop onto a solid surface without any special care, we usually observe a contact angle between the above two extreme values. The hysteresis phenomenon is influenced in a complicated manner by roughness and heterogeneity of the solid surface and also by the irreversible movement of the three-phase contact line. Many authors have extensively investigated the influence of these factors

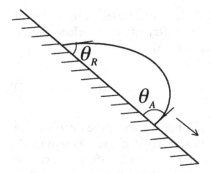

FIG. 2 Contact angle hysteresis observed when a drop rolls down on a slope.

on wetting behavior [6–21]. Although the mechanism of how the macroscopic contact angles are determined is not yet clearly understood, we often use the advancing or receding contact angles as basic quantities indicating the wettability between a liquid and a solid surface. In this section, we call the contact angle defined by Eq. (1) Young's contact angle to distinguish it from the macroscopically observed contact angles θ_A and θ_R.

The correct values of the liquid–vapor interfacial tension and contact angle are necessary to analyze the wetting problems stated at the beginning of this section. Many methods have thus far been proposed for the measurement of these basic quantities. The methods may be classified into two categories, i.e., (1) methods of measuring the dynamic quantities and (2) methods of measuring the geometrical quantities. The Wilhelmy [22] and Du Noüy [23] methods for the measurement of liquid–vapor interfacial tension are typical examples of (1). In these methods, a clean glass plate or ring with zero contact angle is drawn from the liquid and the liquid–vapor interfacial tension acting on the periphery is measured directly using devices such as a microbalance. These methods may also be applied to the measurement of the contact angle by using a test plate, to which the test liquid attaches at certain contact angle, instead of a glass plate [24]. The maximum bubble pressure or drop weight method is another example of (1) for the measurement of interfacial tension [25,26]. Both involve the use of force balance on the bubble or drop due to interfacial tension. The interfacial tension is calculated from the measurement of pressure inside the bubble in the former method and from the drop weight in the latter when it detaches from an orifice. Generally, the Wilhelmy, Du Noüy, and drop weight methods are widely used for the measurement of interfacial tension.

The capillary rise method is a classical example of category (2) for measuring liquid–vapor interfacial tension, in which the liquid height in a capillary tube is measured [27]. On the other hand, the geometrical shape of the liquid–vapor interface, such as drop or bubble, is often used as the object of measurement. It is well known that the geometry of the interface can be determined by the following Laplace equation [28]:

$$\sigma_{LV}\left(\frac{1}{R_1} + \frac{1}{R_2}\right) = \Delta P, \tag{2}$$

where R_1 and R_2 indicate the principle radii of curvature of the interface and ΔP is the pressure difference between gas and liquid phases. Equation (2) indicates the force balance acting on the liquid–vapor interface and can be solved analytically or numerically. If ΔP, such as the static pressure of liquid due to gravitational force, is known, the interfacial tension can be obtained by measuring the geometric specifications of the interface, such as drop height and radius [29–31]. Contact angles of the drop or bubble are often measured directly using a telescope with a goniometer or in photographs [32]. The tilted plate method is another example of direct measurement of the contact angle, in which a plate immersed in liquid is tilted until the liquid meniscus becomes horizontal at the point of contact [33]. The inclined angle of the plate is equal to the contact angle. The method of using a sessile or pendant drop described above can be applied to the measurement of the contact angle since the geometrical shape is also dependent on the contact angle as a boundary condition on the solid surface [34]. Sometimes the capillary rise of the meniscus attached to a vertical plate is measured to obtain contact angles. The geometry of the meniscus is also determined by Eq. (2), and the contact angle can be easily calculated from the measured height [35].

All the above methods are classic for the measurement of the liquid–vapor interfacial tension and the contact angle and have long been used. Detailed reviews of the measurement of interfacial tension and the contact angle have been given in Refs. [5] and [35–37]. Recently, the above methods have been made more sophisticated and precise by using elaborate techniques such as lasers, computers, and graphic data processing [39–52]. Please refer to those references for details.

Each method mentioned above has its own advantage. However, the methods of category (1) require an accurate apparatus for measuring force because the surface tension is usually small. Similarly, an optical or other device is necessary to measure the geometry of a drop or the liquid meniscus in methods of category (2) since they are not measured directly. Although many sophisticated methods have been considered, the measurements of surface tension and contact angle are generally not so easy and somewhat expensive.

In this section, we discuss a new method for measuring the liquid–vapor interfacial tension and the contact angle based on a principle different from those proposed in the past. One can see a liquid meniscus formed under a solid such as a needle when it is drawn from a liquid bath. If the solid is raised further, the meniscus spontaneously drops off the surface. This phenomenon might be the result of some kind of thermodynamic or other instability and might be related to the wetting characteristics between the solid and liquid. If the mathematical relationship could be derived theoretically between the critical height of the solid when the meniscus falls off and the wetting, we could obtain the liquid–vapor interfacial tension or contact angle simply by direct measurement of that height. Below, we first discuss the measurement method based on the thermodynamic instability of the liquid meniscus. For the theoretical consideration of instability, a method of calculating the energy due to wetting behavior when hysteresis of contact angle occurs will be proposed. Secondly, we discuss the feasibility of the method based on the geometrical instability of the liquid meniscus.

II. MEASURING METHOD APPLYING THE THERMODYNAMIC INSTABILITY OF LIQUID MENISCUS

A. Hysteresis and Macroscopic Contact Angle

As stated in Section I, advancing and receding contact angles are observed in the macroscopic wetting behavior. There have been many reports regarding the effect of roughness or inhomogeneity on the hysteresis phenomenon of the contact angle. Johnson and Dettre [7,36] considered the thermodynamic free energy of a system consisting of a spherical drop on a solid surface with concentric regular roughness and heterogeneity. The free energy of the system was calculated, varying the attached surface area of a drop of constant volume. There are numerous metastable states corresponding to the local minima of the system energy. Johnson and Dettre explained the hysteresis phenomenon by suggesting that the extreme contact angles among the metastable states appear when the liquid advances or recedes. Later, Eick et al. [8] and Li and Neumann [17] discussed in detail a problem similar to that treated by Johnson and Dettre, but including gravitational energy. The validity of those analyses was verified experimentally, to some extent, for a surface with regular two-dimensional roughness [10,53]. On the other hand, Joanny and de Gennes [12] and Jansons [13] considered the behavior of the three-phase contact line when it moves past a single defect such as a circular hollow or heterogeneous region. They showed that the state of force balance is different between advancement and retreat of the contact line and also discussed the relation of hysteresis to irreversible dissipation occurring when the contact line passes over the defect region. Although those reports presented an interesting concept, the mechanism of hysteresis is not yet understood completely. For example, Johnson and Dettre et al. discussed the wetting behavior only for the static state, and the results obtained by Joanny and de Gennes et al. are limited to the special case of a single defect. Besides the roughness or heterogeneity stated above, some authors have discussed the effect of line tension on the wetting behavior [3,54–56].

The theoretical considerations mentioned above are based on the assumption that Eq. (1) holds locally on the solid surface, i.e., the inhomogeneous surface is a collection of small homogeneous patches. In this case, the infinitesimal energy change ΔE_W when the liquid wets the solid surface reversibly can be calculated at each local portion as

$$\Delta E_W = -\sigma_{LV} \cos \alpha_Y \Delta s, \tag{3}$$

where Δs indicates the infinitesimal surface area of one homogeneous patch. In Eq. (3), Young's contact angle α_Y is a function of position on the solid surface. Hence it is necessary to integrate Eq. (3) for the calculation of E_W over the entire surface as

$$E_W = -\sigma_{LV} \int \int_s \cos \alpha_Y ds. \tag{4}$$

In the above equation, effects of both the roughness (change of surface area) and heterogeneity (change of α_Y) are included in the integration. It is noted that Eq. (4) holds when the three-phase contact line moves quasi-statically everywhere on the solid surface. We should add extra work if the three-phase contact line moves irreversibly on the solid surface. One can see that the calculation of E_W is complicated and it may be unrealistic to integrate Eq. (4) every time we consider the macroscopic wetting behavior such as mentioned at the beginning of this section.

Wenzel suggested a concept for defining the macroscopic contact angle [57]. He considered a rough surface with constant α_Y. In this case, the energy E_W of Eq. (4) can be written as

$$E_W = -\sigma_{LV} \cos \alpha_Y rS, \tag{5}$$

where r indicates the ratio of the true area of the rough surface to the apparent surface area S. Wenzel's apparent contact angle θ_W is defined as

$$\cos\theta_W = r \cos \alpha_Y, \tag{6}$$

If Wenzel's contact angle actually appears, we can easily calculate the energy change E_W by using measured θ_W as

$$E_W = -\sigma_{LV} \cos \theta_W S \tag{7}$$

We can avoid the difficulty of calculating r by using Eq. (7). In a similar manner, for a smooth but heterogeneous solid surface composed of two regions with different Young's contact angles α_{Y1} and α_{Y2}, Cassie [58] suggested the following apparent contact angle θ_C:

$$\cos \theta_C = A_1 \cos \alpha_{Y1} + A_2 \cos \alpha_{Y2}, \tag{8}$$

where A_1 and A_2 indicate the proportion of surface area with α_{Y1} and α_{Y2}, respectively. We can also calculate E_W if θ_C is used instead of θ_W in Eq. (7). Although the analysis of the wetting problem can be greatly simplified by the use of θ_W or θ_C, unfortunately, practically observed advancing or receding contact angles are usually different from those angles [17,36]. The reason for this may be considered as follows [11,13]. The definitions of θ_W and θ_C are based on the reversible movement of the three-phase contact line since they are derived from Eq. (3) or Eq. (4). However, irreversible motions inevitably occur at some local positions in the actual wetting behavior. For example, one can observe stick-slip motion when the contact line moves on an inhomogeneous solid surface. Also, extra work might be needed to form a residual drop or gas bubble in a trough of the rough surface when the contact line passes over it. These energies may possibly depend on the direction of contact line movement, i.e., advance and retreat. Some authors consider that these irreversible effects may be the cause of contact angle hysteresis, although the details are not yet clear [11–13]. Wenzel's or Cassie's contact angles do not consider the energies due to irreversibility. Hence we cannot use those angles directly to calculate the actual energy change when the liquid wets the solid surface by some apparent area.

In order to calculate E_W in the macroscopic wetting behavior, here we assume the following rather simple relation:

$$E_W = -\sigma_{LV} \cos \theta S \tag{9}$$

$$\Delta E_W = -\sigma_{LV} \cos \theta \Delta S \tag{10}$$

($\theta = \theta_A$ for advancing and $\theta = \theta_R$ for receding).

Equation (10) is the differential form of Eq. (9). The above equation means that the change of energy when the three-phase contact line advances or recedes on the solid surface by apparent area S can be calculated by using the macroscopically observed contact angles θ_A or θ_R. Equation (9) or Eq. (10) suggests that the effect of irreversible energy change stated above could be wholly reflected by the value of the macroscopic contact angles.

The validity of Eq. (9) or Eq. (10) may be justified by a simple fact. Let us consider a plate with roughness or heterogeneity immersed slowly into a liquid bath, as shown in Fig. 3. A liquid meniscus forms on the plate and its height decreases gradually at the beginning of immersion. When the meniscus height reaches a threshold value, it no longer changes and we can observe the constant contact angle, i.e., advancing angle θ_A. The force $-\sigma_{LV} \cos\theta_A$ acts constantly on the plate during immersion. Since the meniscus shape does not

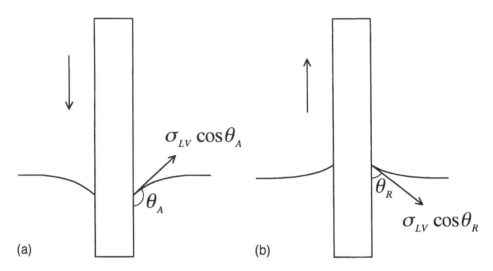

FIG. 3 Liquid–vapor interfacial tension acting on a plate moving quasi-statically: (a) immersion of plate; (b) emersion of plate.

change, the potential energy and the energy related to the liquid–vapor interface area remain constant. Hence the above force would contribute to the work of the liquid wetting the solid plate, i.e., the energy change E_W. In a similar manner, the receding contact angle θ_R can be observed when the plate is slowly drawn from the liquid, and the force $-\sigma_{LV} \cos \theta_R$ produces the energy change E_W. Equations (9) and (10) were verified more rigorously in Ref. [59].

The use of Eq. (9) or Eq. (10) makes it possible to calculate the energy difference E_W from the wetting behavior without touching on the details of the infinitesimal structure of the solid surface or the movement of the three-phase line. In the following, we discuss the instability of the meniscus formed under a solid surface from a thermodynamic viewpoint [60]. The system energy will be estimated based on Eq. (9) or Eq. (10).

B. Wetting Behavior of Two-Dimensional Meniscus Under a Horizontal Plate

One can observe that a meniscus attached to a horizontal plate spontaneously falls off at a certain critical height of the plate. On the other hand, if the plate is immersed into a liquid bath, the liquid spontaneously spreads and wets the entire plate at a critical depth. In this section, we first discuss the unstable phenomenon of a two-dimensional meniscus under a horizontal plate from a thermodynamic viewpoint based on Eq. (10) above. Then, in order to verify the results, the same problem is considered according to the method in the references [7,8,17,36], in which the infinitesimal effects on the solid surface are taken into consideration.

The energy of the system illustrated in Fig. 4 is calculated when the two-dimensional meniscus attaches to the horizontal plate at macroscopic contact angle θ. S, V, and L in the figure indicate the solid, vapor, and liquid phases, respectively. The coordinates x and z are taken to be the horizontal and vertical directions, respectively. The geometry of the meniscus can be obtained from the solution of Laplace equation (2). The radius of curvature is estimated from differential geometry. Considering the static pressure due to gravitation as ΔP, Eq. (2) is rewritten for the two-dimensional case as [2]

$$\sigma_{LV} \frac{\dfrac{d^2z}{dx^2}}{\left\{1 + \left(\dfrac{dz}{dx}\right)^2\right\}^{3/2}} = \Delta\rho g z, \tag{11}$$

where $\Delta\rho$ and g indicate the difference in density between the liquid and gas phases and the gravitational acceleration, respectively. The nondimensionalized form of Eq. (11) can be written as:

$$\bar{z} = \frac{d^2\bar{z}/d\bar{x}^2}{\left\{1 + (d\bar{z}/d\bar{x})^2\right\}^{3/2}}. \tag{12}$$

The quantities with an overbar ($^-$) indicate the length nondimensionalized by the following capillary constant:

$$a \equiv \sqrt{\sigma_{LV}/\Delta\rho g} \tag{13}$$

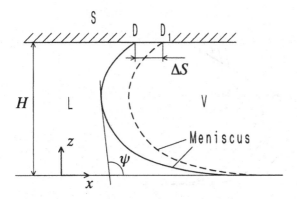

FIG. 4 Two-dimensional meniscus attached to a horizontal plate.

Equation (12) can be solved analytically. Using the inclination angle of the meniscus ψ shown in Fig. 4, the meniscus geometry can be determined by [2]

$$\bar{x} = \left[\ln\left\{ \frac{1 + \sin(\psi/2)}{\cos(\psi/2)} \right\} - 2 \sin \frac{\psi}{2} \right] + C \tag{14}$$

$$\bar{z} = 2 \cos \frac{\psi}{2}, \tag{15}$$

where C is the integral constant.

The infinitesimal energy change is considered when the three-phase contact line shifts by an apparent length ΔS from D to D_1, as seen in Fig. 4. It is noted that ΔS includes roughness or heterogeneity in itself. The contributions to the energy increment are classified into the following three items [8,17]: (1) potential energy of the meniscus ΔE_P, (2) work necessary to increase the liquid–vapor interfacial area ΔE_{LV}, and (3) energy change due to contact line movement ΔE_W. Before the energy of the system in Fig. 4 is discussed, let us calculate energies (1) and (2) for the meniscus attached to an inclined plate, as shown in Fig. 5, for the purpose of generality. Both can be obtained from

$$\Delta E_P = \frac{dE_P}{dS} \Delta S = \frac{d}{dS} \left[\Delta\rho g \left\{ \int_0^H \left(xz - \frac{z^2}{\tan\phi} \right) dz \right\} \right] \Delta S \tag{16}$$

$$\Delta E_P = \frac{dE_{LV}}{dS} \Delta S = \frac{d}{dS} \left[\sigma_{LV} \left\{ \int_0^H \sqrt{1 + (dx/dz)^2}\, dz - \int_0^\infty dx \right\} \right] \Delta S \tag{17}$$

where ϕ and H indicate the angle of plate inclination and the attachment height of the meniscus, as shown in Fig. 5, respectively. E_P and E_{LV} are the potential and liquid–vapor interface energy of the meniscus as a whole, respectively. In the above equations, the horizontal liquid surface is taken as a reference state of energy E_P and E_{LV}. Referring to Li and Neumann [17],

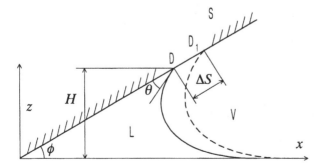

FIG. 5 Two-dimensional meniscus attached to an inclined plate.

the above calculation can be easily obtained by the use of meniscus geometry, i.e., Eqs. (14) and (15). The sum of Eqs. (16) and (17) is written as

$$\Delta E_P + \Delta E_{LV} = \sigma_{LV} \cos \theta \Delta S. \tag{18}$$

Equation (18) can also be applied to the meniscus under a horizontal plate shown in Fig. 4. It is noted that in Eq. (18), we neglect the contribution from the liquid in a trough of small roughness with higher order than ΔS. The energy ΔE_W is dependent on the direction of contact line movement (i.e., the sign of ΔS), as stated by Eq. (10). The values of ΔE_W for liquid advancement and retreat are again written here as

$$\Delta E_W = -\sigma_{LV} \cos \theta_A \Delta S \quad \text{(advancing, } \Delta S > 0) \tag{19a}$$

$$\Delta E_W = -\sigma_{LV} \cos \theta_R \Delta S \quad \text{(receding, } \Delta S < 0). \tag{19b}$$

The total energy change of the system,

$$\Delta E = \Delta E_P + \Delta E_{LV} + \Delta E_W, \tag{20}$$

can be obtained from Eqs. (18), (19a), and (19b), and (dE/dS) is written as

$$\frac{dE}{dS} = \sigma_{LV}(\cos \theta - \cos \theta_A) \quad \text{(advancing)} \tag{21a}$$

$$\frac{dE}{dS} = \sigma_{LV}(\cos \theta - \cos \theta_R) \quad \text{(receding).} \tag{21b}$$

Now the behavior of meniscus as shown in Fig. 4 is discussed using Eqs. (21a) and (21b). The meniscus height $\bar{H}$ shown in Fig. 4 is obtained from Eq. (15) as

$$\bar{H} = \sqrt{2(1 + \cos \theta)}. \tag{22}$$

First, we discuss the case of meniscus height $\bar{H} < \sqrt{2(1 + \cos \theta_A)}$, i.e., $\theta > \theta_A$. The following relation can be written for the differential coefficient of energy from Eq. (21a).

$$\frac{dE}{dS} < 0 \quad \text{(advancing)}$$

The meniscus spreads spontaneously and wets the entire plate since the system energy decreases monotonically in the direction of advance of the contact line ($\Delta S > 0$). In a similar manner, (dE/dS) for retreat can be written from Eq. (21b) as

$$\frac{dE}{dS} < 0 \quad \text{(receding).}$$

Since the energy increases in the direction of retreat ($\Delta S < 0$), the system is stable for the retreat of the contact line.

Next, the case of $\bar{H} > \sqrt{2(1 + \cos \theta_R)}$, i.e., $\theta < \theta_R$, is considered. The following relation can be written for the retreat, from Eq. (21b),

$$\frac{\mathrm{d}E}{\mathrm{d}S} > 0 \quad \text{(advancing)}$$

$$\frac{\mathrm{d}E}{\mathrm{d}S} > 0 \quad \text{(receding)}$$

Since the energy decreases in the direction of retreat ($\Delta S < 0$), the contact line retreats spontaneously, and the meniscus finally falls off the plate.

Lastly, let us consider the system behavior when $\bar{H}$ is in the range of $\sqrt{2(1 + \cos \theta_A)} < \bar{H} < \sqrt{2(1 + \cos \theta_R)}$, i.e., $\theta_A < \theta < \theta_R$. The following inequalities can be derived from Eqs. (23a) and (23b) for the differential coefficient:

$$\frac{\mathrm{d}E}{\mathrm{d}S} > 0 \quad \text{(advancing)}$$

$$\frac{\mathrm{d}E}{\mathrm{d}S} < 0 \quad \text{(receding)}$$

The contact line does not move since the energy increases in the directions of both advance and retreat. This means that the two-dimensional meniscus is stable for macroscopic contact angles between θ_A and θ_R.

It was demonstrated in the above discussion that the meniscus begins to move when the contact angle reaches θ_A or θ_R, but does not move for angles between θ_A and θ_R. This behavior is similar to that of a drop on a plate, described in Section I. In the case of a meniscus under a horizontal plate, however, the unstable phenomenon that the meniscus wets the entire surface or falls off occurs at some critical height of the plate corresponding to advancing or receding contact angles. Those phenomena can easily be observed by the naked eye. This suggests the possibility of obtaining the value of macroscopic contact angles by simple measurement of the critical heights. Since the height of a solid surface can be measured directly, we do not need any optical device, unlike the measurement of drop geometry or capillary height of the meniscus.

The above results regarding the system energy can be schematically summarized, as shown in Fig. 6. The figure shows the change of energy E with apparent displacement S for various values of θ. The energy curves in Fig. 6 are drawn straight according to the following relation:

$$\frac{\mathrm{d}^2 E}{\mathrm{d}S^2} = -\sigma_{LV} \sin \theta \frac{\mathrm{d}\theta}{\mathrm{d}S} = 0. \tag{23}$$

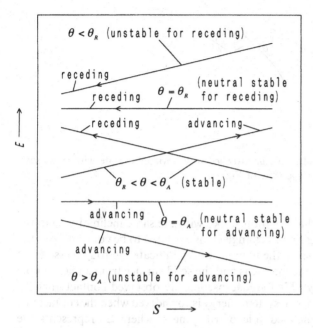

FIG. 6 Schematic of system energy change with the shift of attachment position of meniscus on a horizontal plate: the system becomes unstable when the energy monotonically decreases in the direction of the shift.

The above equation can be derived by the differentiation of Eq. (21) and $(d\theta/dS) = 0$ (the shape of a two-dimensional meniscus does not change with S, as seen in Fig. 4). As shown in Fig. 6, the system becomes neutral stable when $\theta = \theta_A$ or $\theta = \theta_R$.

The above discussion is based on Eq. (9) or Eq. (10) in which the infinitesimal effects on the solid surface are assumed to be wholly reflected by the macroscopic contact angles θ_A and θ_R. The use of Eqs. (9) and (10) makes it simple to analyze the macroscopic wetting behavior of a two-dimensional meniscus under a horizontal plate. In order to confirm the validity of the analysis using Eqs. (9) and (10), the same problem is discussed using the method in which the effect of roughness is taken into consideration. As stated before, there have been many reports concerning the wetting behavior on a rough surface. Unfortunately, the movement of the contact line on a surface with random roughness is not yet understood completely. Here we choose the relatively simple model for two-dimensional roughness treated by Eick et al. [8]. As shown in Fig. 7, we discuss the two-dimensional meniscus under a horizontal plate with a saw-tooth cross section. The surface is made up of a

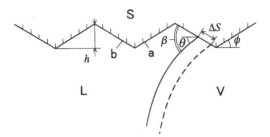

FIG. 7 Two-dimensional meniscus attached to a horizontal plate with small saw-tooth roughness: microscopic view of Fig. 4.

repetition of two small alternation slopes with the same inclination ϕ to the horizontal axis. Young's contact angle α_Y is assumed to be constant over the entire surface. As shown in the figure, h and β indicate the roughness height and the angle between the meniscus and the solid surface from a microscopic viewpoint, respectively. θ is the macroscopically observed contact angle as stated before. The change of system energy is considered when the contact line shifts by Δs on the inclined side of roughness, where Δs represents the infinitesimal distance along the side and is different from the apparent ΔS used in Eqs. (10), (19a), and (19b). The work necessary to change the potential energy and the liquid–vapor interfacial area of the meniscus can be calculated, for the inclined plate, from Eq. (18) as

$$\Delta E_P + \Delta E_{LV} = \sigma_{LV} \cos \beta \Delta s. \tag{24}$$

The energy change ΔE_W due to the movement of the three-phase contact line should be derived using Eq. (3), in this case, as

$$\Delta E_W = -\sigma_{LV} \cos \alpha_Y \Delta s. \tag{25}$$

From Eqs. (24) and (25), the rate of energy change is written as

$$\frac{dE}{ds} = \sigma_{LV}(\cos \beta - \cos \alpha_Y). \tag{26}$$

The above equation must be considered separately for side a and side b shown in Fig. 7. The following relations are obtained for α and β if we substitute the relations $\beta = \theta - \phi$ for α and $\beta = \theta + \phi$ for β into Eq. (26).

$$\left(\frac{dE}{ds}\right)_a = \sigma_{LV}\{\cos(\theta - \phi) - \cos \alpha_Y\} \tag{27}$$

$$\left(\frac{dE}{ds}\right)_b = \sigma_{LV}\{\cos(\theta + \phi) - \cos \alpha_Y\} \tag{28}$$

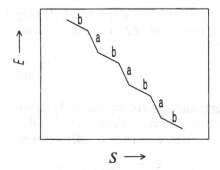

FIG. 8 System energy change with the shift of the meniscus shown in Fig. 7: in the case of $\bar{H} < \sqrt{2\{1 + \cos(\alpha_Y + \phi)\}}$, at which the system is unstable in the advance direction.

First, let us consider the case of apparent contact angle $\theta > \alpha_Y + \phi$ [i.e., the meniscus height $\bar{H} < \sqrt{2\{1 + \cos(\alpha_Y + \phi)\}}$ from Eq. (22)]. The following inequality holds, from Eqs. (27) and (28):

$$\left(\frac{dE}{ds}\right)_b < \left(\frac{dE}{ds}\right)_a < 0. \tag{29}$$

The energy curve can be depicted as shown in Fig. 8 based on the above relation. As is clear in the figure, the energy decreases monotonically for the movement in the positive direction of s. Hence the meniscus spreads and wets the entire plate spontaneously. The system becomes neutral stable at $\theta = \alpha_Y + \phi$, as shown in Fig. 9.

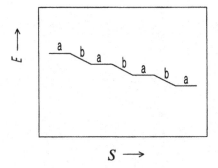

FIG. 9 System energy change with the shift of the meniscus in the case of $\bar{H} = \sqrt{2\{1 + \cos(\alpha_Y + \phi)\}}$, at which the system is neutral stable in the advance direction.

Next, when $\theta < \alpha_Y - \phi$ (i.e., $\bar{H} > \sqrt{2\{1 + \cos(\alpha_Y - \phi)\}}$), the following relation can be derived for the differential coefficient of E in a similar manner as Eq. (29).

$$\left(\frac{dE}{ds}\right)_a > \left(\frac{dE}{ds}\right)_b > 0. \tag{30}$$

According to the above relation, the energy curve can be schematically shown as that in Fig. 10. As seen in the figure, the meniscus is unstable in the receding direction of the contact line and falls off the plate. When $\theta = \alpha_Y + \phi$, the system is at neutral stability, as in Fig. 9.

Finally, when $\alpha_Y - \phi < \theta < \alpha_Y + \phi$ (i.e., $\sqrt{2\{1 + \cos(\alpha_Y + \phi)\}} < \bar{H} < \sqrt{2\{1 + \cos(\alpha_Y - \phi)\}}$), the following inequality holds:

$$\left(\frac{dE}{ds}\right)_a > 0, \quad \left(\frac{dE}{ds}\right)_b < 0. \tag{31}$$

Fig. 11 shows the energy curve obtained according to Eq. (31). The energy exhibits a local minimum at the vertex of the roughness between a and b shown in Fig. 7. The system is stable because there are many positions with minimum energy.

Eick et al. [8] analyzed the wetting behavior of a two-dimensional meniscus attached to a vertical plate with roughness similar to that depicted in Fig. 7. They suggested the following relations for the macroscopically observed contact angles.

$$\theta_A = \alpha_Y + \phi \tag{32a}$$

$$\theta_R = \alpha_Y - \phi \tag{32b}$$

Using Eqs. (32a) and (32b), the above results from Eqs. (29)–(31) can be summarized in Table 1. If Eqs. (32a) and (32b) are valid for the macroscopic

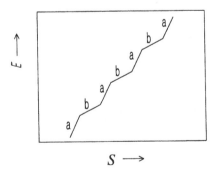

FIG. 10 System energy change with the shift of the meniscus in the case of $\bar{H} > \sqrt{2\{1 + \cos(\alpha_Y - \phi)\}}$, at which the system is unstable in the retreat direction.

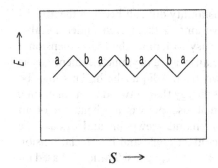

FIG. 11 System energy change with the shift of the meniscus in the case of $\sqrt{2\{1 + \cos(\alpha_Y + \phi)\}} < \bar{H} < \sqrt{2\{1 + \cos(\alpha_Y - \phi)\}}$ at which the system is stable.

contact angles, the results given in Table 1 are the same as those shown in Fig. 6 derived based on Eq. (10). Similar results can be obtained for a smooth plate with two-dimensional heterogeneity of regular shape [60]. Hence it seems to be possible to analyze the macroscopic wetting behavior based on the assumption of Eq. (9) or Eq. (10). The roughness or heterogeneity assumed here is quite simple. The analysis would be more difficult for a general surface with random characteristics if the infinitesimal effects were taken into consideration. It is evident that the method of using Eqs. (9) and (10) is quite simple compared with the above method.

C. Unstable Behavior of Axisymmetric Meniscus

It seems possible to measure contact angles by applying the thermodynamic instability of a two-dimensional meniscus, as stated above. However, there are many metastable positions on the horizontal plate as shown in Fig. 11,

TABLE 1 Wetting Behavior of Two-Dimensional Meniscus Under a Horizontal Plate: The Behavior is Analyzed from a Microscopic Viewpoint

θ	$\bar{H}$	Advance of meniscus	Retreat of meniscus
$\theta > \theta_A$	$\bar{H} < \sqrt{2(1 + \cos\theta_A)}$	Unstable	Stable
$\theta = \theta_A$	$\bar{H} = \sqrt{2(1 + \cos\theta_A)}$	Neutral stable	Stable
$\theta_R < \theta_A < \theta_A$	$\sqrt{2(1 + \cos\theta_A)} < \bar{H} < \sqrt{2(1 + \cos\theta_R)}$	Stable	Stable
$\theta = \theta_R$	$\bar{H} = \sqrt{2(1 + \cos\theta_R)}$	Stable	Neutral stable
$\theta < \theta_R$	$\bar{H} > \sqrt{2(1 + \cos\theta_R)}$	Stable	Unstable

which might make the recognition of instability difficult because the attachment point of the meniscus could move among them even under a stable condition. Also, it is actually not very easy to handle the two-dimensional meniscus under a horizontal plate because an axisymmetric meniscus is usually formed when the plate is drawn from a liquid bath. This may be because the axisymmetric shape has less energy than a two-dimensional one.

In this section, let us consider the macroscopic wetting behavior of an axisymmetric meniscus from a thermodynamic viewpoint and discuss the possibility of the measurement of the contact angle and interfacial tension [61]. As in the analysis stated above, the theoretical consideration is based on the assumption described by Eq. (9).

The axisymmetric meniscus under a conical surface is chosen as the subject, as shown in Fig. 12. The cylindrical coordinates r and z are taken to be the radial and horizontal directions, respectively. If some relations of differential geometry are inserted into the radii of curvature in Laplace equation (2), the profile of the axisymmetric meniscus can be determined by the following differential equation [31,62].

$$r\frac{d^2r}{dz^2} - \left(\frac{dr}{dz}\right)^2 - 1 - \frac{1}{\sigma_{LV}}\Delta\rho grz\left\{1 + \left(\frac{dr}{dz}\right)^2\right\}^{3/2} = 0 \tag{33}$$

The above equation is nondimensionalized by the use of capillary constant a defined by Eq. (13).

$$\bar{r}\frac{d^2\bar{r}}{d\bar{z}^2} - \left(\frac{d\bar{r}}{d\bar{z}}\right)^2 - 1 - \bar{r}\bar{z}\left\{1 + \left(\frac{d\bar{r}}{d\bar{z}}\right)^2\right\}^{3/2} = 0 \tag{34}$$

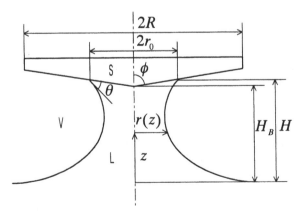

FIG. 12   Schematic of axisymmetric meniscus attached to a downward cone surface.

where $\bar{r} \equiv r/a$ and $\bar{z} \equiv z/a$. The boundary conditions of the meniscus profile shown in Fig. 12 can be written as

$$\bar{z} \to 0: \quad \frac{d\bar{r}}{d\bar{z}} \to -\infty \tag{35}$$

$$\bar{z} = \bar{H}: \quad \bar{r} = \bar{r}_0, \tag{36}$$

where $\bar{H} \equiv H/a$ and $\bar{r}_0 \equiv r_0/a$ indicate the height and radius of the meniscus at the cone surface. In the following, quantities of length are all nondimensionalized by a. Eqs. (34)–(36) were solved numerically [63,64]. Fig. 13 shows a comparison of the meniscus profile calculated numerically with that measured from a photograph. As seen in the figure, the calculated results agree well with the measured profile.

The system energy of Fig. 12 can be estimated by using the above solution for the meniscus profile. As stated in the previous section, we consider the potential energy E_P, the energy of the liquid–vapor interfacial area E_{LV}, and the work done by the three-phase contact line wetting the cone surface E_W. The dry cone surface and the horizontal liquid surface ($z = 0$) are taken for the reference state of system energy. E_P and E_{LV}, which represent the work necessary to form the axisymmetric meniscus shown in Fig. 12, are calculated from

$$E_P = \Delta\rho g \left\{ \int_0^H \pi r^2 z \, dz - \int_{H_B}^H \pi (z - H_B)^2 \tan^2\phi \, z \, dz \right\} \tag{37}$$

$$E_{LV} = \sigma_{LV} \left[\int_0^H 2\pi r \left\{ \sqrt{1 + \left(\frac{dr}{dz}\right)^2} - \frac{dr}{dz} \right\} dz - \pi r_0^2 \right], \tag{38}$$

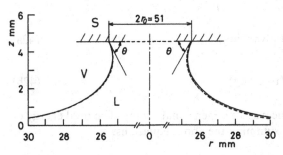

FIG. 13 Comparison of axisymmetric meniscus profile obtained numerically with that measured from a photograph.

where ϕ and H_B indicate the half-vertical angle of the cone and the height of the cone vertex, respectively. Using Eq. (9), the energy change E_W, when the three-phase contact line advances and wets the cone surface to the radius r_0, can be calculated as

$$E_W = -\sigma_{LV} \frac{\pi r_0^2}{\sin \phi} \cos \theta_A \quad \text{(advancing)} \tag{39a}$$

In order to calculate E_W in the receding direction, we should consider that, first, the dry cone surface is completely wetted and then dries to r_0. The energy change during this process can be obtained as

$$E_W = \sigma_{LV} \frac{\pi (R^2 - r_0^2)}{\sin \phi} \cos \theta_R - \sigma_{LV} \frac{\pi R^2}{\sin \phi} \cos \theta_A \quad \text{(receding)}$$

where R is the cone radius shown in Fig. 12. However, it is troublesome to treat the constant quantities in the above equation every time because we are mainly interested in the energy change from one state to another, i.e., the displacement of the meniscus radius r_0. Hence we omit the constant terms in the above equation and write the following expression:

$$E_W = -\sigma_{LV} \frac{\pi r_0^2}{\sin \phi} \cos \theta_R \quad \text{(receding)}, \tag{39b}$$

Equations (37)–(39a,b) are nondimensionalized using the parameter $(\sigma_{LV}^2 / \Delta \rho g)$.

$$\bar{E}_P = \int_0^{\bar{H}} \pi \bar{r}^2 \bar{z} d\bar{z} - \int_{\bar{H}_B}^{\bar{H}} \pi (\bar{z} - \bar{H}_B)^2 \tan^2 \phi \, \bar{z} d\bar{z} \tag{40}$$

$$\bar{E}_{LV} = \int_0^{\bar{H}} 2\pi \bar{r} \left\{ \sqrt{1 + \left(\frac{d\bar{r}}{d\bar{z}}\right)^2} - \frac{d\bar{r}}{d\bar{z}} \right\} d\bar{z} - \pi \bar{r}_0^2 \tag{41}$$

$$\bar{E}_W = -\frac{\pi \bar{r}_0^2}{\sin \phi} \cos \theta_A \quad \text{(advancing)} \tag{42a}$$

$$\bar{E}_W = -\frac{\pi \bar{r}_0^2}{\sin \phi} \cos \theta_R \quad \text{(receding)} \tag{42b}$$

The overbar, such as $\bar{E}_P$, means nondimensional energy. The total energy of the system is calculated by summing the above equations.

$$\bar{E} = \bar{E}_P + \bar{E}_{LV} + \bar{E}_W \tag{43}$$

First, let us discuss the axisymmetric meniscus attached to a vertical circular cylinder, i.e., $\phi = 0°$. For the cylinder, the energy $\bar{E}_W$ is calculated by the following equation instead of Eqs. (42a) and (42b).

$$\bar{E}_W = -2\pi \bar{r}_0 \bar{H} \cos \theta_A \quad \text{(for advancing)} \tag{44a}$$

$$\bar{E}_W = -2\pi \bar{r}_0 \bar{H} \cos \theta_R \quad \text{(for receding)} \tag{44b}$$

Fig. 14a and b shows the calculated energy change with the meniscus height $\bar{H}$ and with the apparent contact angle θ for the advancement and the retreat of the three-phase contact line. The nondimensional cylinder radius $\bar{R} = 14.7$ in the figure corresponds to $R = 40$ mm for water of 25°C. Two energy curves are depicted in each figure corresponding to advancing and receding, for which $\theta_A = 60°$ and $\theta_R = 40°$ are assumed in the calculation of $\bar{E}_W$. As shown in Fig. 14a, the system exhibits a minimum energy at $\theta = \theta_A$ or $\theta = \theta_R$ for both curves. This result is similar to that obtained conventional two-dimensional meniscus attached to a smooth and homogeneous plate [8,17], although here we used the macroscopic apparent contact angles for the calculation of $\bar{E}_W$.

Referring to the results of Fig. 14, let us consider the wetting behavior when the cylinder is immersed into a liquid bath in state A shown in Fig. 14, i.e., the energy minimum at $\theta = 40°$. Since the contact line advances on the cylinder

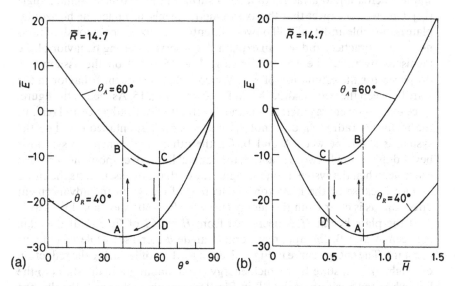

FIG. 14 System energy change with (a) the apparent contact angle θ and (b) nondimensional meniscus height $\bar{H}$, when axisymmetric meniscus attaches to a circular cylinder. The system is stable at states C and A for advance and retreat of meniscus, respectively.

surface, we start from state B on the curve of $\theta_A = 60°$. The meniscus height $\bar{H}$ decreases with the immersion of the cylinder from B because of the reduction of energy from B to C, as seen in Fig. 14b. Once the system reaches C, at which energy is minimum, $\bar{H}$ no longer decreases and is constant. A similar consideration holds for the retreat of the contact line. Inversely, when the cylinder is raised from C, we start at D on the curve $\theta_R = 40°$. Once the system reaches A, $\bar{H}$ remains constant corresponding to $\theta = 40°$. These results are the same as those shown in Fig. 3 in Section I and indicate the general wetting behavior of a meniscus attached to a plate or cylinder observed macroscopically. Here we can explain the mechanism by using the energy curves for advancement and retreat, which were calculated based on the assumption of Eq. (9) that the effects of infinitesimal roughness are entirely reflected in the values of macroscopic contact angles.

Next, the wetting behavior is discussed for an axisymmetric meniscus under a horizontal plate. The calculated results of Eq. (43) for $\phi = 90°$ are shown in Fig. 15 for the plate height $\bar{H} = 1.62$. We assume that $\theta_A = 60°$ and $\theta_R = 40°$, similar to Fig. 14. The abscissa $\bar{r}_0$ in Fig. 15b is the meniscus radius of attachment to the plate. As shown in Fig. 15a, the energy exhibits a maximum value at each contact angle, contrary to the results shown in Fig. 14. These results are quite different from the general understanding that the system is stable when a liquid attaches to a solid surface at an intrinsic contact angle [17]. One would expect that the axisymmetric meniscus under the horizontal plate is unstable and it falls off or wets the entire surface spontaneously. This is not true in practice, and we can explain the observed wetting behavior of the meniscus by using the energy curves in Fig. 15 based on the assumption proposed for the estimation of $\bar{E}_W$. We consider the system behavior in the state between the two maxima A and B shown in Fig. 15. As seen in the figure, since the system energy increases in the directions of both advance and retreat, the meniscus radius $\bar{r}_0$ does not spread or shrink spontaneously, i.e., the system is stable between A and B. On the other hand, when the system is beyond the region between A and B, the contact line moves spontaneously; for example, when the system is on the right side of B, the meniscus wets the entire surface because of the monotonic reduction of energy in the advancement direction, as is clear from the energy curve for $\theta_A = 60°$ shown in Fig. 15b.

If the plate height $\bar{H}$ is decreased from $\bar{H} = 1.62$ of Fig. 15 at the state between A and B, the apparent contact angle θ increases while $\bar{r}_0$ remains constant. The energy curve in Fig. 15b changes its profile during the reduction of height $\bar{H}$, i.e., state B at which energy is maximum gradually shifts to the left, while the positions A and B in Fig. 15a remain unchanged. Finally, the system coincides with state B when θ reaches θ_A at a certain critical height of the plate corresponding to the assumed radius $\bar{r}_0$ between A and B at $\bar{H} = 1.62$. At that time, the system becomes unstable in the direction of advance and the

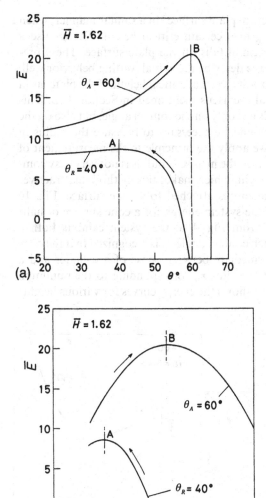

FIG. 15 System energy change with (a) the apparent contact angle θ and (b) non-dimensional meniscus radius $\bar{r}_o$, when axisymmetric meniscus forms under a horizontal plate. The system is stable between states A and B.

liquid spontaneously wets the entire plate surface. In a similar manner, when the plate is raised, θ reaches θ_R at a certain critical height. The meniscus becomes unstable for the retreat and falls off the plate surface. These predictions based on the energy curve describe the actual wetting behavior well. The unstable phenomenon occurs at a certain critical height of the plate, as in the case of the two-dimensional meniscus mentioned in Section II.B. This critical height, however, depends not only on the contact angle, but also on the meniscus radius $\bar{r}_0$. Hence it should be necessary to measure the radius in addition to the plate height if we apply the principle to the measurement of contact angles. The measurement of the meniscus radius is not very easy compared with that of the critical height, which makes this method unattractive.

Lastly, let us consider the meniscus attached to a cone surface. Fig. 16 shows the calculated results of the system energy for a cone surface of half-vertex angle $\phi = 60°$. As seen from Fig. 16a, the system exhibits both a minimum and a maximum at each contact angle. It is recognized that the cone surface has characteristics intermediate between those of a cylinder and a plate. The system is stable at the radius $\bar{r}_0$ corresponding to the minimum energy shown in Fig. 16b. Fig. 17 shows the energy curves for various heights

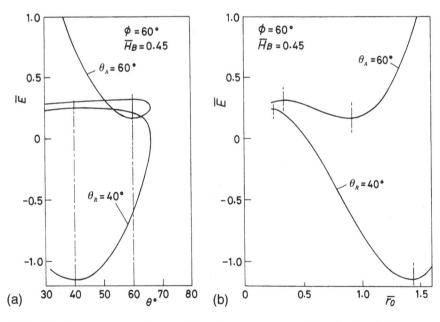

FIG. 16 System energy change with (a) apparent contact angle θ and (b) non-dimensional meniscus radius $\bar{r}_o$, when axisymmetric meniscus attaches to a downward cone surface. The system energy exhibits both a minimum and a maximum at advancing and receding contact angles.

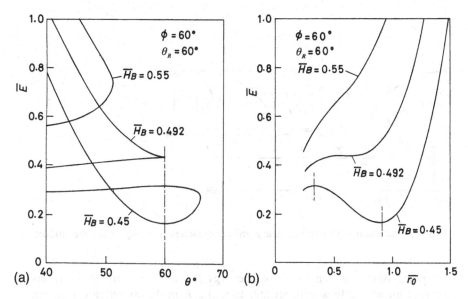

FIG. 17 Energy curve for various cone heights $\bar{H}_B$ shown in Fig. 12. The system does not have energy minimum and becomes unstable in the direction of retreat of meniscus when $\bar{H}_B$ is larger than 0.492.

of the cone vertex $\bar{H}_B$. Here we assume the receding contact angle $\theta_R = 60°$. When the height is relatively low, i.e., $\bar{H}_B = 0.45$, the system energy becomes maximum and minimum at the apparent contact angle $\theta = 60°$, as shown in Fig. 17a. Fig. 17b indicates that when the cone height is raised to $\bar{H}_B = 0.492$, the extreme states disappear from the energy curve and the energy monotonically increases with $\bar{r}_0$. Hence the system becomes unstable in the direction of retreat and the meniscus falls off the cone for $\bar{H}_B$ higher than 0.492. The critical height is dependent only on the receding contact angle θ_R unlike the case of the horizontal plate described above. Moreover, the meniscus radius $\bar{r}_0$ is fixed to that of minimum energy under a stable condition such as $\bar{H}_B = 0.45$ in Fig. 17b, which is different from the two-dimensional meniscus under a horizontal plate where there are many metastable positions on the plate surface. This could make it easy to recognize when instability occurs. It may be possible to obtain θ_R from the measured critical height of $\bar{H}_B$.

A similar result can be obtained for the axisymmetric meniscus attached to an upward cone, as shown in Fig. 18. The coordinate z is taken to be the downward direction, while the other variables are the same as in Fig. 12. $\bar{H}_B$ indicates the depth of the cone vertex. The system energy, instead of $\bar{E}_W$, can be calculated in the same manner as for the downward cone. Here we consider the advance of the three-phase contact line. If we take the dry surface as a

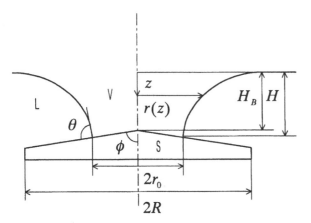

FIG. 18 Schematic of axisymmetric meniscus attached to an upward cone surface.

reference state, the nondimensional energy $\bar{E}_W$ required for the the three-phase contact line to wet the surface to radius $\bar{r}_0$ in the inward direction can be calculated as

$$\bar{E}_W = \frac{\pi(\bar{r}_0^2 - \bar{R}^2)}{\sin \phi} \cos \theta_A,$$

As stated previously in the derivation of Eq. (42b), the constant term is omitted from the above equation for the purpose of simplicity:

$$\bar{E}_W = \frac{\pi \bar{r}_0^2}{\sin \phi} \cos \theta_A. \tag{45}$$

The total energy of the system shown in Fig. 18 is obtained from the sum of Eqs. (40), (41), and (45). Fig. 19 shows the calculated results for the advancing contact angle $\theta_A = 90°$ as an example. The system exhibits both maximum and minimum energies at $\theta = 90°$ for $\bar{H}_B$ smaller than a critical depth, similarly as shown in Fig. 17. When $\bar{H}_B$ is larger than 0.590, the energy has no extreme value and increases monotonically with $\bar{r}_0$, as shown by Fig. 19b. Hence the liquid spreads and wets the entire surface of the upward cone. We can obtain the critical depth $\bar{H}_B$ for each advancing contact angle. It is possible to measure θ_A by the same method as for the receding contact angle if we reverse the cone surface.

D. Measuring Method of Contact Angles and Surface Tension Using the Cone Surface

As stated above, we theoretically discussed the unstable wetting behavior of an axisymmetric meniscus attached to a cone surface. It was suggested that

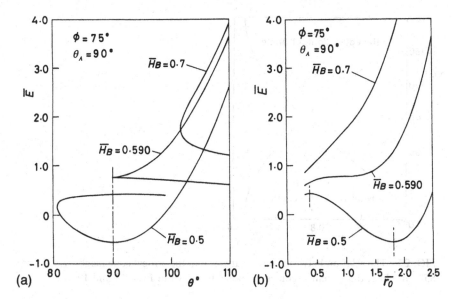

FIG. 19 Energy curve for various cone depths $\overline{H}_B$ shown in Fig. 18. The system has no energy minimum and becomes unstable when $\overline{H}_B$ is larger than 0.590.

contact angles could be obtained by simple measurement of the critical height at which the instability occurs [65]. In this section, the contact angles are actually measured according to the principle suggested here. Fig. 20 shows the functional relation between the critical height of the cone vertex $\overline{H}_{Bcr}$ and contact angles θ_R and θ_A for a cone of $\phi = 85°$. The theoretical curve shown in Fig. 20 was obtained from the height at which the energy has no extreme values, as shown in Fig. 17 or Fig. 19. The contact angles can be calculated from Fig. 20 if the critical height or depth $\overline{H}_{Bcr}$ is measured.

In order to confirm the principle stated above, a rather simple experiment was carried out. The experimental apparatus used here is schematically depicted in Fig. 21. The solid cone surface (1) is drawn from the liquid bath (3) and the critical height at which the meniscus falls off is measured using a micrometer (4). The bottom surface of (3) was chosen as the reference position for the measurement of height. In this experiment, we used a capacitance probe (6) to measure the position of the liquid surface [66]. In order to confirm the method proposed here, the meniscus shape was photographed by a camera (5) from the side of (3) and contact angles were read from the photograph for each measurement of critical height. Water and ethanol solution were used as test liquids. Table 2 shows the properties of the test liquids and the range of temperature in the experiment. The temperature was checked at each measurement of contact angles. The test cone surface was

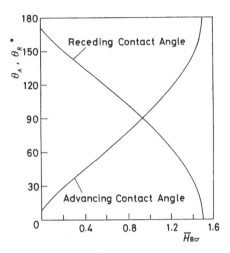

FIG. 20 The relation between contact angles and the critical cone height $\overline{H}_{Bcr}$ ($\phi =$ 85°) at which the system becomes unstable as shown in Figs. 18 and 19.

prepared using a lathe. In this experiment, solid surfaces of different materials were prepared: brass, duralumin, vinyl chloride, and Teflon. Each test solid, except Teflon, was finished by both turning and polishing to vary the surface roughness.

First, the critical height was roughly measured. Then, the cone surface was raised more slowly near the critical position determined above and $\overline{H}_{Bcr}$, at which the instability occurred, was precisely measured. The measurements of $\overline{H}_{Bcr}$'s were repeated five or six times under each set of experimental

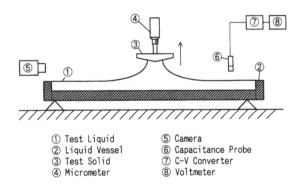

① Test Liquid   ⑤ Camera
② Liquid Vessel   ⑥ Capacitance Probe
③ Test Solid   ⑦ C-V Converter
④ Micrometer   ⑧ Voltmeter

FIG. 21 Schematic of experimental apparatus to measure contact angles using a cone surface.

TABLE 2 Properties of Test Liquids

Liquid	Temperature (°C)	Density (kg/m³)	Surface tension (N/m)
Water	12–19	998–1000	0.0736–0.0743
Ethanol solution	12–19	965–975	0.0368–0.0428

conditions. The scatter of critical height values was about 5/1000 mm, which confirmed the good reproducibility of the measurement.

Fig. 22 shows the results of measurements. The theoretical curve depicted in the figure is the same as that in Fig. 20. Each experimental point is a plot of the measured critical height $\bar{H}_{Bcr}$ and the corresponding contact angle obtained from the photograph. Hence the experimental points would lie on the theoretical curve if the contact angles obtained by the method proposed here coincided with those from the photograph. Fig. 22 shows rough agreement between experimental points and the theoretical curve, which indicates the feasibility of measurement using the method proposed here. The disagreement with the curve may be mainly due to the difficulty of measurement of contact angles from the photograph since the meniscus attaches to a curved cone surface. The agreement indicated by Fig. 22 also shows the validity of the discussion given in the preceding section, where we analyzed the unstable wetting behavior based on Eq. (9) assumed in Section I.

Although only contact angles were measured here, the liquid–vapor interfacial tension could also be measured based on the same principle. A glass cone with zero contact angle with liquids could be used, instead of the test solids described above, in the measurement of interfacial tension. The critical height corresponding to $\theta_R = 0°$ can be obtained as $\bar{H}_{Bcr} = 1.508$, from Fig. 20, for $\phi = 85°$. This relation can be rewritten in the dimensional form, using the capillary constant defined by Eq. (13), as

$$H_{Bcr} = 1.508\sqrt{\frac{\sigma_{LV}}{\Delta\rho g}}. \tag{46}$$

If we measure the critical height H_{Bcr} when the liquid falls off the glass cone surface, the liquid–vapor interfacial tension can be calculated from Eq. (46).

III. METHOD OF APPLYING GEOMETRICAL INSTABILITY OF THE TWO-DIMENSIONAL MENISCUS

In the preceding section, we discussed the method of applying the thermodynamic instability of an axisymmetric meniscus attached to a cone surface. The

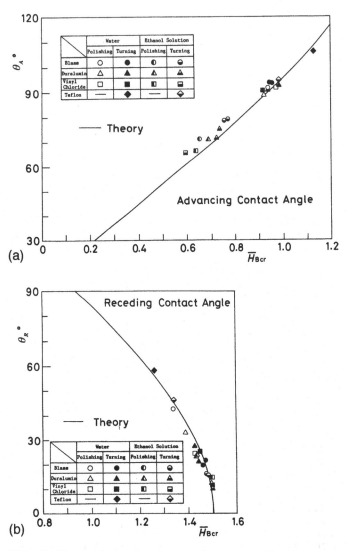

FIG. 22 Comparison of contact angles measured by the proposed method with those from a photograph: (a) advancing contact angle; (b) receding contact angle.

method has the merit of enabling the contact angle or liquid–vapor interfacial tension to be obtained by direct measurement of the critical height. When the method is applied to the measurement of contact angles, however, it may not be convenient to prepare a cone surface as a test piece every time. Practically, it is desirable to use a plate or circular cylinder for the measurement. In this section, we discuss another measuring method of contact angles and interfacial tension, in which the instability of a two-dimensional meniscus is applied in a manner different from that proposed in the preceding section [67].

A. Principle of the Method

Fig. 23 schematically shows the principle of the method of using a circular cylinder as the test solid. S, L, and V in the figure indicate solid, liquid, and vapor phases, respectively. The cylinder held horizontally is first immersed and then slowly drawn from the liquid bath. We can see a pair of two-dimensional menisci formed under the cylinder, as shown in Fig. 23a. As the cylinder is raised to a certain critical height, the waists of the two meniscus curves contact each other and the liquid breaks off from the solid surface. The geometry of the two-dimensional meniscus can be determined from the Laplace equation (11) and the contact angle as a boundary condition, as mentioned in Section I. Hence we could calculate the contact angle using the

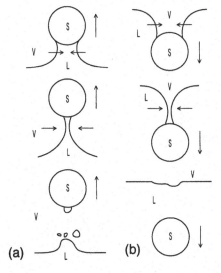

FIG. 23 Principle of the method for the measurement of contact angle using circular cylinder based on geometrical instability of two-dimensional meniscus: (a) receding contact angle; (b) advancing contact angle.

critical height at which the break-off of the meniscus occurs, as shown in Fig. 23a. Since the three-phase contact line recedes on the solid surface as the cylinder is raised in the case of Fig. 23a, the receding contact angle θ_R would be obtained by measuring the critical height. On the other hand, when the cylinder is inversely immersed into a liquid bath, as shown in Fig. 23b, the meniscus curves contact each other at a certain critical depth. In this case, the meniscus is broken and the liquid wets the entire surface of the cylinder. Since the contact line advances on the solid surface as the cylinder is immersed, we can obtain the advancing contact angle from the measurement of the critical depth.

Fig. 24 shows the principle of the method in a similar manner, for a plate used as the test solid. As shown in Fig. 24a, the test plate S_1 is fixed onto the inclined surface of support S_2 and is drawn up from the liquid bath. The two-dimensional meniscus formed under S_1 gradually approaches the support wall as the test plate is raised. The waist of the meniscus curve contacts the wall at a critical height and the meniscus breaks off from the solid surface. In a similar manner to that for the horizontal cylinder, we could calculate the receding contact angle from the measured critical height. Fig. 24b shows the measurement of the advancing contact angle. In this case, the plate and support are reversed, as shown in the figure. The three-phase contact line advances

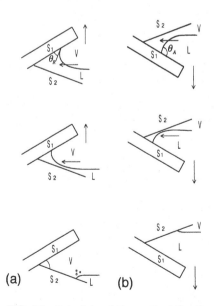

FIG. 24 Principle of the method for the measurement of contact angle using tilted plate based on geometrical instability of two-dimensional meniscus: (a) receding contact angle; (b) advancing contact angle.

gradually on the test plate as S_1 and S_2 are slowly immersed into the liquid. At a critical depth of the plate, the liquid contacts the support wall and wets the entire surface, as shown in the figure. The advancing contact angle could be calculated by measuring the critical depth.

B. Relation Between Critical Height and Contact Angle

As mentioned above, it is, in principle, possible to obtain contact angles by measuring the critical height or depth at which the geometrical instability of a two-dimensional meniscus occurs. Here the relationship between the critical height of cylinder or plate and the contact angles is discussed theoretically. First, let us consider the circular cylinder shown in Fig. 25. In the figure, the x- and z-axes are taken to be the direction of the stationary liquid surface and the vertical direction through the cylinder center, respectively. The height of the cylinder bottom H_B from the stationary liquid surface shown in Fig. 25 is used as the measuring height of the cylinder. In this section, we use the nondimensional solution of the Laplace equation, i.e., Eqs. (14) and (15), which determines the geometry of the two-dimensional meniscus. The unknown integral constant C in Eq. (14) should be determined from the boundary condition. Since the meniscus is in contact with the cylinder of radius R at the receding contact angle θ_R, as shown in Fig. 25, Eq. (14) should satisfy the boundary condition

$$x = x_0 = R\sin\phi \text{ at } \psi = (\phi + \theta_R),$$

where x_0 and ϕ indicate the coordinate at the attachment point of the meniscus and the angle between the z-axis and the cylinder radius at the attachment point, respectively, as seen in Fig. 25. ψ is the inclination of the

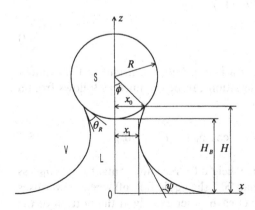

FIG. 25 Schematic of two-dimensional meniscus attached to a circular cylinder.

meniscus curve from the x-axis. Using the capillary constant defined by Eq. (13), the above boundary condition is rewritten in the nondimensional form as

$$\bar{x} = \bar{x}_0 = \bar{R} \sin \phi \text{ at } \psi = \phi + \theta_R.$$

Equation (14) can be rewritten as below if the constant C is calculated from the boundary condition.

$$\bar{x} = \ln \left\{ \frac{1 + \sin\left(\dfrac{\psi}{2}\right)}{\cos\left(\dfrac{\psi}{2}\right)} \right\} - \ln \left\{ \frac{1 + \sin\left(\dfrac{\theta_R + \phi}{2}\right)}{\cos\left(\dfrac{\theta_R + \phi}{2}\right)} \right\}$$

$$+ 2 \left\{ \sin\left(\frac{\theta_R + \phi}{2}\right) - \sin\left(\frac{\psi}{2}\right) \right\} + \bar{R} \sin \phi \tag{47}$$

The angle ϕ in the above equation gradually decreases with increasing cylinder height. Finally, when the waists of two meniscus curves contact each other, the x coordinate at $\psi = \pi/2$, i.e., x_1 shown in Fig. 25, becomes zero. Substituting $\psi = \pi/2$ into Eq. (47) and taking $\bar{x} = \bar{x}_1 = 0$, the following relation holds for the angle ϕ when the meniscus breaks off from the solid surface:

$$\bar{R} \sin \phi - \ln \left\{ \frac{1 + \sin\left(\dfrac{\theta_R + \phi}{2}\right)}{\cos\left(\dfrac{\theta_R + \phi}{2}\right)} \right\} + 2 \sin\left(\frac{\theta_R + \phi}{2}\right)$$

$$+ \ln\left(1 + \sqrt{2}\right) - \sqrt{2} = 0 \tag{48}$$

On the other hand, the meniscus height at the position of attachment to the cylinder can be calculated from Eq. (15) as

$$\bar{H} = 2 \cos\left(\frac{\theta_R + \phi}{2}\right). \tag{49}$$

Using the solution of Eq. (48) as ϕ_R in Eq. (49), the height of the cylinder bottom $\bar{H}_{Bcr}$ under the critical condition can be obtained as follows from a simple geometrical consideration:

$$\bar{H}_{Bcr} = 2 \cos\left(\frac{\theta_R + \phi_R}{2}\right) - \bar{R}(1 - \cos \phi_R) \quad (\theta_R \leq 90°) \tag{50}$$

The above equation gives the critical height for receding contact angles θ_R less than $90°$. If θ_R becomes greater than $90°$, the liquid falls off when the vertexes of the two meniscus curves contact each other exactly at the bottom of the cylinder, as shown in Fig. 26. Since the inclination of the meniscus at the vertex is equal to the receding contact angle θ_R, the critical height of the cylinder can be calculated simply by using θ_R instead of ψ in Eq. (15) as

$$\overline{H}_{Bcr} = 2 \cos\left(\frac{\theta_R}{2}\right) \quad (\theta_R > 90°). \tag{51}$$

The relationship for the advancing contact angle can be obtained in a similar manner as described above. The details are not discussed here to avoid repetition. We can readily obtain the relationship between the critical depth and the advancing contact angle θ_A if we use $(\pi - \theta_A)$ instead of θ_R in Eqs. (48), (50), and (51). Equation (48) is rewritten for θ_A as

$$\overline{R} \sin \phi_A - \ln\left\{\frac{1 + \cos\left(\dfrac{\theta_A - \phi_A}{2}\right)}{\sin\left(\dfrac{\theta_A - \phi_A}{2}\right)}\right\} + 2 \cos\left(\frac{\theta_A - \phi_A}{2}\right)$$

$$+ \ln\left(1 + \sqrt{2}\right) - \sqrt{2} = 0. \tag{52}$$

Using the solution of the above equation as ϕ_A, the following relations between the critical depth and the advancing contact angle can be given as

$$\overline{H}_{Bcr} = 2 \sin\left(\frac{\theta_A - \phi_A}{2}\right) - \overline{R}(1 - \cos \phi_A) \quad (\theta_A \geq 90°) \tag{53}$$

$$\overline{H}_{Bcr} = 2 \sin \frac{\theta_A}{2} \quad (\theta_A < 90°) \tag{54}$$

Next, let us consider the relationship for a plate used as the test solid. As shown in Fig. 27, we take the height of intersection B of test plate S_1 with support S_2 as the measuring height H_B. The z-axis is taken to be the vertical

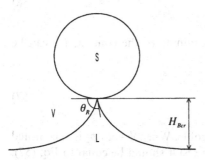

FIG. 26 Critical condition for receding contact angle θ_R larger than 90° at which the meniscus falls off the cylinder surface.

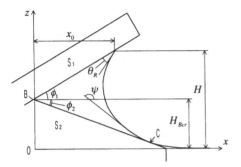

FIG. 27 Schematic of critical condition at which meniscus contacts the surface of the support S_2.

direction through B. Fig. 27 indicates the critical condition just as the meniscus curve contacts the S_2 wall at C. The critical height of B, i.e., $\bar{H}_{Bcr}$, can be written as follows based on a simple geometrical consideration:

$$\bar{H}_{Bcr} = \bar{H} - \bar{x}_0 \tan \phi_1$$
$$= 2 \cos\left(\frac{\theta_R + \phi_1}{2}\right) - \bar{x}_0 \tan \phi_1 \tag{55}$$

Equation (15) was used to obtain the meniscus height $\bar{H}$. As seen in Fig. 27, $\bar{x}_0$ and ϕ_1 indicate the nondimensional coordinate of the meniscus at the attachment position and the inclination of the test plate, respectively. The gradient of the meniscus curve becomes equal to that of the solid wall (i.e., ϕ_2 shown in the figure) at C. Hence the coordinate of the contact, $\bar{z}_C$, can be calculated by using $(\pi - \phi_2)$ instead of ψ in Eq. (15) as

$$\bar{z}_C = 2 \sin\left(\frac{\phi_2}{2}\right). \tag{56}$$

Using the above equation, the $\bar{x}$ coordinate of the contact, $\bar{x}_C$, can be obtained geometrically as

$$\bar{x}_C = \left\{ \bar{H}_{Bcr} - 2 \sin\left(\frac{\phi_2}{2}\right) \right\} \frac{1}{\tan \phi_2}. \tag{57}$$

Now we use Eq. (14) for the meniscus profile. When $\psi = \pi - \phi_2$ is substituted into the right-hand side of Eq. (14), the result should be equal to Eq. (57). Hence we can determine the integral constant C in Eq. (14) under the critical condition. Substituting the calculated C and $\psi = \theta_R + \phi_1$ at the attachment position of the meniscus to the test plate S_1 into Eq. (14) again, $\bar{x}_0$ shown in Fig. 27 can be obtained as follows:

$$
\bar{x}_0 = \ln\left\{\frac{1 + \sin\left(\dfrac{\theta_R + \phi_1}{2}\right)}{\cos\left(\dfrac{\theta_R + \phi_1}{2}\right)}\right\} - \ln\left\{\frac{1 + \cos\dfrac{\phi_2}{2}}{\sin\dfrac{\phi_2}{2}}\right\}
$$
$$
- 2 \sin\left(\frac{\theta_R + \phi_1}{2}\right) + \frac{\bar{H}_{Bcr}}{\tan\phi_2} + \frac{1}{\cos\dfrac{\phi_2}{2}}
$$

(58)

Finally, the critical height $\bar{H}_{Bcr}$ shown in Fig. 27 is written as follows if Eq. (58) is inserted into Eq. (55):

$$
\bar{H}_{Bcr} = 2 \cos\left(\frac{\theta_R + \phi_1}{2}\right) - \frac{2 \tan\phi_1}{\tan\phi_1 + \tan\phi_2}\left[\cos\left(\frac{\theta_R + \phi_1}{2}\right)\right.
$$
$$
-\sin\frac{\phi_2}{2} - \left\{\sin\left(\frac{\theta_R + \phi_1}{2}\right) - \cos\frac{\phi_2}{2}\right.
$$

(59)

$$
\left.\left. - \frac{1}{2}\ln\left(\frac{1 + \sin\left(\dfrac{\theta_R + \phi_1}{2}\right)}{\cos\left(\dfrac{\theta_R + \phi_1}{2}\right)}\right) + \frac{1}{2}\ln\left(\frac{1 + \cos\dfrac{\phi_2}{2}}{\sin\dfrac{\phi_2}{2}}\right)\right\}\tan\phi_2\right]
$$

When $(\theta_R + \phi_1)$, i.e., the inclination of the meniscus from the horizontal at the test plate, becomes greater than $(\pi - \phi_2)$, the meniscus does not contact support S_2 until it reaches B, as shown in Fig. 28. In this case, the three-phase contact line might be trapped at the corner between S_1 and S_2 because the tip of S_2 would be slightly rounded. The meniscus would not break off

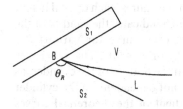

FIG. 28 Critical condition for receding contact angle θ_R larger than $[\pi - (\phi_1 + \phi_2)]$: the three phase line is trapped at the corner between S_1 and S_2.

from the plate even if the plate was raised higher than the critical condition. Hence the measurement applying Eq. (59) should be limited to the following range of receding contact angles:

$$\theta_R < \pi - (\phi_1 + \phi_2) \tag{60}$$

For the measurement of advancing contact angle θ_A, the test plate is immersed inversely into a liquid bath, as shown in Fig. 24b. The critical depth of the test plate can be obtained in the same manner as for the receding contact angle, so we do not discuss the details here. The relationship between the critical depth at which instability occurs and θ_A can be calculated if we use $(\pi - \theta_A)$ instead of θ_R in Eq. (59) as

$$
\bar{H}_{Bcr} = 2 \sin\left(\frac{\theta_A - \phi_1}{2}\right) - \frac{2 \tan \phi_1}{\tan \phi_1 + \tan \phi_2}\left[\sin\left(\frac{\theta_A - \phi_1}{2}\right)\right.
$$

$$
-\sin\frac{\phi_2}{2} - \left\{\cos\left(\frac{\theta_A - \phi_1}{2}\right) - \cos\frac{\phi_2}{2} - \frac{1}{2}\ln\left(\frac{1 + \cos\left(\frac{\theta_A - \phi_1}{2}\right)}{\sin\left(\frac{\theta_A - \phi_1}{2}\right)}\right)\right.
$$

$$
\left.\left.+\frac{1}{2}\ln\left(\frac{1 + \cos\frac{\phi_2}{2}}{\sin\frac{\phi_2}{2}}\right)\right\}\tan \phi_2\right] \tag{61}
$$

According to the same reason as for Eq. (60), the measurement of the advancing contact angle is limited to the following region:

$$\theta_A > \phi_1 + \phi_2 \tag{62}$$

Fig. 29 shows the relationship between θ_R or θ_A and $\bar{H}_{Bcr}$ calculated by Eqs. (50), (51), (53), and (54) for cylinder and by Eqs. (54) and (61) for plate. The calculated results are presented in the figures for several values of cylinder radius $\bar{H}$ and inclinations ϕ_1 and ϕ_2. As seen in the figures, each critical height or depth corresponds to one contact angle, which indicates the validity of the method. In Fig. 29a for a circular cylinder, each curve for various $\bar{R}$'s converges to one curve in the region of $\theta_R > 90°$ or $\theta_A < 90°$. This is because the meniscus falls off the solid surface exactly at the bottom of the cylinder, as described by Fig. 26, and the critical height is not dependent on the cylinder radius. In Fig. 29a, we can see that the gradient of the theoretical curves becomes steep close to $\theta_R = 0°$ and $\theta_A = 180°$ for all cylinder radii. This is due to the fact that the gradient of the meniscus curve becomes nearly horizontal

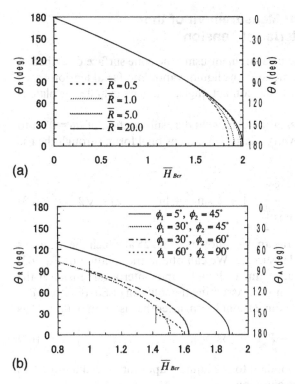

FIG. 29 Theoretical relation between contact angle and nondimensional critical height of (a) cylinder and (b) tilted plate.

at the attachment point for those contact angles. Hence the critical height varies only slightly with the change of contact angle, i.e., $dH/d\theta_R \approx 0$. This might be evident if we calculate $(dH/d\theta)$ at $\theta = 0°$ from Eq. (22), which determines the height of the meniscus attached to a horizontal plate. The above fact indicates an inaccuracy in the measurement around such contact angles. A similar tendency is observed in Fig. 29b for the plate used as the test solid. However, the accuracy can be improved if we use the test plate and support with large inclination in order to make the gradient of the meniscus curve steep at the attachment point. In fact, the theoretical curves of $\phi_1 = 60°$ and $\phi_2 = 90°$ have a gentle gradient near $\theta_R = 0°$ and $\theta_A = 180°$ compared with other curves, as shown in Fig. 29b. However, the region in which measurement is possible becomes more limited as ϕ_1 or ϕ_2 increases, as described by Eq. (60) or Eq. (62). The vertical bar | in Fig. 29b shows the limit of measurement. It would be desirable to prepare some supports with different inclinations to measure various contact angles with sufficient accuracy.

C. Critical Height for Measurement of the Liquid–Vapor Interfacial Tension

In a similar manner to the measurement using the cone surface described in Section II, it is possible to measure the liquid–vapor interfacial tension based on the same principle as for the contact angle, if a glass cylinder or plate is used as the test solid.

In the measurement using the glass cylinder, substitution of $\theta_R = 0°$ into Eq. (48) leads to the following relation in dimensional form, through the use of Eq. (13).

$$\sqrt{\frac{\Delta \rho g}{\sigma_{LV}}} R \sin \phi - \ln \left(\frac{1 + \sin \dfrac{\phi}{2}}{\cos \dfrac{\phi}{2}} \right) + 2 \sin \frac{\phi}{2} + \ln(1 + \sqrt{2}) - \sqrt{2} = 0 \quad (63)$$

The above equation gives the angle ϕ shown in Fig. 25 at which the meniscus breaks off from the cylinder surface. We can obtain the solution of Eq. (63) numerically for an arbitrary value of liquid–vapor interfacial tension if the density of the test liquid is known. Assuming the solution of Eq. (63) to be ϕ_T, Eq. (50) can be rewritten in dimensional form after the insertion of $\theta_R = 0°$ as

$$H_{Bcr} = 2\sqrt{\frac{\sigma_{LV}}{\Delta \rho g}} \cos \frac{\phi_T}{2} - R(1 - \cos \phi_T). \quad (64)$$

The critical height corresponding to the liquid–vapor interfacial tension can be calculated by the above equation.

For the plate, we similarly substitute $\theta_R = 0°$ into Eq. (59) and rewrite the equation in dimensional form. The relation between the liquid–vapor interfacial tension and the critical height H_{Bcr} can be written as

$$\begin{aligned}
\sqrt{\frac{\sigma_{LV}}{\Delta \rho g}} = \frac{1}{2} H_{Bcr} &\left[\cos \frac{\phi_1}{2} - \frac{\tan \phi_1}{\tan \phi_1 + \tan \phi_2} \right. \left\{ \cos \frac{\phi_1}{2} - \sin \frac{\phi_2}{2} \right. \\
&- \left(\sin \frac{\phi_1}{2} - \cos \frac{\phi_2}{2} - \frac{1}{2} \ln \left(\frac{1 + \sin \dfrac{\phi_1}{2}}{\cos \dfrac{\phi_1}{2}} \right) \right. \\
&\left. \left. \left. + \frac{1}{2} \ln \left(\frac{1 + \cos \dfrac{\phi_2}{2}}{\sin \dfrac{\phi_2}{2}} \right) \right) \tan \phi_2 \right\} \right]^{-1}
\end{aligned} \quad (65)$$

Fig. 30 shows the relation between $(\sigma_{LV}/\Delta \rho)$ and H_{Bcr} for the cylinder and the plate. As seen in the figure, the liquid–vapor interfacial tension can be

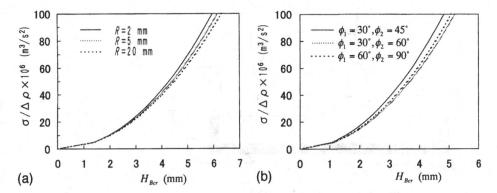

FIG. 30 Theoretical relation between liquid–vapor interfacial tension and critical height: (a) circular cylinder; (b) tilted plate.

obtained by measuring the critical height, in a same manner as for the contact angle. Although there were some regions in which the measurement of the contact angle was inaccurate, as mentioned in the preceding section, it is possible to measure the liquid–gas interfacial tension correctly over its entire range for both cylinders and plates, as shown in Fig. 30.

D. Experiment

The results described in Sections III.B and III.C indicate that it is possible to measure the liquid–vapor interfacial tension as well as the contact angle using the same apparatus if the test cylinder or plate is changed according to the desired measurement. An experiment was carried out to actually measure contact angles and interfacial tensions based on the principle described in the preceding sections. Since the main purpose here is to confirm the validity of the principle and to perform measurement using a simple apparatus, the accuracy is not of high concern in this experiment.

Fig. 31 shows the schematic of the experimental apparatus used. The apparatus is similar to, but slightly modified from, that used for the work described in Section II.D. The test cylinder or plate (4) is attached to the feeding device and drawn from vessel (1) filled with test liquid (2). The critical height H_{Bcr} at which the liquid falls off from the test surface is measured by a micrometer (3) from the bottom surface of (1) as the reference plane. For the measurement of the advancing contact angle, the cylinder or plate is inversely immersed into the test liquid and the critical depth is measured in the same manner. The distance between the liquid surface and the reference plane is measured by the needle contact method [68], which is simple compared with

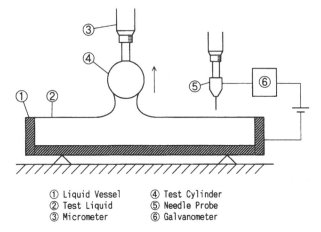

① Liquid Vessel   ④ Test Cylinder
② Test Liquid   ⑤ Needle Probe
③ Micrometer   ⑥ Galvanometer

FIG. 31   Schematic of experimental apparatus for the measurement of contact angle and liquid–vapor interfacial tension.

the capacitance probe method used in Section II.D. A probe with a thin needle is fed by a micrometer. When the tip of the needle contacts the liquid surface or the bottom of the vessel (1), a simple circuit is closed, as shown in Fig. 31. The position of the liquid surface can be measured by the micrometer when a signal from a galvanometer is detected. Although a liquid without conductivity is used, we can determine the position to within 1/100 mm by the naked eye, using a mirror to observe when the needle comes into contact with the liquid surface [35]. The liquid meniscus formed under a solid surface may not remain two-dimensional. Hence, in this experiment, thin plates are fixed to both ends of the test cylinder or plate, as shown in Fig. 32, to maintain the two-dimensionality of the meniscus. The liquid near the ends is fixed by the plates and does not shrink to the middle part of the cylinder. Although the liquid layer may become thick and the meniscus loses two-dimensionality

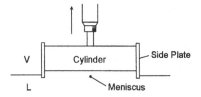

FIG. 32   Side view of the measurement of critical height using circular cylinder: side plates are fixed on the ends of the cylinder to maintain the two-dimensionality of the meniscus.

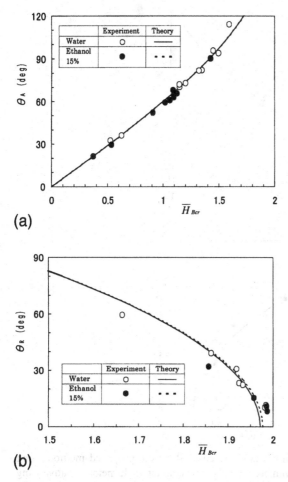

(a)

(b)

FIG. 33 Comparison of contact angles measured by the proposed method using circular cylinder with those from a photograph: (a) advancing contact angle; (b) receding contact angle.

near the end plates, the critical height can be measured correctly because the instability of the meniscus, as described in Figs. 23 and 24, occurs around the central region of the cylinder where the meniscus is two-dimensional. The length of the test cylinder or plate should be about 10 times the meniscus height in order to maintain two-dimensionality in the central region.

Six kinds of solid surfaces were prepared for the measurement of the contact angle: brass, duralumin, stainless steel, vinyl chloride, nylon, and Teflon. In a similar manner as the method using the cone surface described in

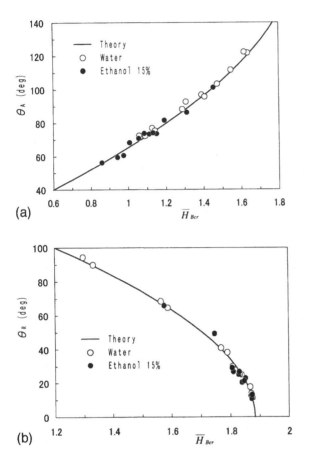

FIG. 34 Comparison of contact angles measured by the proposed method using tilted plate with those by a commercially available contact angle meter: (a) advancing contact angle; (b) receding contact angle.

Section II.D, each material was finished by several methods, such as lathing and polishing, in order to measure contact angles for various surface roughnesses. The test liquids were water and 15% ethanol solution for the measurement of the contact angle and water, four kinds of ethanol solution of different concentrations, and two kinds of machine oils for the measurement of the liquid–vapor interfacial tension.

For comparison, the contact angle was also measured from a photograph for the cylinder and by using a commercially available contact angle meter (Kyowa Kaimen Kagaku CA-A type) for the plate, in which the contact angle of a drop is measured using a telescope with a goniometer. The liquid–vapor interfacial tension was also measured by the Wilhelmy method using a commercially available surface tension meter (Kyowa Kaimen Kagaku CBVP-A3 type).

The measured results of contact angles for the cylinder and for the plate are shown in Figs. 33 and 34, respectively. The experimental points in the figures are plotted in the same manner as in Fig. 22 in Section II.D, i.e., by using the measured critical height $\overline{H}_{Bcr}$ (abscissa) and the contact angle measured from a photograph or using a contact angle meter (ordinate). If the results measured by the two different methods agreed with each other, the experimental points should lie on the theoretical curves obtained in the preceding section, as shown in the figures. There is some scatter of data in Fig. 33b for the receding contact angle measured using a cylinder. This may be due to error in the photograph method because it is difficult to measure contact angles on a

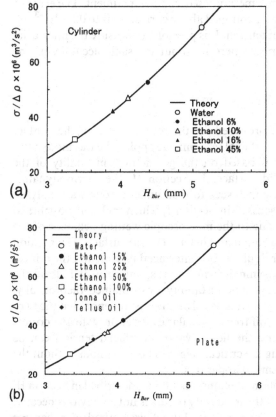

FIG. 35 Comparison of liquid–vapor interfacial tension measured by the proposed method with those by a commercially available surface tensiometer: (a) circular cylinder; (b) tilted plate.

curved surface. For the plate used as the test solid, on the other hand, the theoretical curves fit the data well for both receding and advancing contact angles, as shown in Fig. 34. The discrepancy from the curve is about 2°, which is roughly the same value as the error in the direct method using a goniometer. The error in critical height measured using the apparatus shown in Fig. 31 is about 1/100 mm, which corresponds to less than 10^{-2} in nondimensional height $\bar{H}_{\mathrm{Bcr}}$ and to the error of $\pm 1°$ in the contact angle at $\theta_{\mathrm{R}} = 45°$.

Fig. 35 shows the experimental results of the interfacial tension σ_{LV} measured by the same method as the contact angle, using a glass cylinder or plate as test solid. The experimental points in the figures are plotted using σ_{LV} measured by the Wilhelmy method and critical heights. As shown in the figures, it is possible to measure the interfacial tension by the method proposed here. The discrepancy between experimental points and the theoretical curve is less than 0.2×10^{-3} (N/m), which roughly corresponds to the degree of error in the Wilhelmy method used in this experiment. The error of 1/100 mm in the critical height mentioned above corresponds to 0.2×10^{-3} (N/m) in liquid–vapor interfacial tension. It is possible to measure the interfacial tension by the method proposed here to about the same accuracy as the Wilhelmy method.

IV. DISCUSSION

Two kinds of methods were proposed for the measurement of the contact angle and the liquid–vapor interfacial tension; one applies the thermodynamic instability and the other is based on the geometrical instability of the meniscus attached to a solid surface. In Section II, the thermodynamic instability of an axisymmetric meniscus formed under a cone was analyzed based on the assumption discussed in Section I, which makes it possible to calculate the energy change due to the macroscopic wetting behavior of the three-phase contact line on a rough and heterogeneous surface. The relation between the contact angle or liquid–vapor interfacial tension and the critical height of a cone when the axisymmetric meniscus spontaneously falls off the solid surface due to thermodynamic instability was obtained theoretically. The method proposed in Section III is based on the phenomenon that the two-dimensional meniscus breaks off from a circular cylinder or plate at a critical height. The contact angle and the liquid–vapor interfacial tension can be obtained simply by measuring the critical height in both methods without the need for any optical or dynamical apparatus.

In the method applying thermodynamic instability, described in Section II, the critical condition at which the instability occurs is not so obvious because the movement of the three-phase line is slow. This method also has the disadvantage that a cone surface must be prepared for each test in order to measure contact angles. On the other hand, it is easier to recognize the

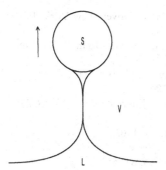

FIG. 36 Schematic of soap-like film when a solution of surface-active materials is used as the test liquid for the measurement using circular cylinder.

meniscus break-off in the method based on the geometrical instability discussed in Section III. In particular, the phenomenon can be seen more clearly when a circular cylinder is used as the test solid. However, when a solution of surface-active materials was used as the test liquid, a pair of two-dimensional menisci did not break under the critical condition shown in Fig. 23 since the liquid forms a thin soap-like film as shown in Fig. 36. The film maintains until it becomes a monolayer. Hence we cannot measure the contact angle by the method using a circular cylinder when the liquid includes surface-active materials. On the other hand, the meniscus formed under the inclined plate still breaks off even if surfactant solution is used since it contacts the wall of support S_2, as illustrated in Fig. 24. The method using a plate has another advantage in that it is possible to compensate the inaccuracy of the measurement near $\theta_R = 0°$ or $\theta_A = 180°$ if we use a support with a different inclination angle, as stated in Section III.B.

For the measurement of the liquid–vapor interfacial tension, we can use the three kinds of glass surfaces stated above, except when surfactant solution is used with a cylinder. The plate would be the most easily available among these surfaces. However, we must take care to clean the support or side plates shown in Figs. 24 and 32, as well as the glass plate, in order to not pollute the test liquid.

REFERENCES

1. Young, T. In *Miscellaneous Works*; Peacock, G., Ed.; J. Murray: London, 1855; Vol. 1, 418.
2. Chappius, J. In *Multiphase Science and Technology*; Hewitt, F., Delhaye, J.M., Zuber, N., Eds.; 1982; Vol. 1, 484.
3. Gaydos, J.; Neumann, A.W. In *Surfactant Science Series*; Neumann, A.W. Spelt, J.K., Eds.; Dekker: New York, 1996; Vol. 63, 169–238.

4. Gibbs, J.W. *The Scientific Papers of J. Willard Gibbs*; Dover: New York, 1961; Vol. 1, 55–371.

5. Adamson, A.W. *Physical Chemistry of Surfaces,* 5th Ed.; John Wiley: New York, 1990.

6. Shuttleworth, R.; Bailey, G.L. Discuss. Faraday Soc. 1948, *3*, 16.

7. Johnson, R.E.; Dettre, R.H. Adv. Chem. Ser. 1964, *43*, 112.

8. Eick, J.D.; Good, R.J.; Neumann, A.W. J. Colloid Interface Sci. 1975, *53*, 235.

9. Hocking, L.M. J. Fluid Mech. 1976, *76*, 801.

10. Bayramli, E.; Mason, S.G. Can. J. Chem. 1981, *59*, 1962.

11. Cox, R.G. J. Fluid Mech. 1983, *131*, 1.

12. Joanny, J.F.; de Gennes, P.G. J. Chem. Phys. 1984, *81*, 552.

13. Jansons, K. J. Fluid Mech. 1985, *154*, 1.

14. Schwartz, L.W.; Garoff, S. J. Colloid Interface Sci. 1985, *106*, 422.

15. Chen, Y.L.; Helm, C.A.; Israelachvili, J.N. J. Phys. Chem. 1991, *95*, 10736.

16. Marmur, A. J. Colloid Interface Sci. 1994, *168*, 40.

17. Li, D.; Neumann, A.W. In *Surfactant Science Series*; Neumann, A.W. Spelt, J.K., Eds.; Dekker: New York, 1996; Vol. 63, 110–168.

18. Extrand, C.W. J. Colloid Interface Sci. 1998, *207*, 11.

19. Sakai, H.; Fujii, T. J. Colloid Interface Sci. 1999, *210*, 152.

20. Lam, C.N.C.; Wu, R.; Li, D.; Hair, M.L.; Neumann, A.W. Advances in Colloid and Interface Science, 2002, *96*, 169.

21. Kamusewitz, H.; Possart, W. Applied Physics A—Materials Science & Processing, 2003, *76*, 899.

22. Wilhelmy, L. Ann. Phys. 1863, *119*, 177.

23. Du Noüy, L. J. Gen. Physiol. 1919, *1*, 521.

24. Neumann, A.W.; Good, R.J. In *Surface and Colloid Science*; Good, R.J. Stromberg, R.R., Eds.; Plenum Press: New York, 1979; Vol. 22, 31–91.

25. Sugden, S. J. Chem. Soc. 1922, *121*, 858.

26. Harkins, W.D.; Brown, F.E. J. Am. Chem. Soc. 1919, *41*, 503.

27. Sugden, S. J. Chem. Soc. 1921, *120*, 1483.

28. de Laplace, P.S. *Mechanique Celeste, Supplement to Book 10*; J. B. M. Duprat: Paris, 1808.

29. Bashforth, F.; Adams, J.C. In *An Attempt to Test the Theory of Capillary Action*; University Press: Cambridge, 1883.

30. Andreas, J.M.; Hauser, E.A.; Tucker, W.B. J. Phys. Chem. 1938, *42*, 1001.

31. Hartland, S.; Hartley, R.W. In *Axisymmetric Fluid–Liquid Interfaces*; Elsevier: Amsterdam, 1976.

32. Fox, H.W.; Zisman, W.A. J. Colloid Sci. 1950, *5*, 520.

33. Adam, N.K.; Jessop, G. J. Chem. Soc. 1925, *1925*, 1863.

34. Bikerman, J.J. Ind. Eng. Chem. Anal. Ed. 1941, *13*, 443.

35. Paddy, J.F. In *Surface and Colloid Science*; Matijevic, E., Eirich, F.R., Eds.; Academic Press: New York, 1969; Vol, 1, 101–149.

36. Johnson, R.E.; Dettre, R.H. In *Surface and Colloid Science*; Matijevic, E., Eirich, F.R., Eds.; Academic Press: New York, 1969; Vol. 2, 85–153.

37. Ambwani, D.S.; Fort, T. In *Surface and Colloid Science*; Good, R.J., Stromberg, R.S., Eds.; Plenum Press: New York, 1979; Vol. 22, 93–119.

38. Spelt, J.K.; Vargha-Butler, E.I. In *Surfactant Science Series*; Neumann, A.W., Spelt, J.K., Eds.; Dekker: New York, 1996; Vol. 63, 379–411.
39. Pallas, N.R. Colloids Surf. 1990, *43*, 169.
40. Suttiprasit, P.; Krisdhasima, V.; Mcguire, J. J. Colloid Interface Sci. 1992, *154*, 316.
41. Holcomb, C.D.; Zollweg, J.A. J. Colloid Interface Sci. 1992, *154*, 51.
42. Hasen, F.K. J. Colloid Interface Sci. 1993, *160*, 209.
43. Hong, K.T.; Imadojemu, H.; Webb, R.L. Exp. Therm. Fluid Sci. 1994, *8*, 279.
44. Lahooti, S.; Rio, O.I.D.; Neumann, A.W.; Cheng, P. In *Surfactant Science Series*; Neumann, A.W., Spelt, J.K., Eds.; Dekker: New York, 1996; Vol. 63, 441–507.
45. Song, B.; Springer, J. J. Colloid Interface Sci. 1996, *184*, 77.
46. Semmler, A.; Ferstl, R.; Kohler, H.H. Langmuir 1996, *12*, 4165.
47. Sedev, R.V.; Petrov, J.G.; Neumann, A.W. J. Colloid Interface Sci. 1996, *180*, 36.
48. Earnshaw, J.C.; Johnson, E.G.; Carrol, B.J.; Doyle, P.J. J. Colloid Interface Sci. 1996, *177*, 150.
49. Frank, B.; Garoff, S. Colloids Surf. 1999, *156*, 177.
50. Graham-Eagle, J.; Neumann, A.W. Colloids Surf. 2000, *161*, 63.
51. Sakai, K.; Mizuno, D.; Takagi, K. Physical Review E. 2001, *63*, Art No. 046302 Part 2.
52. Bateni, A.; Susnar, S.S.; Amirfazli, A.; Neumann, A.W. Colloids and Surfaces-Physicochemical and Engineering Aspects, 2003, *219*, 215.
53. Katoh, K.; Fujita, H.; Yamamoto, M. Trans. Jpn. Soc. Mech. Eng. 1991, *57(B)*, 4124. in Japanese.
54. Boruvka, L.; Gaydos, J.; Neumann, A.W. Colloids Surf. 1990, *43*, 307.
55. Ivanov, I.B.; Dimitrov, A.S.; Nikolov, A.D.; Denkov, N.D.; Kralchevsky, P.A. J. Colloid Interface Sci. 1992, *151*, 446.
56. Gu, Y.G. Colloids and Surfaces-Physicochemical and Engineering Aspects, 2001, *181*, 215.
57. Wenzel, R.N. Ind. Eng. Chem. 1936, *28*, 988.
58. Cassie, A.B.D. Discuss. Faraday Soc. 1948, *3*, 11.
59. Katoh, K.; Azuma, T. Heat Transf. Asian Res. 2001, *30*, 371.
60. Katoh, K.; Fujita, H.; Sasaki, H. Trans. ASME, J. Fluids Eng. 1995, *117*, 303.
61. Katoh, K.; Fujita, H.; Sasaki, H. Trans. ASME, J. Fluids Eng. 1990, *112*, 289.
62. Princen, H.M. In *Surface and Colloid Science*; Matijevic, E., Eirich, F.R., Eds.; Academic Press: New York, 1969; Vol. 2, 1–84.
63. Huh, C.; Scriven, L.E. J. Colloid Interface Sci. 1969, *30*, 323.
64. Rapacchietta, A.V.; Neumann, A.W. J. Colloid Interface Sci. 1977, *59*, 555.
65. Katoh, K.; Fujita, H.; Sasaki, H.; Miyashita, K. Trans ASME, J. Fluids Eng. 1992, *114*, 460.
66. Hirota, M.; Fujita, H.; Katoh, K. Proc. Third Int. Sympo. Multiphase Flow and Heat Transfer; Hemisphere: New York, 1994; 747.
67. Katoh, K.; Tsao, Y.; Yamamoto, M.; Azuma, T.; Fujita, H. J. Colloid Interface Sci. 1998, *202*,54.
68. Katoh, K.; Fujita, H.; Takaya, M. Trans. ASME, J. Fluids Eng. 1994, *116*, 801.

8

Contact Angle Measurements on Fibers and Fiber Assemblies, Bundles, Fabrics, and Textiles

BIHAI SONG   Krüss GmbH, Hamburg, Germany

ALEXANDER BISMARCK   Imperial College London, London, United Kingdom

JÜRGEN SPRINGER   Technical University of Berlin, Berlin, Germany

I. INTRODUCTION

The study of fiber wettability has important implications in detergency [1], water and oil repellency of textiles [2], in textile spinning, optical fiber processing, and in the chemical design of modern fiber reinforced composite systems (such as fiber reinforced polymers or concrete). The degree to which a liquid wets fibers determines how easily the liquid can penetrate fiber assemblages and this is important for both wet processing and for the performance of a textile article. Understanding the nature of the interaction of fibers with various liquids is important to tailor new polymeric composite systems [3]. The measurement of contact angles provides useful information about their wettability. The latter is essential in the modification of the fiber surface state or the adjustment of the rheological properties of the wetting prepolymer (resins) or of a polymer melt. Thermodynamic parameters, such as solid surface tension calculated from measured contact angles and surface polarity, can be used to estimate the expected adhesion between reinforcing fibers and a surrounding matrix [4,5]. It can also lead to a prediction of the compatibility of the components of composite materials.

The wetting behavior of a planar surface is governed by the materials surface chemistry and its local geometry (roughness). The wetting of a fiber of a chemically identical material is, however, significantly different because of the global geometry of the cylindrical shape [6]. The contact angle of a liquid against a fiber is the quantity, which is a fundamental measure [7].

Contact angle measurements provide only information about the outer-most atom layers of a surface (~ 0.5 to 1 nm [8]), depending on the chemistry and the molar volume of the liquid used, whereas spectroscopic techniques, such as X-ray photoelectron spectroscopy (XPS) also known as electron spectroscopy for chemical analysis (ESCA), give information about the elemental composition and the bonding state of the elements originating from a cone typically extending some 3–5 nm deep [9]. Therefore contact angle measurements are most suitable for characterizing the changes in the surface "nature" with a high degree of accuracy.

II. CONTACT ANGLES AND WETTABILITY

Wettability refers to the interactions when a liquid l is brought into contact with a solid surface s initially surrounded by a gas (or the saturated liquid vapor v atmosphere) or another liquid [10]. These interactions can result in spreading without limits of the liquid over the surface displacing the other fluid, or the spreading process might come to an end if the equilibrium state is reached, which is characterized by a contact angle θ between the liquid–fluid and liquid–solid interfaces. This phenomenon is often described by a sessile drop resting on a solid surface (Fig. 1).

Young's equation

$$\gamma_{sv} = \gamma_{sl} + \gamma_{lv} \cos \theta \tag{1}$$

connects the quantities of the liquid vapor surface tension γ_{lv}, the solid vapor surface tension γ_{sv}, the solid–liquid interfacial tension γ_{sl}, and the contact angle θ [11]. Young's equation, however, is only applicable to systems that are in equilibrium. This means that all components have a uniform chemical potential in all (liquid, solid, and vapor) phases. Therefore the liquid must be saturated with the solid, which usually lowers the liquid surface tension, and the vapor and solid surface must be at adsorption equilibrium. The value of the interfacial tension of a solid, covered by a film γ_{sv}, is lower than that for the

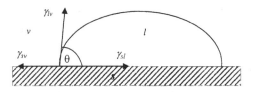

FIG. 1 A drop of liquid l resting on a plane surface of solid s in equilibrium with the liquid's vapor v.

solid-vacuum surface γ_s by the amount of the surface (or spreading or film) pressure π,

$$\pi = \gamma_s - \gamma_{sv}. \tag{2}$$

It is generally accepted that for systems exhibiting contact angles $\theta > 10°$ the surface pressure is negligible [8]. This is usually explained by the fact that a liquid will not adsorb on a solid having a lower surface tension than the liquid [12]. Hence we can write: $\gamma_{sv} \cong \gamma_s$. On the other hand $\gamma_{lv} \cong \gamma_l$, because the interacting energies between the molecules at the surfaces of condensed phases and those in the saturated vapor phase are negligible, compared to the interaction energies acting within the condensed phase [13]. Therefore Young's equation can be written as follows:

$$\gamma_s = \gamma_{sl} + \gamma_l \cos \theta. \tag{3}$$

The only directly measurable quantities in Eq. (3) are the liquid surface tension and the contact angle.

III. DIRECT CONTACT ANGLE MEASUREMENTS FROM THE MENISCUS PROFILES OF DROP-ON-FIBER SYSTEMS

A. Theoretical Background

1. Description of Drop Profile

Like all liquid–fluid interfaces, the equilibrium shape of the interface for a drop-on-fiber system is governed by the gravitational and interfacial tension effects through the classical Laplace equation of capillarity:

$$\Delta p = \gamma \left(\frac{1}{R_1} + \frac{1}{R_2} \right), \tag{4}$$

where Δp is the pressure difference (excess pressure within the drop) across the droplet's liquid–fluid interface, γ is the interfacial tension, and R_1 and R_2 are the principal radii of curvature at a point in the interface.

For a cylinder symmetrical interface as depicted in Fig. 2, R_1 (in the paper plane) and R_2 (perpendicular to the paper plane and to the drop profile $y = f(x)$ depicted in Fig. 2) can be written as

$$\frac{1}{R_1} = \frac{d\phi}{ds} = \frac{d\phi}{dx/\cos \phi} = \frac{d\phi}{dx} \cos \phi, \tag{5}$$

$$\frac{1}{R_2} = \frac{\sin \phi}{x}. \tag{6}$$

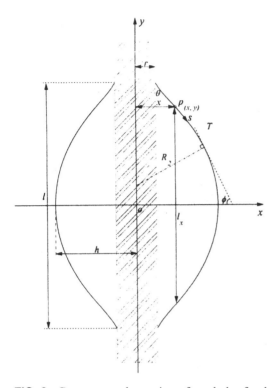

FIG. 2 Geometry and notation of symbols of a drop-on-fiber system.

Eq. (4) may be then rewritten as:

$$\frac{\Delta p}{\gamma} = \cos\phi\,\frac{d\phi}{dx} + \frac{\sin\phi}{x} = \frac{1}{x}\frac{d}{dx}(x\sin\phi). \qquad (7)$$

For very small liquid droplets, the gravitational effect may be negligibly small compared to the interfacial tension, so the term $\Delta p/\gamma$ on the left side of Eq. (7) can be considered constant along the whole interface. Eq. (7) can be then integrated to:

$$x\sin\phi = \frac{1}{2}\frac{\Delta p}{\gamma}x^2 + C_2 = \frac{1}{2}C_1 x^2 + C_2, \qquad (8)$$

where $C_1\,(=\Delta p/\gamma)$ and C_2 are the system constants, which are restricted to the following boundary conditions

$$x_{\phi = \pi/2} = h,$$
$$x_{\phi = \pi/2 - \theta} = r. \qquad (9)$$

Combination of Eqs. (8) and (9) gives:

$$x \sin \phi = \left(\frac{h - r \cos \theta}{h^2 - r^2}\right) x^2 + \left(\frac{h \cos \theta - r}{h^2 - r^2}\right) h \cdot r. \tag{10}$$

Eq. (10), together with the relationship:

$$\frac{dy}{dx} = -\tan \phi = -\frac{\sin \phi}{\sqrt{1 - \sin^2 \phi}}, \tag{11}$$

allows now the gradient of the drop profile, dy/dx, to be expressed as a function of x, which, after substitution for $\sin\phi$ in Eq. (11) with the expression of Eq. (10) followed by expansion and simplification as described in detail by Carroll [19], can be written as:

$$-\frac{dy}{dx} = \frac{x^2 + a \cdot h \cdot r}{[(h^2 - x^2)(x^2 - a^2 r^2)]^{\frac{1}{2}}}, \tag{12}$$

where a is defined as:

$$a = \frac{h \cdot \cos \theta - r}{h - r \cdot \cos \theta}. \tag{13}$$

With the following transformation:

$$x^2 = h^2(1 - k^2 \sin^2 \varphi), \tag{14}$$

where

$$k^2 = \frac{h^2 - a^2 \cdot r^2}{h^2}, \tag{15}$$

Eq. (12) can be rewritten as

$$dy = \pm \frac{a \cdot r + h(1 - k^2 \sin^2 \varphi)}{(1 - k^2 \sin^2 \varphi)^{\frac{1}{2}}} d\varphi, \tag{16}$$

which can further be integrated to give

$$y = \pm [a \cdot r \cdot F(\varphi, k) + h \cdot E(\varphi, k)], \tag{17}$$

where $F(\varphi,k)$ and $E(\varphi,k)$ are the Legendre standard incomplete elliptic integrals of the first and second kind, respectively, and are defined as

$$F(\varphi, k) = \int_0^\varphi \frac{d\vartheta}{\sqrt{1 - k^2 \sin^2 \vartheta}}, \tag{18}$$

$$E(\varphi, k) = \int_0^{\varphi} \sqrt{1 - k^2 \sin^2 \vartheta} \, d\vartheta. \tag{19}$$

Equations (12)–(15) and (17) give the expression of the drop profile in a drop-on-fiber system where the gravitational effect is negligibly small compared to the interfacial tension. To make it easier for the later generalization, it is useful to transfer these equations into their corresponding dimensionless forms by dividing all the used dimensional variables by the radius of the fiber r_f involved in the system:

$$a = \frac{H \cdot \cos \theta - 1}{H - \cos \theta}, \tag{13a}$$

$$X^2 = H^2 (1 - k^2 \sin^2 \varphi), \tag{14a}$$

$$k^2 = 1 - a^2 / H^2, \tag{15a}$$

$$Y = \pm [a \cdot F(\varphi, k) + H \cdot E(\varphi, k)], \tag{17a}$$

where X, Y, and H are the dimensionless counterparts of x, y, and h, respectively (in this text we use capital and lowercase letters to describe the dimensionless and dimensional variables, respectively). As can be seen from these relations, a drop profile, as represented by a collection of (X, Y)-coordinates, may be expressed as a function of H and θ and can be obtained through the numeric evaluation of the elliptic integrals F and E in Eq. (14a) and (16a). The Legendre's standard elliptic integrals F and E can be transformed to the Carlson's elliptic integrals [14,15] of the first and the second kind, R_F and R_D, which are more suitable for the numerical evaluation [16,17]:

$$F(\varphi, k) = \sin \varphi \cdot R_F(\cos^2 \varphi, 1 - k^2 \sin^2 \varphi, 1), \tag{20}$$

$$E(\varphi, k) = \sin \varphi \cdot R_F(\cos^2 \varphi, 1 - k^2 \sin^2 \varphi, 1) - \frac{1}{3} \cdot k^2 \cdot \sin^3 \varphi$$
$$\cdot R_D(\cos^2 \varphi, 1 - k^2 \sin^2 \varphi, 1). \tag{21}$$

B. Diverse Characteristic Drop Parameters

(a)   Drop Length.   Combining Eq. (17a) with Eqs. (20) and (21), one obtains the dimensionless length L of a droplet with maximum height H and contact angle θ:

$$L = 2 \left[(a + H) \cdot \sin \varphi \cdot R_F - \frac{1}{3} \cdot H \cdot k^2 \cdot \sin^3 \varphi \cdot R_D \right] = f(H, \theta). \tag{22}$$

The right side in Eq. (22) is a function of H and θ only [as denoted by $f(H,\theta)$], which means that the contact angle θ in a drop-on-fiber system can be unambiguously determined from its length L and height H.

(b)   Drop Volume.   The (net) liquid volume v of a droplet is calculated from the total volume enclosed under the drop–vapor interface by subtraction of the fiber volume inside the droplet:

$$v = \int_{-\frac{l}{2}}^{\frac{l}{2}} \pi \cdot (x^2 - r^2)dy = 2\pi \int_0^{\frac{l}{2}} (x^2 - r^2)dy. \tag{23}$$

Using Eqs. (12) and (14) the above expression can be integrated to give:

$$v = \frac{2\pi h}{3}\left[(2a^2r^2 + 3arh + 2h^2)E(\varphi,k) - a^2r^2F(\varphi,k) + \frac{r}{h}(h^2 - r^2)^{\frac{1}{2}} \right.$$
$$\left. (r^2 - a^2r^2)^{\frac{1}{2}} \right] - \pi r^2 l \qquad \text{(with } \sin^2\varphi = (h^2 - 1)/h^2k^2) \tag{24}$$

Eq. (24) may be transformed to the following dimensionless form:

$$V = \frac{2\pi H}{3}\left[(2a^2 + 3aH + 2H^2)E(\varphi,k) - a^2F(\varphi,k) + \frac{1}{H}(H^2 - 1)^{\frac{1}{2}} \right.$$
$$\left. (1 - a^2)^{\frac{1}{2}} \right] - \pi r^2 L \qquad \text{(with } \sin^2\varphi = (H^2-1)/H^2k^2) \tag{24a}$$

(c)   Contact Areas.   There are two interfacial contact areas involved in the system: solid–liquid a_{sl} and liquid–fluid (usually liquid–gas) a_{lv}. The fiber-liquid contact area can be deduced from the length of the droplet l at once $a_{sl} = 2\pi r l$. The liquid-fluid contact area can be calculated through surface integration of the drop interface:

$$a_{lv} = \int_{-\frac{l}{2}}^{\frac{l}{2}} 2\pi \cdot x ds = 4\pi \int_0^{\frac{l}{2}} x ds = 4\pi \int_0^{\frac{l}{2}} x \frac{dx}{\cos\phi}$$
$$= 4\pi \int_0^{\frac{l}{2}} x \left[1 + \left(\frac{dy}{dx}\right)^2 \right]^{\frac{1}{2}} dx. \tag{25}$$

Using the relationship of Eq. (12) and the transformation (14), Eq. (25) can be integrated to give:

$$a_{lv} = 4\pi h(ar + h)E(\varphi,k), \tag{25a}$$

which may be further transformed into the following dimensionless form:

$$A_{lv} = 4\pi H(a + H)E(\varphi,k), \tag{25b}$$

where φ is given by Eq. (14a) with $X = 1$.

(d) *Excess Pressure.* The excess pressure in the droplet Δp can be deduced from Eqs. (8) and (10):

$$\frac{\Delta p}{\gamma} = \frac{2(h - r \cos \theta)}{h^2 - r^2}. \tag{26}$$

(e) *Inflection Angle.* The liquid–fluid interface of a droplet may show an inflection point as it approximates the solid–liquid interface. The inflection point occurs when the gradient of the drop profile, dy/dx, reaches a maximum, i.e., where $d^2y/dx^2 = 0$. By substitution of Eq. (12) into the above formulation and after rearrangement, we have:

$$\frac{d^2y}{dx^2} = 0 \quad \Rightarrow \quad x = \pm\sqrt{ahr} \quad \text{or} \quad X = \pm\sqrt{aH}. \tag{27}$$

Using the relationship of the inflection angle θ_i to the gradient of the drop profile $\tan(90-\theta_i) = dy/dx$, and Eqs. (12) and (27) lead to the following expression of θ_i:

$$\tan \theta_i = \frac{H - a}{2\sqrt{aH}}. \tag{28}$$

The relationship given in (Eq. (27) must be, however, restricted to the fact that in a real droplet-on-fiber system the value of X can never be smaller than 1, which leads to:

$$X^2 = aH \geq 1 \quad \Rightarrow \quad \frac{2H}{H^2 + 1} \leq \cos \theta \leq 1 \quad \text{or} \quad H \geq \frac{1 + \sin \theta}{\cos \theta}. \tag{29}$$

Eqs. (29) and (27) indicate that the existence of an inflection point and its location for a given system with a certain contact angle θ is dependent of the droplet size H. It is in general to be observed by droplets with a small contact angle (e.g., $< 30°$), whereas for droplets of small size and large contact angles it may be missed.

It can be verified that for a droplet having a reduced drop height of H, its maximum possible inflection angle α_{max} is equal to its maximum possible contact angle under the fulfillment of Eq. (29) and its minimum possible inflection angle α_{min} occurs at $\theta = 0$:

$$\alpha_{max} = \text{arc sin}\left(\frac{H^2 - 1}{H^2 + 1}\right) \tag{30}$$

$$\alpha_{min} = \text{arc tan}\left(\frac{H - 1}{2\sqrt{H}}\right) \tag{31}$$

C. Maximum Drop Length (L)–Height (H) Method

As pointed out above, the right side of Eq. (22) is a function of H and θ only, which suggests that the contact angle θ can be unambiguously determined from the maximum drop length L and height H. Based on this principle, Yamaki and Katayama [18] and Carroll [19] have introduced this method for determining the contact angle of a liquid droplet on a cylindrical fiber surface.

However, Eq. (22) cannot be transformed into an explicit form of contact angle in dependence of H and L. To evaluate the contact angle from the measured H and L, Eq. (22) must be solved numerically in an iterative way. The resolutions are available both in graphical [18,19] and table forms (see table in Appendix A). Because of the complexity of the dependence (see curve a in Fig. 3 [20]), the graphical relationship allows the contact angle generally to be found to within about 5° [16], while the tables provided by Carroll [19] later giving L as a function of H and θ in steps of 0.01 in H and of 1° in θ do make a highly precise determination possible.

To make the determination more accurate than using the graphical relationship and more convenient than using the tables, a computer program

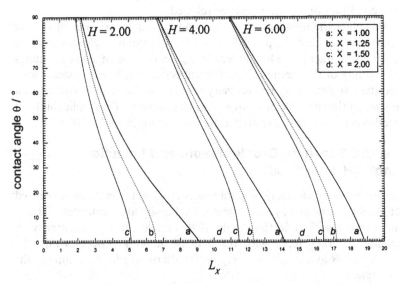

FIG. 3 Contact angle as a function of droplet length at a given height.

may be used to solve Eq. (22) numerically [16]. There are many ways to do it. One of them is by minimizing the following objective function E_{LH}

$$E_{LH} = \left| L - 2\left[(a+H) \cdot \sin \varphi \cdot R_F - \frac{1}{3} \cdot H \cdot k^2 \cdot \sin^3 \cdot R_D\right]\right|$$
$$= |L - f(H, \theta)| \tag{32}$$

at $X = 1$.

E_{LH} is by given L and H, a one-dimensional function of θ. Any one-dimensional minimization routine may be therefore suitable for solving this problem. For example, the Brent's method in one-dimension as described in Ref. 17 was found to be quite suitable for this purpose [20].

D. Methods Based on the Drop's Inflection Angle

Like the maximum droplet height (H) and length (L), the inflection angle (α) of a droplet is a characteristic parameter, which can be used in combination with other parameters to determine the contact angle. For drops exhibiting a visible (or measurable) inflection angle, the direct measurement of inflection angle may be easier and more accurately practicable than the contact angle itself.

As pointed out above, not all droplets exhibit an inflection angle that must be located within the range of $1 \leq X \leq H$. For the occurrence of an inflection angle the condition of Eq. (29) must be fulfilled.

Fig. 4 displays the relationship of α_{max} and α_{min} in dependence of H. For droplets with a reduced drop height of H, an inflection angle can only be observed if its contact angle lies under the α_{max} curve, which also determines the applicability of the methods based on inflection angle. For systems with larger contact angles it may be necessary to increase the droplet volume to a certain size, so that an inflection angle can be observed. The applicability of these methods is, in general, restricted for contact angle under $60°$.

E. Method Based on Droplet Height and Inflection Angle (H, α Method)

Among the droplet characteristic parameters H, L, α, the maximum reduced droplet height H can be in general determined easily and accurately. Therefore for droplets with inflection angle, if this angle can be measured with appropriate accuracy (e.g., with the help of imaging processes [3]), the combination of H and α will be a very useful alternative for determining the contact angle.

Fig. 5 shows the relationship between the contact angle θ and inflection angle α of a droplet at a given H. It can be seen that the contact angle is very

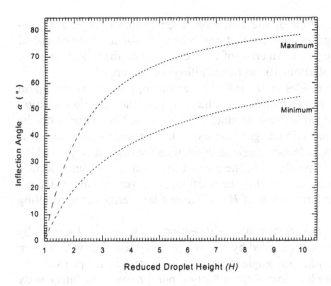

FIG. 4 Dependence of minimal and maximal inflection angle α on H.

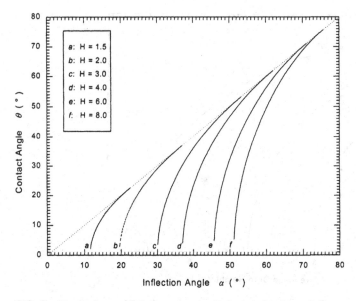

FIG. 5 Dependence of the contact angle θ on inflection angle α at a given H.

sensitive to the variation of the inflection angle, which is especially pronounced in the range of small contact angles ($\theta < 10°$ for small droplets and $\theta < 20°$ for large droplets). An error of $1°$ in determining the inflection angle can cause an error of about $10°$ in the resulting contact angle.

This sensitivity subsides as the inflection angle approaches its maximum value at each given H. Nevertheless, at the same time the inflection position moves closer to the three-phase contact (TPC) line and coincides with the TPC point as the inflection angle reaches its maximum, which makes the determination of the inflection angle as difficult as the contact angle itself.

Fig. 6 shows the dependence of the contact angle on H at given inflection angles α. For small contact angles, the relationship is very sensitive; a small uncertainty in the determination of H may cause a large error in the resulting contact angle.

Fig. 7 displays the variation of the inflection angle and its location by changing the droplet size (implied by H) for a given system (i.e., with a given contact angle). The inflection angle increases with increasing droplet volume. At the same time the location of the inflection point moves absolutely away from, but relatively closer to, the solid surface, because the reduced maximum droplet height increases far more rapidly than the movement of the location of the inflection point.

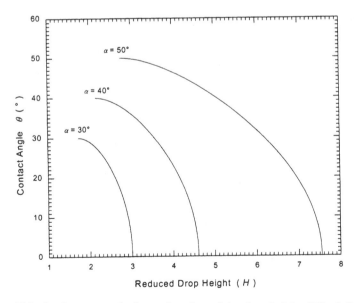

FIG. 6 Contact angle θ as a function of the drop height H for inflection angles α.

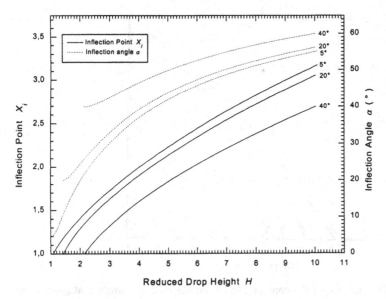

FIG. 7 Inflection angle α and inflection point X_i as a function of the drop size (as indicated by droplet height H) for given contact angles θ.

As a result, the method based on the measurement of H and α is, in general, too error-sensitive to have a wide applicability. Nevertheless, the method can be quite interesting when the inflection angle can be determined with high accuracy, for example, using some appropriate profile analysis routines [2].

F. Method Based on Droplet Length and Inflection Angle (L, α Method)

Similar to the method based on droplet height and inflection angle (H,α method), the contact angle may be determined by measuring the droplet length L and the inflection angle (L,α method). Fig. 8 shows the dependence of the contact angle on the inflection angle by some given drop lengths.

It can be seen from Fig. 8 that the contact angle is not so sensitive to the variation of inflection angle as in case of the H,α method, whereas the value of drop length L is, in general, more difficult to measure precisely than H.

Fig. 9 shows the dependence of the contact angle on L by given inflection angles. For small droplets of small contact angles, the value of the contact

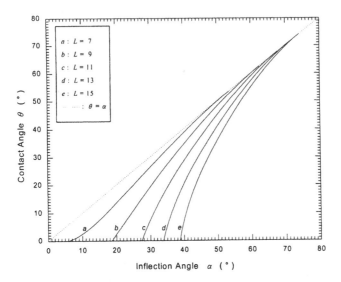

FIG. 8 Dependence of the contact angle θ on the inflection angle α at given drop lengths L.

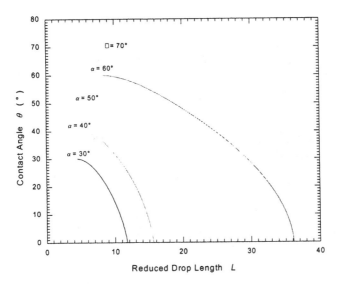

FIG. 9 Dependence of the contact angle θ on the drop length L for given inflection angles α.

angle is much more sensitive to possible uncertainties in the determination of L than for larger droplets with large contact angles.

G. Method Based on Droplet Length, Height, and Inflection Angle (L/H, α Method)

One more imaginable method based on the measurement of the inflection angle is to combine all three droplet characteristic parameters H, L, and α for determining the contact angle. Fig. 10a–c shows the relationship of the inflection angle in dependence of L/H-ratio by given contact angles in steps of 2°. Using L/H-ratio instead of absolute values of L or H replaces the need to determine the fiber radius r_f. This is particularly advantageous because of the fact that the equations have to be transferred into their corresponding dimensionless forms by dividing all the used dimensional variables involved in the system by r_f having a limited accuracy.

H. A Generalized Drop Length (L$_x$)–Height (H) Method

The maximum L,H method works accurately, insofar as the L- and H-values can be determined with sufficient precision. This is often not the case, however, because the curve of a liquid drop surface changes steeply near the three-phase contact line [18] and the determination of the drop dimensions (especially the L-values) is often associated with large uncertainty and subject to subjective influences. This is particularly true for drops of small contact angles and for cases where the drop dimensions are measured with video images instead of photographs [21]. For fibers of very small diameter (in the order of 1 μm), the formation of "precursor manchon" [22] or "protruding foot" [23] may further complicate the determination of the contact points. Because the value of the contact angle is quite sensitive to the values of L and H, a small uncertainty in the determination of contact points may introduce large errors in the resultant contact angle values and make the determination of contact angles in some cases more or less subjective.

To partially overcome the disadvantages of the maximum L,H method, Song et al. [20] have recently extended the concept and introduced a generalized drop length–height method. Instead of using only the maximum drop length L, together with the maximum drop height H, for calculating the contact angle, a general drop length L_x was introduced. This is the length (i.e., diameter) of a drop measured at the drop plane $X = X$ (see the dimensional representation in Fig. 2). According to this definition we have $L = L_1$, i.e., the maximum drop length is the drop length measured at $X = 1$. As can be seen from Eqs. (14a) and (17a), any combination of L_x and H of a drop profile can

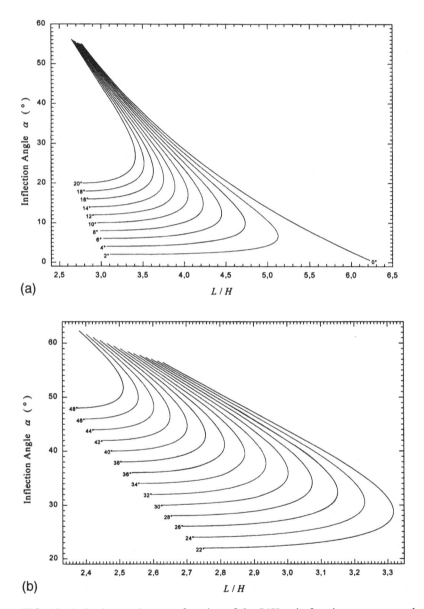

FIG. 10 Inflection angle α as a function of the L/H ratio for given contact angles θ.

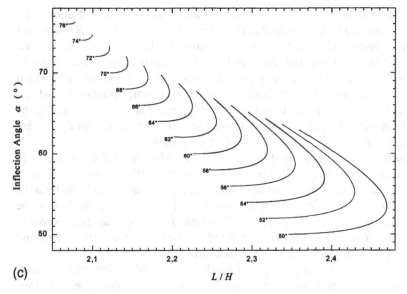

(c)

L / H

FIG. 10 Continued.

now be used to calculate the contact angle θ. Similarly to the maximum L,H method θ is now determined by minimizing the following objective function $E_{L_x,H}$:

$$E_{L_x,H} = \left| L_x - 2\left[(a+H) \cdot \sin \varphi \cdot R_F - \frac{1}{3} \cdot H \cdot k^2 \cdot \sin^3 \cdot R_D \right] \right|$$

$$= |L_x - f(H, \theta)|$$

(33)

at $X = X$.

Fig. 3 shows, the calculated relationship of contact angles to L_x-values for given drop heights $H = 2, 4,$ and 6. The value of the contact angle is increasingly sensitive to the value of L_x with increasing X, particularly for contact angles smaller than $10°$. This means that with the same extent of error in L_x-values the error in the resultant contact angle value increases with increasing X. Therefore this new procedure will only work well if the L_x-values at $X > 1.0$ can be determined more accurately than the maximum drop length itself.

One advantage of this generalized method over the common maximum L,H method lies in the fact that one is no more restricted to measuring the maximum drop length in order to determine the contact angle. Alternatively, it is possible to measure the drop length at any drop levels and use the measured values to calculate the contact angle. As pointed out above, the determination of the maximum drop length is generally associated with uncertainty and subjective influences, because the three-phase contact points are often smeared out and may not be determined with sufficient precision. However, the drop lengths at higher X-levels (e.g., $X > 1.5$) are well defined and may be measured more accurately, which may compensate or overwhelm the weakness of sensitivity by using L_x,H-values with $X > 1$ for computing the contact angle. Compared to the L-value, the maximum drop height H can be, in general, measured with more satisfactory accuracy.

Furthermore, this generalized method provides us with a procedure to determine the contact angle from a number of pairs of L_x,H-values around a drop profile, so that their resultant average value can be more reliable than the value obtained only from the single pair of L,H-values as usually done in the literature. In this way a large part of the whole drop profile can be involved in the determination process, which will reduce the random statistical error and thus improve the accuracy of the measurement.

Principally L_x,H-values at any X-levels can be used to determine the contact angles. However, with increasing X-value, L_x-value decreases, and thus the relative error in its determination rises. The meaningful X-values therefore usually range from 1 to about $H-2$ (for drops with $H > 3$) [20].

1. Theoretical Simulations

Theoretical simulations have been carried out in order to examine the applicability and reliability of the generalized drop length–height L_y,H method as described above. Dimensionless theoretical drop profiles are generated by numerically solving the following equations:

$$X^2 = H^2(1 - k^2 \sin^2 \varphi), \tag{14a}$$

$$Y = \pm [a \cdot F(\varphi,k) + H \cdot E(\varphi,k)]. \tag{17a}$$

The theoretical profiles (i.e., X-coordinates of the profiles) are then overlapped with random errors in order to simulate the experimentally obtained drop profiles. As a typical example we have shown in Fig. 11 a simulated drop profile with $H = 3$ and $\theta = 15°$.

Contact angles were then evaluated from the profiles using the program described in Ref. 20. The result of applying the L_y,H method to the profile shown in Fig. 11 is plotted in Fig. 12.

Generated profiles were not the same, as they were all overlaid with random errors and, therefore, the calculation for every $H-\theta$ combination included in the table was repeated 40 times. Table 1 contains some of the results, the average values, and the standard deviation for each $H-\theta$-combination. The maximum extent of the overlaid random error amounts to 0.025, which is

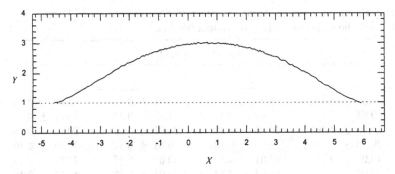

FIG. 11 A simulated (dimensionless) drop profile overlapped with random errors with a maximum content of 0.025.

close to the maximum statistical errors of real drop profiles extracted using digital image processing.

2. Experimental Measurements

The L_x,H method was successfully applied to study a variety of drop-on-fiber systems [20,24] and also to characterize the wetting behavior of (carbon) fiber–polymer melt systems [25–27]. Fig. 13 shows exemplarily a drop of polyimide (Pl, Aurum PD 400) melt on a carbon fiber (Toray T700SC) at 420°C in air atmosphere together with the extracted drop profile. The carbon fibers studied have diameters of approx. 7 μm.

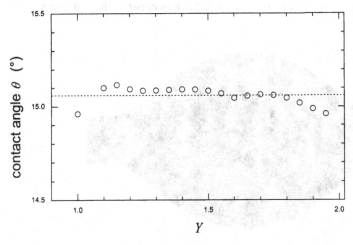

FIG. 12 Results of the L_x,H method applied to the drop profile shown in Fig. 11.

TABLE 1 Results of Contact Angle Determination for the Simulated Drop Profiles Using the Generalized Length–Height Method[a]

	Maximum drop height H^b			
θ^2	2	3	5	7
5	5.10 ± 0.82	5.03 ± 0.65	4.95 ± 0.75	5.06 ± 0.88
10	10.05 ± 0.50	9.99 ± 0.37	9.98 ± 0.39	10.02 ± 0.51
20	19.95 ± 0.40	20.00 ± 0.20	20.02 ± 0.23	20.02 ± 0.28
30	29.96 ± 0.16	29.99 ± 0.18	30.01 ± 0.12	30.03 ± 0.30
40	40.03 ± 0.27	40.01 ± 0.18	40.01 ± 0.15	40.01 ± 0.18
50	49.94 ± 0.61	50.04 ± 0.14	50.01 ± 0.12	50.01 ± 0.16

[a] The y-coordinates of the theoretical dimensionless drop profiles were overlapped with random errors ranging from −0.025 to 0.025. The results are shown here as average value ± standard derivation, which were evaluated from 40 simulations (cf. text).
[b] The listed H- and θ-values are the values used to generate the theoretical dimensionless drop profiles.
Source: From Ref. 20.

Carbon fibers were tightened to a metal frame using dental cement. Droplets of polymer melts (as shown in Fig. 13) can be prepared by first dipping a tightened fiber into the polymer powder, followed by slowly heating the "powdered" fiber in a hot stage to melt the polymer. Droplets form on the fiber from a receding movement and, therefore, the measured contact angles are (or approaching) their receding angles. The drop images were recorded using a microscope equipped with a digital camera. The images shown here were taken from above (i.e., top view) and it was ensured that all fibers were

FIG. 13 Image of a polyimide melt droplet resting on a carbon fiber at 420°C.

lying horizontally, perpendicular to the microscope lenses. The drop profiles were extracted from the images using a subpixel routine, as for instance described in Ref. 35. The extracted profiles (Fig. 14) were then used for calculating the contact angles.

The results of the L_y, H method applied to the real carbon fiber/Pl drop profile are shown in Fig. 15. Both sides of the drop profile as well as their averaged profile were used to calculate the contact angles.

Other than the theoretical simulations described above the contact angle values calculated using L_x, H-values in case of the experimentally obtained drop profile are quite sensitive to the X-levels by X approaching 1 (Fig. 15). Nevertheless, as X increases to values around 1.5–3.0 the calculated contact angle values are less dependent on X. The sensitivity of the calculated θ-values against X by small X-values reveals the difficulty in determining the L_x-values near the liquid–fiber contact surface. This might be partially due to the weakness of the program employed. However, the fact that the three-phase contact points in a drop-on-fiber system are often smeared out to some extent is certainly a dominant factor which usually makes an accurate determination of the L_x-value near the contact line rather difficult. This sensitive range is generally more significant for drops having small contact angles as compared to those with large contact angles. The results suggest that contact values determined using the common maximum drop length-height L, H method might deviate significantly from their real values. The L_x, H-values in the X-

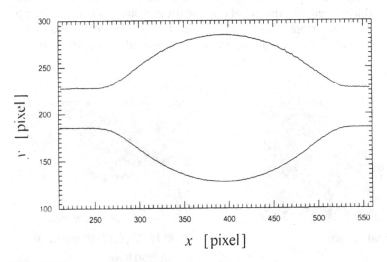

FIG. 14 Extracted profile of the polyimide melt droplet (Fig. 13) resting on a carbon fiber at 420°C.

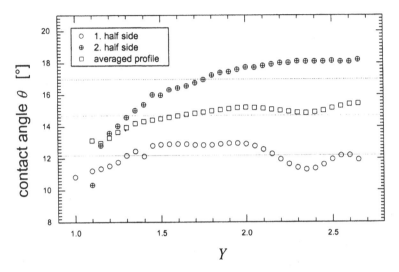

FIG. 15 Results of the L_x, H method applied to the PI melt droplet profile shown in Fig. 14.

range of 1.5 to 3.0 usually produce more reliable contact angle values. Using average-side profiles rather than simply the single-side profiles of a droplet improves the reliability of calculation even further. The difference between the contact angles obtained from the two-side profiles of a droplet is not due to the possible influence of the gravity on the drop profiles, because the drop images are recorded from above (i.e., top view). It is more likely due to the geometrical asymmetry (see Fig. 16) and/or the chemical heterogeneity of the fibers.

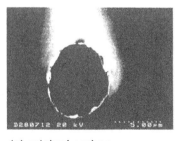

(a) original carbon

(b) in HNO_3 for 24h oxidized carbon fiber

FIG. 16 SEM micrographs of carbon fiber cross sections (fiber diameter).

For experimental applications, care should be taken that optical aberrations and distortions of the drop images are considered (i.e., corrected) in the calculation as far as possible. Image distortions have a rather strong influence on the resulting values because the calculated contact angles are very sensitive to the drop dimensions (see also Fig. 3). Usually, the total image distortions will be summarized into an aspect ratio factor, which has to be determined for each experimental set-up using, for instance, the procedure described in Ref. 35. Further care should be taken to ensure that the fibers are positioned at right angles to the microscope lenses to avoid extra image distortions.

IV. INDIRECT CONTACT ANGLE MEASUREMENTS

Direct methods for the determination of the contact angles from profiles of droplets resting on fibers are also quite difficult to perform, which is related to the tricky handling of thin fibers and small liquid volumes. They are also restricted to fibers with almost perfect circular cross sections. Furthermore, the accuracy of some of the aforementioned methods is directly related to the accuracy of determining the exact position of the three-phase contact point (compare Fig. 17) and the fiber diameter.

The contact angle range that is usually accessible for direct measurements of the contact angle in symmetric drop-on-fiber systems is $0° < \theta < 60°$ [28]. However, if it is possible to control the droplet size, it should be possible to create droplets of a suitable size for optimum direct measurements of the contact angle from drop profiles. For higher contact angles and smaller liquid

FIG. 17 Image of a polyimide melt droplets resting on a carbon fiber at 420°C. Who would be able to determine exactly the three-phase contact point?

volumes the clam-shell (or asymmetric) configuration of the droplets on fibers becomes more favorable [6,29], which makes it even more difficult to determine the contact angles accurately. In cases where the asymmetric droplet conformation is preferred, it is essential to ensure that the fiber has been rotated to present the exact meridional profile of the droplet to the observer. In cases where this cannot be guaranteed, the determined apparent contact angles will deviate significantly from the real contact angles [1].

In the following section the focus is on the methods that allow the characterization of the wetting behavior indirectly: the contact angle is derived by measuring the wetting force exerted from a meniscus of a liquid on a vertical fiber or by determining the penetration rates of liquids in fiber assemblies.

A. The Modified Wilhelmy Method

The modified Wilhelmy technique [7,30,31] for measuring contact angles θ between fibers and liquids with known surface tensions γ_{lv} or for determining the fiber or filament perimeter and the increase of the fiber perimeter by swelling [7] in a test liquid was introduced several decades ago. This method is very sensitive for measuring contact angles between fibers and low-viscose liquids and is most widely used in all areas where fiber wetting is of interest. The method has been reviewed several times [32,33].

A schematic setup of the Wilhelmy balance is shown in Fig. 18. The fiber is attached to the arm (or using a sample carrier) of a high-precision ultra-

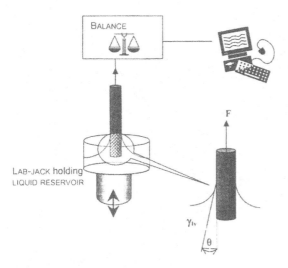

FIG. 18 Schematic of a Wilhelmy balance for measuring contact angles on fibers.

microbalance placed on an antivibration table in an enclosed measuring chamber (in order to avoid undesirable interference in the measured signal due to percussion, drafts, and adsorption of environmental contaminants). The fiber(s) can be immersed into/withdrawn from the test liquid using a reversible elevator platform carrying the beaker containing the liquid, and the mass change detected at the balance will be recorded using a computer. The Wilhelmy method is very accurate, because a balance can be exactly calibrated using calibration weights.

Using the Wilhelmy technique the wetting tension ($\gamma_{lv} \cdot \cos\theta$) exerted by a liquid meniscus on a vertical fiber can be measured directly, using the balance:

$$\gamma_{lv} \cdot \cos\theta = \frac{\Delta m \cdot g}{\pi \cdot d_f} \tag{34}$$

where $\Delta m \cdot g$ is the measured "apparent wetting force" (Δm is the change in "mass" before and after the fiber is immersed in the liquid), g is the acceleration due to gravity and d_f is the fiber diameter. In the case of small fibers with diameters $d_f \leq 50$ μm the buoyant weight of the immersed fiber is negligible and only the capillary term accounts. However, the accuracy of the measurement depends strongly on the accuracy of the determined fiber diameter (or more exactly the determined fiber perimeter). The diameter can be measured easily with the Wilhelmy method using a test liquid which is known to wet the fiber completely (usually n-alkanes*), i.e., $\cos\theta = 1$, but only if penetration of the liquid into the fiber can be excluded. In the most practical cases the total absence of contact-angle hysteresis is a convenient test for complete wetting. In this case Eq. (34) becomes:

$$d_f = \frac{\Delta m \cdot g}{\pi \cdot \gamma_{lv}}. \tag{35}$$

The accuracy of the diameter determination depends only on the accuracy of the liquid surface tension γ_{lv}, which can be measured with high precision using the pendent drop method [34,35].

Alternatively, fiber diameters can be measured using scanning electron microscopy (SEM) [36] or optical microscopy equipped with a calibrated

---

*When using alkanes it is important to ensure that all surface-active impurities are removed from the hydrocarbon. Alkanes can be easily purified by passing them several times through adsorbent columns (i.e., in the case of hydrocarbons the most efficient adsorbent will be basic alumina). The purity of the hydrocarbon can be checked by measuring the hydrocarbon–water interfacial tension [Aveyard R.; Haydon, D.A. Trans. Faraday Soc. 1965, *61*, 2255], which is very sensitive to those impurities [Goebel A.; Lunkenheimer, K. Langmuir 1997, *13*, 369].

eyepiece [37] (but these techniques are only applicable for fibers that do not swell in the test liquid [38]) by averaging the measured diameters for several fibers.

If the receding contact angle is zero (for instance, for water receding contact angles on wood fibers [39]), the following equation has been used to determine advancing contact angles θ_a:

$$\cos \theta_a = \frac{F_a}{F_r} \tag{36}$$

where F_a and F_r are the advancing and receding dynamic wetting forces, respectively. The above equation suggests that the advancing contact angle can be measured without knowing the fiber perimeter [38,40].

When measuring contact angles on very small single fibers, the detectable mass change Δm due to capillary rise on a "monofilament" will be very small. The accuracy of the measurement can be improved by using more than one monofilament. Thus disturbing effects through surface roughness and -heterogeneities and varying fiber perimeters will be averaged out [36,38]. However, care has to be taken that all fibers are placed perpendicular with respect to the liquid surface and parallel to each other on to an appropriate sample holder. The holder should be preferably made of a low-energy material, such as Teflon, in order to avoid problems related to vapor adsorption.

The modified Wilhelmy technique can be used to measure either static or dynamic contact angles. In the static mode the fiber is immersed into the test liquid only by a few millimeters in order to avoid possible end-effects. Afterwards, the fiber is held stationary until a constant mass is measured so that the contact angle can be calculated. In the dynamic mode, however, the mass change is recorded during the whole fiber immersion–emersion cycle at a constant stage velocity. Advancing θ_a and receding θ_r contact angles can be calculated using the Wilhelmy equation from the mass changes, which were detected during the immersion and emersion, respectively, of the fibers into and from the test liquid [Eq. (34)]:

$$\cos \theta = \frac{\Delta m \cdot g}{\pi d_f \cdot \gamma_{lv}} \tag{37}$$

A typical example of the measured weight changes during dynamic contact angle measurements between carbon fibers and water as well as diiodomethane (DIM) is shown in Fig. 19. No buoyancy slope can be observed, even when using more fibers because of the small fiber diameter (around 7 μm) [41].

The ability to measure dynamic contact angles has a major advantage. It enables us to characterize a large surface area of fibers and, therefore, to

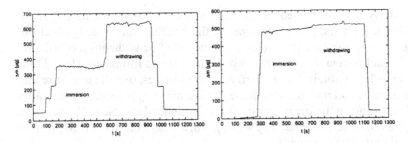

FIG. 19 Typical weight recordings as a function of time during fiber immersion and emersion procedure for four carbon fibers in water (left) and DIM (right).

detect local changes in the wetting behavior. When measuring dynamic advancing and receding contact angles it is possible to study the (chemical) surface heterogeneity [42] and roughness* or kinetic effects such as liquid penetration into the fibers by means of the ubiquitous contact angle hysteresis (for further details see Refs. [43–46]) and, therefore, study fiber swelling, adsorption–desorption processes [47], or in the case of polymer fibers, polymer surface dynamics (due to overturning of molecular segments at the surface). Because of all omnipresent inhomogeneities of fiber surfaces, the measurements have to be repeated several times in order to check the reproducibility of obtained contact angles.

As the solid surface tension γ_s cannot be measured directly, contact angles are, in many cases, used to estimate (calculate) the solid surface tension. Different approaches, all based on the Young equation but on different models for obtaining solid–liquid-interfacial tension γ_{sl} from their surface tension properties (γ_{sv} and γ_{lv} as well as their components), are available to calculate solid surface tensions from measured contact angles [48,49]. The debate still continues over the usage of different approaches in estimating solid surface tensions from experimental contact angle values [50–54]. The majority of these approaches require more than one contact angle to be measured between a studied solid and several test liquids of known surface tension components (or parameters). Test liquids with known surface tension components and parameters can be found in the literature [4,55–59].

---

*Differently sized glass fibers were characterized using atomic force microscopy and contact angle measurements, it was concluded that roughness plays only a minor roll in causing hysteresis whereas the chemical heterogeneity plays the major part [Wolff, V.; Perwuelz, A.; El Achari, A.; Caze C.; Carlier, E. J. Mater. Sci. 1999, 34, 3821].

If one is only interested in the solid surface tension and its dispersive (London forces) and polar components, these values can be easily obtained, using the tensiometric method (based on the modified Wilhelmy method) and a two-liquid system (i.e., different hydrocarbons and formamide) [60]. However, in order to obtain "true" surface tension values, one has to ensure that no thin wetting film of the low dense (hydrocarbon) phase will remain after passing through it, before immersing into the higher dense phase (formamide) and vice versa. Such a remaining wetting film acts as a barrier, which prevents a true liquid–fiber contact [60].

1. Some Examples

The modified Wilhelmy method has been frequently used to determine the wetting behavior of fiber materials, ranging from natural fibers [61] to all kinds of synthetic fibers [62]. The method was used (and still is) for the characterization of the changes in the surface chemistry of carbon [63–67], glass [68,69], and natural [70–72] fibers after modifying the fibers, applying sizings and coatings [73] and grafting processes [74–76] as well as for characterizing all kinds of polymer fibers [77,78] regarding their solid surface tension.

Besides wettability the modified Wilhelmy technique used in the dynamic mode also offers the opportunity to study and quantify the swelling properties of modified (polyacrylic acid grafted) and unmodified regenerated cellulose fibers [79]. Measurements of single-fiber swelling give accurate values of the fiber swelling capacity. Such measurements performed on single fibers exclude information about liquid adsorption between the fibers that occur in fiber assemblies. As mentioned above, the accuracy of the contact angles measured using the Wilhelmy method strongly depends on the accuracy of the fiber perimeters. Therefore it was stated that the Wilhelmy method is hardly applicable in the case of natural fibers considering their wide variation in perimeters and shapes [1,80]. The surface tension of cellulose-type fibers can, however, be determined using a simple floating technique using polar (water–methanol) and nonpolar (1-methylnaphthalene/octane) liquid mixtures avoiding contact angle measurements [80].

V. STUDY OF FIBER SURFACE CHEMISTRY BY CONTACT ANGLE MEASUREMENTS

Hüttinger et al. [81] suggested that the acidity and basicity of (carbon) fiber surfaces could be characterized by determining the work of adhesion/pH diagram (Fig. 20). Advancing contact angles are measured using the modified Wilhelmy technique between fiber and acidic and basic aqueous solutions, with varying pH values ranging from 1 to 14. The method is based on the fact that the surface tension and the Lifshitz (L)/van der Waals (W) component of

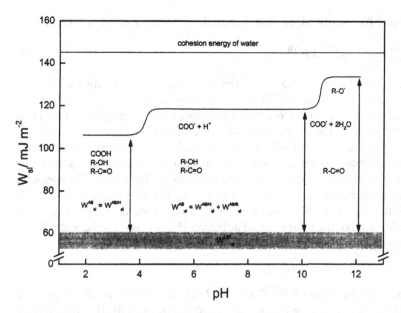

FIG. 20 Schematic work of adhesion/pH diagram including a model for interpreting such a diagram according to Ref. [82].

the surface tension γ_l^{LW} of pure water are not, or only slightly, affected by the addition of small amounts of inorganic acids and bases. This behavior is explained by the strong self-association of the water molecules due to hydrogen bonds. These bonds are practically unaffected by additional ion–dipole interactions. Contact angle measurements as a function of pH can be a valuable tool for analyzing surface functionalities having different pK_a values [82].

The work of adhesion between the solid and aqueous solution W_{sl} depends on the pH value and is composed of a constant contribution due to dispersion (or Lifshitz–van der Waals) interactions W_{sl}^{LW} and different contributions deriving from acid (A)–base (B) interactions W_{AB}^{sl} between the functional groups at the solid–water interface:

$$W_{sl} = W_{sl}^{LW} + W_{sl}^{AB} \tag{38}$$

The acid–base interactions include the formation of Brønsted acid–base complexes ($W_{sl}^{AB/B}$) but also hydrogen bonds ($W_{sl}^{AB/H}$):

$$W_{sl}^{AB} = W_{sl}^{AB/H} + W_{sl}^{AB/B} \tag{39}$$

W_{sl}^{AB} may be calculated, or at least explained, by the following expression:

$$W_{sl}^{AB} = -f \cdot n_i^{AB} \cdot \Delta H_i^{AB} \tag{40}$$

where ΔH_i^{AB} is the enthalpy of the complex or adduct formation of a Brønsted acid–base pair, n_i^{AB} is the number of the acid–base pairs i and f is a correlation factor to correct enthalpy values to free energy values (≈ 1), as the work of adhesion is equivalent to a free energy.

The W_{sl}^{LW} component of the surfaces under investigation can be determined by contact angle measurements, using a nonpolar liquid as the interaction of a nonpolar liquid is restricted to dispersive forces [83]. The nonpolar liquid should have a high surface tension (such as DIM) to avoid the spreading of the liquid. In the case of negligible acid–base interactions (i.e., for nonpolar solids), the total work of adhesion W_{sl} between a solid and a liquid corresponds to W_{sl}^{LW}:

$$W_{sl} = W_{sl}^{LW} = \gamma_1(1 + \cos \theta) \tag{41}$$

Knowing the value of W_{sl}^{LW} and the London–van der Waals component γ_l^{LW} of the liquid surface tension, the same component for the solid surface tension γ_s^{LW} can be calculated using the geometric mean expression:

$$W_{sl}^{LW} = 2\sqrt{\gamma_s^{LW} \cdot \gamma_l^{LW}} \tag{42}$$

If the adsorption of the probe liquid vapor on the solid surface is negligible, γ_s^{LW} can also be determined, in analogy to the acid–base approach for estimating solid surface tensions [84], by measuring the contact angle of a neutral, non-self-associating liquid (i.e., DIM, squalene, or n-alkanes) using the following expression:

$$\gamma_s^{LW} = \frac{\gamma_l^{LW}(1 + \cos \theta)^2}{4}. \tag{43}$$

Brønsted acid–base complexes may be formed selectively between surface functional groups and aqueous solutions with different pH values. The free energy of the acid–base complex formation [Eq. (40)] causes an increase in W_{sl} and therefore a step in the W_{sl}/pH diagram. The increased W_{sl} causes the contact angles to decrease. However, it was found that the overall W_{sl} is mainly determined by hydrogen bonds and to a smaller content by well-defined Brønsted acid–base complexes [82]. At very low pH values the hydrogen bonds form between carboxyl-, hydroquinone-, hydroxyl-, phenol-, and carbonyl groups on fiber surfaces and water [82]. With increasing pH values Brønsted acid–base complexes will form. The increase in W_{sl} cannot be large, because the contribution of the hydrogen bonds formed by carboxyl groups at lower pH values is simultaneously reduced by the formation of carboxylate anions. An analogous interpretation can be given for the second step in the basic range. Brønsted acid–base complexes will be formed between phenolic-OH functional groups and the basic solution medium.

The acidic and basic aqueous solutions to be used in these measurements can simply be prepared by adding hydrochloric acid and sodium hydroxide to water, but other acids and bases might be used alternatively.

Zielke et al. [83] concluded that neither temperature-programmed desorption of functional groups nor quantitative X-ray photoelectron spectroscopy can give absolute information about the exterior surface functionalities on a fiber surface, because both techniques include subsurface contributions. Interactions between the functionalities on the outer fiber surface and test liquids (including aqueous solutions with varying pH values), however, affect the measurable contact angles [85].

VI. CHARACTERIZING FIBER–SURFACTANT INTERACTIONS BY CONTACT ANGLE MEASUREMENTS

Contact angle measurements can also be used to obtain information about adsorption of surfactants on solid surfaces (in our case on fiber materials). In such a case contact angles have to be measured as a function of the surfactant concentration c. Under certain circumstances, but generally if the liquid–vapor surface tension γ_{lv} is larger than the solid surface tension γ_s, no adsorption of the liquid vapor on the solid surface takes place, and therefore the solid surface tension will stay constant with changing surfactant concentration. The Young equation [Eq. (3)] becomes therefore:

$$-\frac{d\gamma_{sl}}{dc} = \frac{d(\gamma_{lv} \cos \theta)}{dc}. \tag{44}$$

It follows the Gibbs equation for solid–liquid interfaces:

$$\Gamma_{sl} = \frac{1}{RT} \frac{d(\gamma_{lv} \cos \theta)}{d\ln c}, \tag{45}$$

where Γ_{sl} is the excess concentration at the solid surface, T is the temperature, and R is the gas constant.

Information about the surfactant adsorption at solid–liquid interfaces can now be obtained by measuring the contact angle as a function of the surfactant concentration [31,86]. However, in many practical applications the direct measurable results are sufficient enough.

The modified Wilhelmy method has been used to follow time-dependent adsorption processes of surfactants from aqueous solutions onto modified and untreated glass fibers [30] and to study interactions between fibers and surfactants [31,38,87,88].

VII. WETTING OF FIBERS BY POLYMER MELTS

The interfacial properties and the flow of polymer melts are important in many technological processes such as polymer blending, fiber, and wire coating as well as in composite manufacturing. In the area of fiber reinforced polymers the nature of the fiber–polymer melt (or prepolymer solution) interface plays a fundamental role, as the adhesive strength of the desired composite material depends strongly on the interphase between both materials. Load stress must be transferred from the matrix through the interphase into the reinforcing fiber. Therefore the wettability of reinforcing fibers by polymer melts or prepolymers (such as resins and vinylesters) is a crucial issue. Poor wetting of the fibers by the liquid matrix will lead to the formation of voids in the fiber–solidified polymer matrix interphase, which will reduce the final strength of the final composite. According to adhesion theories, a thermodynamic prerequisite for good adhesion is that the liquid (the adhesive, i.e., the liquid matrix material) wets the adherend (i.e., the reinforcement). In order for this to happen, the surface tension of the fiber has to be higher than that of the liquid. The fiber surface tension is indirectly accessible by "standard" contact angle measurements vs. test liquids. The wetting kinetic of fibers by the liquid matrix or polymer melts can be characterized by wetting force measurements using the Wilhelmy method [89,90]. However, it is still difficult to obtain directly experimental data about the kinetics of the formation and the interactions in a fiber–high viscose polymer melt interphase [91,92].

Until the present day, only very few methods are available for the study of the wetting behavior of fibers directly by polymer melts. The modified Wilhelmy technique [89,93] can be used to measure directly the wetting tension ($\gamma_{lv} \cdot \cos \theta$). When the surface tension of the polymer melt is known, then the contact angle can be calculated [94]. In addition, direct measurement methods can be used. Such methods enable the calculation of the contact angle of drop-on-fiber systems from the drop shape [18–20].

The first attempts to measure contact angles between a thermoplastic melt (polypropylene) and graphite-, Kevlar-, and glass fibers using the modified Wilhelmy technique were made by Chang and coworkers [95]. Since then, several developments have taken place to improve the Wilhelmy equipment for measuring either the surface tension of polymer melts [89] or the contact angle between single fibers and polymer melts [93,94,96]. The latest description of a new refined device (Fig. 21) to measure the wetting properties of fibers by polymer melts was given by Grundke et al. [93,97]. The experimental setup allows the characterization of polymer melts and the wetting behavior of thin fibers (with diameters in the range 8 μm < d_f < 100 μm) up to 300°C in an inert gas (Ar) atmosphere [98]. This device is furthermore equipped with two windows that enable parallel imaging of the fiber polymer melt meniscus using a CCD camera.

The major problem when measuring the contact angles between polymer melts and fibers (besides the difficulties of thermal convection currents affecting the balance signal, polymer oxidation, degradation, or crosslinking) is the high melt viscosity. Therefore not only surface tension (capillary) but also hydrodynamic effects have to be considered [89,99]. Two factors determine the rate and extent of the equilibrium of the fiber–melt menisci [89]. First, when using small fibers the meniscus will equilibrate more rapidly (the meniscus height is directly proportional to the fiber diameter on small diameter fibers [89]). The second effect, related to the first, is due to the relative rate of impurities adsorbing at the moving, equilibrating contact line. If small fibers (d_f < 100 μm) are used as the Wilhelmy probe the dimensions of the wetting meniscus between fiber and polymer melt are reduced, which accelerates the equilibration rate [94]. Sauer [94] measured the time needed for a meniscus to equilibrate. The measurements were performed at 23°C using dry glass fibers having different diameters (8.1 μm ≤ d_f ≤ 155 μm) and poly(dimethylsiloxane)

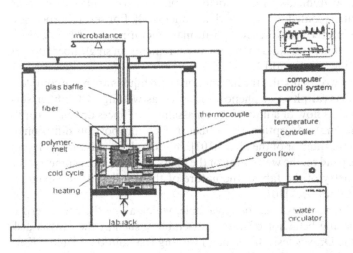

FIG. 21 Schematic of an apparatus allowing the determination of surface tensions of polymer melts by immersing fibers based on the modified Wilhelmy principle [98].

(PDMS) with different viscosities (from 50 mPas to 5 Pas) and found that equilibrium state is established in less than 200 sec between fibers with diameters of $d_f \leq 155$ µm and a high molecular weight PDMS with a viscosity of 5 Pas. In the case of polymeric melts, the time to reach the equilibrium can take up to several minutes for a polypropylene melt on approx. 60 µm glass fibers at temperatures above $T > 160°C$ [93].

VIII. CONTACT ANGLE MEASUREMENTS ON FIBROUS ASSEMBLIES (WOVEN FABRICS, MAT, AND BUNDLES)

Because fibers are in most cases gathered either in rovings (fiber bundle) or as woven textiles, fabrics or mats, and also for these fibrous assemblies, the interaction with liquids, their wetting behavior, is important for manufacturing processes (e.g., dyeing and finishing, sizing, impregnation in composite making) as well as performance or maintenance in use (e.g., washing processes). However, when characterizing the wetting behavior of fabrics, it is not clear whether the wetting should be measured using single fibers taken from the fabric or on fibrous assemblies itself [100]. If the fibers have an irregular surface, in case of natural fibers, both the fiber diameter and the contact angle will vary making it inapplicable to use for instance the Wilhelmy-technique [1]. Furthermore, contact angles measured on single fibers are influenced by the uneven distribution of coatings and sizings after applying fiber treatments. Single-fiber measurements should not be applicable when fiber assemblies have been treated, coated, or sized with a binder [100]. For several processes, such as fiber reinforced polymer composite manufacturing, a number of fibers are wetted simultaneously and therefore measurements on single fibers do not always reflect the "real world."

In the case of woven textiles the interactions with liquids are more complex. They involve several physical phenomena, such as wetting of the fiber surfaces, spontaneous flow of a liquid into an assembly of fibers driven by capillary forces (wicking), adsorption on the fiber surface, and possibly diffusion of the liquid into the interior of the fibers [101]. In general, fiber assembly–liquid interactions depend on the wettability of the fibers, their surface geometry, the capillary geometry of the fabric, the nature of the test liquid, and external forces (applied pressure) [101].

There are different test methods available, which allow the characterization of the wicking behavior of fibrous assemblies, such as the canvas disk wetting test, the Draves test, the demand wetting test, and the Lennox–Kerr test or sinking float test. Kissa [101] reviewed these techniques and the theoretical basics of wetting and wicking of such fiber assemblies. Therefore we want to concentrate here only on methods that permit the measurement of at least advancing contact angles or the characterization of the fiber assemblies with regard to their surface tension.

A. Contact Angle Measurements and Retention Properties of Woven Fabrics

Hsieh and Yu [102] describe a method to measure simultaneously the liquid wetting and retention characteristics of single-component woven textile fabrics. For the measurements, defined strips (in size and fiber alignment) have to be cut out of the fabric using a die cutter. Each specimen has to be weighted before being attached to a film tab. The tab contains a hole to attach the fabric to the arm of an electronic balance. A commercial tensiometer or a balance (Fig. 18) can be used. The weight of the sample (including the holder) is tarred on the balance. Afterwards, the surface of liquid contained in a reservoir will be raised slowly until it almost touches the sample edge (about 1 mm), then the measurement is started. Upon the first force detection the liquid level is to be stopped. The weight-force recording is to continue until a steady state in the measurement can be established. Then, the fabric will be withdrawn from the test liquid at a higher stage speed until the entire specimen is pulled out of the liquid. After the fabric is completely separated from the liquid a residual weight can be detected, indicating the total liquid retention (W_t), i.e., the amount of liquid retained in the vertical hung fabric. During the measurement, wetting and wicking processes will be detected simultaneously. Therefore the wetting component has to be decoupled from the measurement. The steady state is a combined result of both these processes. Knowing W_t the wetting force F_w can simply be decoupled from the steady state or stabilized balance reading (ΔB_{st}):

$$F_w = (\Delta B_{st} - W_t) \cdot g, \tag{46}$$

where g is the gravitational acceleration and F_w is the vertical attraction force exerted on the fabric as it immerses into the liquid and is expressed as:

$$F_w = \gamma_{lv} \cdot P \cdot \cos \theta \tag{47}$$

where P is the perimeter of the liquid–fabric interface or its wetted length, which can be measured, as described above, using a liquid that will totally wet the fibers (such as alkanes).

$$P = \frac{F_w}{\gamma_{alkanes}} \tag{48}$$

Therefore if the dimensions of the fabric, its wetted length, P, and the liquid surface tension γ_{lv} are known, the advancing contact angle between the fabric and the test liquid θ_a can be calculated.

The alkane retention is used as a measure for the liquid retention capacity of the fabrics (C_m). C_m can also be expressed by the weight ratio of the absorbed liquid over the dry fabric weight [g/g]:

$$C_m = \frac{\rho_l}{\rho_f} \frac{\Phi}{1 - \Phi},\tag{49}$$

where ρ_l is the liquid density, ρ_f is the fiber density, and Φ is the porosity of the fabric. Φ is defined as the fraction of void space in a porous medium:

$$\Phi = 1 - \frac{\rho_b}{\rho_f},\tag{50}$$

here ρ_b is the "fabric bulk density," which can be calculated as follows:

$$\rho_b = \frac{\text{fabric weight}}{\text{thickness}} \left[\frac{\text{g/cm}^2}{\text{cm}}\right].\tag{51}$$

In order to validate their measuring procedure, Hsieh and Yu [102] performed measurements using 100% cotton fabrics of different geometry (warp and werf direction, fabric size, and wetting perimeter) at varying depth of immersion of the fabric into water. The amount of liquid rise in a fibrous assembly depends not only on the wetting property of the fibers but also on the geometric configurations of the pores (capillaries) in the assembly. Their measured data, however, confirmed that the water contact angles are independent of the actual wetted length, i.e., the liquid-substrate perimeters. There was no evidence either that liquid meniscus affects the measured contact angles on fabrics (including woven fabrics, yarns, and fibers). They compared the measured contact angles on a variety of fabrics (natural and synthetic textile fibers) with the wetting properties of single fibers. They found that the contact angles obtained for fabrics and single fibers were identical and that neither the existence nor the magnitude of liquid retention (W_t) interferes with the determination of the contact angle from fabrics [103]. The error range was broader for contact angles in the case of contact angle measured on single fibers extracted from a fabric. Dimensional irregularities along the cotton fiber axis and in cross-sectional shapes complicate the measurements and thus affect the contact angles measured for single fibers. Greater variations in the measured contact angle might also be due to the inhomogeneous character of the fiber surfaces and therefore varying fiber surface wettability.

This method was also applied in the study of the influence of scouring and bleaching on the wetting behavior of slivers, yarns, and plain and satin weave fabrics [104]. In summary, the method introduced by Hsieh and Yu should be applicable for measuring contact angles between liquids and different kinds of fiber assemblies and for estimating the liquid retention behavior as well as the "porosity" of such fabrics.

B. Capillary-Penetration Method or Rising-Height- or Imbibition Technique

The modified Washburn or capillary-rise method can be used to characterize the wetting behavior not only of porous solids [105], powders [106,107] but also of fiber assemblies [108–110], such as rovings or yarns, fabrics, and mats, or at least their surface tension. Different experimental techniques are available for characterizing fiber assemblage. All of them are based on measuring either the weight gain due to liquid penetration as a function of time [111–114] or the penetration time needed for a liquid to rise to a certain height (known as thin-layer wicking technique) [115,116].

C. The Rising-Height or Capillary-Rise Method

Fiber bundles or rovings can be characterized with respect to their wettability or, more or less directly, their surface tension using the rising-height method, which was introduced for porous solids or powders. Apart from the preparation of reproducible fiber "bundles," the same capillary penetration method for powder beds can be used [113]. In this case, a precisely weighted quantity of fibers has to be inserted reproducibly into a sample holder. The packed sample holder will be attached to an electrobalance or a commercial tensiometer and brought into contact with the test liquid. As soon as the sample touches the liquid surface, the immersion movement has to be stopped. Upon the first force detection the liquid level is stopped at almost no immersion of the holder. Then, the penetration velocity of the test liquid into the fiber bundle is determined by recording the weight gain (which is proportional to the capillary height of the liquid front) as a function of time using a computer until a steady state in the measurement is established. Depending on the nature of both the fibers and the test liquid, the measurement can be stopped after a few tens of seconds (typically after 10 to 60 sec). After this time, the deviation of the experimental data from the expected theoretical weight-time dependence increases drastically. This renders any additional data useless with respect to the data evaluation, according to the Washburn equation. The deviations are caused by 1) gravity, which hinders the liquid from penetrating into the solid with increasing·penetration height; and 2) by evaporation of the test liquid (if the liquid is volatile enough) [113]. After completing the measurement, the sample holder can be withdrawn from the test liquid and residual weight indicating the total liquid retention can be determined.

The evaluation of the measured data is based on modifications [111] of the Washburn equation [117] for a single capillary, which arises from the combination of expression for the Laplace pressure and the Hagen–Poiseuille equation for steady flow conditions.

$$\gamma_{lv} \cos \theta = \left[\frac{2}{A^2 r}\right] \left[\frac{\eta}{\rho^2}\right] \left[\frac{m^2}{t}\right] \tag{52}$$

where A is the cross-sectional area and r is the radius of the capillary, η is the liquid viscosity, ρ is the liquid density, m is the weight of the liquid which penetrates into the capillary, and t is the time.

However, in the case of short fiber beds or fiber bundles the geometry of the capillary system is unknown and therefore the factor $[2/A^2 r]$ has to be replaced by an unknown factor $1/C$, which leads to:

$$C \cdot \gamma_{lv} \cos \theta = \left[\frac{\eta}{\rho^2}\right] \left[\frac{m^2}{t}\right]. \tag{53}$$

The factor $[\eta/\rho^2]$ reflects the properties of the test liquid and $[m^2/t]$ is measured using a balance or tensiometer.

This modified Washburn equation can be used under the following assumptions: 1) laminar flow is predominating in the pore spaces; 2) gravity is negligible; and 3) the geometry of the porous solid–fiber bundle is constant [112].

As experimental result we obtain the weight increase m as a function of the square root of time $\sqrt{t}$ (or the weight increase m^2 as a function of time t). By determining the slope of the linear part of these plots (see also the example shown in Fig. 22), we can obtain the needed experimental quantity $[m^2/t]$.

If this quantity is measured for a series of different test liquids, the surface tension of the solid (fiber bundle, powder, or porous solid) can be determined. Because $[m^2/t][\eta/\rho^2]$ equals $C \cdot \gamma_{lv} \cos \theta$ according to Eq. (53), $C \cdot \gamma_{lv} \cos \theta$ can be plotted vs. the liquid-surface tension γ_{lv} of the test liquids used. This plot (see Fig. 23) will show a maximum that is analogous to Zisman's critical solid–vapor surface tension γ_c of the investigated material [108,113].

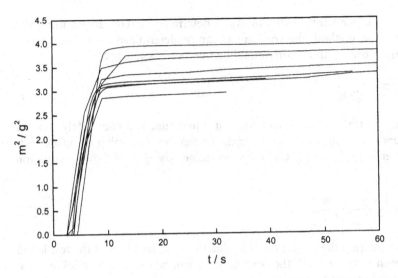

FIG. 22 Weight rise curves of N,N-dimethyl formamide into cellulose fibers as a function of time. (From Ref. 118.)

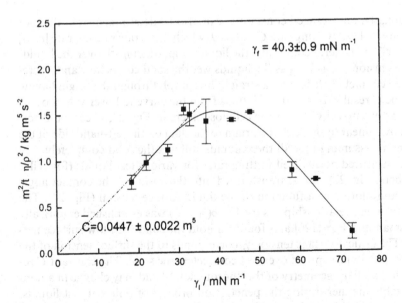

FIG. 23 Plot of the normalized wetting rates as a function of the test liquid surface tension for pure cellulose. (From Ref. 118.)

On the other hand, if one needs to determine contact angles using the capillary-rise method, the constant C can be determined assuming a mean capillary radius r_c and a corresponding number of capillaries n_c

$$C = \frac{\pi^2}{2} \cdot r_c^5 \cdot n_c^2 \tag{54}$$

with an "extra" measurement using a liquid that will completely wet the solid under investigation, as for instance hexane. Assuming complete wetting, therefore $\cos\theta = 1$, C can be determined using the following equation [119]

$$C = \left[\frac{\eta}{\gamma_{lv} \cdot \rho^2}\right]\left[\frac{m^2}{t}\right]. \tag{55}$$

After measuring the weight increase due to the penetration of the test liquid of interest into the solid, the contact angle can be calculated as follows, assuming C is really a constant:

$$\cos\theta = \left[\frac{1}{C\gamma_{lv}}\right]\left[\frac{\eta}{\rho^2}\right]\left[\frac{m^2}{t}\right] \tag{56}$$

Additionally, an averaged constant C can also be determined from the plot of the normalized wetting rate $C \cdot \gamma_{lv} \cos\theta$, which is the raw wetting rate $[m^2/t]$ multiplied by the term containing the liquid properties $[\eta/\rho^2]$, over the liquid-surface tension γ_{lv}; as long as the liquids wet the solid completely and, therefore, the contact angle is zero, a straight line graph through the origin having the slope C results. If $\theta > 0$, deviations from the curve to lower values occur [108]. As exemplarily shown for cellulose fibers in Fig. 23, C can be determined by a linear fit through the origin to the data on the left-hand side of the maximum, assuming that all these liquids still wet the solid completely.

The measured normalized wetting rates for various test liquids (for cellulose fibers; Fig. 23) can be transformed into the cosine of the contact angle ($\cos\theta$) and plotted as a function of the liquid surface tension (Fig. 24). The resulting linear relationship $\cos\theta = 1 - b(\gamma_l - \gamma_c)$ was established empirically by Zisman and Fox [120] and found to hold for solid with low surface tensions. The critical surface tension γ_c corresponds to the surface tension of the liquid that will just spread over/wet completely the solid. The constant C reflects the capillary geometry of the "porous solid" and may change in a nonpredictable manner during the penetration process of different test liquids. It was concluded from the experiments performed [112] that there is no need to determine the constant C in order to obtain solid-surface tensions, because the position of the maximum in the $C \cdot \gamma_{lv} \cos\theta$ vs. γ_{lv} plot, which is expected to

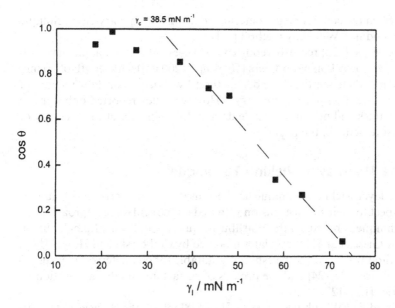

FIG. 24 Zisman plot for estimating the cellulose fiber surface tension (the cosines of the contact angles were extracted from the measured normalized wetting rates) using the averaged constant $\bar{C}$ determined from Fig. 23. (From Ref. 118.)

reflect the solid-surface tension, is not affected by the sample geometry. However, Chibowski and Perea-Carpio [121] state that this "apparently interesting approach should be carefully considered, because for low energy nonpolar (or weakly polar) solid many approaches give reasonable agreement just because the surface energy is low."

D. Some Examples

The capillary-rise technique [108] as described above was successfully applied to characterize differently sized glass fibers [91] and modified jute fibers [70] with respect to their surface tensions. Gassan et al. [70] found that the packing density of the fibers influences only slightly the measured contact angle. However, in the case of hydrophilic natural fibers the measured contact angle is affected by the diffusion of the test liquid into the fibers. The measured $[m^2/t]$ function contains not only information about the wetting process but also about the absorption of the liquid, the penetration of the test liquid into the fibers. Swelling takes place and therefore the fiber diameter increases, which again influences the geometry factor C, which influences the contact angle. Significant differences were found comparing the surface-tension values calculated from the contact angles determined using the capillary-rise technique and the modified Wilhelmy method [70].

Bubert et al. [122] recently reported on successful contact angle measurements between carbon nano-fibers ($50 < d_f < 200$ nm) with length of 40 mm and test liquids using the "powder-method by which a constant volume of nanofibers are given into a capillary." However, they reported only surface tension values and no contact angle data and nothing about the accuracy of the measurements in the paper.

E. The Thin-Layer Wicking Technique

The thin-layer wicking technique has been used to characterize textile fabrics with respect to their surface tension γ_s (based on the acid–base approach [84]). The technique is based on the Washburn equation and was originally introduced by Giese et al. [115] and later modified by Chibowski and Holysz [116]. The method was found to be especially suitable for characterizing (strips of) textile fabrics [123,124] and the process of surfactant and dye adsorption by the fabrics [125–127].

The modified Washburn equation [116,128] relates the changes in the surface free energy ΔG connected to the replacement of the solid–gas by a solid–liquid interface during the wicking process:

$$x^2 = \frac{r_p \cdot t}{2\eta} \cdot \Delta G. \tag{57}$$

In a typical experiment the time t that a liquid with a viscosity η needs to travel a certain distance x through a textile having an "effective pore radius" r_p has to be measured at a constant temperature [116,129].

Chibowski and Holysz [116,130] considered four cases, in which ΔG has different values:

(i)   A low surface tension liquid, such as n-alkanes, completely wets a surface that is equilibrated with the liquid's saturated vapor for a sufficiently long time (a duplex film has formed), then $\Delta G = \gamma_{lv}$ and the original form of the Washburn equation can be used:

$$x^2 = \frac{r_p \cdot t}{2\eta} \cdot \gamma_{lv} \tag{58}$$

The measured function $x^2 = f(t)$ allows determining the "effective pore radius" r_p parameter.

(ii) $\Delta G = W_{sl} - W_{ll}$ in the case of using the same set of liquids as described in case (i) now spreading/wetting the "bare" surface (not precontacted with the liquid vapor). $W_{sl}[= \gamma_{lv}(1 + \cos\theta)]$ and $W_{ll}(= 2\gamma_{lv})$ are the thermodynamic work of adhesion between the liquid and the solid and the liquid's work of cohesion, respectively.

Using the formulations of the acid–base approach of interfacial interactions [84,131] we find:

$$\Delta G = 2\sqrt{\gamma_s^{LW}\gamma_{lv}^{LW}} + 2\sqrt{\gamma_s^+\gamma_{lv}^-} + 2\sqrt{\gamma_s^-\gamma_{lv}^+} - 2\gamma_{lv} \qquad (59)$$

where γ^{LW} is the Lifshitz–van der Waals component of the surface tension, γ^+ is the electron acceptor/proton donor (acidity parameter) and γ^- is the electron donor/proton acceptor (basicity parameter) according to van Oss et al. [84]. The subscripts s and l always indicate solid and liquid, respectively.

(iii) A liquid is not completely wetting the solid that was equilibrated with the liquid vapor, and a dynamic contact angle θ, which is not the Young contact angle on a flat solid [132], appears at the liquid penetration front and $\Delta G = \gamma_{lv} \cos \theta$.

(iv) $\Delta G = \gamma_{lv} \cos \theta + W_{sl} - W_{ll}$, if the same liquid as in case (iii) is wetting the "bare" nonpreequilibrated solid surface. Using the acid–base approach to estimate the solid surface tension we obtain:

$$\Delta G = \gamma_{lv} \cos \theta + 2\sqrt{\gamma_s^{LW}\gamma_{lv}^{LW}} + 2\sqrt{\gamma_s^+\gamma_{lv}^-} + 2\sqrt{\gamma_s^-\gamma_{lv}^+} \qquad (60)$$

Once the "effective pore radius" parameter r_p was determined [Eq. (58)] in a first experiment using a liquid with known surface tension γ_{lv} and viscosity that wets the solid completely, it will be possible, according to Refs. [116,128, 130], to use the appropriate ΔG-values from cases (iii) and (iv) to determine the solid surface tension components.

All nonpolar liquids (n-alkanes or DIM) interact with solid surface only via long range, Lifshitz–van der Waals forces γ_{lv}^{LW}, and, therefore, applying the acid–base approach formulations Eq. (59) becomes:

$$\Delta G = W_{sl} - W_{ll} = 2\sqrt{\gamma_s^{LW}\gamma_{lv}^{LW}} - 2\gamma_{lv} \qquad (61)$$

Now, using "bare" samples in an n-alkane wicking experiment, the driving force for liquid penetration, according to case (ii), should be $\Delta G = W_{sl} - W_{ll}$. The Lifshitz–van der Waals component of the solid surface tension γ_s^{LW} can be calculated by combining Eqs. (57) and (61). Another way to determine γ_s^{LW} is by using high surface tension nonpolar liquids, e.g., DIM. We then measure the functions $x^2 = f(t)$ for the "bare" and preequilibrated solid considering the cases (iii) and (iv). The value of $\Delta G = \gamma_{lv} \cos \theta$ from case (iii) will be determined and introduced into the equation of case (iv), assuming that γ_{lv}

$\cos \theta$ is equal in the two cases. Solving the equations for cases (iii) and (iv) simultaneously allows the determination of γ_{s}^{LW}.

The electron acceptor γ^{+} and electron donor γ^{-} parameters of the solid (and, therefore, the acid–base component $\gamma^{AB} = 2\sqrt{\gamma^{+}\gamma^{-}}$) can be determined using polar liquids with known surface tension components and parameters (see for instance Refs. [58,59,131,133]), as well as viscosities for the liquid penetration experiments and considering the proper experimental conditions (i) to (iv). Again, the functions $x^{2} = f(t)$ for the "bare" and preequilibrated solid have to be measured separately, and ΔG_{bare} as well as ΔG_{pre} has to be determined according to Eq. (58). The following Eq. (62) needs to be solved to estimate the γ^{+} and γ^{-} parameters:

$$\Delta G_{bare} - \Delta G_{pre} = 2\sqrt{\gamma_{s}^{LW}\gamma_{lv}^{LW}} + 2\sqrt{\gamma_{s}^{+}\gamma_{lv}^{-}} + 2\sqrt{\gamma_{s}^{-}\gamma_{lv}^{+}} - 2\gamma_{lv} \tag{62}$$

The experiments can be performed as follows [116,123,124,130]: The fabric of interest is to be placed on a glass plate. The strip must be slightly strained to guarantee uniform porosity for the penetrating liquids. In order to obtain the solid surface tension components and parameters from such thin-layer wicking experiments, the latter have to be conducted not only on bare (cleaned and dried) fabric strips but also on strips that were preequilibrated with the test liquid's vapor. The equilibration of the fabric strips with the saturated liquid vapors can be performed for instance in desiccators containing the appropriate liquid on the bottom for a certain period of time at the measuring temperature.

The thin-layer wicking experiments have to be performed in a closed sandwich chamber in which the glass plate (containing the fiber strip) is to be placed in an exactly horizontal position. The chamber lid (if not the whole chamber) should be made from glass, so that the moving boundary of the liquid penetration front in the strip is visible. On the edges of the sandwich chamber there has to be a scale, which allows the recording of the penetration time. The test liquid is transported from the bottles to the "porous layer" using a flannel wick. The liquid surface has to be on the same level as the testing plate in the sandwich chamber. One end of the flannel wick is dipped into the liquid and the other will be attached to the end of the plate when starting the penetration experiment.

All the above-described forms of the Washburn equation [cases (i) to (iv)] should result in a straight-line dependence for $x^{2} = f(t)$, having different slopes depending on the true free energy changes accompanying the liquid penetration into the fabric (or any porous solid medium) [116]. Deviations from the linearity of the $x^{2} = f(t)$ relationship could occur if the (apparent) contact angle changes during the penetration process or if a nonuniform liquid front develops due to a nonuniform distribution of the pore size of the fabric [134]. However, it was recently shown by a mathematical analysis of the Washburn equation that most of discordance found from its application is due to the inadequate use of this equation [135]. It was proposed to use the Washburn equation as a polynomial expression instead of its linearized form. In the proposed way the deviation from linearity will disappear [135].

Note again: The contact angle appearing under these dynamic conditions is not expected to be the equilibrium Young contact angle [132]. Therefore it is not possible to calculate a "Young" contact angle using the modified Washburn equation [case (iii)]:

$$\cos \theta = \frac{x^2}{t} \frac{2\eta}{r_p \cdot \gamma_{lv}} \tag{63}$$

for a liquid with known surface tension and viscosity, but not completely wetting the solid, penetrating a porous medium with known "effective pore radius" [116,136].

IX. OTHER METHODS TO MEASURE CONTACT ANGLES ON FIBERS

In addition to the aforementioned techniques, most commonly used for the characterization of fibers, fiber assemblies, and fabrics in terms of their wetting behavior or surface tension, there are also other methods available: the determination of the contact angle from the equilibrium meniscus near a floating fiber [80], the reflection method initially developed by Jones and Porter [137] (reviewed in Refs. 1 and 32), the tilted fiber or rotating stage technique [138,139], and the solidification front method (recently reviewed in Ref. 140).

A comparison between contact angles of various test liquids (glycerol and water) on carbon- and Kevlar fibers measured using the solidification front technique and the modified Wilhelmy method can be found in Ref. 141. The results obtained using both techniques are consistent with each other, which illustrates that both techniques will produce reliable data for small-diameter fibers. However, the solidification technique offers a number of potential advantages compared to the modified Wilhelmy method [140]. This is because it is not possible to measure the fiber diameter and the contact angle simultaneously with the Wilhelmy method. Also, in cases where the fibers have a nonuniform character it becomes even more difficult. The fiber geometry may also influence the results, so that a convoluted fiber cross section can cause wicking effects and therefore lead to incorrect balance readings [140].

APPENDIX: Contact Angle (θ) as a Function of the Maximum Drop Height (H) and Length (L)

H	0°	5°	10°	15°	20°	25°	30°	35°	40°
1.50	7.7750	6.8605	5.9460	5.1230	4.4240	3.8465	3.3725	2.9825	2.6590
1.55	7.9170	7.0655	6.2020	5.4065	4.7145	4.1305	3.6435	3.2370	2.8955
1.60	8.0590	7.2610	6.4405	5.6710	4.9880	4.4010	3.9030	3.4825	3.1265
1.65	8.1990	7.4460	6.6645	5.9195	5.2465	4.6585	4.1530	3.7210	3.3510
1.70	8.3380	7.6240	6.8770	6.1545	5.4925	4.9050	4.3940	3.9520	3.5700
1.75	8.4765	7.7965	7.0795	6.3780	5.7270	5.1415	4.6260	4.1760	3.7840
1.80	8.6140	7.9650	7.2740	6.5920	5.9515	5.3695	4.8510	4.3940	3.9930
1.85	8.7510	8.1280	7.4615	6.7975	6.1675	5.5885	5.0685	4.6060	4.1965
1.90	8.8880	8.2880	7.6425	6.9955	6.3755	5.8010	5.2795	4.8125	4.3960
1.95	9.0225	8.4435	7.8190	7.1870	6.5770	6.0065	5.4850	5.0135	4.5915
2.00	9.1575	8.5970	7.9905	7.3730	6.7720	6.2065	5.6845	5.2105	4.7830
2.05	9.2915	8.7485	8.1580	7.5540	6.9620	6.4005	5.8795	5.4030	4.9705
2.10	9.4250	8.8980	8.3220	7.7305	7.1470	6.5900	6.0700	5.5910	5.1545
2.15	9.5575	9.0445	8.4830	7.9025	7.3275	6.7750	6.2560	5.7755	5.3350
2.20	9.6900	9.1900	8.6410	8.0715	7.5040	6.9560	6.4385	5.9565	5.5130
2.25	9.8215	9.3335	8.7965	8.2370	7.6770	7.1335	6.6170	6.1340	5.6875
2.30	9.9525	9.4755	8.9500	8.3995	7.8465	7.3070	6.7925	6.3090	5.8595
2.35	10.084	9.6170	9.1010	8.5595	8.0130	7.4775	6.9645	6.4805	6.0290
2.40	10.214	9.7560	9.2500	8.7170	8.1765	7.6455	7.1340	6.6495	6.1960
2.45	10.343	9.8950	9.3975	8.8720	8.3380	7.8105	7.3010	6.8160	6.3605
2.50	10.472	10.033	9.5430	9.0250	8.4965	7.9730	7.4650	6.9805	6.5230
2.55	10.601	10.169	9.6875	9.1760	8.6530	8.1330	7.6270	7.1425	6.6835
2.60	10.729	10.305	9.8300	9.3255	8.8075	8.2910	7.7870	7.3020	6.8420
2.65	10.857	10.439	9.9715	9.4730	8.9605	8.4470	7.9445	7.4600	6.9990
2.70	10.984	10.573	10.112	9.6195	9.1110	8.6010	8.1005	7.6160	7.1540
2.75	11.111	10.706	10.251	9.7640	9.2605	8.7535	8.2545	7.7705	7.3070
2.80	11.238	10.838	10.389	9.9075	9.4080	8.9040	8.4065	7.9230	7.4590
2.85	11.364	10.970	10.527	10.050	9.5540	9.0535	8.5575	8.0740	7.6095
2.90	11.490	11.101	10.663	10.191	9.6990	9.2010	8.7065	8.2240	7.7580
2.95	11.615	11.231	10.798	10.330	9.8425	9.3470	8.8545	8.3720	7.9055
3.00	11.740	11.361	10.932	10.468	9.9850	9.4920	9.0010	8.5190	8.0520
3.05	11.865	11.490	11.066	10.606	10.126	9.6355	9.1460	8.6645	8.1970
3.10	11.989	11.619	11.199	10.743	10.266	9.7785	9.2900	8.8090	8.3410
3.15	12.113	11.747	11.331	10.880	10.405	9.9195	9.4330	8.9525	8.4840
3.20	12.237	11.875	11.463	11.015	10.543	10.060	9.5745	9.0945	8.6255
3.25	12.361	12.002	11.594	11.149	10.681	10.200	9.7150	9.2355	8.7665
3.30	12.484	12.128	11.724	11.283	10.817	10.338	9.8550	9.3755	8.9060
3.35	12.607	12.255	11.853	11.415	10.952	10.476	9.9935	9.5150	9.0450
3.40	12.729	12.381	11.983	11.547	11.087	10.612	10.132	9.6530	9.1830
3.45	12.852	12.506	12.111	11.678	11.221	10.748	10.269	9.7905	9.3200
3.50	12.974	12.631	12.239	11.809	11.354	10.883	10.405	9.9270	9.4565
3.55	13.096	12.756	12.367	11.940	11.487	11.017	10.540	10.063	9.5920
3.60	13.217	12.880	12.494	12.069	11.619	11.151	10.675	10.198	9.7270
3.65	13.339	13.004	12.620	12.198	11.750	11.284	10.809	10.333	9.8610

45°	50°	55°	60°	65°	70°	75°	80°	85°	90°
2.3875	2.1565	1.9585	1.7865	1.6360	1.5025	1.3830	1.2760	1.1785	1.0900
2.6070	2.3605	2.1475	1.9620	1.7985	1.6540	1.5240	1.4075	1.3015	1.2045
2.8225	2.5610	2.3340	2.1355	1.9605	1.8045	1.6650	1.5390	1.4240	1.3195
3.0330	2.7580	2.5185	2.3075	2.1210	1.9545	1.8050	1.6700	1.5470	1.4345
3.2395	2.9520	2.7000	2.4780	2.2805	2.1040	1.9450	1.8010	1.6700	1.5500
3.4420	3.1430	2.8795	2.6465	2.4390	2.2525	2.0845	1.9320	1.7930	1.6655
3.6405	3.3310	3.0570	2.8135	2.5960	2.4005	2.2235	2.0630	1.9160	1.7810
3.8355	3.5155	3.2320	2.9785	2.7515	2.5470	2.3620	2.1930	2.0390	1.8970
4.0265	3.6975	3.4045	3.1420	2.9060	2.6930	2.4995	2.3230	2.1615	2.0125
4.2140	3.8770	3.5750	3.3040	3.0595	2.8380	2.6365	2.4530	2.2840	2.1285
4.3985	4.0535	3.7435	3.4640	3.2115	2.9820	2.7730	2.5820	2.4060	2.2440
4.5800	4.2275	3.9100	3.6225	3.3625	3.1255	2.9090	2.7105	2.5280	2.3595
4.7580	4.3995	4.0745	3.7795	3.5120	3.2675	3.0440	2.8390	2.6500	2.4750
4.9335	4.5685	4.2370	3.9350	3.6605	3.4090	3.1785	2.9670	2.7715	2.5905
5.1065	4.7355	4.3975	4.0890	3.8075	3.5495	3.3125	3.0945	2.8930	2.7060
5.2770	4.9005	4.5565	4.2420	3.9535	3.6890	3.4460	3.2215	3.0135	2.8210
5.4450	5.0635	4.7140	4.3930	4.0985	3.8280	3.5785	3.3480	3.1345	2.9360
5.6105	5.2245	4.8695	4.5430	4.2425	3.9660	3.7105	3.4740	3.2545	3.0510
5.7740	5.3840	5.0235	4.6915	4.3855	4.1030	3.8415	3.5995	3.3750	3.1655
5.9355	5.5410	5.1765	4.8390	4.5275	4.2390	3.9725	3.7245	3.4945	3.2800
6.0950	5.6970	5.3275	4.9855	4.6685	4.3745	4.1025	3.8495	3.6140	3.3945
6.2530	5.8510	5.4775	5.1305	4.8085	4.5095	4.2320	3.9735	3.7330	3.5085
6.4090	6.0035	5.6260	5.2740	4.9475	4.6435	4.3610	4.0975	3.8520	3.6225
6.5635	6.1550	5.7730	5.4170	5.0855	4.7770	4.4890	4.2210	3.9705	3.7360
6.7160	6.3045	5.9190	5.5590	5.2230	4.9095	4.6170	4.3440	4.0885	3.8495
6.8675	6.4530	6.0640	5.7000	5.3595	5.0415	4.7445	4.4665	4.2065	3.9630
7.0175	6.6000	6.2080	5.8395	5.4950	5.1725	4.8710	4.5885	4.3240	4.0760
7.1660	6.7460	6.3505	5.9785	5.6300	5.3035	4.9970	4.7105	4.4415	4.1890
7.3130	6.8910	6.4920	6.1165	5.7640	5.4335	5.1230	4.8320	4.5585	4.3020
7.4590	7.0345	6.6325	6.2540	5.8975	5.5625	5.2480	4.9530	4.6755	4.4145
7.6040	7.1770	6.7725	6.3905	6.0300	5.6915	5.3730	5.0735	4.7920	4.5270
7.7475	7.3185	6.9115	6.5260	6.1625	5.8195	5.4970	5.1935	4.9080	4.6390
7.8905	7.4590	7.0490	6.6605	6.2935	5.9475	5.6210	5.3135	5.0240	4.7510
8.0320	7.5990	7.1860	6.7950	6.4245	6.0745	5.7445	5.4330	5.1395	4.8630
8.1725	7.7375	7.3225	6.9280	6.5545	6.2010	5.8675	5.5525	5.2550	4.9745
8.3120	7.8755	7.4580	7.0610	6.6840	6.3275	5.9900	5.6715	5.3705	5.0860
8.4510	8.0125	7.5925	7.1930	6.8130	6.4530	6.1120	5.7900	5.4850	5.1970
8.5885	8.1485	7.7265	7.3240	6.9415	6.5780	6.2340	5.9080	5.6000	5.3085
8.7255	8.2840	7.8600	7.4550	7.0690	6.7025	6.3550	6.0260	5.7145	5.4190
8.8620	8.4185	7.9925	7.5850	7.1965	6.8270	6.4760	6.1435	5.8285	5.5300
8.9970	8.5525	8.1245	7.7145	7.3230	6.9505	6.5970	6.2610	5.9425	5.6405
9.1320	8.6855	8.2555	7.8435	7.4495	7.0740	6.7170	6.3780	6.0565	5.7510
9.2660	8.8180	8.3865	7.9715	7.5750	7.1970	6.8370	6.4950	6.1700	5.8615
9.3990	8.9500	8.5165	8.0995	7.7005	7.3195	6.9565	6.6115	6.2830	5.9715

(*Continued on next page*)

APPENDIX (Continued)

H	0°	5°	10°	15°	20°	25°	30°	35°	40°
3.70	13.460	13.127	12.746	12.327	11.881	11.416	10.942	10.466	9.9945
3.75	13.580	13.251	12.872	12.455	12.011	11.547	11.075	10.599	10.127
3.80	13.701	13.374	12.998	12.583	12.140	11.678	11.206	10.731	10.259
3.85	13.821	13.496	13.123	12.710	12.269	11.809	11.338	10.863	10.391
3.90	13.942	13.619	13.247	12.837	12.397	11.938	11.468	10.994	10.522
3.95	14.062	13.741	13.371	12.963	12.525	12.068	11.598	11.125	10.652
4.00	14.181	13.863	13.495	13.088	12.653	12.196	11.728	11.254	10.782
4.05	14.301	13.984	13.619	13.214	12.780	12.325	11.857	11.384	10.911
4.10	14.420	14.105	13.742	13.339	12.906	12.452	11.986	11.513	11.040
4.15	14.539	14.226	13.865	13.463	13.032	12.580	12.114	11.641	11.168
4.20	14.658	14.347	13.987	13.587	13.158	12.706	12.241	11.769	11.295
4.25	14.777	14.467	14.109	13.711	13.283	12.833	12.368	11.896	11.423
4.30	14.895	14.588	14.231	13.834	13.408	12.959	12.495	12.023	11.549
4.35	15.014	14.708	14.353	13.958	13.533	13.084	12.621	12.150	11.676
4.40	15.132	14.827	14.474	14.080	13.657	13.209	12.747	12.276	11.802
4.45	15.250	14.947	14.595	14.204	13.781	13.334	12.872	12.401	11.927
4.50	15.368	15.066	14.716	14.325	13.904	13.459	12.997	12.526	12.052
4.55	15.486	15.185	14.837	14.448	14.027	13.583	13.122	12.651	12.177
4.60	15.603	15.304	14.957	14.569	14.150	13.706	13.246	12.776	12.301
4.65	15.720	15.423	15.077	14.690	14.272	13.830	13.370	12.900	12.425
4.70	15.838	15.542	15.197	14.811	14.395	13.953	13.493	13.024	12.549
4.75	15.955	15.660	15.317	14.933	14.516	14.075	13.617	13.147	12.672
4.80	16.072	15.778	15.436	15.053	14.638	14.198	13.739	13.270	12.795
4.85	16.188	15.896	15.555	15.173	14.759	14.320	13.862	13.393	12.917
4.90	16.305	16.014	15.674	15.293	14.880	14.441	13.984	13.515	13.040
4.95	16.421	16.131	15.792	15.414	15.001	14.563	14.106	13.637	13.162
5.00	16.537	16.248	15.911	15.532	15.121	14.684	14.227	13.758	13.283
5.05	16.653	16.366	16.029	15.651	15.241	14.805	14.349	13.880	13.404
5.10	16.769	16.483	16.148	15.771	15.361	14.925	14.470	14.001	13.525
5.15	16.885	16.599	16.266	15.890	15.481	15.046	14.591	14.122	13.646
5.20	17.001	16.716	16.383	16.008	15.601	15.166	14.711	14.243	13.766
5.25	17.117	16.833	16.501	16.128	15.720	15.286	14.831	14.363	13.887
5.30	17.232	16.950	16.618	16.246	15.839	15.405	14.951	14.483	14.007
5.35	17.348	17.066	16.736	16.364	15.958	15.525	15.071	14.603	14.126
5.40	17.463	17.182	16.853	16.482	16.077	15.644	15.191	14.723	14.246
5.45	17.578	17.298	16.970	16.599	16.195	15.763	15.310	14.842	14.365
5.50	17.693	17.414	17.086	16.717	16.313	15.882	15.429	14.962	14.484
5.55	17.808	17.530	17.203	16.834	16.432	16.001	15.548	15.080	14.603
5.60	17.923	17.645	17.320	16.952	16.549	16.119	15.667	15.199	14.722
5.65	18.037	17.761	17.436	17.069	16.667	16.237	15.785	15.318	14.840
5.70	18.152	17.876	17.552	17.185	16.785	16.355	15.904	15.436	14.958
5.75	18.266	17.991	17.668	17.302	16.902	16.473	16.022	15.554	15.076
5.80	18.381	18.107	17.784	17.419	17.019	16.590	16.139	15.672	15.194
5.85	18.495	18.222	17.900	17.535	17.136	16.708	16.257	15.790	15.311
5.90	18.609	18.336	18.015	17.651	17.253	16.825	16.375	15.907	15.429

45°	50°	55°	60°	65°	70°	75°	80°	85°	90°
9.5315	9.0815	8.6460	8.2270	7.8255	7.4415	7.0760	6.7275	6.3965	6.0815
9.6635	9.2120	8.7750	8.3540	7.9500	7.5635	7.1950	6.8435	6.5090	6.1910
9.7950	9.3420	8.9035	8.4800	8.0740	7.6850	7.3135	6.9590	6.6220	6.3010
9.9255	9.4715	9.0310	8.6060	8.1975	7.8060	7.4315	7.0745	6.7345	6.4105
10.056	9.6005	9.1585	8.7315	8.3205	7.9265	7.5495	7.1900	6.8465	6.5200
10.186	9.7290	9.2855	8.8565	8.4435	8.0470	7.6675	7.3050	6.9590	6.6290
10.315	9.8570	9.4120	8.9810	8.5660	8.1670	7.7850	7.4195	7.0710	6.7380
10.443	9.9845	9.5380	9.1055	8.6880	8.2865	7.9020	7.5340	7.1825	6.8470
10.571	10.112	9.6635	9.2290	8.8095	8.4060	8.0190	7.6485	7.2940	6.9560
10.699	10.238	9.7885	9.3525	8.9310	8.5250	8.1355	7.7625	7.4055	7.0645
10.826	10.364	9.9135	9.4755	9.0520	8.6440	8.2520	7.8765	7.5170	7.1730
10.953	10.490	10.038	9.5980	9.1725	8.7625	8.3680	7.9900	7.6280	7.2815
11.079	10.615	10.162	9.7200	9.2930	8.8805	8.4840	8.1035	7.7390	7.3900
11.205	10.740	10.285	9.8420	9.4130	8.9985	8.6000	8.2165	7.8495	7.4980
11.330	10.864	10.408	9.9635	9.5325	9.1165	8.7150	8.3300	7.9600	7.6065
11.455	10.989	10.531	10.085	9.6520	9.2335	8.8305	8.4425	8.0705	7.7145
11.579	11.112	10.654	10.206	9.7715	9.3510	8.9455	8.5555	8.1810	7.8220
11.704	11.235	10.776	10.327	9.8900	9.4675	9.0600	8.6680	8.2910	7.9300
11.827	11.358	10.897	10.447	10.009	9.5845	9.1745	8.7800	8.4010	8.0375
11.951	11.481	11.019	10.567	10.127	9.7010	9.2890	8.8925	8.5110	8.1450
12.074	11.603	11.140	10.687	10.245	9.8170	9.4030	9.0045	8.6210	8.2525
12.197	11.725	11.261	10.806	10.363	9.9330	9.5170	9.1165	8.7305	8.3600
12.319	11.847	11.381	10.925	10.481	10.049	9.6310	9.2280	8.8400	8.4670
12.441	11.968	11.501	11.044	10.598	10.164	9.7445	9.3395	8.9495	8.5745
12.563	12.089	11.621	11.163	10.715	10.280	9.8580	9.4510	9.0585	8.6815
12.684	12.210	11.741	11.281	10.832	10.395	9.9710	9.5620	9.1675	8.7880
12.805	12.330	11.860	11.399	10.948	10.509	10.084	9.6730	9.2765	8.8950
12.926	12.450	11.979	11.517	11.064	10.624	10.197	9.7840	9.3855	9.0020
13.047	12.570	12.098	11.634	11.180	10.738	10.310	9.8950	9.4945	9.1085
13.167	12.689	12.217	11.752	11.296	10.853	10.422	10.005	9.6030	9.2150
13.287	12.809	12.335	11.869	11.412	10.967	10.535	10.116	9.7115	9.3215
13.407	12.928	12.453	11.986	11.528	11.081	10.647	10.226	9.8200	9.4280
13.526	13.047	12.571	12.102	11.643	11.195	10.759	10.336	9.9280	9.5345
13.646	13.165	12.689	12.219	11.758	11.308	10.871	10.446	10.036	9.6405
13.765	13.284	12.806	12.335	11.873	11.422	10.983	10.557	10.144	9.7470
13.884	13.402	12.924	12.451	11.988	11.535	11.094	10.666	10.252	9.8530
14.002	13.520	13.041	12.567	12.103	11.648	11.206	10.776	10.360	9.9590
14.121	13.638	13.157	12.683	12.217	11.761	11.317	10.886	10.468	10.065
14.239	13.755	13.274	12.799	12.331	11.874	11.428	10.995	10.576	10.170
14.357	13.872	13.390	12.914	12.445	11.987	11.539	11.105	10.683	10.276
14.474	13.989	13.507	13.029	12.559	12.099	11.650	11.214	10.791	10.382
14.592	14.106	13.623	13.144	12.673	12.211	11.761	11.323	10.898	10.488
14.709	14.223	13.739	13.259	12.787	12.324	11.872	11.432	11.006	10.593
14.826	14.340	13.854	13.374	12.900	12.436	11.982	11.541	11.113	10.699
14.943	14.456	13.970	13.488	13.013	12.548	12.093	11.650	11.220	10.804

(Continued on next page)

APPENDIX (Continued)

H	0°	5°	10°	15°	20°	25°	30°	35°	40°
5.95	18.723	18.451	18.131	17.768	17.370	16.942	16.492	16.025	15.546
6.00	18.837	18.566	18.246	17.884	17.486	17.059	16.609	16.142	15.663
6.05	18.951	18.680	18.361	17.999	17.602	17.176	16.726	16.259	15.779
6.10	19.064	18.795	18.476	18.115	17.719	17.292	16.843	16.375	15.896
6.15	19.178	18.909	18.591	18.230	17.835	17.409	16.959	16.492	16.012
6.20	19.291	19.023	18.706	18.346	17.950	17.525	17.076	16.608	16.129
6.25	19.405	19.137	18.821	18.462	18.066	17.641	17.192	16.725	16.245
6.30	19.518	19.251	18.935	18.576	18.182	17.757	17.308	16.841	16.361
6.35	19.631	19.365	19.050	18.691	18.297	17.873	17.424	16.957	16.476
6.40	19.745	19.479	19.164	18.806	18.412	17.988	17.540	17.073	16.592
6.45	19.858	19.592	19.278	18.920	18.528	18.104	17.655	17.188	16.708
6.50	19.971	19.706	19.392	19.035	18.643	18.219	17.771	17.304	16.823
6.55	20.083	19.819	19.506	19.150	18.758	18.334	17.886	17.419	16.938
6.60	20.196	19.933	19.620	19.264	18.872	18.449	18.001	17.534	17.053
6.65	20.309	20.046	19.734	19.379	18.987	18.564	18.116	17.649	17.168
6.70	20.422	20.159	19.848	19.493	19.102	18.679	18.231	17.764	17.282
6.75	20.534	20.272	19.961	19.606	19.216	18.794	18.346	17.879	17.397
6.80	20.646	20.385	20.075	19.721	19.330	18.908	18.461	17.993	17.511
6.85	20.759	20.498	20.188	19.835	19.444	19.023	18.575	18.108	17.626
6.90	20.871	20.611	20.301	19.949	19.558	19.137	18.690	18.222	17.740
6.95	20.983	20.723	20.415	20.062	19.672	19.251	18.804	18.337	17.854
7.00	21.096	20.836	20.528	20.176	19.786	19.365	18.918	18.451	17.968
7.05	21.208	20.949	20.641	20.289	19.900	19.479	19.032	18.565	18.082
7.10	21.320	21.061	20.754	20.402	20.014	19.593	19.146	18.679	18.195
7.15	21.431	21.173	20.866	20.515	20.127	19.707	19.260	18.792	18.309
7.20	21.543	21.286	20.979	20.629	20.240	19.820	19.374	18.906	18.422
7.25	21.655	21.398	21.092	20.741	20.354	19.934	19.487	19.019	18.536
7.30	21.767	21.510	21.204	20.855	20.467	20.047	19.601	19.133	18.649
7.35	21.878	21.622	21.317	20.967	20.580	20.160	19.714	19.246	18.762
7.40	21.990	21.734	21.429	21.079	20.693	20.274	19.827	19.359	18.875
7.45	22.101	21.846	21.541	21.193	20.806	20.387	19.940	19.472	18.988
7.50	22.213	21.958	21.654	21.305	20.919	20.500	20.053	19.585	19.101
7.55	22.324	22.070	21.766	21.418	21.032	20.613	20.166	19.698	19.213
7.60	22.435	22.181	21.878	21.530	21.144	20.725	20.279	19.811	19.326
7.65	22.547	22.293	21.990	21.642	21.257	20.838	20.392	19.924	19.438
7.70	22.658	22.404	22.102	21.755	21.369	20.951	20.505	20.036	19.551
7.75	22.769	22.516	22.213	21.867	21.481	21.063	20.617	20.149	19.663
7.80	22.880	22.627	22.325	21.978	21.594	21.176	20.730	20.261	19.775
7.85	22.991	22.739	22.437	22.091	21.706	21.288	20.842	20.373	19.887
7.90	23.102	22.850	22.548	22.203	21.818	21.400	20.954	20.486	19.999
7.95	23.213	22.961	22.660	22.314	21.930	21.512	21.066	20.598	20.111
8.00	23.323	23.072	22.771	22.426	22.042	21.624	21.178	20.710	20.223
8.05	23.434	23.183	22.883	22.538	22.154	21.736	21.290	20.822	20.334
8.10	23.545	23.294	22.994	22.649	22.266	21.848	21.402	20.933	20.446
8.15	23.655	23.405	23.105	22.761	22.377	21.960	21.514	21.045	20.558

45°	50°	55°	60°	65°	70°	75°	80°	85°	90°
15.060	14.572	14.085	13.602	13.126	12.660	12.203	11.759	11.327	10.910
15.177	14.688	14.200	13.717	13.239	12.771	12.313	11.868	11.434	11.015
15.293	14.804	14.315	13.830	13.352	12.883	12.424	11.976	11.541	11.120
15.409	14.919	14.430	13.944	13.465	12.994	12.534	12.085	11.648	11.225
15.525	15.035	14.545	14.058	13.578	13.106	12.643	12.193	11.755	11.331
15.641	15.150	14.659	14.172	13.690	13.217	12.753	12.301	11.862	11.436
15.757	15.265	14.774	14.285	13.802	13.328	12.863	12.410	11.968	11.541
15.872	15.380	14.888	14.398	13.915	13.439	12.973	12.518	12.075	11.646
15.988	15.495	15.002	14.511	14.027	13.550	13.082	12.626	12.182	11.751
16.103	15.610	15.116	14.624	14.139	13.660	13.191	12.734	12.288	11.855
16.218	15.724	15.230	14.737	14.250	13.771	13.301	12.841	12.394	11.960
16.333	15.839	15.343	14.850	14.362	13.881	13.410	12.949	12.501	12.065
16.448	15.953	15.457	14.963	14.474	13.992	13.519	13.057	12.607	12.170
16.562	16.067	15.570	15.075	14.585	14.102	13.628	13.165	12.713	12.274
16.677	16.181	15.683	15.187	14.696	14.212	13.737	13.272	12.819	12.379
16.791	16.295	15.796	15.300	14.808	14.322	13.846	13.380	12.925	12.483
16.905	16.408	15.909	15.412	14.919	14.432	13.954	13.487	13.031	12.588
17.019	16.522	16.022	15.524	15.030	14.542	14.063	13.594	13.137	12.692
17.133	16.635	16.135	15.636	15.141	14.652	14.172	13.702	13.243	12.797
17.247	16.748	16.247	15.747	15.251	14.762	14.280	13.809	13.349	12.901
17.361	16.862	16.360	15.859	15.362	14.871	14.389	13.916	13.454	13.005
17.474	16.975	16.472	15.971	15.473	14.981	14.497	14.023	13.560	13.110
17.588	17.088	16.585	16.082	15.583	15.090	14.605	14.130	13.666	13.214
17.701	17.200	16.697	16.193	15.694	15.199	14.713	14.237	13.771	13.318
17.814	17.313	16.809	16.305	15.804	15.309	14.821	14.344	13.877	13.422
17.927	17.426	16.921	16.416	15.914	15.418	14.929	14.450	13.982	13.526
18.040	17.538	17.032	16.527	16.024	15.527	15.037	14.557	14.088	13.630
18.153	17.650	17.144	16.638	16.134	15.636	15.145	14.664	14.193	13.734
18.266	17.763	17.256	16.749	16.244	15.745	15.253	14.770	14.298	13.838
18.379	17.875	17.367	16.859	16.354	15.854	15.361	14.877	14.404	13.942
18.491	17.987	17.479	16.970	16.464	15.962	15.468	14.983	14.509	14.046
18.604	18.099	17.590	17.080	16.573	16.071	15.576	15.090	14.614	14.150
18.716	18.211	17.701	17.191	16.683	16.180	15.683	15.196	14.719	14.254
18.828	18.322	17.812	17.301	16.792	16.288	15.791	15.302	14.824	14.358
18.940	18.434	17.923	17.412	16.902	16.397	15.898	15.408	14.929	14.461
19.052	18.546	18.034	17.522	17.011	16.505	16.005	15.515	15.034	14.565
19.164	18.657	18.145	17.632	17.120	16.613	16.113	15.621	15.139	14.669
19.276	18.768	18.256	17.742	17.230	16.721	16.220	15.727	15.244	14.772
19.388	18.880	18.366	17.852	17.339	16.830	16.327	15.833	15.349	14.876
19.499	18.991	18.477	17.962	17.448	16.938	16.434	15.939	15.453	14.979
19.611	19.102	18.588	18.071	17.557	17.046	16.541	16.045	15.558	15.083
19.722	19.213	18.698	18.181	17.666	17.154	16.648	16.151	15.663	15.186
19.834	19.324	18.808	18.291	17.774	17.262	16.755	16.256	15.768	15.290
19.945	19.435	18.919	18.400	17.883	17.369	16.862	16.362	15.872	15.393
20.056	19.545	19.029	18.510	17.992	17.477	16.968	16.468	15.977	15.497

(Continued on next page)

APPENDIX (Continued)

H	0°	5°	10°	15°	20°	25°	30°	35°	40°
8.20	23.766	23.516	23.216	22.872	22.489	22.072	21.626	21.157	20.669
8.25	23.876	23.627	23.328	22.983	22.600	22.183	21.738	21.268	20.780
8.30	23.987	23.738	23.439	23.095	22.712	22.295	21.849	21.380	20.892
8.35	24.097	23.848	23.550	23.206	22.823	22.406	21.961	21.491	21.003
8.40	24.207	23.959	23.660	23.317	22.935	22.518	22.072	21.603	21.114
8.45	24.318	24.069	23.771	23.428	23.046	22.629	22.183	21.714	21.225
8.50	24.428	24.180	23.882	23.539	23.157	22.740	22.295	21.825	21.336

REFERENCES

1. Carroll, B.J. Colloids Surf. A 1993, *74*, 131.
2. Coulson, S.R.; Woodward, I.S.; Badyal, J.P.S.; Brewer, S.A.; Willis, C. Chem. Mater. 2000, *12*, 2031.
3. Rebouillat, S.; Letellier, B.; Steffenino, B. Intern. J. Adhesion Adhes. 1999, *19*, 303.
4. Wu, S. Polymer Interface and Adhesion. Marcel Dekker: New York, 1980.
5. Clint, J.H. Curr. Opin. Colloid Interface Sci. 2001, *6*, 28.
6. McHale, G.; Newton, M.I. Colloids Surf. A 2002, *206*, 79.
7. Collins, G.E. J. Text. Inst. 1947, *38*, T73.
8. Morra, M.; Occiello, E.; Garbassi, F. Adv. Colloid Interface Sci. 1990, *32*, 79.
9. Paynter, R.W. Surf. Interface Anal. 1998, *26*, 674.
10. Berg, J.C. Preface. In *Wettability*; Berg, J.C., Ed.; Marcel Dekker: New York, 1993.
11. Aveyard, R.; Haydon, D.A. *An Introduction to the Principles of Surface Chemistry*; University Press: Cambridge, 1973.
12. Johnson, R.E. Jr.; Dettre, R.H. Surfactant Science. In *Wettability*; Series; Berg, J.C Ed.; Marcel Dekker: New York, 1993; Vol. 49, 1.
13. Fowkes, F.M. Attractive Forces at Solid–Liquid Interfaces. In *Wetting*; S.C.I. Monograph No. 25: London, 1967.
14. Carlson, B.C. SIAM J. Math. Anal. 1977, *8*,231.
15. Carlson, B.C. Numer. Math. 1979, *33*, 1.
16. Wagner, H.D. J. Appl. Phys. 1990, *67*, 1352.
17. Press, W.H.; Teukolsky, S.A.; Vetterling, W.T.; Flannery, B.P. *Numerical Recipes in C (The Art of Scientific Computing)*, 2nd Ed.; Cambridge University Press: Cambridge, 1992.
18. Yamaki, J.; Katayama, Y. J. Appl. Polym. Sci. 1975, *19*, 2897.
19. Carroll, B.J. J. Colloid Interface Sci. 1976, *57*, 488.
20. Song, B.; Bismarck, A.; Tahhan, R.; Springer, J. J. Colloid Interface Sci. 1998, *197*, 68.
21. Gilbert, A.H.; Goldstein, B.; Marom, G. Composites 1990, *21*, 408.

45°	50°	55°	60°	65°	70°	75°	80°	85°	90°
20.167	19.656	19.139	18.619	18.100	17.585	17.075	16.573	16.081	15.600
20.278	19.767	19.249	18.728	18.209	17.693	17.182	16.679	16.186	15.703
20.389	19.877	19.359	18.838	18.317	17.800	17.289	16.785	16.290	15.807
20.500	19.987	19.469	18.947	18.426	17.908	17.395	16.890	16.395	15.910
20.611	20.098	19.578	19.056	18.534	18.015	17.502	16.996	16.499	16.013
20.722	20.208	19.688	19.165	18.642	18.123	17.608	17.101	16.603	16.116
20.832	20.318	19.798	19.274	18.751	18.230	17.714	17.206	16.708	16.220

22. Brochard, F. J. Chem. Phys. 1986, *84*, 4664.
23. De Gennes, P.G. Rev. Mod. Phys. 1985, *57*, 827.
24. Bismarck, A.; Kumru, M.E.; Song, B.; Springer, J.; Moos, E.; Karger-Kocsis, J. Composites A 1999, *30*, 1351.
25. Bismarck, A.; Pfaffernoschke, M.; Song, B.; Springer, J. J. Appl. Polym. Sci. 1999, *71*, 1893.
26. Bismarck, A.; Richter, D.; Wuertz, C.; Springer, J. Colloids Surf. A 1999, *159*, 341.
27. Bismarck, A.; Richter, D.; Wuertz, C.; Kumru, M.E.; Song, B.; Springer, J. J. Adhesion 2000, *73*, 19.
28. Carroll, B.J. J. Adhesion Sci. Technol. 1992, *6*, 938.
29. Carroll, B.J. Langmuir 1986, *2*, 248.
30. Neumann, A.W.; Tanner, W. Proceedings of the Vth Int. Congress on Surface Active Substances; Barcelona, 1968; Vol. II, 727–733.
31. Neumann, A.W. Chem. Ing. Techn. 1970, *42*, 969.
32. Neumann, A.W.; Good, R.J. In *Surface and Colloid Science*; Good, R.J. Stromberg, R.R., Eds.; Plenum Press, New York, 1979; Vol. 11, 31 pp.
33. Bascom, W.D. In *Modern Approaches to Wettability: Theory and Applications*; Schrader, M.E. Loeb, G., Eds.; Plenum Press: New York, 1992; 359 pp.
34. Song, B.; Springer, J. J. Colloid Interface Sci. 1996, *184*, 64.
35. Song, B.; Springer, J. J. Colloid Interface Sci. 1996, *184*, 77.
36. Bismarck, A.; Kumru, M.E.; Springer, J. J. Colloid Interface Sci. 1999, *210*, 60.
37. González-Benito, J.; Baselga, J.; Aznar, A.J. J. Mater. Proc. Technol. 1999, *92–93*, 129.
38. Deng, Y.L.; Abazeri, M. Wood Fiber Sci. 1998, *30*, 155.
39. Hodgson, K.T.; Berg, J.C. Wood Fiber Sci. 1988, *20*, 3.
40. Shen, W.; Sheng, Y.J.; Parker, I.H. J. Adhesion Sci. Technol. 1999, *13*, 887.
41. Bismarck, A. Chemische Modifizierung von Carbonfasern: Elektrokinetische und oberflächenenergetische Charakterisierung/Einfluß auf die Adhäsion zu thermoplastischen Polymeren. dissertation.de: Berlin, 1999.
42. Tagawa, M.; Ohmae, N.; Umeno, M.; Gotoh, K.; Yasukawa, A.; Tagawa, M. Colloid Polym. Sci. 1989, *267*, 702.
43. Johnson, R.E. Jr.; Dettre, R.H. In *Contact Angle, Wettability, and Adhesion*.

Advances in Chemistry Series; American Chemical Society: Washington, D.C., 1964, *43*, 112.

44. Dettre, R.H.; Johnson, R.E. Jr. In Contact Angle, Wettability, and Adhesion, Advances in Chemistry Series; American Chemical Society: Washington, D.C., 1964, *43*, 136.

45. Garbassi, F.; Morra, M.; Occhiello, E. *Polymer Surfaces; From Physics to Technology*; John Wiley and Sons: Chichester, 1994; 171 pp.

46. Extrand, C.W.; Kumagai, Y. J. Colloid Interface Sci. 1997, *191*, 378.

47. Uyama, Y.; Inoue, H.; Ito, K.; Kishida, A.; Ikada, Y. J. Colloid Interface Sci. 1991, *141*, 275.

48. Morra, M.; Occhiello, E.; Garbassi, F. Adv. Colloid Interface Sci. 1990, *32*, 79.

49. Kloubek, J. Adv. Colloid Interface Sci. 1992, *38*, 99.

50. Fowkes, F.M.; Riddle, F.L.; Pastore, W.E.; Weber, A.A. Colloids Surf. 1990, *43*, 367.

51. In: *Applied Surface Thermodynamics*; Neumann, A.W., Spelt, J.K., Eds.; Marcel Dekker: New York, 1996.

52. Lyklema, J. Colloids Surf. A 1999, *156*, 413.

53. Della Volpe, C.; Siboni, S. In *Encyclopedia of Surface and Colloid Science*; Hubbard, A., Ed.; Marcel Dekker: New York, 2002; 17 pp.

54. Morra, M. In *Encyclopedia of Surface and Colloid Science*; Hubbard, A., Ed.; Marcel Dekker: New York, 2002; 74 pp.

55. Hammer, G.E.; Drzal, L.T. Appl. Surf. Sci. 1980, *4*, 340.

56. Comyn, J. Adhesion Science; RSC Paperbacks: Cambridge, 1997.

57. Good, R.J.; van Oss, C.J. In *Modern Approaches to Wettability*; Schrader, M.E. Loeb, G.I., Eds.; Plenum Press: New York, 1992; 1 pp.

58. Lee, L.-H. Langmuir 1996, *12*, 1681.

59. Della Volpe, C.; Siboni, S. J. Colloid Interface Sci. 1997, *195*, 121.

60. Schultz, J.; Cazeneuve, C.; Shanahan, M.E.R.; Donnet, J.B. J. Adhesion 1981, *12*, 221.

61. Felix, J.M.; Gatenholm, P. Polym. Composites 1993, *14*, 449.

62. Lu, W.; Fu, X.; Chung, D.D.L. Cem. Concr. Res. 1998, *28*, 783.

63. Tsutsumi, K.; Ishida, S.; Shibata, K. Colloid Polym. Sci. 1990, *268*, 31.

64. Gotoh, K.; Yasukawa, A.; Ohikita, M.; Obata, H.; Tagawa, M. Colloid Polym. Sci. 1995, *273*, 1144.

65. Lee, J.-S.; Kang, T.-J. Carbon 1997, *35*, 209.

66. Bismarck, A.; Springer, J. Colloids Surf. A 1999, *159*, 331.

67. Bismarck, A.; Wuertz, C.; Springer, J. Carbon 1999, *37*, 1019.

68. Wu, H.F.; Dwight, D.W.; Huff, N.T. Composites Sci. Technol. 1997, *57*, 975.

69. Bismarck, A.; Egia-Ajuriagojeaskoa, E.; Springer, J.; Habel, W.R. J. Chim. Phys. 1999, *96*, 1269.

70. Gassan, J.; Bledzki, A.K.; Gutowski, V.S. Materialprüfung 1998, *40*, 93.

71. van de Velde, K.; Kiekens, P. In *Proceedings 2nd International Wood and Natural Fiber Composites Symposium*, Kassel, Germany, 1999; 7-1-7-12 pp.

72. de Meijer, M.; Haemers, S.; Cobben, W.; Militz, H. Langmuir 2000, *16*, 9352.

73. Dilsiz, N.; Wightman, J.P. Carbon 1999, 37, 1105.
74. Iroh, J.O.; Yuan, W. Polymer 1994, 37, 4197.
75. Bismarck, A.; Pfaffernoschke, M.; Springer, J. J. Appl. Polym. Sci. 1999, 71, 1175.
76. Saïhi, D.; El-Achari, A.; Ghenaim, A.; Cazé, C. Polym. Test 2002, 21, 615.
77. Tagawa, M.; Gotoh, K.; Yasukawa, A.; Ikuta, M. Colloid Polym. Sci. 1990, 268, 589.
78. Tate, M.L.; Kamath, Y.K.; Wesson, S.P.; Ruetsch, S.B. J. Colloid Interface Sci. 1996, 177, 579.
79. Karlsson, J.O.; Andersson, M.; Berntsson, P.; Chihani, T.; Gatenholm, P. Polymer 1998, 39, 3589.
80. van Hazendonk, J.M.; van der Putten, J.C.; Keurentjes, J.T.F.; Prins, A. Colloids Surf. A 1993, 81, 251.
81. Hüttinger, K.J.; Höhmann-Wien, S.; Krekel, G. Carbon 1991, 29, 1281.
82. Zielke, U.; Hüttinger, K.J.; Hoffman, W.P. Carbon 1996, 34, 1007.
83. Zielke, U.; Hüttinger, K.J.; Hoffman, W.P. Carbon 1996, 34, 999.
84. van Oss, C.J. Interfacial Forces in Aqueous Media. Marcel Dekker: New York, 1994.
85. Zielke, U.; Hüttinger, K.J.; Hoffman, W.P. Carbon 1996, 34, 1015.
86. Yaminsky, V.V.; Yaminskaya, K.B. Langmuir 1995, 11, 936.
87. Rhee, H.; Young, R.A.; Sarmadi, A.M. J. Text. Inst. 1993, 84, 394.
88. Okamura, Y.; Gotoh, K.; Kosaka, M.; Tagawa, M. J. Adhesion Sci. Technol. 1998, 12, 639.
89. Sauer, B.B.; Diapaolo, N.V. J. Colloid Interface Sci. 1991, 144, 527.
90. Sauer, B.B.; Kampert, W.G. J. Colloid Interface Sci. 1998, 199, 28.
91. Mäder, E.; Jacobasch, H.-J.; Grundke, K.; Gietzelt, T. Composites A 1996, 27, 907.
92. Morra, M.; Occiello, E.; Garbassi, F. Polymer 1993, 34, 736.
93. Grundke, K.; Uhlmann, P.; Gietzelt, T.; Redlich, B.; Jacobasch, H.-J. Colloids Surf. A 1996, 116, 93.
94. Sauer, B.B. J. Adhesion Sci. Technol. 1992, 6, 955.
95. Chang, H.W.; Smith, R.P.; Li, S.K.; Neumann, A.W. Polymer Sci. Technol. Ser. In Molecular Characterization of Composite Interfaces; Ishida, H. Kumar, G., Eds.; Plenum Press: New York, 1985; Vol. 27, 413.
96. Kern, T. LaborPraxis 1998, 22, 48.
97. Jacobasch, H.-J.; Grundke, K.; Augsburg, A.; Gietzelt, T.; Schneider, S. Progr. Colloid Polym. Sci. 1997, 105, 44.
98. Jenschke, W.; Michel S.; Arnold, D. "Anlage zur Bestimmung der Oberflächenspannung von Polymerschmelzen mittels eintauchender Faser nach dem modifizierten Wilhelmy-Prinzip" link in: http://www.ipfdd.de/frameset.html?./research/equip/equipment.html (accessed on 11.02.2003).
99. Giannotta, G.; Morra, M.; Occhiello, E.; Garbassi, F.; Nicolais, L.; D'Amore, A. J. Colloid Interface Sci. 1992, 148, 571.
100. Dahlbäck, L.M., Lundström T.S. Proceedings of ICCM-10, Whistler, B.C., 1995; III-293–III-300 pp.

101. Kissa, E. J. Text. Res. 1996, *66*, 660.
102. Hsieh, Y.-L.; Yu, B. Text. Res. J. 1992, *62*, 677.
103. Hsieh, Y.-L. Text. Res. J. 1994, *64*, 552.
104. Hsieh, Y.-L.; Thompson, J.; Miller, A. Text. Res. J. 1996, *66*, 456.
105. Labajos-Broncano, L.; González-Martin, M.L.; Bruque, J.M.; González-Garcia, C.M. J. Colloid Interface Sci. 2001, *233*, 356.
106. Siebold, A.; Nardin, M.; Schultz, J.; Walliser, A.; Oppliger, M. Colloids Surf. A 2000, *161*, 81.
107. Grundke, K.; Augsburg, A. J. Adhesion Sci. Technol. 2000, *14*, 765.
108. Etmanski, B.; Bledzki, A.K.; Fuhrmann, U. Kunststoffe 1994, *84*, 46.
109. Perwuelz, A.; Mondon, P.; Caze, C. Text. Res. J. 2000, *70*, 333.
110. Park, S.-J.; Jin, J.S. J. Polym. Sci. B, Polym. Phys. 2003, *41*, 55.
111. Grundke, K.; Boerner, M.; Jacobasch, H.-J. Colloids Surf. 1991, *58*, 47.
112. Grundke, K.; Bogumil, T.; Gietzelt, T.; Jacobasch, H.-J.; Kwok, D.Y.; Neumann, A.W. Progr. Colloid Polym. Sci. 1996, *101*, 58.
113. Tröger, J.; Lunkwitz, K.; Grundke, K.; Bürger, W. Colloids Surf. A 1998, *134*, 299.
114. Grundke, K.; Augsburg, A. J. Adhes. Sci. Technol. 2000, *14*, 765.
115. Giese, R.F.; Costanzo, P.M.; van Oss, C.J. Phys. Chem. Miner. 1991, *17*, 611.
116. Chibowski, E.; Holysz, L. Langmuir 1992, *8*, 710.
117. Washburn, E.W. Phys. Rev. 1921, *17*, 374.
118. Aranberri-Askargorta, I.; Lampke T.; Bismarck, A.; J. Colloid Interface Sci. 2003, *263*, 580.
119. Krüss Users Manual. K121 Contact Angle- and Adsorption Measuring System, Version 2.1, Part C, Krüss GmbH, 1996; 11 pp.
120. Fox, H.W.; Zisman, W.A. J. Colloid Sci. 1950, *5*, 514.
121. Chibowski, E.; Perea-Carpio, R. Adv. Colloid Interface Sci. 2002, *98*, 245.
122. Bubert, H.; Ai, X.; Haiber, S.; Heintze, M.; Brüser, V.; Pasch, E.; Brandl, W.; Marginean, G. Spectrochim. Acta, Part B: Atom. Spectrosc. 2002, *57*, 1601.
123. Ontiveros-Ortega, A.; Espinosa-Jiménez, M.; Chibowski, E.; González-Caballero, F. J. Colloid Interface Sci. 1998, *199*, 99.
124. Espinosa-Jiménez, M.; Giménez-Martin, E.; Ontiveros-Ortega, A. J. Colloid Interface Sci. 1998, *207*, 170.
125. Espinosa-Jiménez, M.; Ontiveros-Ortega, A.; Giménez-Martin, E. J. Colloid Interface Sci. 1997, *185*, 390.
126. Chibowski, E.; Espinosa-Jiménez, M.; Ontiveros-Ortega, A.; Giménez-Martin, E. Langmuir 1998, *14*, 5237.
127. Chibowski, E.; Ontiveros-Ortega, A.; Espinosa-Jiménez, M.; Perea-Carpio, R.; Holysz, L. J. Colloid Interface Sci. 2001, *235*, 283.
128. Chibowski, E.; González-Caballero, F. Langmuir 1993, *9*, 330.
129. Chibowski, E.; Bolivar, M.; González-Caballero, F. J. Colloid Interface Sci. 1992, *154*, 400.
130. Holysz, L.; Chibowski, E. Langmuir 1992, *8*, 717.
131. van Oss, C.J. Colloids Surf. A 1993, *78*, 1.
132. van Remoortere, P.; Joos, P. J. Colloid Interface Sci. 1991, *141*, 348.

133. Lee, L.-H. J. Adhesion 1998, *67*, 1.
134. Marmur, A. Adv. Colloid Interface Sci. 1992, *39*, 13.
135. Labajos-Broncano, L.; González-Martin, L.; Jańczuk, B.; Bruque, J.M.; González-Garcia, C.M. J. Colloid Interface Sci. 1999, *211*, 175.
136. Chibowski, E.; Holysz, L. J. Adhesion Sci. Technol. 1997, *11*, 1289.
137. Jones, W.C.; Porter, M.C. J. Colloid Interface Sci. 1967, *24*, 1.
138. Adam, N.K.; Jessop, G. J. Chem. Soc. 1865.
139. Grindstaff, T.H. Text. Res. J. 1969, *39*, 958.
140. Li, D.; Neumann, A.W. Surfactant Science Series. In *Applied Surface Thermodynamics*; Neumann, A.W. Spelt, J.K., Eds.; Marcel Dekker: New York, 1996; Vol. 63, 557 pp.
141. Li, S.K.; Smith, R.P.; Neumann, A.W. J. Adhesion 1984, *17*, 105.

9
Bubble Nucleation and Detachment

STEVEN D. LUBETKIN   Eli Lilly & Company, Greenfield, Indiana, U.S.A.

I. INTRODUCTION

Bubbles are a rather spectacular embodiment of the forces produced by surface tension. Their spherical shape is a testimony to the isotropic nature of the gas/liquid interfacial tension. Of all the forces involved in shaping and making bubbles, surface tension is preeminent, and this dominant position is reinforced by the appearance of the surface tension to the third power in the exponential in the rate expression for bubble nucleation as shown in Eq. (6). In this review, the author has chosen to emphasize a property of gas bubbles that has been noted in the past, but whose implications have not been fully worked out—that the surface tension is a function of the bubble size. This is not true for cavitation or boiling in unary liquids.

Bubbles play a bigger part in our lives than is often realized. An example is that most fundamental of processes: boiling. Boiling is nothing more than the formation of bubbles in a liquid where the vapor pressure is above ambient pressure. When the liquid is water, boiling is the way in which steam is formed, and because that in turn is the first step in most electricity production, the efficiency of the process is important. Gasoline separation from crude oil also depends on bubble generation (distillation), so two of the key sources of energy, the pillars of modern technological society, rest firmly on bubble generation. Bubbles play a crucial role in determining the efficiency of these processes, and so ultimately in the cost of energy. That is the most obvious example, but it by no means exhausts the list, because boiling or gas bubble evolution also occur in many other situations, almost if not equally important: Heat pumps, heat exchangers, refrigerators, and electrolysis cells are industrially important examples. The formation of bubbles is an essential step in the formation of many foamed plastics, and these materials play a key part in our lives—principally as insulation foams and structural foams. If we open the discussion, and consider processes where bubbles play a significant part, then a representative list becomes more extensive:

1. Efficiency of boilers (steam, refrigerators, heat exchangers).
2. "The bends" or bubble formation in the tissues of divers.
3. The eruption of volcanoes and geysers.
4. Steel making.
5. Bubble chambers for detection of subatomic particles.
6. How does transpiration occur in plants more than 10 m tall?
7. Ships screws and other cavitation effects, including spillways, valves, and other hydraulic equipment.
8. Electrochemical cells and gas evolution, and the sound of electrolysis.
9. The long-range attraction between hydrophobic surfaces.
10. Bubble jet printers.
11. Beer gushing and oil well gushing.
12. Cell damage in fermentation processes.
13. Ecology of the ocean.
14. Insulation, structural, and other foams.
15. Electrostatic (spark) ignition of volatile vapors.
16. Sonoluminescence.
17. Disaster at Lake Nyos and the release of subterranean gases.
18. Cloud formation, and ultimately the global pattern of rainfall.
19. Measurement of surface tension/contact angle.
20. Strength of liquids.
21. Champagne, beer, and carbonated drinks manufacture.
22. Flotation for ore benefication.
23. "Cracking" knuckles.
24. Theories of cosmology.
25. Mechanical failure in nuclear fuel rods.
26. Antibumping granules.
27. Holes in Swiss cheese, bread, and other food foams.

This review is primarily directed at bubbles of gas or vapor surrounded by a bulk liquid phase, and in what follows, the word "bubble" will be taken to have that meaning unless the context implies otherwise. Bubbles can be of three essentially different types, although only two of these are commonly met. The two common forms are: First, the soap bubble, which consists of an approximately spherical surfactant bilayer bounded on each side by a gas phase. The second is a volume of gas, also approximately spherical, surrounded by a liquid. The third, much less usual is the "negative bubble," which occasionally appears when a drop of liquid falls through the interface of a liquid in the presence of a surfactant species. The drop can acquire a "coat" of the gas phase through which it fell, and this thin gas coat is in effect a bubble, filled with liquid. All three types can be brought into a single definition, if instead of specifying a liquid or gas phase, we can accept as a definition: "An approximately spherical or part spherical gas-filled cavity partially or fully surrounded by a fluid phase."

This definition could also encompass various other possibilities, some of which will be addressed here in more detail than others. These include the fact that bubbles in water greater than a certain size (a radius $r > ~1$ mm) will not be spherical, and as the size gets bigger, so too does the deviation from sphericity. Also included in this definition are bubbles attached to an interface, be it liquid or solid or indeed that of another bubble. In each case, the presence of a second interface will generally cause distortion from the roughly spherical shape, or at least truncate the sphere.

The size of the bubble is significant at the small end of the range as well as at the large. There is a theoretical absolute lower limit on the size of bubbles. For a given set of thermodynamic conditions, the critical radius, r^*, is the smallest size of bubble that is in an unstable equilibrium with the supersaturated solution. Bubbles of smaller size than this radius will spontaneously dissolve or collapse. In this context, smaller means even one molecule smaller. The gain of one molecule increases the radius past the critical value, and the larger bubble is now more likely to grow than collapse. The loss of one molecule from the critical bubble leaves the radius in the subcritical range where the Laplace pressure is greater than the vapor pressure in the bubble, which thus tends to collapse. At the critical size, the probability of collapse or growth is approximately equal. It is a corollary of the fact that there is an excess pressure in the bubble, that any gas- (or vapor-) filled bubble cannot indefinitely persist unless the solution surrounding it is supersaturated. Thus all bubbles in a saturated solution are unstable. Another way to look at this is that $r^* = \infty$ for a saturated solution, so that *all* bubbles are below the critical size in a saturated solution. Yet another way to view this is that the presence of a single bubble is sufficient to supersaturate an otherwise saturated solution. These may be important considerations when the role of bubbles [1] in the long-range interaction between hydrophobic surfaces is discussed.

II. HOMOGENEOUS NUCLEATION

There are many treatments of nucleation processes in the literature, although those addressing bubble nucleation are far less common than those dealing with crystal or droplet nucleation. The so-called classical nucleation theories (which can trace their lineage directly back to Volmer and Weber [92] and Becker and Döring [21]) are conceptually simpler than the more modern theories. They have also been outstandingly successful at predicting some nucleation phenomena. For these reasons, the present review only encompasses this type of theory. There are a number of reviews available on bubble nucleation, and the treatment here draws on those in Refs. 2–8, 26 and Hirth (1969).

For a phase change to occur in a homogeneous system, a necessary condition is that the system must be unstable, in the sense that the new (daughter) phase must have a lower chemical potential μ than the old (mother) phase, μ'.

Referring to Fig. 1, the full line labeled "binodal" represents the phase boundary, and is the vapor-pressure vs. temperature curve. In raising the temperature, so passing from B to A, the system starts at B with a relatively stable liquid phase, and a relatively unstable vapor phase. As it reaches the binodal, the system is at equilibrium, with equality of chemical potential for

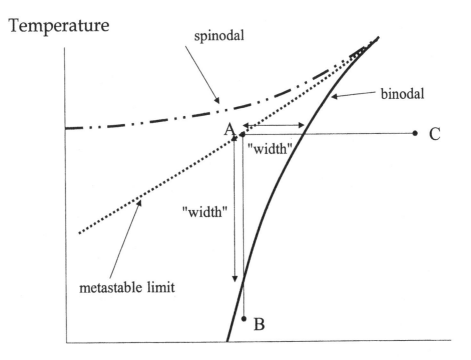

FIG. 1 At point "A," bubbles begin to appear. Two pathways to point "A" are shown. BA represents raising the temperature at constant pressure, whereas CA causes boiling by reducing the pressure at a constant temperature. The portions of the line BA above the full curve, or of CA to the left of the full curve until they reach "A" are sometimes referred to as the "widths" of the metastable zone. Ostwald's metastable limit is the kinetic limit of stability, and is shown here as the dotted line. The spinodal represent the thermodynamic limit of instability.

each phase: There is no drive to change state. As soon as it moves above the full line, toward A, the vapor phase is more stable than the liquid, and bubbles may form. The probability of this happening is a strong function of the chemical potential difference, $\Delta\mu$ ($\Delta\mu = \mu - \mu'$) as will be discussed below. Displaced from the full line by some distance, and crudely parallel to it (Ref. 70), is the dotted curve on which the point A lies, representing the "metastable limit." This line has no absolute theoretical significance. It is an empirical limit set by experiment, and much beyond which (further toward the line labeled "spinodal") is a region that is usually experimentally inaccessible, because of the effectively instantaneous bubble nucleation that then takes place. In practice, increasing the drive by moving further above the point A, has little effect. At the spinodal line, there ceases to be any barrier to nucleation—the new phase will very rapidly appear; the limiting rate of bubble formation is set by transport of matter (diffusion) or of heat, or both, with these terms appearing in the preexponential of the nucleation equation. In this region, the large value of $\Delta\mu$ ensures that the exponential term is very close to 1. Under these conditions, the kinetics are determined by the preexponential alone. Strictly speaking, at the spinodal, nucleation ceases, and spontaneous phase change takes place—it is impossible for the super-heated state to persist.

The further above or to the left of the binodal curve one goes, thus increasing $\Delta\mu$, the greater the rate at which the phase change will occur. At the binodal, the rate is zero, but as regions further above or to the left of the coexistence curve are reached, this rate rises to a high value as the departure from equilibrium increases. The magnitude of the departure is measured by the chemical potential difference $\Delta\mu$, between the liquid and vapor states under the prevailing conditions of temperature, T and pressure, P.

Of course, the "width" of the metastable zone is not a precisely defined quantity. The concept of having "no nucleation" is reasonably clear (to become completely clear, a time for which the observer is willing to wait to see a nucleation event becomes important as noted by Fisher [9]). In contrast, what constitutes an observable or worse yet, a "rapid" rate of nucleation is open to discussion. Conventionally, an observable rate of homogeneous nucleation (taking place in the bulk of the mother phase, well away from the influence of any surface) is taken to be 1 nucleation event in 1 cubic centimeter per second ($J = 1 \text{ cm}^{-3} \text{ sec}^{-1}$, which corresponds to $10^6 \text{ m}^{-3} \text{ sec}^{-1}$). A rapid rate would then perhaps be 3–4 orders of magnitude greater. For heterogeneous nucleation (taking place at or on a surface), the rate is defined in terms of events per unit area per unit time, and J has units of $\text{cm}^{-2} \text{ sec}^{-1}$; for this case, defining the area is not straightforward because usually the surface on which nucleation takes place is itself heterogeneous, chemically or topographically or both, and the nucleation event thus takes place at preferred or active sites. The presence of the surface has a number of potential effects, which are discussed in more detail below.

However, the key point is that the change from essentially zero rate to a very large rate occurs sharply, within a very small range of $\Delta\mu$. This is illustrated in Fig. 2.

The sharpness experimentally appears as a limiting superheat before which nothing happens, but beyond which very rapid boiling takes place. Calculating this rate is the province of nucleation theory, and the so-called "classical" theory is outlined below.

A.   The Rate of Homogeneous Bubble Nucleation

The free energy required for the formation of a bubble in a single component system is a function of the bubble size. To see why this is so, it is sufficient to note that the surface area of the bubble is proportional to r^2, and thus for each unit of interfacial area created by the new bubble, energy has to be expended in proportion to that surface area, and in proportion to the surface tension, γ:

$$\Delta G_{\text{area}} = 4\pi r^2 \gamma$$

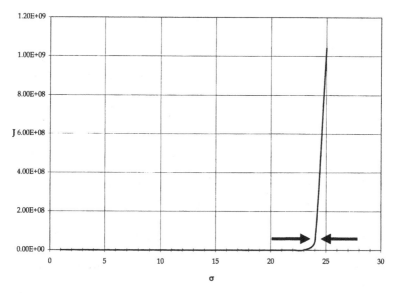

FIG. 2   The nucleation rate is indistinguishable from the X axis for most of the range. In a narrow interval of supersaturation, σ, it dramatically rises to very large values. It is this sudden change from essentially zero to a large rate that gives rise to the appearance of a kinetic "metastable limit." The metastable limit is crossed somewhere between the two arrows.

If the system is superheated and is thus is above the full line (the binodal) in Fig. 1, then the vapor phase is of lower chemical potential than the liquid. Consider a bubble exposed to an applied, external (hydrostatic) pressure P', and having an internal pressure, p. The decrease in free energy is proportional to the volume of the new phase produced, and is thus proportional to $-r^3$:

$$\Delta G_{volume} = -\frac{4}{3}\pi r^3 (p - P')$$

The total free energy change for the formation of this bubble is then given by:

$$\Delta G_{total} = \Delta G_{area} + \Delta G_{volume} = 4\pi r^2 \gamma - \frac{4}{3}\pi r^3 (p - P') \tag{1}$$

Regardless of the details of the functional form of the ΔG terms, it can be seen that the total free energy must go through a maximum as r increases, because the positive term ΔG_{area} grows as r^2, and thus dominates the total at low r, while the negative term ΔG_{volume} grows as r^3, and thus dominates as r gets larger. The maximum occurs at some characteristic value of the size, r^*, which strongly depends on the distance ($\Delta\mu$) the system has moved to the left or above the equilibrium full curve. The position of the maximum can be evaluated as shown below. The graph of ΔG_{total} as a function of the bubble size is shown in Fig. 3.

The maximum in the curve is identified as the height of the kinetic barrier, ΔG^*_{total}, which has to be surmounted for the phase change to take place. The free energy increase on the path to the summit is in effect the "activation energy" for the "reaction" leading to the appearance of the new bubble. By analogy with the conventional Arrhenius expression for the rate of a reaction, we might write for J, the rate of nucleation of bubbles:

$$J = C\exp\left[\frac{-\Delta G^*_{total}}{kT}\right] \tag{2}$$

The formal similarity to the conventional Arrhenius equation is deceptive; in the Arrhenius equation, E_a is essentially a constant, or at worst a weak function of the chief experimental variable, T:

$$R = A\exp\left[\frac{-E_a}{kT}\right]$$

but the situation with ΔG^*_{total} is different. It depends very strongly on the experimental variable, $(p - P')$, as we will see below. The preexponential, C in common with the A in the Arrhenius rate expression, represents an encounter frequency. The details will be discussed below, but for the present discussion, it can be treated as approximately constant.

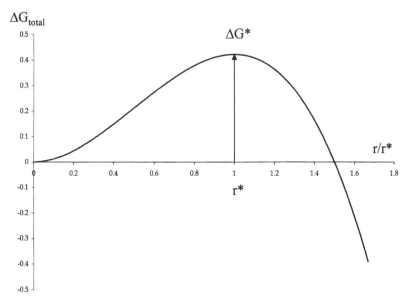

FIG. 3 In this diagram, the Y axis represents ΔG_{total}, and the X axis is the radius of the bubble, in units of r^*, the critical bubble size. The height of the curve at the point where $x = 1$, or $r = r^*$ is the height of the kinetic barrier the system must overcome to undergo the phase change, ΔG^*. The curve crosses the X axis at the point $r = 1.5r^*$, and this is where the surface term exactly equals the volume term, and $\Delta G_{total} = 0$.

The position of the maximum in the free energy curve can be obtained by differentiating expression (1) with respect to the radius

$$\frac{\partial \Delta G_{total}}{\partial r} = 0 = 8\pi r \gamma - \frac{12}{3}\pi r^2(p - P')$$

or:

$$\frac{\partial \Delta G_{total}}{\partial r} = 0 = 2\gamma - r(p - P')$$

therefore

$$r^* = \frac{2\gamma}{(p^* - P')} \tag{3}$$

$$\Delta G^*_{total} = 4\pi r^2 \gamma \left(1 - \frac{2}{3}\frac{r}{r^*}\right)$$

when $r = r^*$: $\Delta G^*_{total} = 4\pi r^{*2}\gamma[1 - (2/3)] = (4/3)\pi r^{*2}\gamma$

$$J = C\exp\left[\frac{-4\pi r^{*2}\gamma}{3kT}\right] \tag{4}$$

Now

$$r^* \approx \frac{2\gamma}{\sigma P'} \quad \text{(Fig. 4)} \tag{5}$$

Because

$$r^* = \frac{2\gamma}{(p^* - P')} \approx \frac{2\gamma}{p - P'} = \frac{2\gamma}{P'(\alpha - 1)} = \frac{2\gamma}{\sigma P'}$$

So

$$\Delta G^*_{\text{total}} = \frac{4}{3}\pi\gamma \frac{4\gamma^2}{(\sigma P')^2}$$

or

$$\Delta G^*_{\text{total}} = \frac{16\pi\gamma^3}{3(\sigma P')^2}$$

Because

$$J = C\exp\left[\frac{-\Delta G^*_{\text{total}}}{kT}\right]$$

Substituting for $\Delta G^*_{\text{total}}$:

$$J = C\exp\left[\frac{-16\pi\gamma^3}{3kT(\sigma P')^2}\right] \tag{6}$$

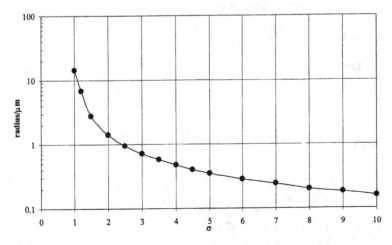

FIG. 4 Log plot of the radius of the critical bubble as a function of the supersaturation, σ. The size increases very rapidly as the supersaturation decreases.

Note that $(\sigma P')$ is an approximation for the full expression accounting for nonidealities, etc.

$$\left(\frac{\eta p_\infty}{v_V} + \frac{cP'}{c_\infty v_S} - P' \right)$$

This is more fully discussed later.

B. Bubble Nucleation by Raising the Temperature (Boiling)

The Clausius–Clapeyron equation gives the slope of the (P,T) curve at any point:

$$\frac{dP}{dT} = \frac{H_f}{T(v^V - v^L)} = \frac{H_f \rho^V \rho^L}{T(\rho^L - \rho^V)}$$

Given that $\rho^V = MP/RT$ (assuming ideal gas behavior), and that $\rho^V/\rho^L \ll 1$, integrating gives:

$$\ln\frac{P_\infty}{P^L} = \frac{H_f M}{R} \frac{T^L - T_{sat}}{T^L T_{sat}}$$

Using the Kelvin equation at the critical size, r^*:

$$T^L - T_{sat} = \frac{2\gamma T_{sat}}{\rho^V \Delta H_f r^*}$$

$$J = C' \exp\left[\frac{-4\pi r^{*2} \gamma}{3} \right]$$

Taking the conventional detectable rate of nucleation to be $J = 1 \text{ cm}^{-3} \text{ sec}^{-1}$, and substituting in the expression for J

$$\ln(1) = 0 = \ln(C') + \frac{-4\pi r^{*2} \gamma}{3}$$

$$\ln(C') = \frac{4\pi r^{*2} \gamma}{3}$$

$$\frac{3\ln(C')}{4\pi\gamma} = r^{*2}$$

$$r^* = \left[\frac{3\ln(C')}{4\pi\gamma} \right]^{\frac{1}{2}}$$

$$T^L - T_{sat} = \frac{2\gamma T_{sat}}{\rho^V \Delta H_f} \left[\frac{4\pi\gamma}{3\ln(C')} \right]^{\frac{1}{2}}$$

$$\frac{T^L - T_{sat}}{T_{sat}} = \frac{1}{\rho^V \Delta H_f} \left[\frac{16\pi\gamma^3}{3kT^L \ln(C')} \right]^{\frac{1}{2}} \tag{7}$$

The calculated superheat using Eq. (7) is compared with the measured values of Kenrick et al. [10] for various simple liquids in Table 1A. T_{sat}, T^L are the temperatures corresponding to the saturation pressure, and the actual temperature in the liquid, respectively. A typical early experiment of Kenrick et al. [10] involved immersing a U tube, open to the atmosphere, and partially filled with the freshly distilled liquid to be studied, in a thermostatic bath that had been set to the test temperature. The U tube was immersed and given sufficient time (5 sec) for the tube and liquid to thermally equilibrate. If boiling did not take place within that time, the thermostatic bath temperature was raised, and the experiment was repeated, until a temperature (T^L) was established at which boiling did take place within the allotted time. The general agreement of the theory with the data is remarkably good, given the simplicity of the experimental methods.

More sophisticated and more accurate methods for examining the boiling behavior of pure liquids have been developed. Subdividing the liquid phase into many small droplets reduces the probability of any given drop containing a nucleation catalyzing surface. As a procedure, it also has the great advantage of removing the surface of the container as a potential site for heteronucleation. Trefethen [11] appears to have been the first to use this method, although Turnbull [12] had earlier used the same technique to look at nuc-

TABLE 1A Calculated and Early Experimentally Measured Values of the Superheat Required for Boiling

Liquid	Experimental boiling temperature (K)	Calculated from Eq. (7)	Theory/measured ratio
Water	543	539	0.99
Methanol	453	435	0.90
Ethanol	474	444	0.85
Diethyl ether	416	400	0.89
Benzene	480	476	0.98
Chlorobenzene	523	534	1.04
n-Pentane	419	419.5	1.00
n-Hexane	455	453.3	0.99
n-Heptane	484	485.8	1.01
i-Pentane	411	410.1	0.99
Cyclopentane	453	451	0.99
Methylcyclopentane	473	470	0.99
Cyclohexane	489	492.3	1.02
Water	543	539	0.99
Mean ± SD			0.97 ± 0.05

TABLE 1B Calculated and More Recently Measured Values of the Superheat Required for Boiling

Substance	Boiling point (1 bar)	Measured superheat limit	Calculated superheat limit	Theory/measured ratio
Methane	−161.5		−107.5	
Ethane	−88.6	−4	−3.5	0.88
Flourethene	−72.2	16.9		
Sulfur dioxide	−10.	50.		
Propene	−47.7	52.4	50.3	0.96
Propane	−42.1	53.0	55.3	1.04
1,1-Difluoroethane	−24.7	70.4		
Propadiene	−34.5	73.		
Cyclopropane	−32.9	77.5		
Propyne	−23.2	83.6	88.2	1.06
2-Methylpropane	−11.8	87.8	87.7	1.00
Chloromethane	−24.2	93.0		
2-Methylpropene	−6.9	96.4	99.3	1.03
1-Butene	−6.3	97.8	100.2	1.02
Chloroethene	13.9	100.9		
1,3-Butadiene	−4.4	104.1		
Butane	−0.5	105.	105.2	1.00
Trans 2-butene	0.9	106.5		
Perfluoropentane	27.0	108.3	108.9	1.01
Cis 2-butene	3.7	112.2		
2,2-Dimethylpropane	9.5	113.4		
Ethyl chloride	12.3	126.		
Perfluorohexane	50.9	136.6	137.4	1.01
2-Methylbutane	27.9	139		
1-Pentene	30.0	144.	141.9	0.99
Diethyl ether	34.5	147.	145.	0.99
Pentane	36.0	147.8	148.3	1.00
Perfluoroheptane	70.9	161.6	161.2	1.00
Carbon disulfide	46.3	168.		
Chloroform	61.7	173.		
2,3-Dimethylbutane	58.0	173.2	175.9	1.02
Acetone	56.2	174.		
Cyclopentene	44.2	173.2		
Perfluorooctane	94.8	183.8	185.2	1.01
Cyclopentane	49.3	183.8	173.2	0.94
Hexane	68.7	184.	184.3	1.00
Methanol	65.0	186.0	186.5	1.00
Ethanol	78.5	189.5	191.8	1.01
1-Hexyne	71.3	192.		

TABLE 1B Continued

Substance	Boiling point (1 bar)	Measured superheat limit	Calculated superheat limit	Theory/measured ratio
Hexafluorobenzene	74.5	194.7	195.4	1.00
Methylcyclopentane	71.8	202.9		
Perfluorononane	114.5	205.3	205.7	1.00
Heptane	98.0	214.	214.5	1.00
2,2,4-Trimethylpentane	99.2	215.3	214.9	1.00
Cyclohexane	80.7	219.6	216.3	0.98
Perfluorodecane	133.0	223.9	223.1	1.00
Benzene	80.1	225.3		
1,3-Dimethylbenzene	139.1	235.		
1-Octene	121.3	237.1		
Methylcyclohexane	100.9	237.2	232.0	0.98
Octane	125.7	239.8	242.7	1.01
Chlorobenzene	132	250.		
Bromobenzene	156.	261.		
Aniline	184.1	262.		
Nonane	150.8	265.3	262.0	0.99
Decane	174.1	285.1	282.8	0.99
Cyclooctane	148.5	287.5		
Mean ± SD				0.997 ± 0.03

Source: Refs. 3 and 29.

leation of the solid phase from liquid mercury. Trefethen's methods were improved by Wakeshima and Takata [13]. Skripov and Sinitsyn [14] essentially used the same method to look at boiling in the presence or absence of ionizing radiation. Apfel [15] adapted the method by using a standing acoustic wave to halt the rise of the drop at a chosen level, and thus to more accurately control the degree of superheating. Finally, Skripov and Pavlov [16] used pulse heating methods, where very rapidly heated surfaces, typically platinum wires (with heating rates in excess of 10^6K sec^{-1}) result in explosive boiling. Using resistance thermometry allows the temperature at which boiling takes place to be accurately defined.

C. Bubble Nucleation by Lowering the Pressure (Cavitation)

Essentially the same theory as above can be applied with no modification to the case of cavitation. The experimental variable is now the applied (hydro-

static) pressure, not the temperature. Therefore a more convenient form of
Eq. (7) is sought, expressed in terms of $(p^* - P)$. With very minor modification in the preexponential factor, [C in Eq. (6) is changed to C''] the same
equation can be used:

$$J = C'' \exp\left[\frac{-16\pi\gamma^3}{3kT(\sigma P')^2}\right] \tag{8}$$

Consulting Fig. 1, it can be seen that in traversing the path CA, the destination
(A) is the same as before, while the route has changed. It is of course well
known that reducing the applied pressure (e.g., atmospheric pressure) results
in the lowering of the boiling point of water. At sufficient low applied pressure, the water would boil at room temperature, thus reaching the line
representing the kinetic superheat limit, though not now at the point A, but
further toward the Y axis. Thus boiling and cavitation are part of a continuum
of behavior, and it is not surprising that they share closely similar mathematical descriptions.

Experimentally, applying negative pressures to liquids to cause cavity formation (cavitation) is not simple. Reducing the pressure in the vapor space
above a liquid using a vacuum pump is ineffectual in causing cavitation in
low vapor pressure liquids. Means of applying substantial negative pressures
are needed. Early experiments, e.g., those of Meyer [17] in this area were
performed by filling and sealing glass tubes at high temperatures, and then
cooling them in a controlled fashion. The differential shrinkage rate on
cooling of the glass and the liquid resulted in negative pressure (tension) being applied. Knowing the temperature at which cavitation occurred, knowledge of the coefficients of expansion of the glass and the liquid allowed a
calculation of the tension. Other methods for putting liquids under known
tension include the use of metal bellows [18], centrifugal force applied to
tubes containing the experimental liquid [19], and acoustic methods (see, e.g.,
Ref. 20).

Using the centrifugal method, Briggs obtained the data in Table 2,
consisting of experimental fracture tensions for various liquids. These values
are compared with the theoretical calculation based on the full Eq. (10).

The predictions are generally in reasonable agreement with experiment,
although mercury appears to be an exception, and was not used to calculate
the ratio, average, or standard deviation.

The good agreement of theory and experiment in the case of boiling, as seen
by inspection of Tables 1A and 1B, contrasts with the somewhat worse
agreement for cavitation at room temperature (Table 2) and the generally very
poor agreement for gas bubble formation in liquids, the data for which are
given in Tables 6–8, and are discussed below. This may be a reflection of the

TABLE 2 Calculated and Theoretical Fracture Tensions for Various Liquids at 20°C

Liquid	Theoretical fracture tension/atmosphere	Measured fracture tension/atmosphere	Theory/measured ratio
Water	1380	270	5.11
Chloroform	318	290	1.10
Benzene	352	150	2.35
Acetic acid	325	288	1.13
Aniline	625	280	2.23
Carbon tetrachloride	315	275	1.15
Mercury	23100	425	54.35
Mean ± SD			2.18 ± 1.55

Source: Ref. 57.

fact that boiling is an effective means of reducing the dissolved gas content of liquids, while cavitation is not equally so. However, most cavitation experiments start with careful degassing and distillation (which of course involves boiling) of the test liquid. It may also be relevant that many cavitation experiments are focused on the first nucleation event, rather than on generating massive numbers of bubble nuclei. The effects of gas on bubble nucleation are discussed below. However, it is unlikely that dissolved gas accounts for the very poor concordance for mercury's experimental and theoretical fracture tensions.

D. A Closer Look at the Preexponential

Up to this point, the preexponential has been taken to be a constant, independent of the temperature and pressure. An empirical justification for this is that the nucleation rate depends on the exponential of the free energy of formation of the critical nucleus, and that this strong dependence effectively swamps any variation in the preexponential. An illustration of the relative importance of the two terms is shown in Fig. 5.

Here values typical of boiling have been used, and a change of 5% in the exponent gives the same order of magnitude change in J as does a 1000% change in the preexponential. In terms of the sudden onset of nucleation experimentally observed, the exponential is dominant. Generally speaking, it is safe to ignore the variability of C. This does not mean that it *must* be so; where the surface tension is very low, or very close to the critical point, or where the spinodal is closely approached, then the kinetics may be dominated by the preexponential.

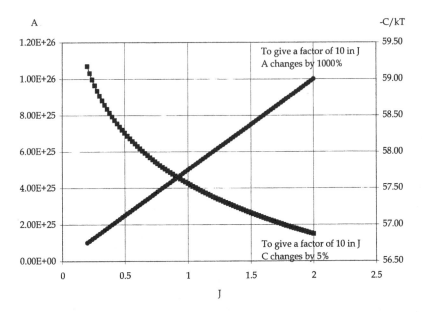

FIG. 5 Composite plot of the values of the exponential and preexponential terms needed to cause a change of 1000% (from 0.2 to 2) in the rate of nucleation, J, which is plotted along the X axis. The left-hand Y axis is the preexponential term, which changes from 1×10^{25} to 1×10^{26}—a change of 1000%. The right hand Y axis (the exponential term) changes from about 56.9 to 59.2—about 5%.

C can be broken down into two components: a concentration term relating to the total number, n, of possible sites at which a monomer (atom, molecule) can join the growing cluster, and a frequency factor, f, relating to the probability of a successful incorporation:

$$C = nf$$

The number of molecules per unit volume in the liquid state is denoted by n, and the frequency factor, f, can be estimated from the theory of absolute rates.

$$f = \left(\frac{kT}{h}\right) \exp\left[\frac{-E_D}{kT}\right]$$

Thus we obtain for the complete expression for the rate of nucleation:

$$J = n\left(\frac{kT}{h}\right) \exp\left(-\frac{E_D + \frac{4}{3}\pi r^{*2}\gamma}{kT}\right)$$

by rewriting this equation, we can emphasize the structure of the preexponential

$$J = n\left(\frac{kT}{h}\right)\exp\left(-\frac{E_D}{kT}\right)\exp\left(-\frac{4\pi r^{*2}\gamma}{3kT}\right) \tag{9}$$

The quantity $\exp(-E_D/kT)$, where E_D is the activation energy for diffusion through the liquid, emphasizes the importance of the diffusion in the liquid state. Volmer [2] derived a very similar rate expression, where the activation energy was related to the evaporation of a molecule rather than its diffusion in the liquid state:

$$J = n\left(\frac{6\gamma}{m(3-b)}\right)^{1/2}\exp\left(-\frac{H_v + \frac{4}{3}\pi r^{*2}\gamma}{kT}\right)$$

or:

$$J = n\left(\frac{6\gamma}{m(3-b)}\right)^{1/2}\exp\left(-\frac{H_v}{kT}\right)\exp\left(-\frac{4\pi r^{*2}\gamma}{3kT}\right) \tag{10}$$

here b is $(p^* - P)/p^*$. For the situation where $b = 3$, Eq. (10) does not hold. The discontinuity arises as a result of an approximation used in the derivation. The interested reader is referred to the original reference for details.

E. Zeldovich Factor

The theories represented by the Eqs. (9) and (10) are based on the assumption that critical size nuclei are built up by a sequence of additions of monomers, with each step in the process being in a quasi equilibrium with the step before (and after):

$$A + A \leftrightarrows A_2$$
$$A + A_2 \leftrightarrows A_3$$
$$A + A_3 \leftrightarrows A_4$$
$$\cdots\cdots\cdots\cdots\cdots$$
$$A + A_{i-1} \leftrightarrows A^*$$
$$A + A^* \rightarrow A_{i+1} \tag{11}$$

The critical nucleus A^* contains i^* monomer units. At this critical size, the capture of a single monomer unit causes the critical nucleus to become free growing, descending the curve to the right of the maximum in Fig. 3. To be able to treat this distribution as quasi steady state, a conceptual procedure has to be imagined where any nucleus that gets larger than the critical size is returned to the mother phase as monomer units. The distribution can then be approximated as being at a steady state. It was the avoidance of this approximation that prompted Becker and Döring [21] and Zeldovich [22] to solve the set of equations making allowance for the reduction below the steady state concentration due to both promotion above the critical size, and by the rate of the reverse reaction leading to the critical nucleus. The solution introduces a "nonequilibrium factor," Z, usually having an order of magnitude of about 10^{-2}, and given by:

$$Z = \left(\frac{\Delta G^*_{total}}{3\pi k T i^{*2}} \right)$$

A further correction makes allowance for thermal nonaccommodation. The condensation coefficient, β, is generally close to unity for bulk phases, but there are exceptions, as noted by Hirth and Pound [5]. For the case of monomer units impinging on an isolated subcritical nucleus entirely in the vapor phase, and consisting of relatively few molecules, disposing of the excess thermal energy becomes an issue. This accounts for a reduction in the value of β significantly below one. However, for the case of boiling, thermal accommodation is most unlikely to be an issue, bearing in mind that the mother phase is here the condensed phase, and all subcritical bubbles are expected to be fully thermally accommodated with their surrounding mother phase. Taking these factors into account, the full expression for the rate of nucleation becomes:

$$J = \beta Z n \left(\frac{6\gamma}{m(3-b)} \right)^{1/2} \exp\left(-\frac{H_\nu}{kT} \right) \exp\left(-\frac{4\pi r^{*2}\gamma}{3kT} \right) \tag{12}$$

F. Non-Steady-State Nucleation

Imagine a uniform liquid system at equilibrium. Suddenly, the conditions are changed so that now the system is superheated. The system does not instantaneously arrive at the steady state represented by Eq. (11). The equilibria take a finite amount of time to develop, and during this time, the nucleation rate is below the calculated steady state rate.

This delay or time lag, τ, can be very considerable. Dunning and Shipman [23] found a time lag of about 100 hr for concentrated sucrose solutions. For typical gaseous mother phases, τ is usually of the order of microseconds. For boiling liquids, τ is likely to be of the order of tens of milliseconds, but this depends on the viscosity. For foam formation in molten polymers (this is the basis of the manufacture of insulation foams), these time lags can become comparable with or exceed the processing time, and hence may become the dominant kinetic step in bubble nucleation.

Kantrowitz [24] showed that the rate of nucleation as a function of time, $J(t)$ was related to the steady state rate, J_0 by:

$$J(t) = J_0 \exp\left(-\frac{\tau}{t}\right)$$

with

$$\tau = \frac{i^{*2}}{g}$$

Here g is the rate of growth of the critical nucleus, of i^* monomer units. Typical plots are shown in Fig. 6.

G. Two Component (Gas/Liquid) Systems

The discussion so far has centered on the behavior of pure liquids when the liquid phase was destabilized by changing the external conditions. In this section, in addition to the liquid, a second component is considered. One of the commonest and also most interesting cases is where the second component is a gas. A number of possibilities arise, including the case where the gas forms an essentially ideal but rather weak solution. In this case, the gas provides a

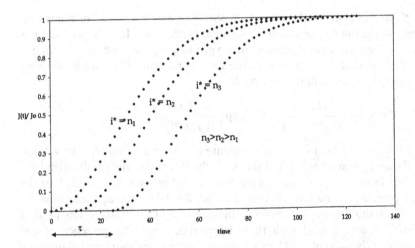

FIG. 6 Plots of the non-steady-state nucleation rate for increasing sizes of critical nucleus. The time lag (τ) before any appreciable nucleation takes place is indicated on the plot for the case of the n_3 nucleus. On occasion, this lag can be a substantial proportion of the overall delay before nucleation—for concentrated sucrose crystallization, it has been estimated at 100 hr. Even when the time lag has elapsed, there may be further significant delays as the rate of time-dependent nucleation, $J(t)$, gradually builds up to the steady state value, J_0.

contribution to the pressure in the critical bubble, thus allowing it to be smaller than would otherwise be the case. We have already seen that it is easier to produce a small critical nucleus than a large one, so the conclusion here is that the presence of the gas acts to reduce the thermodynamic drive needed to produce the phase change. If the gas/liquid solution is not ideal in either the vapor phase or in the liquid phase, then even at low levels, the effects of the dissolved gas can be enhanced. We will discuss the theory in more detail below.

The whole new field that opens up is where the gas is present in relatively large amounts, and now the nucleation and phase change represent not the ebullition of the liquid, which in this case mainly acts as a carrier solvent for the dissolved gas, but the appearance of bubbles of the previously dissolved gas itself, with a lesser contribution from the vapor pressure of the solvent. Of course, there is a continuum of behavior from pure boiling in a single component system consisting of the liquid, through the bubble evolution of essentially the pure gas from solution—and this continuity is reflected in the mathematical description of the process, with essentially one equation describing both extremes.

H. Dilute Gas Solution

The point of departure for this discussion is a dilute solution of a gas in a liquid, where the solution and vapor phases are ideal. In this case, the gas provides some additional pressure in the growing bubble, and this in turn has the effect of reducing the size of the critical nucleus. This can be seen by examining the denominator in Eq. (6)

$$J = C \exp\left[\frac{-16\pi\gamma^3}{3kT(p^* - P')^2}\right] \sim C \exp\left[\frac{-16\pi\gamma^3}{3kT(\sigma P')^2}\right]$$

$(\sigma P')$ is an approximation to the pressure in the critical bubble. We will see that the approximation has four parts: (1) that the vapor and the dissolved gas are ideal in both vapor phase in the bubble and in the solution, (2) that the vapor pressure in the critical nucleus is unaffected by the applied pressure, $p^* + P'$, (3) that the surface tension is independent of the pressure, and (4) that $p^* \sim \alpha P'$. We now deal with these approximations. The presence of gas dissolved in the solvent, and hence having some characteristic partial pressure in the nucleus, supports a part of the compression due to the surface tension. Therefore the simplest possible model [25] gives the nucleation rate in the presence of a dissolved gas as:

$$J = C'' \exp\left[\frac{-16\pi\gamma^3}{3kT(p^* + p_2 - P')^2}\right] \tag{13}$$

Here the pressure of the dissolved gas in the nucleus is p_2. This extra pressure results in the critical nucleus being smaller than without the gas being present, and this in turn implies that the nucleation is easier. For the single component system, we have seen that the radius of the critical nucleus is given by Eq. (3):

$$r^* = \frac{2\gamma}{(p^* - P')}$$

The addition of the gas pressure in the nucleus gives the expression:

$$r^* = \frac{2\gamma}{(p^* + p_2 - P')}$$

Because p_2 is always positive, r^* for the second case is always smaller than for the first. Our assumption of a weak solution equates to the statement that p^* and p_2 are of comparable magnitude.

Take now the case where allowance is made for the effect of applied pressure on the vapor pressure of the solvent. The calculation depends on the Kelvin equation, and expresses the relationship between the saturation vapor pressure, p_∞ and the actual vapor pressure under the applied hydrostatic pressure, P'. The factor η is given by Ward et al. [4]:

$$\eta = \exp\left[\frac{\Omega(P' - p_\infty)}{kT} - \frac{c'_2}{c'_1}\right]$$

$$p_2 = \eta p_\infty$$

This gives an expression for the radius of the critical nucleus, r^* in terms of the concentration of the dissolved gas, c, and the equilibrium concentration, c_∞:

$$r^* = \frac{2\gamma}{\left(\eta p_\infty + \dfrac{cP}{c_\infty} - P\right)} \tag{14}$$

The definition of the saturation ratio, α, is that $\alpha = c/c_\infty$ and because $k_Hp = c$, combining these two definitions and using the expression for the supersaturation, $\sigma = \alpha - 1$ allows for some simplification:

$$r^* = \frac{2\gamma}{(\eta p_\infty + \alpha P' - P')} = \frac{2\gamma}{(\eta p_\infty + \sigma P')}$$

Comparing with Eq. (5), it is immediately apparent that r^* is smaller than before, and that nucleation will therefore be easier. For relatively involatile solvents, with gases dissolved in them, η is close to 1, or somewhat less than 1; also, p_∞ is relatively small. It is on this basis that the approximation used in Eq. (6) is justified.

The nonideality of the gas mixture in the nucleus, and of the solution, can be accounted for by modifying Eq. (14):

$$r^* = \frac{2\gamma}{\left(\dfrac{\eta p_\infty}{v_V} + \dfrac{cP'}{c_\infty v_S} - P'\right)} \tag{15}$$

Here v_V and v_S are the activity coefficients in the vapor and in the solution, respectively. For a full derivation, the interested reader is referred to Ref. 4.

Finally, the surface tension is generally a function of the dissolved gas pressure, in accordance with an isotherm such as the Gibbs adsorption isotherm. This was pointed out by Hirth and Pound [5] and Hirth et al. [26]. It is a known, but somewhat obscure, fact that gases can act as surfactants (Refs. 27, 28, 79–87, 89–91). The more soluble gases appear to be more effective than the less soluble. The mechanism of action of gases as surfactants is not clear. The classical picture of a surfactant as a molecule with hydrophilic and hydrophobic parts is clearly inapplicable to most gases, and completely inappropriate for gaseous elements, whether atomic or molecular. Nonetheless, the positive adsorption of these species is a well-established experimental fact, as is the resulting lowering of the interfacial tension, as reported for example by Lubetkin and Akhtar [28]. Thus dissolved gases act to lower the surface tension, and the extent of lowering depends upon the adsorbed amount.

I. Higher Gas Concentrations

The reduction in surface tension is highly significant for the rate of nucleation of bubbles, because the surface tension appears to the third power in the exponent in the expression for the nucleation rate, Eq. (6). With dilute gas solutions, although this effect is important, it may not be dominant. However, as the gas concentration increases, so the importance of the surface tension term is increased.

Based on the information in Table 3, the size of the first coefficient, **b**, of the reduction in surface tension with pressure strongly depends on the nature of the gas, ranging from helium at one extreme, to n-C_4H_{10} at the other. The common atmospheric gases N_2 and O_2 have relatively modest coefficients, with the surface tension falling about 0.1 mN m^{-1} atm^{-1}, whereas for n-C_4H_{10}, the corresponding reduction is nearly 3 mN m^{-1} atm^{-1}. Similar data has been reported for ethene in cyclohexane by Lubetkin and Akhtar [28], where the coefficient **b** was about 1.7 mN m^{-1} atm^{-1}.

At low pressures, the surface tension decreases in a roughly linear fashion, so that to a reasonable approximation, the surface tension can be represented by the equation $\gamma = \gamma_0 + (d\gamma/dp)p$, and $(d\gamma/dp) = $ **b**, where **b** is the first coefficient shown in Table 3, and γ_0 is the surface tension against air at atmospheric pressure. For helium, the available data suggests that **b** is very close to zero, but probably a negative quantity. The small size of the effect for helium makes it difficult statistically to distinguish **b** from zero. For all other gases so far examined, **b** is a negative quantity, showing that the gases are positively adsorbed, and reduce the surface tension. At the very least, this reduction in surface tension will result in a significant reduction in the size of the critical nucleus, as seen from Eq. (3) and its variants. That equation may be written:

$$r^* = \frac{2\left(\gamma_0 + \frac{d\gamma}{dp}p^*\right)}{(p^* - P')} = \frac{2(\gamma_0 + bp^*)}{(p^* - P')} = \frac{2\gamma}{\sigma P'} + \frac{2b\alpha}{\sigma}$$

The coefficient **b** is negative, thus the second term represents a reduction in the size of the classical critical nucleus (which is given by the first term). Note that because the dissolved gas is the main contributor to the pressure in both subcritical and critical size bubbles, the necessary precondition for the lowering of the surface tension (the presence of a substantial pressure of surface active gas) is fulfilled. The assumption that the gas concentration is large is equivalent to the statement that the denominator in the nucleation equation, while retaining the same form as before, now represents the dominance of the dissolved gas in the critical nucleus. Neglecting the term

TABLE 3 Variation in Surface Tension of Water with Dissolved Gas Content, Expressed as $\gamma = \gamma_0 + bp + cp^2 + dp^3$, with p in Bar

Gas	b	c	d
He	0.0000	—	—
H_2	−0.0250	—	—
O_2	−0.0779	+0.000104	—
N_2	−0.0835	+0.000194	—
Ar	−0.0840	+0.000194	—
CO	−0.1041	+0.000239	—
CH_4	−0.1547	+0.000456	—
C_2H_4	−0.6353	+0.00316	—
C_2H_6	−0.4376	−0.00157	—
C_3H_8	−0.9681	−0.0589	—
N_2O	−0.6231	+0.00287	−0.000040
CO_2	−0.7789	+0.00543	−0.000042
$n\text{-}C_4H_{10}$	−2.335	−0.591	—

Source: Ref. 64.

in ηp_∞ in comparison with $\sigma P'$ becomes a better approximation as the proportion of gas increases.

At the critical size, the bubble is in (unstable) equilibrium, with the pressure in the critical bubble given by p^*. Therefore the reduction in the size of the critical nucleus could be very significant and approximately may be given by:

$$r^* \approx \frac{2(\gamma_0 + b\alpha P')}{\sigma P'} = \frac{2\gamma_0}{\sigma P'} + \frac{2\mathbf{b}(\sigma + 1)}{\sigma}$$

Below the critical size, the pressure in the incipient bubble is p, which is greater than p^* or $\sigma P'$. Thus below the critical size, the adsorption of the gas and the corresponding reduction in surface tension (and presumably, the increased growth rate to the supercritical size) could be even greater than for the case of the critical nucleus. The extent of the reduction in surface tension is dependent on the kinetics of adsorption as balanced by the rate of collapse or growth of the subcritical bubbles. These kinetic details are not known at present. It seems very likely that this reduction in surface tension is implicated in the very poor agreement between theory and experiment in the case of gas bubble nucleation. The $\mathbf{b}$ coefficient for He gas is very close to zero. If He is used in a bubble nucleation experiment, is it anticipated that the predictions of classical nucleation would be verified? Not entirely, because it is not only the adsorption of the gas reducing the surface tension that matters. The pressure of gas in the critical bubble will reduce its size, and will also affect the vapor pressure of the liquid, and corrections for this can be applied. These aspects are discussed in more detail below. Nonetheless, it is not surprising to see He at the top of the list of supersaturations experimentally measured (see Table 7), with a ratio *theory/measured* close to that measured for water cavitation, Table 2, and noticeably smaller than for all other gases. In other words, He is not acting as a surfactant, so that the nucleation process is similar to that found in unary bubble nucleation systems.

III. HETERONUCLEATION

Up to this point, it has been assumed that the nucleation, whether caused by cavitation, boiling, or dissolved gas evolution, was taking place in the bulk of the mother phase, without the influence of any heterosurface. Generally speaking, the presence of such a surface makes nucleation easier, so that unless special precautions are taken, nucleation will preferentially take place at surfaces. However, the number of sites upon which such heteronucleation may take place is limited, whereas the number of sites for homonucleation is essentially unlimited. Thus as the supersaturation increases, homogeneous nucleation may come to dominate the rate of phase change, although heteronucleation is simultaneously taking place. This does not contradict the very useful generalization that most nucleation is indeed heteronucleation. Even when careful precautions have been taken, it is nearly impossible to eliminate all possibility of hetero effects. When a surface is involved in the nucleation step, the texture of the surface itself may be relevant. With very few exceptions, solid surfaces have sites of differing adsorption energies or them, very often associated with chemical inhomogeneities. In addition, topographical features including steps, pits, and scratches are of significance. For this reason, our discussion starts with idealized smooth solid surfaces and liquid surfaces.

A. Heteronucleation at a Smooth Solid Surface

Different surfaces have different catalytic potencies for nucleation, and of course, the catalytic efficacy depends on the nature of the liquid, the gas, and the surface. Generally speaking, and considering only atomically smooth, clean solid surfaces, the greater the contact angle of the liquid on the solid, the better the surface is in catalyzing the bubble nucleation. This is the opposite of the case for the formation of liquid droplets from a supersaturated vapor, where a well-wetted solid surface can encourage condensation at or at least very close to the binodal, or put another way, at very low supersaturations. In pores of suitable size, condensation can take place below $\alpha \, (= p/p_0) < 1$. This is called capillary condensation. A counterpart would be bubbles with an overall negative curvature, and such bubbles may have longer lifetimes. This may well be relevant to the experimental observation of apparently kinetically stable bubbles attached to hydrophobic surfaces, and which are implicated in long-range attractive forces between such surfaces, as discussed by Christenson and Claesson [1].

Recall that the contact angle is conventionally measured through the liquid (condensed) phase. Fig. 7 shows the important conceptual point that the cap-shaped bubble is part spherical, and that ϕ is the fraction that the cap represents of the volume of the whole sphere. Taking the sphere to represent the volume of the critical nucleus under these conditions, if the nucleation is homogeneous (or if the contact angle is zero, which amounts to the same thing), then of course the critical nucleus is the whole sphere. As the surface becomes a better and better catalyst, or in other words, as the contact angle gets larger, the fraction, ϕ, represented by the cap, of the volume of the whole sphere shrinks. We have already seen that a decrease in size of the critical nucleus corresponds to an increase in the rate of nucleation. Thus the factor ϕ is a multiplier that appears in the expression for the rate of nucleation, in the exponential, as seen by inspection of Eq. (17).

Geometrically, it is rather easy to see that ϕ is the ratio of the volumes of the cap-shaped nucleus to that of the whole sphere (Fig. 8).

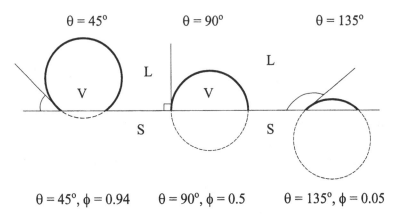

$$V_{bubble} = \phi V_{sphere}$$

$$\phi = [2 + 3\cos\theta - \cos^3\theta]/4$$

$\theta = 45°$ $\theta = 90°$ $\theta = 135°$

$\theta = 45°, \phi = 0.94$ $\theta = 90°, \phi = 0.5$ $\theta = 135°, \phi = 0.05$

FIG. 7 The relative volume ϕ of the spherical bubble cap and the total sphere changes as the contact angle changes, and becomes rapidly smaller as the contact angle increases above 90°. The work of formation of the nucleus correspondingly decreases, so that bubble nucleation becomes easier as the contact angle gets larger.

The volume of the whole sphere is given by:

$$V_{total} = \frac{4}{3}\pi r^3$$

The volume of the spherical cap ABDC is:

$$V_{cap} = \frac{\pi h^2}{3}(3r - h)$$

$$\frac{V_{cap}}{V_{total}} = \frac{h^2}{4r^3}(3r - h)$$

$$\sin(\theta - 90) = \frac{r}{r - h}$$

Eliminating h gives

$$\frac{V_{cap}}{V_{total}} = \frac{[2 + 3\cos\theta - \cos^3\theta]}{4}$$

$$\phi(\theta) = \frac{V_{cap}}{V_{total}} = \frac{[2 + 3\cos\theta - \cos^3\theta]}{4} \qquad (16)$$

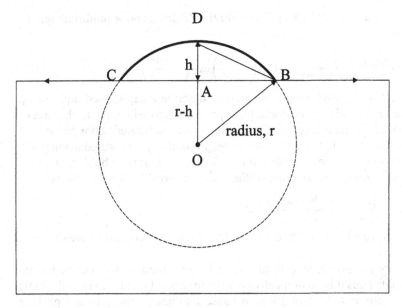

FIG. 8 The geometry of the spherical cap ACDB, showing the cap height, h. It is easy to show that $h = R(1 + \cos\theta)$, and from this to deduce that $V_{cap}/V_{total} = \phi = (2 + 3\cos\theta - \cos^3\theta)/4$. For liquid drops nucleating on a surface, $V_{cap}/V_{total} = \phi = (2 - 3\cos\theta + \cos^3\theta)/4$.

The simplified expression for the nucleation rate for a bubble on a flat, smooth solid surface, upon which the contact angle is θ, is given by:

$$J = C'''' \exp\left[\frac{-16\pi\gamma^3\phi(\theta)}{3kT(\sigma P')^2}\right] \tag{17}$$

For more details, see Ref. 29. The preexponential, C'''' is modified from the previous values [see, e.g., Eq. (8)] in two significant respects. First, the value of n (the number of molecular positions in the liquid per unit volume) is modified. For homogeneous nucleation, n is approximately $(1/\Omega)$, where Ω is the molecular volume. For heterogeneous nucleation, the number of possible sites is greatly reduced, and because a surface rather than a volume is involved, the dimensionality has to be reduced. A useful approximation is to take $n^{2/3}$, which is approximately the number of molecular positions per unit area of heterosurface. This change is reflected in the fact that the unit of J in this case is $cm^{-2}\,sec^{-1}$, rather than the $cm^{-3}\,sec^{-1}$ for Eq. (8). The second is the appearance of the factors $2s = [1 + \cos(\theta)]$, and ϕ, defined by Eq. (16), both of which involve the contact angle of the bubble on the nucleating surface. Allowance can also be made for the Zeldovich nonequilibrium factor, as before.

$$J = n^{2/3} s \left(\frac{6\gamma}{m\phi(3-b)} \right)^{1/2} \exp\left(-\frac{H_v}{kT} \right) \exp\left(-\frac{4\pi r^{*2}\gamma}{3kT} \right)$$

Because Fig. 7 could equally be used to describe a cap-shaped liquid drop nucleus on a solid surface, the same type of analysis will apply to the case of heterogeneous nucleation of a liquid from a supersaturated vapor phase. The only difference is that now the contact angle (as always, measured through the liquid) is equal to the complement ($= 180 - \theta$) of that for a bubble, and thus the expression for the fraction of the whole sphere has its signs reversed:

$$V_{cap} = \frac{[2 - 3\cos\theta + \cos^3\theta]}{4} V_{total}$$

The plots for $\phi(\theta)$ for both drops (circles) and bubbles (squares) are shown in Fig. 9.

Because, in principle, $\phi(\theta)$ can be arbitrarily small, so too, can be the exponential. Thus it is theoretically possible for a solid surface to be sufficiently catalytic for bubble nucleation to cause bubbles to form at any positive supersaturation. To sufficiently lower the exponential term to explain some of

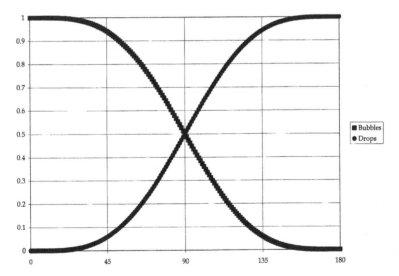

FIG. 9 The two functions ϕ (for bubbles, shown as squares, and for drops, shown as circles). The nucleation of drops gets easier as the liquid becomes better wetting (contact angle ~0), while the nucleation of bubbles gets easier as the liquid becomes increasingly poorly wetting. Contact angles are measured through the liquid phase (or more generally, through the denser phase).

the experimental observations (see Section V) the contact angle would have to be of the order of 98% of perfect nonwetting ($\theta \sim 176°$). This is not likely to be a widespread occurrence. In fact, perfect nonwetting is probably more of an idealization than is perfect wetting ($\theta = 0°$).

B. Heteronucleation at a Liquid/Liquid Interface

Apfel [30] and Jarvis [31] have analyzed the thermodynamics of the bubble formed at an interface between two immiscible liquids. The situation is shown in Fig. 10:

Following Blander [3] in defining m_A and m_B as

$$m_A = \cos(\pi - \theta) = \frac{\gamma_A^2 + \gamma_{AB}^2 - \gamma_B^2}{2\gamma_A \gamma_{AB}}$$

$$m_B = \cos(\pi - \phi) = \frac{\gamma_B^2 + \gamma_{AB}^2 - \gamma_A^2}{2\gamma_B \gamma_{AB}}$$

$$\Phi = \frac{1}{4\gamma_A^3} \left[\gamma_A^3 (2 - 3m_A + m_A^3) + \gamma_B^3 (2 - 3m_B + m_B^3) \right]$$

$$J = C'''' \exp\left[\frac{-16\pi\gamma^3 \Phi}{3kT(\sigma P')^2} \right] \tag{18}$$

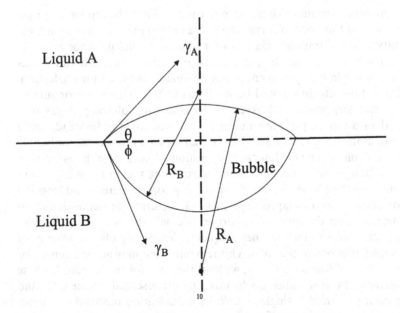

FIG. 10 A bubble at the interface between two immiscible liquids, showing the parameters needed to calculate the free energy of formation of the critical bubble.

In addition to the modifications to Eq. (16), the preexponential, C'''' includes a factor of ϕ, as it did for the case of the smooth solid surface. Equation (18) is clearly predicated on the assumption of thermodynamic equilibrium when calculating the work of formation of the critical nucleus. This is equally true for other treatments that involve the contact angle as a variable. Given that all nucleation takes place well away from equilibrium, it is important to keep in mind the possibility of misleading interpretations from this cause. Specifically, this stricture applies to the use of Eq. (18) for liquid drops that are superheated while rising through a second immiscible liquid.

Finally, Eq. (18) specifically applies to the case of mercury as the second liquid phase, and this may be important when attempting to identify the locus of the nucleation event for electrolytic bubbles produced at liquid mercury electrodes. Some generalizations can be made based on the relative values of the various interfacial tensions: (1) If $\gamma_B \geq \gamma_A + \gamma_{AB}$, then it requires less energy to form a bubble homogeneously in A than at the interface; thus the nucleation will be homogeneous. (2) If $\gamma_A \geq \gamma_B + \gamma_{AB}$, the bubbles are most stable in liquid B, and tend to detach from the interface. (3) If both conditions $\gamma_A < \gamma_B + \gamma_{AB}$ and $\gamma_B < \gamma_A + \gamma_{AB}$ hold, then bubbles are most stable at the interface, and the nucleation will be heterogeneous.

C. Heteronucleation at Rough Solid Surfaces

Rough surfaces introduce two new complications. First, the topography now directly impacts the work of formation of the critical nucleus and second, the possibility arises of trapping gas or vapor in sites of suitable shape and size. This is important because the presence of any supercritical amount of gas phase anywhere in the system effectively eliminates the need for nucleation. Included in this category would be small gas bubbles either free or surface resident, and any so-called Harvey nuclei consisting of trapped gas in a crevice, depression, or pit. Such nuclei are believed to be widespread, and a number of authors (Ref. 55) have claimed that they are responsible for bubble "nucleation" phenomena at low supersaturations. The case for this is far from proven, although there is good reason for expecting this to be an important contributor to the phase change at low driving forces. Here we address the question of how various shaped steps, pits, cracks, scratches, depressions, and prominences affect the nucleation barrier. We will also see that as bubbles detach from a nucleation site, they frequently leave a significant amount of gas or vapor attached to the solid. This remaining gas may act as a center for the production of further bubbles, without the need for nucleation. In these circumstances, these so-called nucleation sites are essentially "one-off," and having each produced a single bubble by a nucleation mechanism, subsequently act as repetitive sites for the evolution of further bubbles. They may become deactivated, in which case another nucleation event would be required to reactivate them. Often, sites will behave in this way, but under some circumstances (where the contact angle is either zero, or close to it), detachment may not leave any gas at the site, and this may be common. The question is considered in more detail elsewhere in this review.

D. Surface Topography

As examples, consider the conical projection shown on the left in Fig. 11a and the conical pit shown on the right, Fig. 11b. Other geometries have been examined in several other publications (e.g., Ref. 3 or 32) but here these will be taken as representative. Take first the conical pit of half-angle β illustrated in Fig. 11b.

The reversible work of formation, W, of the bubble in this geometry is evaluated as:

$$W = \gamma_{LV} A_{LV} + (\gamma_{SV} - \gamma_{SL}) A_{SV} - (P_V - P_L) V_V + \int_{P_V}^{P_S + P_V} V dP$$

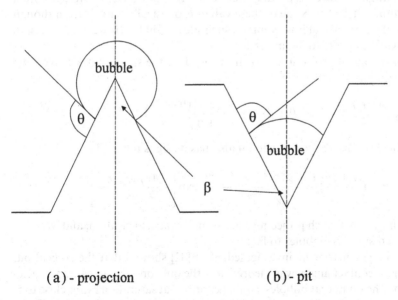

(a) - projection   (b) - pit

FIG. 11 A bubble forming at a conical projection (a, on the left) or in a conical cavity (b on the right). The half angle of the cone is β in each case.

Using Young's equation, $\gamma_{LV} \cos \theta = \gamma_{SV} - \gamma_{SL}$, and geometrically evaluating the areas, A_A and A_B, and volumes, V_A and V_B, together with the condition for equality of chemical potential at the critical size:

$$W = 4\pi r^2 \gamma \phi_{pit} - \frac{4}{3}\pi r^3 (P_V - P_L)\phi_{pit} : \text{ here } \gamma = \gamma_{LV}$$

$$\phi_{pit} = \frac{1}{4}\left(2 - 2\sin(\theta - \beta) + \frac{\cos\theta\cos^2(\theta - \beta)}{\sin\beta}\right)$$

At the critical size, $r = r^*$, the terms can be expanded in a Taylor series. Taking the leading two terms in the Taylor series leads to the full equation for the nucleation rate, J, in a conical pit of half angle, β:

$$J = n^{2/3}\left(\frac{1 - \sin(\theta - \beta)}{2}\right)\left(\frac{2\gamma}{\pi m \phi_{pit}}\right)^{1/2}\exp\left[-\frac{4\pi\gamma\phi_{pit}r^{*2}}{3kT}\right] \tag{19}$$

It will be seen that both preexponential and exponential terms are affected, but as noted earlier, changes in the preexponential are relatively unimportant. The main result is that the function ϕ_{pit} appears as a multiplier in the exponential, and that ϕ_{pit} in turn depends on both the contact angle, θ and the cone half angle, β. Wilt [32] showed that with a contact angle of 94°, a cone of half angle 4.7° would produce nucleation in a carbonated beverage with a saturation ratio of 5. None of these values is physically unrealistic, although one could argue about how common such sites might be. However, fizziness in Coca Cola appears to be universal.

For a conical projection, shown in Fig. 11a, the same analysis gives two new functions.

$$\phi_{cone} = \frac{1}{4}\left(2 + 2\sin(\theta + \beta) + \frac{\cos\theta\cos^2(\theta + \beta)}{\sin\beta}\right)$$

and the term in the preexponential also has its signs reversed:

$$J = n^{2/3}\left(\frac{1 + \sin(\theta + \beta)}{2}\right)\left(\frac{2\gamma}{\pi m \phi_{cone}}\right)^{1/2}\exp\left[-\frac{4\pi\gamma\phi_{cone}r^{*2}}{3kT}\right]$$

Wilt showed that such projections are stable toward nucleation, and were not preferred sites for bubbles to form.

Of the geometries he investigated, Wilt [32] showed that the conical pit, with high contact angle, was theoretically the only one able to catalyze nucleation to the point that bubbles would be formed at saturation ratios close to 5. This was important because this is typical of carbonated drinks, and the fact is that such drinks do indeed produce massive numbers of bubbles when the pressure is released (i.e., when the supersaturation is established). Simple theoretical calculations show that without catalytic assistance for homogeneous nucleation of CO_2 in water, the supersaturation required approaches 1400. Thus there was a very significant difference of more than 2 orders of magnitude between theory and experiment. The geometry of the pit is one aspect necessary for the Wilt theory to be successful. The other is the contact angle. Experiments have been undertaken in an attempt to test this theory.

E. Importance of the Contact Angle

Pyrex glass usually has a high density of silanol groups on its surface exposed to atmospheric (moist) air. These groups will react with chloromethyl silanes, thus replacing some of the silanol groups with methyl silane. These groups, have their -(CH$_3$) groups exposed to the solution, and thus render the surface hydrophobic [33]. The extent of the hydrophobicity can be controlled by the concentration of the chloromethyl silane solution exposed to the Pyrex glass. The parameter describing the hydrophobicity is the contact angle. Some typical data [34] are shown in Fig. 12.

Both receding and advancing contact angles can be adjusted in the region from about 0° to about 90°, and the advancing angles are generally about

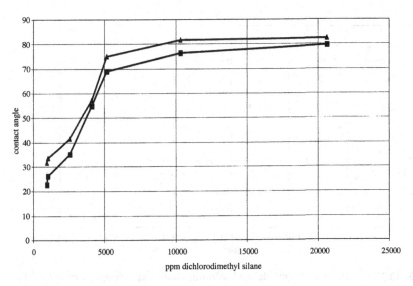

FIG. 12 The contact angle for a Pyrex glass surface treated with various concentrations of dichlorodimethyl silane. The advancing and receding angles are shown. The Pyrex surface is close to complete coverage at 1000 ppm.

between 5° and 10° greater than the receding angles. The measured angles decline with the passage of time, by about $0.15°$ hr^{-1}. This is due to the gradual hydrolysis of the Si–O–Si bond.

Nucleation rates of distilled water/CO_2 solutions were measured on two surfaces. First, a fully silanized surface, where the contact angle was close to 90°, and second, a carefully cleaned Pyrex glass surface, where the contact angle was close to zero.

The procedure was to equilibrate the solvent with the CO_2 gas at the chosen pressure overnight. The pressure vessel was opened, thus rapidly allowing the pressure to drop to ambient, whereupon the vessel was quickly closed. As bubbles formed and burst, the pressure rose in the sealed pressure vessel. The nucleation rate was measured as the rate of pressure rise, allowing for adiabatic cooling, and the data shown in Fig. 13 were plotted according to Eq. (6). The methods and apparatus have been fully described in Ref. 35. The plots should be straight lines if the theory embodied in Eq. (6) holds. Clearly, they are not straight lines. The hydrophobic nucleating heterosurface gives a more rapid nucleation rate than the hydrophilic surface at a given supersaturation. The hydrophobicity of silanized Pyrex is greater than that of clean Pyrex. Thus there is a qualitative agreement with expectations regarding the

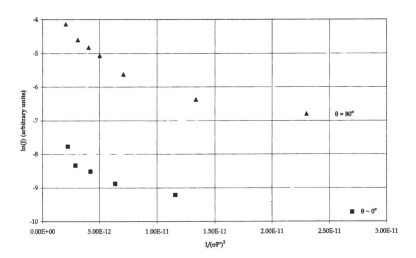

FIG. 13 The rate of CO_2 bubble nucleation measured as a function of $(\sigma P)^{-2}$ for two conditions. The upper curve (triangles) is for silanized Pyrex, with a contact angle of close to 80°. The lower curve (squares) is for clean Pyrex, with a contact angle close to 0°.

ease of nucleation on heterosurfaces. No significance attaches to the absolute values of the ordinates—the units are arbitrary.

F. The Geometry of the Nucleation Site

Attempts to exploit the geometry of the sites to control bubble nucleation have not been particularly successful. An early attempt was that of Griffith and Wallis [36]. The authors used a copper metal surface carefully finished with fine emery paper, and then deliberately pricked with a gramophone needle, which itself had been sharpened. The opening of the approximately conical cavities was about 25 μm radius. The experimental results showed that the presence of the pits did indeed reduce the superheat necessary for boiling at relatively low superheats. Unexpectedly, the pits did not reduce the superheat needed to cause nucleation to the extent predicted: experimentally 20°F vs. the prediction of 3°F. The radius of curvature of the tip of the gramophone needle was about 2.5 μm, and the opening was about 50 μm; the depth is not mentioned, but the cone half angle was about 9°. The authors attributed the larger than expected superheating results to the fact that the mean wall temperature did not coincide with the temperature in an active conical pit.

A different approach was adopted by Lubetkin [37], who made conical pits in a polymer, CR39. This was performed by first exposing the polymer to high energy alpha particles, and then etching the polymer along the tracks of those alpha particles. According to the conditions used, and the energy of the particles, the tracks can be designed to have a range of shapes and sizes. A computer simulation corresponding to the conditions used in the study is shown in Fig. 14. By controlling both the chemical conditions and the etching time (shown on the figure) pits with various profiles could be achieved, including various conical pits of known (calculated) diameters at their mouths, and known half angles, through curved sided approximately conical pits, with arbitrarily small half angles, through round-bottomed pits. All are shown in Fig. 14.

Once these pits had been formed, they were exposed to chlorotrimethyl silane, and contact angles as high as 118° were achieved [38]. The resulting surfaces with their pits were examined as nucleating surfaces for supersaturated CO_2 solutions. The predictions of the theory embodied in Eq. (19) are that this combination of very high contact angle, and small cone half angles should be an effective catalyst for bubble nucleation at a supersaturation of about 4. The results were unexpected: The rate of nucleation as measured by the pressure rise method (used above) was much lower than for glass or stainless steel. It was surmised that the high contact angle caused bubble detachment to be much more difficult than on glass or stainless steel, and

0.15 HRS
0.30 HRS
0.45 HRS
0.60 HRS
0.75 HRS
0.90 HRS
1.05 HRS
1.20 HRS
1.35 HRS

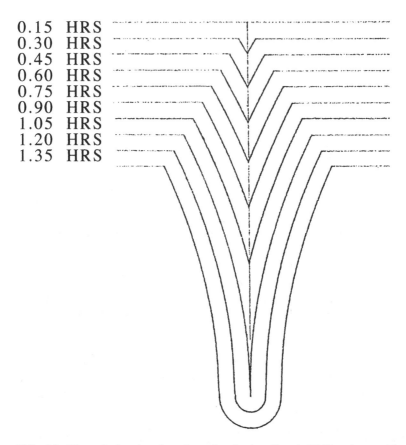

FIG. 14 The calculated etch trajectories for irradiated CR39 polymer. The simulation was performed using conditions designed to represent those actually used in the study (see text). It can be seen that by stopping the etch at various times (9–81 m are shown) a range of half angles can be produced. It is possible to produce essentially straight-sided conical pits, steeply curved pits or round-ended pits, but these geometries cannot be achieved independently of the half angle. By using a controlled, collimated radiation source, it is possible in principle to produce pits where the axis intersects the polymer surface at different angles, giving openings of different sizes and geometries (e.g., ellipses of various eccentricities), but this was not performed in the study referred to in the text.

that while the bubbles remained attached to the surface, no further nucleation was possible there. In other words, detachment had become the rate-limiting step in bubble evolution. This question is dealt with in more detail later.

Volanschi et al. [39] have performed elegant experiments with cavities of well-defined geometry. An inverted square pyramidal cavity of accurately known size (depth and half-angle) was made by etching silicon. Electrochemical bubbles were generated in this site, and were observed with an optical system giving a magnification of about X400. Overpotential and capacitance measurements were made, and this allowed studies of individual bubble events. The results led the authors to postulate what they call the "concentrator" effect: essentially, that the concentration of liberated electrolytic gas species is greater at the base of the cavity than elsewhere. Given that such sites are already favored for nucleation, this added effect could go some way to explaining the much lower than expected supersaturation needed to cause electrolytic bubble nucleation.

Marto and Sowersby [40] used glass cavities of known geometry, in a variety of aqueous solutions, and photographically studied the motion of the gas/liquid interface during bubble formation and release during boiling, from the cavities. The authors observed a liquid wave traveling down the walls of the cavity after bubble departure. If this wave reached the bottom of the cavity, the cavity often became deactivated as a nucleation site. On the other hand, if the liquid evaporated before reaching the base of the cavity, it remained active as a nucleation site.

When bubbles detach from a surface under gravity, a common state of affairs is that a proportion of the bubble remains attached to the surface, or in the cavity. The amount remaining is predictable from the physical properties of the system. Bearing in mind that from point of view of the bubble shape, the detachment of a vapor or gas bubble is the same physical problem as the detachment of a liquid drop in the same vapor or gas (see, e.g., Ref. 41), the treatment of detachment, at least under quasi equilibrium conditions is well understood. The relevant point is that the detachment of one bubble (however formed) very often leaves enough gas phase behind to avoid the need for a fresh nucleation event.

IV. PREEXISTING NUCLEI

There are three types of preexisting nuclei we need to consider: (1) free approximately spherical bubbles, which generally will be coated to a greater or lesser extent with surface active molecules, (2) similar bubbles, attached to a surface, and (3) portions of gas or vapor trapped in crevices, cracks, pits, and other geometrically nonspherical cavities. Such bubbles may be filled with air (or other gases) or with the vapor of the liquid (usually water), and we will see that the composition of the vapor phase can be important.

A. Free Bubbles

An important distinction needs to be made between free and attached bubbles. Ignoring gravity, free spherical bubbles have a single radius of curvature. Recall that the curvature of an interface is defined as:

$$\kappa = \left(\frac{1}{r_1} + \frac{1}{r_2} \right) \tag{20}$$

where r_1 and r_2 are the principal radii of curvature. For a sphere, $r_1 = r_2 = r$, and the curvature is then $2/r$. The Laplace equation can be written as

$$\Delta P = \kappa \gamma \tag{21}$$

so for a sphere, $\Delta P = 2\gamma/r$, as before. When a bubble is attached to a surface, not only can r_1 and r_2 be different, but the possibility of a negative curvature exists, and this may have important consequences. For this reason, we start with the simplest case, where the bubble is free, and assumed to be spherical.

Bubbles over about 1 μm radius in water will rise under gravity, eventually intersect the free liquid surface, and burst. The 1 μm limit is not completely arbitrary. The demarcation between kinetic stability and instability is defined by the Péclet number, Pe, which is the ratio of the gravitational force to the Brownian force:

$$Pe = \frac{\Delta \rho g V r}{kT}$$

$\Delta \rho$ is the density difference between the liquid and gas phases, g is the gravitational acceleration. The bubble has a volume V, a radius r, while k is the Boltzman constant, and T is the temperature. If $Pe \gg 1$, then gravity will overcome the randomizing tendency of Brownian collisions, and the bubble will rise. If $Pe \ll 1$, then the bubble may remain suspended for an indefinite period, assuming no other instabilities supervene. The dividing size is that for which $Pe \sim 1$, which for water at room temperature occurs close to $r = 1$ μm. A lower limit on size is set by the critical radius under the conditions imposed on the system. Between these two sizes, free bubbles might *in principle* survive.

What factors make survival unlikely? Clearly, the most serious is Ostwald ripening (see, e.g., Ref. 42). Gases have enough solubility in most liquids to make Ostwald ripening a significant mechanism for bubble size redistribution to occur on reasonable time scales. For the case of gas bubbles in water, this mechanism will be quite rapid. The original theory for crystalline solids was outlined by Lifshitz and Slyozov [43], and applies equally to gas bubbles. In an assembly of bubbles, where the assembly can consist of any number greater than one, and where there is more than one size represented, the larger bubble(s) will grow at the expense of the smaller. In principle, it is easy to see why the pressure inside the smaller bubble is greater than that in the bigger by inspection of Eq. (21). Given that solubility is directly proportional to pressure (Henry's law), the gas in the smaller bubble is more soluble than that in the larger, and the latter will grow at the expense of the former. This is true regardless of absolute size, although the kinetics will be faster for smaller bubbles. A similar situation holds for assemblies of more than two bubbles. Eventually, any assembly of bubbles will be reduced to a single bubble; if this single bubble has a radius greater than about 1 μm, the Péclet condition, $Pe >$ 1, ensures that it will rise and burst. This is simply another way of expressing the underlying truth that suspensions (of crystals, droplets, or bubbles) are thermodynamically unstable, and will phase separate given time. As a lower bound, the rate at which a single bubble in a saturated (but no supersaturated) solution will dissolve, has been evaluated by Epstein and Plesset [44]. Their expression for the time, t, required for disappearance from an initial size r_0 in a medium where the surface tension is γ,

$$t = \frac{r_0^2 k_H}{3DRT} \left[\frac{r_0(P_L - P_V)}{2\gamma} + 1 \right]$$

The diffusivity is D, and k_H is the Henry law constant. P_L and P_V are the pressure in the liquid and in the vapor, respectively. Typical calculations are shown in Table 4.

A subsidiary question is what difference does the composition of the vapor in the bubble make? The answer, by analogy with the situation with emulsions [45], is that this may be significant in terms of the kinetics of ripening. With either gaseous component present on its own, the Ostwald ripening will proceed as described by Lifshitz and Slyozov [43]. However, when two gases

TABLE 4 Time Required for Disappearance of a Bubble in a Saturated Solution

r_0	Dissolution time	Rise rate (U mm sec^{-1})	Reynolds number, Re
1 μm	8 msec	0.0022	4×10^{-6}
10 μm	6.6 sec	0.22	4×10^{-3}
100 μm	5,880 sec	22	~1
1 mm	5,800,000 sec	330	≫ 1

The rise rates quoted in the third column are all estimated from Stokes law, and thus will be inaccurate for sizes above about 100 μm. Since the condition $Re = \Delta\rho U d/\mu < 1$ for Stokes Law to be reliable, the rise rate for the 1 mm bubble was estimated from Ref. 77.

(or more) are mixed, then interesting behaviors including the development of bimodal distributions become possible. As far as the author is aware, these behaviors have not been reported nor does there appear to be any data on Ostwald ripening of bubbles in general.

When the discussion focuses on the detachment of a succession of bubbles from a source, the composition of the remaining gas in the reservoir will change with time. This point is taken up later. Finally, as was the case for nucleation, the presence of the gas increases the internal pressure sustainable in the bubble, and this may be important to the stability of trapped bubbles.

At low bubble densities in the bulk liquid, the probability of collisions and hence of coalescence is rather small, but the underlying theory of coalescence of emulsions, or aggregation of suspensions, would be expected to equally apply to bubbles. It seems likely that coatings of surfactant would exercise their main influence on stability by prevention of such coalescence. There is a second possibility and that is that the adsorbed surfactant layer becomes compressed as the (partially soluble) gas contents of the bubble diffuse out, as pointed out by Yount [46]. In doing so, it becomes less permeable to the gas, and at the same time, more rigid (see, e.g., Ref. 47). For such compression to become a reality, the pressure in the bubble would have to be less than would normally be accommodated in a bubble of that radius. For it to then function as a nucleation center, gas or vapor would have to diffuse into the interior, thus relieving the compression of the surfactant "skin." There would clearly be an activation energy barrier to the onward growth.

In summary, then, free bubble suspensions are thermodynamically unstable, and phase separation will occur, but the time scale for this depends on the detailed circumstances.

B. Bubbles Attached to a Smooth Surface

The situation changes when the bubble is attached to a smooth surface. Depending on the contact angle, we have already seen that in the absence of gravitational distortion, the bubble will adopt a part spherical shape. The pressure in the cap is still dictated by the curvature, not by the volume, and so remains essentially the same as for a full spherical bubble of the same radius. The significant change is that when the bubble grows (as it will in a supersaturated environment, allowing for the excess pressure in the bubble owing to the curvature), the buoyancy forces will grow, too. A point will come where the bubble detaches, as discussed in more detail below. Often when the bubble detaches, some gas remains at the previous site of attachment of the now rising bubble. Provided that this new spherical cap has a radius $R > r^*$, it will act as a site for a new bubble to grow, without the need for nucleation. In these circumstances, heterogeneous nucleation will be a "one-off" process, followed by bubble growth and detachment. At sufficiently low contact angles, or at sites with composite nature, this mechanism may not operate, and this is discussed below.

Note that attached, part spherical bubbles are subject to the same Ostwald ripening process as free bubbles, and will therefore disappear in a saturated or even a mildly supersaturated solution of the gas phase, or when the temperature is sufficiently lowered. It is a matter of experimental observation that bubbles that detach from surfaces are generally of a size such that $Pe > 1$, and bubbles released from surfaces generally rise rapidly through the solution and burst at the free liquid surface.

C. Sources of Gas Bubbles at a Rough Surface

Next we consider roughness on a scale that allows bubbles to have a three-dimensional surface of attachment, rather than the two-dimensional circle of attachment typical of a smooth surface. An appropriate length scale would be the bubble diameter. The new feature of such attachments is that negative curvatures become possible, and this has important consequences. We start by considering the question, where do these bubbles come from?

We will consider only three types of cavities: a crevice (an extended V shape, such as might be produced by a scratch) with a triangular cross section, a conical pit, and a reservoir with an arbitrary, but reentrant shape. These cases are schematically shown in Fig. 15.

For the crevice, this same diagram can also be used to illustrate a conical depression, a case that will be discussed in a later section.

Taking the crevice first, consider how an advancing thin sheet of water will enter the crevice. The situation is illustrated in Fig. 16. In A, the contact angle, θ, is less than or equal to 2β, the wedge angle: Strictly speaking, the relevant contact angle is the advancing angle, θ_a. This nomenclature derives from the advance of a liquid over a "dry" surface. In these circumstances, the liquid will penetrate to the apex of the wedge before meeting the far wall, thus filling the entire cavity. There will be no air trapped in the cavity by the advancing water. In B, on the other hand, where $\theta > 2\beta$, the advancing water will trap some air. As a generalization, one would expect that the greater θ is, in comparison to 2β, the larger the volume of air trapped in a given cavity. A previously dry surface containing both types of cavity, when exposed to water or other liquids, might be expected to show different extents of air trapping. Summarizing the predictions of Bankoff [48] gives Table 5.

The prediction is that steep sided cavities, with large contact angles (i.e., poor wetting) are likely to be the sources of bubble production. In well-wetted conditions, cavities are likely to be unimportant.

In practice, the advancing water will generally not be parallel to the solid surface nor will it be configured as a thin sheet. The model serves only to point out the possible importance of the size and shape of such cavities, and the

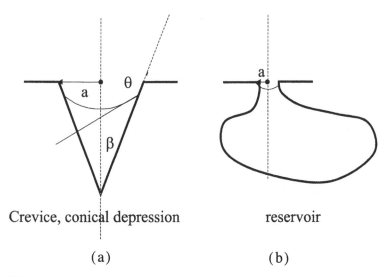

Crevice, conical depression   reservoir

(a)   (b)

FIG. 15 Three possible gas reservoirs: On the left (a), a crevice (extended perpendicular to the plane of the page) or conical pit. On the right (b), a small-mouthed reentrant cavity. In each case, the opening radius is a.

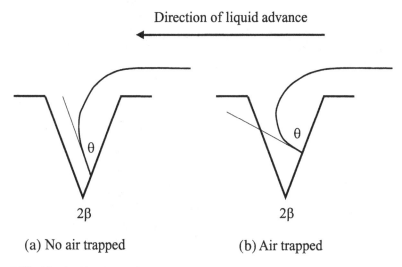

(a) No air trapped   (b) Air trapped

FIG. 16 A thin sheet of liquid advancing from right to left encounters a crevice. If $\theta < 2\beta$, then the crevice would fill completely, but if $\theta > 2\beta$, then air would be trapped.

TABLE 5 Correlation of Cavity Type, Contact Angle and Trapped Phase

Type of Cavity	Wetting Conditions	Phase Trapped
Steep	Poor	Gas only
Steep	Good	Liquid only
Shallow	Poor	Gas and liquid
Shallow	Good	Neither

possibility that some sites are better at trapping air pockets than others, and to emphasize the role of the contact angle, as discussed by Bankoff [48], Cole [49], Atchley and Prosperetti [50], and Carr [51]. Certain conclusions can also be drawn from this analysis about the behavior of such crevices in supersaturated and undersaturated conditions, and this question is considered below. This includes the possibility of successive cycles of heating and cooling causing such cavities to produce or eliminate gas/vapor pockets. In this connection, it should be noted that each successive bubble detaching from a reservoir takes with it a portion of the gas in the cavity, leaving an increased proportion of vapor. The composition of the contents of the cavity is thus a function of time.

Next, let us consider the case of the reservoir cavity illustrated in Fig. 15B. There is no obvious property relating to contact angle that can be used to predict the filling of such cavities. The expectation is that the filling would be governed more by hydrodynamics than thermodynamics, in the absence of cycling of temperature or pressure, or both. We will examine the effect of such cycling on the reservoir cavity below. The thermodynamic approach also hides the importance of kinetics in the filling, emptying, and bubble release processes. It is this aspect that has been examined by Atchley and Prosperetti.

Of course, both types of cavity can be filled with vapor by a nucleation event followed by growth. In the case of the conical cavity or wedge, it has already been established that such reentrant sites are preferred for bubble nucleation, and with suitable roughness inside the reservoir cavity, this too, would act as a preferred place for nucleation. The chief difference is that nucleation can only fill the cavities with vapor not with noncondensable gases such as N_2 or O_2.

D. The Effect of Supersaturation or Undersaturation on a Bubble Trapped in a Conical Pit

For the sake of simplicity, we will consider only the behavior of a conical pit containing a bubble. This case and others have been the subject of an extensive literature, which includes Bankoff [48], Griffith and Wallis [36], Apfel [52], Cole [49], Winterton [53], Trevena [54], and Atchley [50]. We will first look at a particularly simple case (and one that probably does not exist in practice) where the contact angle is held constant at 90°, as examined by Griffith and Wallis [36].

The curvature ($= 1/r$) of the interface is plotted against the volume of gas phase for each of the positions of the interface, r_1 though r_4 (Fig. 17). The case where $r = r^\#$ is of particular interest. This value of r may be called the "threshold" radius. Some authors use the term "critical radius" but this allows confusion with the critical nucleus. The radius is both the radius of the bubble at the point where the bubble shape is hemispherical (for the fixed contact angle of 90°) and equal to the radius of the opening of the conical pit. At this point, the curvature is at a maximum, and the pressure difference, ΔP, is also a maximum. If the internal pressure is at least this great, a bubble will form, grow, and eventually be released from the cavity. Should the pressure be lower, the meniscus will retreat back into the cavity. For boiling, to get to this threshold size, the superheat is the minimum needed to cause the growth of a bubble in this particular cavity. A set of cavities, each with the same opening, $r^\#$, would all become centers for bubble release at the same, sharply defined value of the superheat:

$$\Delta P = \frac{2\gamma}{r^\#}$$

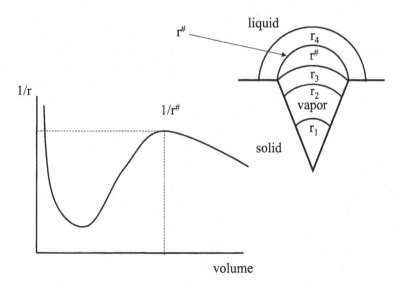

FIG. 17 The curvature is plotted vs. the volume enclosed for a constant contact angle of 90°. The local maximum is where the curvature is equal to the radius of the opening of the cavity.

Now using the Clausius Clapeyron equation,

$$\frac{\Delta P}{\Delta T} = \frac{\Delta H_V}{TV_V}$$

$$r^{\#} = \frac{2\gamma T_W V_V}{\Delta H_V (T_W - T_S)} \tag{22}$$

where $(T_W - T_S)$ is the superheat (the difference between the temperature of the heated wall, T_W, and the saturation temperature, T_S). Thus the smaller the opening of the cavity, the larger the required superheat to generate a bubble.

The following discussion is based on Apfel [52] and Trevena [54]. Denote the advancing contact angle by θ_a, and the receding contact angle by θ_r and consider the case of cavitation, where negative pressures are applied, and let the contact angle not be restricted to 90°. Referring to Fig. 18a, a conical cavity is shown where the applied pressure is greater than the pressure in the incipient bubble, while in the case B, the applied pressure is lower than that in the bubble. In the former case, where the meniscus is being forced into the

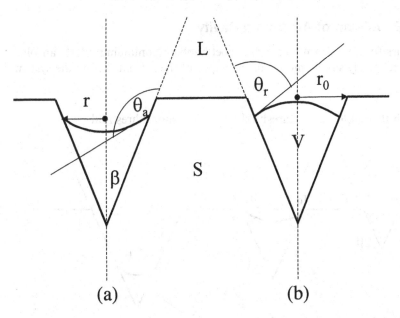

(a)   **(b)**

FIG. 18 Because the advancing contact angle, θ_a, is generally larger than the receding angle, θ_r, the geometry of the meniscus is different when the hydrostatic pressure is forcing the liquid into the cavity, or when the pressure is reduced so that the gas expands outward from the cavity.

cavity, the contact angle rises until it becomes equal to the advancing contact angle, θ_a, whereupon the meniscus advances down the cavity, maintaining that angle.

When the applied pressure is reduced, as in the latter case, the meniscus withdraws from the cavity. The relevant contact angle is now θ_r, the receding angle, as shown in Fig. 18b.

The concept of the threshold size is also applicable here. The pressure inside and outside the bubble are related through the Laplace equation:

$$\Delta P = (P - P_V) = \frac{2\gamma}{r}$$

In the case where the contact angle is exactly θ_a when the external pressure P is imposed, the cavity is a threshold cavity, and

$$r = r^{\#} = \frac{2\gamma(\cos(\theta_a - \beta))}{(P - P_V)}$$

The situation is shown in Fig. 19. The opening of the conical pit has a radius r_0. The conditions (contact angle, radius of contact) are such that the cavity is "subthreshold," (A), "threshold," (B) or "superthreshold."

E. Expulsion of Air from a Cavity

Cavities freshly filled with air can expel bubbles (containing mostly air plus some relatively small amount of the vapor of the solvent) before the system

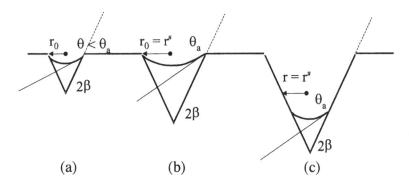

FIG. 19 The definition of a "threshold" cavity. In the literature, the term "critical" cavity is used. To avoid confusion with the critical bubble the term, "threshold" is used in the text.

becomes superheated. Cold tap water standing in a glass will often release bubbles. It is very likely that some of these bubbles are of this type, although clearly there is a contribution from the supersaturation of the air dissolved in the water in this case. For a freshly filled reservoir type of cavity, containing only air, one can compare the expansion ($V_1/T_1 = V_2/T_2$) due to heating from main water temperature up to room temperature T_2. In the example of tap water left to warm up to room temperature, T_1 would be about 10°C, and room temperature would be say 25°C. To calculate a maximum effect, we can use say 99°C for T_2, with the calculated departure volume of a bubble from the orifice. This would allow one to predict whether air bubbles would be emitted from such a cavity before any nucleation could possibly take place:

$$V_2/V_1 = T_2/T_1 = 372/273 = 1.36$$

In other words, a volume equal to 36% of the cavity volume is available to form a bubble large enough to become detached. Assume that a bubble has to be about 100 μm in diameter to detach (see below), this would then give an estimate of the equivalent radius of the cavity needed to expel a bubble of air from water (just) before boiling as 70 μm. The larger the volume of the cavity relative to the detachment volume of the bubble, and the narrower the opening, the lower the temperature required to expel the air bubble(s). Conversely, if the volume of the cavity is less than about three times greater than the volume of the detaching bubble, then an air bubble will not be released before the system becomes superheated with respect to the solvent. This calculation does not account for the change in solubility of the dissolved gas with temperature (supersaturation of the solute), which will have the effect of decreasing the size of the cavity needed.

F. Deactivation of Cavities

As each bubble detaches from a cavity, it carries away a portion of the gas/vapor mixture from the cavity. This mechanism is also at least in part responsible for the removal of the dissolved gases from solution by boiling. Thus the composition of the contents of the cavity changes with time asymptotically approaching pure vapor of the solvent. To a good approximation, the cavity will be filled only with vapor at long times. As the proportion of noncondensable gas in the cavity decreases, the total pressure for a given curvature of the cavity also decreases. This decrease affects the ability of the cavity to withstand compression. Not only will the cavity become more easily deactivated as the external pressure is raised, but as the temperature is reduced, the water may reenter the cavity. If it fully reenters the cavity, the site no longer acts as a Harvey nucleus, and a fresh nucleation event is necessary to reactivate it. The probability of such a nucleation event may be very much less than the probability of a bubble release from the active Harvey site. The deactivation process is the main reason for antibumping granules having to be freshly added to the liquid to be boiled; a cycle of boiling and cooling will quickly result in the antibumping granules becoming inactive.

G. How Important are Cavities (Harvey Nuclei)?

Jones et al. [55] are representative of a significant school of thought that postulates that under low supersaturations, there is no classical nucleation, and bubble release is only a matter of growth from a preexisting gas filled cavity. Depending on the radius of curvature of the interface in the cavity, no nucleation step is involved, and thus it is irrelevant what the nucleation theory predicts. This latter approach has been hinted at a number of times in the literature. The presence of any supercritical amount of any gas phase anywhere in contact with the liquid phase of the system is theoretically enough to remove the need for nucleation. There are undoubtedly cases where this is the main mechanism of bubble evolution. The regular emission of strings of bubbles from beers and champagnes, originating at a single site, are very likely to be of this type; see, e.g., Liger-Belair et al. [56].

Once again, it is important to separate the cases of a single component system (boiling, cavitation) from that of the supersaturated gas solution. For a reservoir cavity, illustrated in Fig. 15b, consider first the case of a supersaturated gas solution. As bubbles form and leave, the composition of the gas in the reservoir will approach the composition (gas + vapor) appropriate to steady state described by the curvature of the opening under the existing conditions.

Diffusion from the supersaturated bulk gas solution will ensure that the pressure in the reservoir climbs to a level great enough to cause a next bubble to detach. As time goes by, and the supersaturation of the solution decreases, the emission of bubbles becomes slower and slower, and finally stops. The process has been described by Jones et al. [55]. This is in contrast to the state of affairs with a single component system. Bubbles gradually detaching rid the cavity of all noncondensable gas, and the reservoir is then filled with the vapor of the boiling liquid. In the absence of liquid in the cavity, the only drive to cause a bubble to leave the cavity is heating of the vapor above ambient temperatures. It clearly would be expected that the continuous regular strings of bubble characteristic of beer, wine, and carbonated soft drinks will be less common in boiling or cavitation.

V. EXPERIMENTAL TESTS OF BUBBLE NUCLEATION THEORY

Finally, we collect here some representative experimental results, and the calculations used to interpret them.

A. Boiling

The early experiments of Kenrick et al. [10] involved submerging filled but open tubes of the study liquid in a thermostatic bath at a given temperature, waiting for thermal equilibration (5 sec) and noting if there was boiling in that time. For various reasons, it is probable that there were adventitious nuclei in these early experiments (Table 1A). However, comparing these data with the later experiments, summarized by Blander, as shown in Table 1B, there is a somewhat greater variability in the earlier results, but overall the agreement with the classical theory is very satisfactory. The later results, mostly obtained using the rising drop technique, largely avoid the possibility of Harvey nuclei, but suffer from the use of an equilibrium expression [Eq. (18)], for the evaluation of the work of formation of the critical nucleus. The agreement between theory and experiment is nonetheless very satisfactory.

Factors other than Harvey nuclei may affect the results. High energy radiation is one such influence. It has been shown that radiation does not greatly affect gas bubble nucleation, but clearly it does affect boiling, this being the basis of the bubble chamber used to detect subatomic events. The effect of radiation on the kinetics of boiling nucleation is well illustrated by the data in Fig. 20 from Ref. 6. The line AB represents the steep decrease in the time (τ) before the droplet boils as the temperature increases. The waiting time decreases from about 550 to 0.5 sec (three orders of magnitude) as the temperature is raised from A (145.6°C) to B (146.0°C). Measurable rates of bubble nucleation occur on the branch DC at much lower superheats (the temperature at D is low as 129.5°C), when the superheated liquid is exposed to radiation (Fig. 20).

B. Cavitation

A sample of data for the fracture tension of liquids is shown in Table 2. The data were obtained by Briggs [57] using a centrifugal method to apply tension to the samples. It can be seen that in all cases, the measured fracture tension is smaller than the prediction. There is a suspicion that heterogeneous sites at the walls of the containing tubes may have influenced the results. The ratio cavitation *theory/measured* is not as large as for gas bubble nucleation, but is larger than for the case of boiling (Table 3).

C. Gas Evolution

1. Electrolytic

The pressure in a small, newly formed bubble (at an electrode, for example) is by definition, rather high, and close to p^*. For gas molecules to diffuse into this bubble and make it grow, the solution surrounding this bubble must be highly supersaturated, and have σ approximately equal to the critical super-

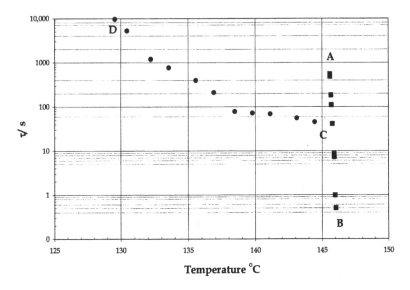

FIG. 20 A plot of experimental data from Skripov [6] for the homogeneous boiling of pentane. The part ACB is the normal superheat limit. As the temperature is increased, so the time required for explosive boiling of liquid drops rapidly decreases. In going from 145.6 to 146.0°C, the waiting time drops from 550 to about 0.5 sec. If the superheated liquid droplets are exposed to radiation, very much lower superheats are needed to cause bubble nucleation. Temperatures as low as 129.5°C can cause bubble formation.

saturation needed to cause nucleation. Conversely, a mature bubble near departure has only a small excess pressure, and thus needs a correspondingly small supersaturation to cause growth. Using the rate of growth of bubbles at or near electrodes as a means of measuring supersaturations in the vicinity is thus fraught with difficulties. It seems probable that this factor accounts for at least some of the radically different estimates for the supersaturations shown in Table 6A,B.

2. Nonelectrolytic Bubbles in Aqueous Solutions

Two main techniques other than electrolytic methods have been used to obtain data on the nucleation of gas bubbles in aqueous media. These are pressure release (as used by Lubetkin, and Hemmingsen and coworkers) and chemical (as used by Noyes, Bowers, and Hey among others). In the former technique, water is saturated with the study gas at an elevated pressure, then the pressure is released down to a target pressure (P') often atmospheric. Bubbles form, rise, and burst and the gas is released into the headspace. This

Table 6A Experimental and Theoretical Values for Electrolytic Supersaturation

Author(s)	$\sigma(c_e/c_b)$ or $(c_e - c_b)^a$	Notes
Sides [65]	~100	A maximum value
Westerheide and Westwater [66]	8–24	
Dapkus and Sides [67]	9–16	On clean, smooth mercury
Shibata [68]	7–70 (1,5... > 100 mA)	From transients
Glas and Westwater [69]	See Part B of this table	From growth

[a]The dimensionless version (c_e/c_b) is preferred in the present review because it directly relates to α, the saturation ratio, but some authors use the dimensional version $c_e - c_b$.

Table 6B

H_2	O_2	CO_2	Cl_2
1.54	1.36	1.08	1.018
19.9	15.4	1.64	1.324

release is studied by the pressure rise (if the vessel is immediately resealed) or by measuring the amount of gas evolved, or by acoustically counting bubbles [38]. In the chemical generation of bubbles, the supersaturation is evaluated by measuring the total gas released, usually sonicating the solution to ensure complete gas evolution (Table 7).

3. Nonelectrolytic Bubbles in Organic Liquids

N_2 gas bubbles were formed by pressure release from solutions in various simple organic liquids, by Kwak and Kim [58]. Similar experiments had been performed on nitrogen bubbles nucleating in ether by Wismer [59], and more recently, by Forest and Ward [60] (Table 8).

4. Organic Gas Bubbles from an Organic Liquid

Data from Lubetkin and Akhtar [28] are shown for the nucleation of bubbles of ethene from a supersaturated solution in cyclohexane, plotted according to Eq. (6). The Y axis is in arbitrary units. The two curves are for nucleation on a Pyrex glass surface and a stainless-steel surface. The corresponding advancing contact angles were about 5° and 25°, respectively (Fig. 21).

5. Example Calculation of Bubble Nucleation Rate

A calculation is presented in Table 9 based on the case of a supersaturated gas nucleating in an aqueous solution at 30°C and atmospheric pressure. The

TABLE 7 Supersaturation Needed to Cause Bubble Nucleation (Mostly Heterogeneous) in Water or Aqueous Solutions. Supersaturations Generated by Nonelectrochemical Means

Gas	Measured	Reference/theory[†]	Theory/measured ratio
He	230–320	(h)/1400	6–4
Ar	110	(i)/1400	13
N_2	100–160	(e)/1400	14–9
O_2	100–150	(i)/1400	14–9
O_2	95–127	(c)/1400	15–11
H_2	80–100	(g)/1400	17–14
CO	80	(f)/1400	18
CH_4	80	(i)/1400	18
N_2	20–30	(g)/1400	70–47
O_2	16–50	(c)/1400	87–28
NO	16	(g)/1400	88
CO_2	10–20	(i)/1400	140–70
CO_2	5.4–7.6	(b)/1400	260–184
Cl_2	5	1400	280
CO_2	4.62	(d)/1400	303
CO_2	1.3–2.1	(a)/1400	1077–667
Mean ± SD			89 ± 107

[†] Calculation for a gas dissolved in water—see Table 9.
(a) Jones et al. [55]; (b) Lubetkin [37,38]; (c) Bowers et al. [70]*; (d) Hey et al. [71]; (e) Hemmingsen [72]*; (f) Smith et al. [73]*; (g) Rubin et al. [74]*; (h) Hemmingsen [75]*; (i) Bowers et al. [76].
References marked with* are claimed to be homogeneous nucleation. The mean and standard deviation do not include the last entry for CO_2.

TABLE 8 Homogeneous N_2 Bubble Nucleation in Nonaqueous Liquids

Liquid	Theory	Measured	Theory/measured ratio
Methanol	250	90	3
Ethanol	252	80	3
Chloroform	330	70	5
Carbon tetrachloride	326	53	6
Benzene	361	35	10
n-Hexane	184	56	3
Mean ± SD			5.0 ± 2.76

Source: Ref. 58. The authors claim that the nucleation is homogeneous.

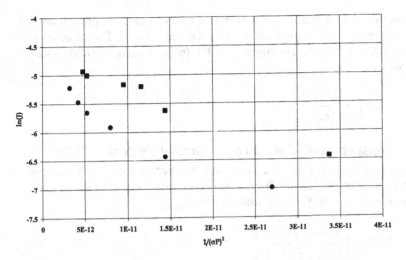

FIG. 21 Experimental data for the nucleation of ethene from cyclohexane, plotted according to Eq. (6). The upper data (squares) are for stainless steel, while the lower curve (circles) is for Pyrex glass. The contact angles are roughly 25° and 5°, respectively.

TABLE 9 Calculation of the Rate of Nucleation Using Eq. (6)

Variable	Unit	Magnitude	Comment
C	$m^{-3} sec^{-1}$	1×10^{35}	Arbitrary
γ	$N\ m^{-1}$	70×10^{-3}	For water/air[a]
T	K	303	
k	$J\ K^{-1}$	1.38×10^{-23}	
σ	Dimensionless	1400	The required σ for ~ 1 event $cm^{-3}\ sec^{-1}$. This value is used in the text as theory for gas bubble nucleation
J	$m^{-3}s^{-1}$	10^6	Equivalent to 1 event $cm^{-3}\ sec^{-1}$
P	Pa	101,325	1 atm
$16\pi\gamma^3$	$N^3\ m^{-3}$	1.72×10^{-2}	A
$3kT(\sigma P)^2$	$N^3\ m^{-3}$	2.52×10^{-4}	B
$-16\pi\gamma^3/3kT(\sigma P)^2$	Dimensionless	-68	$-A/B$
$\exp[-16\pi\gamma^3/3kT(\sigma P)^2]$	Dimensionless	2.17×10^{-30}	$\exp(-A/B)$
$J = C\exp[-16\pi\gamma^3/3kT(\sigma P)^2]$	$m^{-3}\ sec^{-1}$	2.17×10^5	~ 1 event $cm^{-3}\ sec^{-1}$

[a] Note that, in general, the presence of the gas will lower the interfacial tension. This effect has not been included in the calculation here.

value of the supersaturation, σ, was chosen to give a rate of approximately 1 event per cm^{-3} sec^{-1}, equivalent to 10^6 events m^{-3} sec^{-1}. The calculation is based on the simplifying assumptions of Eq. (6), and makes no allowance for nonideality:

$$J = C\exp\left[\frac{-16\pi\gamma^3}{3kT(\sigma P')^2}\right]$$

6. Comparison with Calculation of Droplet Nucleation Rate

For comparison purposes, a calculation is performed for the rate of homogeneous droplet nucleation from a supersaturated vapor (Table 10). The equation is that from Dunning and Shipman [23]:

$$J = C'''\exp\left[\frac{-16\pi\gamma^3 M^2 N_A}{3R^3 T^3 \rho^2 (\ln\alpha)^2}\right]$$

7. Reconciliation of Theory and Experiment

In the case of boiling, the data shows that the theory and experiment agree remarkably well. The data from Table 1B show an average value of the superheat calculated from the theory vs. the measured superheat, the ratio boiling *theory/measured* = 0.997, with a standard deviation of $\pm$ 0.03. For cavitation, the comparable figure for the theoretical and measured fracture

TABLE 10 Calculation of the Homogeneous Water Droplet Nucleation from the Vapor Phase

Variable	Unit	Magnitude	Comment
C (or g_k)	m^{-3} sec^{-1}	1×10^{35}	The original equation uses g_k, equivalent to C here
γ	N m^{-1}	7×10^{-2}	
M	kg	1.8×10^{-2}	Water
N_A	Dimensionless	6.023×10^{23}	
R	J K^{-1} mol^{-1}	8.314	
T	K	2.93×10^2	
ρ	kg m^{-3}	1×10^3	
σ	Dimensionless	4	
$16\pi\gamma^3 M^2 N_A$	N^3 m^3 kg^2	3.36×10^{18}	A
$3R^3 T^3 \rho^2$	J^3 kg^2 m^{-6}	5.23×10^{16}	B
$-16\pi\gamma^3 M^2 N_A/3R^3 T^3\rho^2$	Dimensionless	-7.76×10^1	$-A/B$
$\exp[-16\pi\gamma^3 M^2 N_A/3R^3 T^3\rho^2 \ln^2\sigma]$	Dimensionless	1.21×10^{-28}	$\exp(-A/B)$
$J = C \exp[-16\pi\gamma^3 M^2 N_A/3R^3 T^3\rho^2 \ln^2\sigma]$	m^{-3} sec^{-1}	1.21×10^7	Roughly 1 event cm^{-3} sec^{-1}

tensions are cavitation *theory/measured* = 2.18 ± 1.55 (if the values for mercury are discarded) and 9.63 ± 19.77 (if they are included). The comparable figures for the nucleation of gas bubbles in water are somewhat less easily assembled, the chief difficulties being distinguishing homogeneous from heterogeneous nucleation, or indeed whether bubble nucleation is involved at all in a couple of cases, and in the effect of gas adsorption on the surface tension. Ignoring these potential complications, the theoretical calculation performed as shown in Table 9 gives a theoretical value of gas in water *theory* of about 1400 for gases dissolved in water at ambient temperature of 25°C. Using this as a basis, the values of the supersaturation calculated as shown above and measured by various means (pressure release, bubble train timing, and chemical generation) in situ for various gases give a ratio of gas in water *theory/measured* = 89 ± 107. For bubbles of N_2 in various organic solvents, a much smaller value of the ratio results: gas in organics *theory/measured* = 5 ± 2.76.

The evidence strongly supports the hypothesis that dissolved gases, particularly at relatively high concentrations in water, cause substantial lowering of the interfacial tension in bubbles, and that this lowering is the main reason for the fact that gas bubbles nucleate at low supersaturations, and thus give such high values of the ratio *theory/measured*.

Boiling is an effective means of reducing or eliminating the gas content of a liquid, and this removal of gas probably accounts for the rather good agreement of the data with the predictions of the theory in the case of boiling.

The experimentalists who measured the fracture tension of liquids (by cavitation) were acutely aware that preexisting nuclei (e.g., Harvey nuclei) would strongly influence their results, and so took pains to attempt to eliminate dissolved and suspended gas bubbles before the experiment. The test liquids were carefully purified by distillation before use, and were degassed. This care, while not completely effective, probably was enough to reduce the gas content to the point where the gas could be considered to be a minor impurity, rather than a major contributor to the results. Thus cavitation data are relatively reliable, although not providing quite such a good agreement with theory as in the case of boiling.

By their nature, the experiments on gas bubble nucleation, whether electrolytic, chemically generated, or by pressure release, all have high concentrations of gas in solution during the bubble nucleation process, and thus show the most pronounced effects of surface tension reduction. Two aspects of this observation are particularly noteworthy: first, that He of all the gases so far tested shows the smallest ratio of *theory/measured*—and is also the only gas with a **b** coefficient of ~ 0. Second, that organic liquids with N_2 gas dissolved in them (including the case of diethyl ether) show relatively modest ratios *theory/measured*. In the case of diethyl ether, the liquid is relatively close to its critical point (as are all the simple organic liquids, if water is used as the yardstick), and so the surface tensions are relatively low to start with. Thus gases thus cannot reduce the surface tension as dramatically for organic liquids as for water. In brief, the most dramatic effects of dissolved gas are expected in aqueous gas solutions, and in confirmation of this, the highest values of the ratio theory/measured are indeed found for these systems. Furthermore, the more soluble gases produce the largest effects: The upward trend in the ratio theory/measured as one goes down Table 7 follows the general trend in increased solubility going down the table. The **b** coefficient also follows the same trend, with **b** increasing with the solubility.

One final observation is that Eq. (6) or any of its variants depends only on the properties of the liquid, not on those of the gas. Theory predicts that all gases should nucleate at the same supersaturation, σ. Table 7 shows a factor of about 250:1 in the supersaturation needed to cause nucleation for different gases. Even allowing for the fact that some of these results may be derived from homogeneous and others from heterogeneous nucleation experiments (Table 11), the classical nucleation theory provides no clues to understanding the systematic variation of the ratio with both the solubility of the gas and the **b** coefficient, which together form a most surprising result. However, it has a simple interpretation in the light of gas adsorption as a mechanism for reducing the interfacial tension, and thus lowering the supersaturation needed for nucleation. When a single gas is examined in a variety of liquids as seen in Table 8, the experimental data show a rather small variability, and much lower values of the ratio theory/measured. Once again, this is consistent with the gas adsorption mechanism, but unexpected based on classical nucleation theory.

TABLE 11 Summary Comparison of Theoretical Prediction and Measured Drives Required for Bubble Nucleation

System	Ratio	Comment
Boiling (single component system)	0.997 ± 0.03	Heterogeneous (liquid/liquid interface)
Cavitation (single component system)	2.18 ± 1.55	Homogeneous (?)
Gas evolution (pressure release, chemical reaction, etc.) from water	89 ± 107	Heterogeneous (mostly)
Gas evolution (N_2 from organic liquids)	5 ± 2.76	Homogeneous (?)

VI. THE IMPORTANCE OF DETACHMENT IN NUCLEATION KINETICS

It is not the intention to thoroughly review the detachment literature here. The approach will be to mention very briefly the principles of calculating the detachment volume of a bubble, and to apply these principles to a discussion of the interaction of nucleation and detachment.

A. Quasi Static Detachment

Detachment must by definition be a nonequilibrium process, but it is a convenient fiction to assume that the forces are the same (or at least, close enough to being the same) as for a genuine equilibrium state. The errors involved become smaller as the system more closely approximates to equilibrium conditions. This is one reason why detachment should be approached as slowly as possible during experimental investigations of drop/bubble detachment volumes.

B. Tate's Law

The weight, w, of a detaching drop of wetting liquid of surface tension, γ, formed on a circular tube in air with an orifice of radius r is approximately given by:

$$w = 2\pi r\gamma$$

Tate actually expressed his "law" as follows: "Other things being equal, the weight of a drop of liquid is proportional to the diameter of the tube in which it is formed." In practice, a lesser weight, w', is obtained. The reason is that only a proportion of the total drop actually detaches—as much as 40% of the liquid may remain attached to the orifice. Thus the actual weight is given by the expression $2\pi r\gamma$ multiplied by a correction factor, f:

$$w' = 2\pi r\gamma f$$

The values of f were experimentally derived by Harkins and Brown [61]. For a given drop weight, it is possible by using the correction factor to calculate what proportion of the drop weight (and hence volume) is retained on a given orifice. In calculations involving the interface between two immiscible fluids α and β, the shape of the interface, and properties such as the enclosed volume are available, and it only needs an alteration of sign if the two fluid phases are exchanged. Thus water in air (α in β) is the same mathematical problem as air in water (β in α), but with a sign reversal. What holds for drops, generally speaking, holds for bubbles, too. However, as far as the author is aware, there has been no experimental verification of Tate's law and the Harkins Brown f factor for the case of detaching bubbles.

Recourse to empirical factors such as the Harkins Brown f factor can be avoided by calculations based on the volumes of axisymmetric drops (bubbles) as discussed by Boucher [78].

The significance of this for bubble nucleation is that, in principle, it is possible to calculate the remaining amount of gas at a nucleation site, and thus to be able to answer unambiguously the question "Is a fresh nucleation event needed for the next bubble?" It is worth mentioning in this context that when the inner fluid wets the orifice, then the appropriate radius for detachment is the outer radius of the orifice, but when the inner fluid does not wet the orifice, it is the inner radius that is relevant. In the presence of water as the outer fluid phase, and the generally reasonably good wetting reported in most nucleation experiments, it seems that often, the nonwetting bubble will be detaching leaving a minimal amount of gas behind. Composite nucleation sites, with both hydrophilic and hydrophobic parts, could be an exception.

C. Axisymmetric Profiles

A useful program was written by Boucher et al. [62] to calculate axisymmetric profiles (meridian curves) for holms, bridges, pendent, and sessile drops and bubbles. This program has been rewritten in Quick Basic, and is available from the author. In combination with the theoretical volume of a bubble deduced from the known interfacial tension, using the program, and the theory outlined by Boucher [62], the remaining volume can be estimated. An example of the results for the calculated profiles for bubbles with various shape factors, H (shown alongside the curves), are given in Fig. 22.

D. Detachment vs. Nucleation: The $\mathcal{L}$ Number

It has been noted above that heterogeneous nucleation is energetically favored over homogeneous nucleation, and for this reason, surfaces are usually implicated in nucleation processes. The kinetics of release of bubbles in the presence of a surface are different from the kinetics in its absence, the reason being that the surface introduces a new kinetic step, detachment, into the overall rate of release. The ratio, $\mathcal{L}$, of the rate of nucleation to that of detachment is a significant quantity because it determines which is rate determining in the overall release kinetics. For heterogeneous nucleation, the three most important experimental variables are the saturation ratio, α, the contact angle, θ, and the interfacial tension, γ. For detachment, the selection of important variables depends upon the treatment of the detachment process. In the case of (quasi) equilibrium detachment, a quantitative

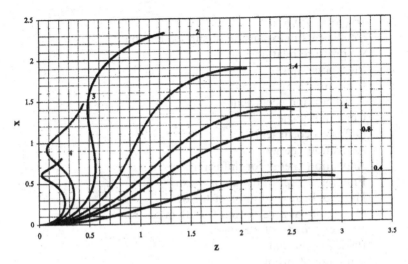

FIG. 22 Plots of bubble profiles for various values of the shape parameter, H, obtained from the QuickBasic program written by Boucher et al. [62].

treatment exists which allows a prediction of the regions in which detachment dominates or is dominated by nucleation [37]:

$$\mathcal{L} = \frac{J}{\lambda}\left(\frac{3\gamma\sin(\theta)(2 - \cos(\theta))}{2g\Delta\rho}\right)^{1/2}$$

The fraction J/λ represents the ratio of the rate of nucleation J, to the rate of growth λ, of the bubble, while the term in brackets accounts for the effect of the contact angle. The rate of growth of a bubble has been addressed by many authors, and adequate theories are available; a recent example is given by Bisperink and Prins [63]. Therefore, in principle, it is possible to assess regions in which classical nucleation theory will apply, and those where it is not expected to hold. It emerges that the contact angle plays an important role, because small contact angles make detachment easy, while inhibiting nucleation, thus making nucleation the rate-determining step. On the other hand, large contact angles make detachment difficult while promoting heterogeneous nucleation, thus making detachment the rate-determining step.

SYMBOLS

A	Preexponential
A_j	Cluster of j monomer units
$A_{l,m}$	Area of interface between phases l,m

b	$= (p^* - P')/p^*$
b, c, d	Coefficients in expansion of surface tension vs. pressure relationship
C	Preexponential
c_j	Concentration
β	Condensation coefficient
β	Cone half angle
D	Diffusion coefficient
E_a	Activation energy
E_D	Activation energy for diffusion
η	Fractional vapor pressure $\eta = \exp\left[\frac{\Omega(P'-p_\infty)}{kT} - \frac{c'_2}{c'_1}\right]$
f	Harkins Brown adjustment factor
g	Gravitational acceleration ($\sim$9.81 m sec^{-2})
g	Rate of growth of nucleus
h	Planck's constant
H_f	Enthalpy of fusion
H_v	Enthalpy of vaporization
i^*	Number of monomer units in critical nucleus
J	Rate of nucleation
$J(t)$	Time-dependent nucleation rate
J_0	Steady state nucleation rate
k	Boltzmann constant
k_H	Henry's law constant
λ	Rate of growth of a bubble
$\mathcal{L}$	Ratio of rate of nucleation to detachment rate
M	RMM
m	Molecular mass
μ	Viscosity
τ	Nucleation time lag
P	Pressure
P'	Imposed hydrostatic pressure, or target pressure
p	Pressure in bubble
p_∞	Saturation vapor pressure (over a flat surface)
p^*	Pressure in critical bubble
p_2	Partial pressure of gas in bubble
P_L	Pressure in liquid (hydrostatic pressure)
Pe	Péclet number
R	Universal gas constant
r	Radius of bubble
$r^\#$	Threshold radius for cavity
r^*	Critical radius
r_0	Initial radius

Re	Reynolds number, $= \Delta\rho U d/\mu$
T	Absolute temperature
T_L	Temperature in the liquid
T_{sat}	Temperature corresponding to the saturated vapor pressure at standard pressure
T_W	Temperature of the (heated) wall
U	Rise rate of a bubble
V	Volume
w	Theoretical maximum weight of a detaching drop
w'	Actual weight of a detaching drop
Z	Zeldovich nonequilibrium factor
ΔG	Change in Gibbs free energy
$\Delta\mu$	Change in chemical potential
Φ	Function of the two contact angles at a liquid/liquid interface
Ω	Molecular volume
α	Saturation ratio (p/P')
ϕ	Function of the contact angle
ϕ_{cone}	A function of the contact angle and cone half-angle
ϕ_{pit}	A function of the contact angle and pit half-angle
$\phi(\theta)$	A function of contact angle for a spherical cap
γ	Interfacial tension
γ_0	Surface tension of the pure liquid in air at a pressure of 1 atm
γ_{LV}	Liquid/vapor interfacial tension
γ_{SV}	Solid/vapor interfacial tension
κ	Curvature $(= 1/r)$
μ	Chemical potential
ν	Activity coefficient
θ	Contact angle
θ_a	Advancing contact angle
θ_r	Receding contact angle
ρ	Density
σ	Supersaturation $(= \alpha - 1)$

REFERENCES

1. Christenson, H.K.; Claesson, P.M. Adv. Colloid Interface Sci. *91*, 391.
2. Volmer, M. *Kinetik der Phasenbildung*; Steinkopff Verlag: Dresden, 1939.
3. Blander, M. Adv. Colloid Interface Sci. 1979, *10*, 1.
4. Ward, C.A.; Balakrishnan, A.; Hooper, F.C. J. Basic Eng. 1970, *92*, 695. And the discussion following 701–704.

5. Hirth, J.P.; Pound, G.M. Condensation and evaporation—nucleation and growth kinetics. *Progress in Materials Science*; Pergamon Press: Oxford, 1963; Vol. 11.

6. Skripov, V.P. *Metastable Liquids*; Wiley: New York, 1974.

7. Lubetkin, S.D. Bubble nucleation and growth. In *Controlled Particle, Droplet and Bubble Formation*; Wedlock, D.J., Ed.; Butterworth Heinemann: Oxford, 1994; pp 159–190.

8. Debenedetti, P.G. *Metastable Liquids*; Princeton University Press: Princeton, NJ, 1996.

9. Fisher, J.C. J. Appl. Phys. *19*, 1062.

10. Kenrick, F.B.; Gilbert, C.S.; Wismer, K.L. J. Phys. Chem. 1924, *28*, 1297.

11. Trefethen, L. J. Appl. Phys. 1957, *28*, 923.

12. Turnbull, D. J. Chem. Phys. 1952, *20*, 411.

13. Wakeshima, H.; Takata, K. J. Phys. Soc. Jpn. 1958, *13*, 1398.

14. Skripov, V.P.; Sinitsyn, E.N. Sov. Phys., Usp. 1964, *7*, 887.

15. Apfel, R.E. Nature 1972, *238*, 63.

16. Skripov, V.P.; Pavlov, P.A. High Temp. USSR 1970, *8*, 782.

17. Meyer, J. *Zur Kenntnis des negativen Druckes in Flüssigkeiten*; Knapp Halle: Halle, 1911.

18. Vincent, R.S. Proc. Phys. Soc. Lond. 1941, *53*, 126.

19. Briggs, L.J. J. Appl. Phys. 1950, *21*, 721.

20. Knapp, R.T.; Daily, J.W.; Hammitt, F.G. *Cavitation*; McGraw-Hill: New York, 1970.

21. Becker, R.; Döring, W. Ann. Phys. 1935, *24*, 719.

22. Zeldovich, J. J. Exp. Theor. Phys. 1942, *12*, 525.

23. Dunning, W.J.; Shipman, A.J. Proc. Intern. Ind. Agr. Alimento 1954, *2*, 1448.

24. Kantrowitz, A. J. Chem. Phys. 1951, *19*, 1097.

25. Mori, Y.; Hijikata, K.; Nagatani, T. Int. J. Heat Mass Transfer 1976, *19*, 1153.

26. Hirth, J.P.; Pound, G.M.; St. Pierre, G.R. Metall. Trans. 1969, *1*, 939.

27. Kundt, A. Weid. Ann. 1881, *12*, 538.

28. Lubetkin, S.D.; Akhtar, M. J. Colloid Interface Sci. 1996, *180*, 43.

29. Blander, M.; Katz, J.L. AIChE J. 1975, *21*, 833.

30. Apfel, R.E. J. Chem. Phys. 1970b, *54*, 62.

31. Jarvis, T.J.; Donohue, M.D.; Katz, J.L. J. Colloid Interface Sci. 1975, *50*, 359.

32. Wilt, P.M. J. Colloid Interface Sci. 1986, *112*, 530.

33. Sindorf, D.W.; Maciel, G.E. J. Am. Chem. Soc. 1983, *105*, 3767.

34. Clifton, B.B. Sc thesis, Bristol, 1991.

35. Lubetkin, S.D.; Blackwell, M.R. J. Colloid Interface Sci. 1988, *126*, 610.

36. Griffith, P.; Wallis, J.D. Chem. Eng. Prog. Symp. Ser. #30 1960, *56*, 49.

37. Lubetkin, S.D. J. Chem. Soc., Faraday Trans. I 1989a, *85*, 1753.

38. Lubetkin, S.D. J. Appl. Electrochem. 1989b, *19*, 668.

39. Volanschi, A.; Nijman, J.G.H.; Olthuis, W.; Bergveld, P. Sens. Mater. 1997, *9*, 223.

40. Marto, P.J.; Sowersby, R.L. Role Nucleation Boiling Cavitation, Symp. ASME 70-HT-16, 1970.

41. Hartland, S.; Hartley, R.W. *Axisymmetric Fluid Liquid Interfaces*; Elsevier Amsterdam: Amsterdam, 1976.

42. Schmeltzer, J; Schweitzer, F. J. Non-Equilib. Thermodyn. 1987, *12*, 255.

43. Lifshitz, I.M.; Slyozov, V.V. J. Phys. Chem. Solids 1961, *19*, 35.

44. Epstein, P.S.; Plesset, M.S. J. Chem. Phys. 1950, *18*, 1505.

45. Binks, B.P.; Clint, J.H.; Fletcher, P.D.I.; Rippon, S.; Lubetkin, S.D.; Mulqueen, P.J. Langmuir 1998, *14*, 5402 and *ibid* 1999, *15*, 4497.

46. Yount, D.E. J. Acoust. Soc. Am. 1979, *65*, 1429.

47. Fox, F.E.; Herzfeld, K.F. J. Acoust. Soc. Am. 1954, *26*.

48. Bankoff, S.G. AIChE. J. 1958, *4*, 24.

49. Cole, R. Adv. Heat Transf. 1974, *10*, 85.

50. Atchley, A.A.; Prosperetti, A J. Acoust. Soc. Am. 1989, *86*, 1065.

51. Carr, M.W. PhD thesis, Bristol, 1990.

52. Apfel, R.E. J. Acoust. Soc. Am. 1970a, *48*, 1179.

53. Winterton, R.H.S. J. Phys. D Appl. Phys. 1977, *10*, 2041.

54. Trevena, D.H. *"Cavitation and tension in liquids"*; Adam Hilger: Bristol, 1987.

55. Jones, S.F.; Evans, G.M.; Galvin, K.P. Adv. Colloid Interface Sci. 1999, *80*, 27 and *ibid* 1999, *80*, 51.

56. Liger-Belair, G.; Vignes-Adler, M.; Voisin, C.; Robillard, B.; Jeandet, P. Langmuir 2002, *18*, 1294.

57. Briggs, L.J. J. Appl. Phys. 1953, *24*, 488.

58. Kwak, H.Y.; Kim, Y.W. Int. J. Heat Mass Transfer 1997, *41*, 757.

59. Wismer, K.L. J. Phys. Chem. 1922, *26*, 301.

60. Forest, T.W.; Ward, C.A. J. Chem. Phys. 1978, *69*, 2221.

61. Harkins, W.D.; Brown, F.E. J. Am. Chem. Soc. 1919, *41*, 499.

62. Boucher, E.A.; Evans, M.J.B.; Jones, T.G.J. Adv. Colloid Interface Sci. 1987, *27*, 43.

63. Bisperink, C.G.J.; Prins, A. Colloids Surf., A 1994, *85*, 237.

64. Massoudi, R.; King, A.D. Jr. J. Phys. Chem. 1974, *78*, 2262.

65. Sides, P.J. Mod. Aspects Electrochem; White, R.E. Bockris, J.O.'M., Conway, B.E., Eds; Plenum: New York, 1986; *18*, 303.

66. Westerheide, D.E.; Westwater, J.W. AIChE J. 1961, *7*, 357.

67. Dapkus, K.V.; Sides, P.J. J. Colloid Interface Sci. 1986, *111*, 133.

68. Shibata, S. Electrochim. Acta 1978, *23*, 619.

69. Glas, J.P.; Westwater, J.W. Int. J. Heat Mass Transfer 1964, *7*, 1427.

70. Bowers, P.G.; Hofstetter, C.; Letter, C.R.; Toomey, R.T. J. Phys. Chem. *99*, 9632; also Bowers, P.G.; Hofstetter, C.; Ngo, H.L.; Toomey, R.T. J. Colloid Interface Sci. 1999, *215*, 441.

71. Hey, M.J.; Hilton, A.M.; Bee, R.D. Food Chem. 1994, *51*, 349.

72. Hemmingsen, E.A. J. Appl. Phys. 1975, *46*, 213; see also Gerth, W.A.; Hemmingsen, E.A. J. Colloid Interface Sci. 1980, *74*, 80.

73. Smith, K.W.; Noyes, R.M.; Bowers, P.G. J. Phys. Chem. 1983, *87*, 1514.

74. Rubin, M.B.; Noyes, R.M.; Smith, K.W. J. Phys. Chem. 1987, *91*, 1618.

75. Hemmingsen, E.A. Nature 1977, *267*, 141.
76. Bowers, P.G.; Bar-Eli, K.; Noyes, R.M. J. Chem. Soc., Faraday Trans. 1996, *92*, 2843.
77. Moore, D.W. J. Fluid Mech. 1965, *23*, 749.
78. Boucher, E.A.; Kent, H.J. J. Colloid Interface Sci. 1978, *67*, 10.
79. Defay, R; Prigogine, I; Bellemans, A. Surface Tension and Adsorption; Translated by Everett, D.H.; Longmans: Green London, 1966; 87–88 pp.
80. Freundlich, H *Kapillarchemie*; Akad. Verlag: Leipzig, 1922; pp 113.
81. Herrick, C.S.; Gaines, G.L., Jr. J. Phys. Chem. 1973, *77*, 2703.
82. Hough, E.W.; Heuer, G.J.; Walker, J.W. Trans. A.I.M.E. 1959, *216*, 469.
83. Hough, E.W.; Rzasa, M.J.; Wood, B.B., Jr. Trans. A.I.M.E. 1951, *192*, 57.
84. Hough, E.W.; Wood, B.B., Jr.; Rzasa, M.J. J. Phys. Chem. 1952, *56*, 996.
85. Jho, C.; Nealon, D.; Shogbola, S.; King, A.D., Jr. J. Colloid Interface Sci. 1978, *65*, 141.
86. Massoudi, R.; King, A.D., Jr. *Colloid Interface Sci.*; Kerker, M., Ed.; Academic Press: New York, 1976, *3*, 33.
87. Massoudi, R.; King, A.D., Jr. J. Phys. Chem. 1975, *79*, 1670.
88. Masterton, W.L.; Bianchi, J.; Slowinski, E.J., Jr. J. Phys. Chem. 1963, *67*, 615.
89. Richards, T.W.; Carver, E.K. J. Am. Chem. Soc. 1921, *43*, 827.
90. Rusanov, A.I.; Kochurova, N.N.; Khabarov, V.N. Dokl. Akad. Nauk. SSSR 1972, *202*, 380.
91. Slowinski, E.J., Jr.; Gates, E.E.; Waring, E.E. J. Phys. Chem. 1957, *61*, 808.
92. Volmer, M.; Weber, A. Z. Phys. Chem. 1925, *119*, 277.

10

Thermodynamics of Curved Interfaces in Relation to the Helfrich Curvature Free Energy Approach

JAN CHRISTER ERIKSSON and STIG LJUNGGREN   Royal Institute of Technology, Stockholm, Sweden

I. INTRODUCTION

In 1878, Gibbs [1] published his celebrated "Theory of Capillarity," the standard reference of surface thermodynamics ever since. In a rather compact—yet exhaustive and profound—manner, Gibbs treated fluid–fluid, as well as solid–fluid, interfaces and their equilibrium properties while representing the interfacial region in an Euclidean manner by a single dividing interface, preferably the so-called surface of tension. For this particular dividing surface, the standard Laplace (or Young–Laplace) equation [2]:

$$\Delta P = 2H\gamma \tag{1}$$

holds exactly for a majority of cases. Here $H = (c_1 + c_2)/2$ denotes the mean curvature, γ is the interfacial tension, and ΔP is the pressure jump at the interface, and c_1 and c_2 are the principal curvatures of the surface of tension. Moreover, for any given interface, the interfacial tension γ attains a minimum value when the surface of tension is chosen to be the dividing surface, as may readily be verified.

Half a century later, Verschaffelt [3] and Guggenheim [4] initiated formal investigations where the interface was attributed a certain thickness by means of using *two* dividing surfaces instead of just one, and, after World War II, Tolman [5] took up the study of the (first-order) effect of curvature on the surface tension. In basically the same vein, Hill [6] and Buff [7] contributed to the development of the localized pressure tensor (hydrostatic) description of interfaces, and began generalizing the Gibbs treatment of curvature effects, as did likewise Ono and Kondo [8].

It is only during the last couple of decades, however (i.e., about a century after the appearance of the Gibbs treatise), that curvature effects have come into focus in surface thermodynamics. On one hand, the mathematical tools developed long ago have gotten more widely appreciated; on the other hand, the existence of a large variety of strongly curved surfactant aggregates, microemulsion droplets, and bilayer vesicles has been experimentally documented. In this largely novel context, the original Gibbs approach employing the surface of tension has turned out to be less suitable. The reason is fairly obvious. In the formalism of Gibbs, the effect of curvature is not considered explicitly, but expresses itself as a shift of the surface of tension relative to the physical interface—a shift that is often neither readily quantified nor easily interpreted (cf. Fig. 1).

Hence, looking back, there was definitely a need of progressing along a different route and, in 1973, Helfrich [9] started off an amazingly fruitful development, simply by invoking the following quadratic *ansatz* for the local curvature (bending) free energy f_{curv} per unit area:

$$f_{curv} = 2k_c(H - H_0)^2 + \bar{k}_c K \tag{2}$$

where $K = c_1 c_2$ is the Gaussian curvature and H_0 is the spontaneous curvature [where, for a cylindrical interface ($K = 0$), the curvature free energy f_{curv} has a minimum equal to zero]. The constants k_c and $\bar{k}_c$ are usually called bending (elasticity) constants or moduli, although, more often than not, they do refer to an interface that in reality constitutes an open system, and the curvature changes contemplated take place while keeping the chemical potentials fixed. In such a case, f_{curv} merely accounts for the effect of curvature on the ordinary interfacial tension γ, in line with the expression:

$$\gamma = \gamma_0 + 2k_c(H - H_0)^2 + \bar{k}_c K \tag{3}$$

which, by the way, implies that the interfacial tension of the plane interface (γ_∞) equals $\gamma_0 + 2k_c H_0^2$, with γ_0 denoting the value of γ for a cylindrical interface of mean curvature H_0.

A few years later, in 1977, Boruvka and Neumann [10] published the first *rigorous* thermodynamic treatment of fully equilibrated, curved interfaces, resulting in a generalized Laplace equation of the kind:

$$\Delta P = 2H\gamma - C_1(2H^2 - K) - 2C_2 HK - \frac{1}{2}\nabla_s^2 C_1 - K\nabla_s^* \cdot (\nabla_s C_2) \tag{4}$$

which holds for an *arbitrary* dividing surface; in other words, it is valid for any dividing surface condition related with the actual properties of the physical interface. Note that the last two curvature-dependent terms of Eq. (4) containing the two-dimensional operators ∇_s^2, ∇_s^*, and ∇_s will vanish for a

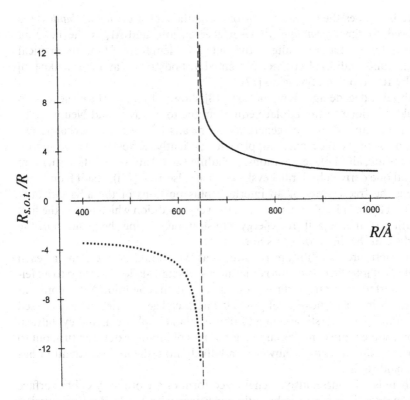

FIG. 1 The ratio between the radius of the surface of the tension $R_{s.o.t.}$ and the radius of the hydrocarbon–water contact surface R for a Winsor II water-in-oil microemulsion droplet. The calculation is based on the general equation [Eq. (160)], and assumes the following curvature-dependent interfacial tension: $\gamma = 1.00 \times 10^{-6} - 1.28 \times 10^{-13} R^{-1} + 1.65 \times 10^{-20} R^{-2}$ J m^{-2}, which has a minimum for $R = 257.8$ nm. The discontinuity occurs for $R = 64.0$ nm where the pressure inside the droplet is the same as the external pressure. Note that negative values for $R_{s.o.t.}$ are obtained below this radius, which are lacking physical significance.

surface of constant curvature. Moreover, the second and third terms can be made to disappear by choosing, as Gibbs did, the surface of tension as the dividing surface.

As will be discussed in detail below, a Laplace equation of even broader generality was derived by Evans and Skalak [11], relying on the old Kirchhoff–Love [12,13] mechanical theory of shells. It can be cast in the following tensorial form [14,15]:

$$\Delta P = \mathbf{b} : \sigma_s - \nabla_s \nabla_s : \mathbf{M}_s \tag{5}$$

where **b** denotes the curvature tensor, σ_s is the tensor of the *mechanical* (as opposed to the *thermodynamic*) surface tension, and M_s is the bending moment tensor. Later on, alternative but equivalent forms of this mechanical equilibrium condition have been presented by Podstrigach and Povstenko [16] and by Rusanov and Shchekin [17].

About a decade ago, Kralchevsky [18] showed that a surface shear term should be added to the Laplace equation due to Boruvka and Neumann to attain the same degree of generality by means of the thermodynamic approach as by the mechanical approach. Evidently, choosing the thermodynamic route, all Laplace equations, including Eq. (4) above and the even more general one derived by Kralchevsky [18] (see also Ref. [19]), result from optimizing the free energy of an interface plus portions of the adjacent bulk phases. Hence, a Laplace equation is the very condition which makes the first variation of the overall free energy vanish upon varying the position of the interface in the direction of its normal.

Furthermore, equilibrium requires that ΔP should be the same at each point of an interface. By approximating a surfactant-laden interface to a Helfrich interface (i.e., an interface obeying Eq. (3)), the condition ΔP = constant will result in a classification of possible interfacial geometries to be discussed in Section IV. Interestingly, in addition to planar, spherical, and cylindrical shapes, some more complex curved interfaces of infinite extension turn out to be permissible, to which, however, relatively little theoretical attention has been paid so far.

There is, of course, also a generalized form of the ordinary Gibbs surface tension equation, which is broadly consistent with the Helfrich expression (Eq. (3)) above, albeit of an entirely different model-independent nature. At constant temperature, it reads [20]:

$$d\gamma = - \sum_i \Gamma_i d\mu_i + C_1 dH + C_2 dK \tag{6}$$

which shows that the surface tension γ of an arbitrary dividing surface depends not just on T and all the chemical potentials μ_i but, in addition, on the curvature parameters H and K characterizing the dividing surface, thus providing definitions of the curvature coefficients C_1 and C_2 participating in Eq. (4). It is worth noting already at this point, however, that for a geometrically closed aggregate surface that is not connected to any three-phase contact line, this kind of Gibbs–Duhem condition is of a somewhat limited use in practice, simply because γ is not a readily accessible experimental parameter for such aggregates.

This book chapter is complementary to a previous review article on the mechanics and thermodynamics of curved interfaces by Kralchevsky and the present authors [15]. It is organized as follows. In Section II, we discuss the minimization of the free energy of interfaces of arbitrary curvature in general terms, whereas Section III is devoted to a number of illuminating micromechanical expressions for surface excess quantities. In Section IV, we employ some of the general results obtained to elucidate the formation of microemulsions, vesicles, and surfactant aggregates.

II. THERMODYNAMICS OF ARBITRARILY CURVED INTERFACES

A. The Generalized Laplace Equation

The reader no doubt recognizes that the proper way to proceed when deriving the full equilibrium conditions for an interface consists in minimizing the free energy $\mathcal{F}$ of the interface plus its ambient bulk phases. The exact choice of this free energy $\mathcal{F}$ will depend on the actual properties of the system and is not in general equal to the Helmholtz free energy F as discussed below in Section II.B.

For an ordinary fluid two-phase system, minimization of the free energy at constant temperature first of all results in the well-known chemical potential condition for diffusive equilibrium with respect to the soluble components present. Furthermore, the mathematical extremum condition for the free energy contains the pressure difference ΔP across the interface as a parameter. Solving for ΔP, this condition takes the form of a generalized Laplace equation. Whether or not this equation signifies a *stable* equilibrium is, however, often a rather complex issue where the detailed system properties may enter in a crucial manner.

In the present part, we consider various forms of the generalized Laplace equation, which are valid for interfaces of *variable curvature* and for an *arbitrary dividing surface* between two homogeneous bulk phases. As already indicated in Section I, this rather complete surface-thermodynamic background is needed, especially when treating microemulsion droplets, micellar aggregates, biomembranes, and infinite interfacial structures for which the curvature free energy plays a key role. The pioneering works in this direction of Buff [7], Murphy [21], Melrose [22], and Eliassen [23] culminated in the general treatment derived by Boruvka and Neumann [10], which was later on further extended by Kralchevsky [18], who added the possible effect of elastic surface shear.

To correspond to a stable equilibrium situation, the overall free energy $\mathcal{F}$ of an interface plus the adjacent bulk phases must have a minimum, implying, to begin with, that the first variation $\delta^{(1)}\mathcal{F}$ should be equal to zero when the interface is subject to an infinitesimal deformation from its equilibrium shape. Let the deformation vector generally be given by:

$$\delta R = \mathbf{s} + \psi \mathbf{n} \tag{7}$$

where $\mathbf{s} = s^1 \mathbf{a}_1 + s^2 \mathbf{a}_2$ is in the tangent plane of the surface and $\mathbf{a}_1$ and $\mathbf{a}_2$ are the corresponding basis vectors, while $\psi \mathbf{n}$ is in the direction of the surface normal $\mathbf{n}$. After a complicated series of deduction steps, Kralchevsky [loc. cit.] arrived at the following expression for the variation of the free energy of a *closed* interfacial system with fixed overall volume comprising the interface and the adjacent phases, held at constant temperature, and for an arbitrary choice of the dividing surface:

$$\delta \mathcal{F} = \Delta P \iint_S \psi \mathrm{d}S + \iint_S (\gamma \delta \alpha + \zeta \delta \beta + B \delta H + \theta \delta D) \mathrm{d}S \tag{8}$$

The first integral simply represents the overall PV work and, apart from the surface shear term $\zeta \delta \beta$, the integrand of the second integral is the same as the expression for the infinitesimal work of surface deformation employed by Gibbs [1]. The above expression is actually valid both for the variation of the grand Ω-potential of an *open* interface, equilibrated at constant chemical potentials, and the variation of the Helmholtz free energy of an interface consisting of *insoluble* adsorbents, provided that the chemical potentials of the components of the interface are constant throughout the interface so that there is no ongoing surface diffusion, and, finally, also for the mixed case, as is explicitly verified in Appendix A.

For completely rigid solid surfaces, Eq. (8) reduces to:

$$\delta \mathcal{F} = \iint_S (\gamma \delta \alpha + \zeta \delta \beta) \mathrm{d}S \tag{9}$$

because then we have to set ψ, δH, and δD equal to zero everywhere. This special case was treated earlier by Shuttleworth [24], Eriksson [25], and Rusanov [26], whereby the fundamental distinction, already emphasized by Gibbs, between the work of *stretching* an interface (γ) and the work of *forming* an interface (ω_s), was found to be of crucial significance.

In Eq. (8) above, ΔP is the pressure difference between the two sides of the interface: $H = (c_1 + c_2)/2$, the mean curvature, and $D = (c_1 - c_2)/2$, the so-called *deviatoric* curvature.

All the variations $\delta \alpha$, $\delta \beta$, δH, and δD are *local*; that is, they are functions of the curvilinear surface coordinates $(u^1$ and $u^2)$ with the following meaning: $\delta \alpha$ is the relative dilation of the element $\mathrm{d}S$ of the dividing surface, $\delta \beta$ is the shearing deformation, whereas δH and δD are the local variations of H and D, respectively. These four modes of surface deformation are illustrated in Fig. 2.

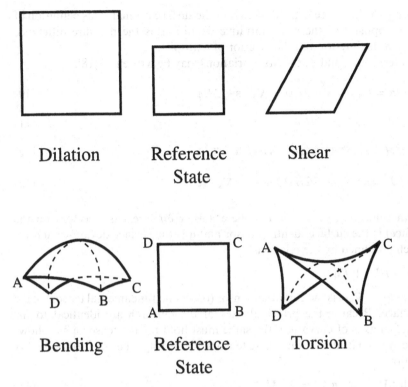

FIG. 2 The four independent modes of surface deformation.

Furthermore, γ is the ordinary (thermodynamic) surface tension, ζ is the surface density of shearing free energy ("shearing tension"), whereas B and Θ are quantities related to the local bending moments of the interface, which are defined below.

The sign conventions for the curvatures c_1 and c_2, and the pressure difference ΔP between the two sides of the interface are related to the direction of the surface normal $\mathbf{n}$, which, as a rule, is assumed to be the outward normal in the case of a surfactant aggregate. This means that the surface normal is directed toward the bulk phase on the (predominantly) convex side of the interface. Furthermore, because, according to differential geometry, we have $2H = -\nabla_s \cdot \mathbf{n}$, where $\nabla_s = a^\alpha(\partial/\partial u^\alpha)$ is the surface gradient operator, it follows that H is normally negative for a surfactant aggregate. More generally, we adopt the convention that the curvature is negative when the corresponding principal curves of normal section are convex toward the phase to which the surface normal is pointing, and we define ΔP as the pressure jump when crossing the interface in the direction of the surface normal. This definition of ΔP is opposite to the more intuitive definition as the pressure difference between the interior and the exterior of a droplet.

In terms of $\mathbf{s}$ and ψ, the first variations may be written as [18]:

$$\delta^{(1)}\alpha = \mathbf{U}_s : \nabla_s\mathbf{s} - 2H\psi = \nabla_s \cdot \mathbf{s} - 2H\psi \tag{10}$$

$$\delta^{(1)}\beta = \mathbf{q} : \nabla_s\mathbf{s} - 2D\psi \tag{11}$$

$$\delta^{(1)}H = \mathbf{s} \cdot \nabla_s H + (H^2 + D^2)\psi + \frac{1}{2}\nabla_s^2\psi \tag{12}$$

$$\delta^{(1)}D = \mathbf{s} \cdot \nabla_s D + 2HD\psi + \frac{1}{2}\mathbf{q} : \nabla_s\nabla_s\psi \tag{13}$$

Furthermore, $\mathbf{U}_s = \mathbf{a}^1\mathbf{a}_1 + \mathbf{a}^2\mathbf{a}_2$ (where $\mathbf{a}_\alpha = \partial\mathbf{r}/\partial u^\alpha$ are basis vectors on the surface) is the surface identity tensor and $\mathbf{q}$ is the surface deviatoric tensor, which is defined by the relation:

$$\mathbf{b} = H\mathbf{U}_s + D\mathbf{q} \tag{14}$$

where $\mathbf{b} = -\nabla_s \cdot \mathbf{n}$ is the curvature tensor (or second fundamental tensor) of the interface. Because the principal axes of the tensor $\mathbf{b}$ are identical to the principal axes of curvature, the same must hold for the tensor $\mathbf{q}$ as follows directly from its definition. These tensors, by the way, obey the orthogonality relations:

$$\mathbf{U}_s : \mathbf{U}_s = 2 \quad \mathbf{q} : \mathbf{q} = 2 \quad \mathbf{U}_s : \mathbf{q} = 0 \tag{15}$$

Because $\mathbf{b}$ is diagonal in the basis of the principal curvatures (which are, in fact, the diagonal components of the curvature tensor $\mathbf{b}$), Eq. (14) tells us that the tensor $\mathbf{q}$ is also diagonal in the basis of the principal curvatures. This may constitute a certain limitation of the generality of Eq. (8) because the latter is based on the assumption that the rate of deformation tensor $\mathbf{d}$ can be written in the form:*

$$\mathbf{d}\mathrm{d}t \equiv \nabla_s\delta R \equiv \nabla_s\mathbf{s} - \psi\mathbf{b} = \frac{1}{2}\delta\alpha\mathbf{U}_s + \frac{1}{2}\delta\beta\mathbf{q} \tag{16}$$

which, in turn, requires $\mathbf{d}$ to be diagonal in the basis of the principal curvatures. For symmetry reasons, however, this requirement is automatically satisfied for cylindrical and spherical geometries. Furthermore, we note that Eqs. (10) and (11) follow from the definitions $\delta\alpha = \mathbf{U}_s:\mathbf{d}\mathrm{d}t$ and $\delta\beta = \mathbf{q}:\mathbf{d}\mathrm{d}t$. Eqs. (12) and (13) were derived by Eliassen [23] (Eq. (13) implicitly).

---

*For a definition of $\mathbf{d}$, which is more general than Eq. (16), see Eq. (80) below.

The equilibrium condition $\delta^{(1)}\mathcal{F}=0$ at constant T and overall volume V can now be written (irrespective of whether the system is closed, partially closed, or completely open):

$$\delta^{(1)}\mathcal{F} = \iint_S [\boldsymbol{\gamma}_s : \nabla_s \mathbf{s} + (B\nabla_s H + \Theta\nabla_s D)\cdot \mathbf{s}]dS$$

$$+ \iint_S [\Delta P - 2H\gamma - 2D\zeta + (H^2 + D^2)B + 2HD\Theta]\psi dS \qquad (17)$$

$$+ \iint_S \frac{1}{2}(BU_s + \Theta\mathbf{q}) : \nabla_s\nabla_s\psi dS = 0$$

where we have introduced the tensor of the surface tension defined by:

$$\boldsymbol{\gamma}_s = \gamma U_s + \zeta\mathbf{q} \qquad (18)$$

which is likewise diagonal in the basis of the principal curvatures.

According to the ordinary procedure of variational calculus, ψ and $\nabla_s\nabla_s\psi$ (or $\mathbf{s}$ and $\nabla_s\mathbf{s}$) cannot be regarded as being independent of one another. On the contrary, a procedure equivalent to partial integration has to be undertaken in order to convert the integrals in the expression for $\delta^{(1)}\mathcal{F}$ into integrals that contain only ψ (or $\mathbf{s}$) itself. Considering ψ and its derivatives as being independent would be similar to setting the terms containing the factors $\partial f/\partial y$ and $\partial f/\partial y'$ separately equal to zero in the expression for the variation of I obtained when minimizing the integral $I = \int f(x,y,y')dx$. For this very reason, the argument invoked by Neogi et al. [27] is false and their resulting ΔP expression is of restricted validity. The same criticism is implicitly valid also for the derivation of an expression for ΔP by Melrose [22] using parallel displacements of the dividing surface.

Making use of a divergence theorem for curved surfaces due to Green, Gauss, and Ostrogradsky (cf. Weatherburn [28]), we obtain:

$$\iint_S \boldsymbol{\gamma}_s : \nabla_s \mathbf{s}dS = \int_C \mathbf{m}\cdot\boldsymbol{\gamma}_s\cdot \mathbf{s}dl - \iint_S [\nabla_s\cdot\boldsymbol{\gamma}_s + 2H\mathbf{n}\cdot\boldsymbol{\gamma}_s]\cdot \mathbf{s}dS \qquad (19)$$

where C denotes the boundary curve, which limits the interface, and where the last term in the last integral containing $2H$ vanishes because $\boldsymbol{\gamma}_s$ is a tensor in the tangent plane of the surface. The unit vector $\mathbf{m}$ is perpendicular both to the tangent of the boundary curve and to the normal vector of the surface at the boundary, and it is directed outward. Thus:

$$\iint_S \boldsymbol{\gamma}_s : \nabla_s \mathbf{s}dS = -\iint_S (\nabla_s\cdot\boldsymbol{\gamma}_s)\cdot \mathbf{s}dS + \int_C \mathbf{m}\cdot\boldsymbol{\gamma}_s\cdot \mathbf{s}dl \qquad (20)$$

We next introduce the surface-bending moment tensor:

$$\mathbf{M}_s = \frac{1}{2}(B\mathbf{U}_s + \Theta\mathbf{q}) \tag{21}$$

which is also supposed to be diagonal in the basis of the principal curvatures. The corresponding eigenvalues are denoted by M_1 and M_2. Using another surface divergence theorem for curved surfaces, we obtain:

$$\iint_S \frac{1}{2}(B\mathbf{U}_s + \Theta\mathbf{q}) : \nabla_s\nabla_s\psi dS \equiv \iint_S \mathbf{M}_s : \nabla_s\nabla_s\psi dS = \iint_S \psi\nabla_s\nabla_s : \mathbf{M}_s$$
$$+ \int_C \mathbf{m} \cdot \mathbf{M}_s \cdot \nabla_s\psi dl - \int_C \psi(\nabla_s \cdot \mathbf{M}_s) \cdot \mathbf{m} dl \tag{22}$$

and, finally, we get:

$$\delta^{(1)}F = \iint_S [(B\nabla_s H + \Theta\nabla_s D) - \nabla_s \cdot \boldsymbol{\gamma}_s] \cdot \mathbf{s} dS + \iint_S \psi[\nabla_s\nabla_s : \mathbf{M}_s$$
$$+ \Delta P - 2H\gamma - 2D\zeta + (H^2 + D^2)B + 2HD\Theta] dS \tag{23}$$
$$+ \int_C \mathbf{m} \cdot \mathbf{M}_s \cdot \nabla_s\psi dl - \int_C \psi(\nabla_s \cdot \mathbf{M}_s) \cdot \mathbf{m} dl + \int_C \mathbf{m} \cdot \boldsymbol{\gamma}_s \cdot \mathbf{s} dl = 0$$

The expression within brackets in the first integral of Eq. (23) represents the tangential force acting on the surface element dS, and the expression within brackets in the second integral represents the corresponding normal force. These expressions are, in fact, nothing else than the functional derivatives $\delta\mathcal{F}/\delta\mathbf{s}(\mathbf{r})$ and $\delta\mathcal{F}/\delta\psi(\mathbf{r})$, respectively, with $\mathbf{r}$ denoting the coordinate of the surface element.

The last three contour integrals on the left-hand side of Eq. (23) will vanish, of course, for a geometrically closed interface as the length of the perimeter in such a case shrinks to zero. Likewise, there will be no contributions from these integrals insofar as one can assume that $\nabla_s\psi$, ψ, and $\mathbf{s}$ all vanish on the boundary. It turns out that, for the most part, we can actually invoke this assumption without any loss of generality. Nevertheless, such an assumption does not right away appear to be a valid one for, for example, bicontinuous microemulsions and so-called L_3 phases with interfacial structures resembling infinite periodical minimal surfaces. There is, hence, an unresolved issue about the modes of external closure of such structures for which the detailed implications of the above contour integrals might well turn out to be of importance. In this context, it is worth noting that the coefficients in front of the variables $\nabla_s\psi$, ψ, and $\mathbf{s}$ in the integrands represent the forces and the bending moment actually acting on the boundary.

The scalar surface tension γ is related to the tensor $\boldsymbol{\gamma}_s$ in the following way:

$$\gamma = \frac{1}{2} \mathbf{U}_s : \boldsymbol{\gamma}_s = \frac{1}{2} Tr(\boldsymbol{\gamma}_s) = \frac{1}{2} (\gamma_1 + \gamma_2) \tag{24}$$

where γ_1 and γ_2 (or, more precisely, γ_1^1 and γ_2^2) are the principal components of $\boldsymbol{\gamma}_s$. In an analogous way, the shearing tension is given by the relation:

$$\zeta = \frac{1}{2} \mathbf{q} : \boldsymbol{\gamma}_s = \frac{1}{2} (\gamma_1 - \gamma_2) \tag{25}$$

From Eq. (23), we directly obtain the Euler–Lagrange equations for the variational problem with respect to the variations $\mathbf{s}$ and ψ. Firstly, using Eq. (18):

$$\nabla_s \gamma + \nabla_s \cdot (\zeta \mathbf{q}) - B \nabla_s H - \Theta \nabla_s D = 0 \tag{26}$$

which is the condition for lateral equilibrium (i.e., the condition for the chemical potentials to be constant within the interface), and, second:

$$\Delta P = 2H\gamma + 2D\zeta - (H^2 + D^2)B - 2HD\Theta - \nabla_s \nabla_s : \mathbf{M}_s \tag{27}$$

which is the condition for equilibrium in the direction perpendicular to the interface. This equation agrees with the generalized Laplace equation derived by Kralchevsky et al. [15] and Kralchevsky [18].

It is worth noting that Eq. (27) for ΔP is nothing else than an *extremum condition for the free energy*. To calculate the equilibrium shape of the interface under given conditions, in addition, the numerical value of ΔP must be known. Employing the terminology of Ou-Yang and Helfrich [29], Eq. (27) can also be regarded as a general shape equation. We stress that assigning a value to ΔP is an integral part of the minimization of the free energy and not an extra, unrelated condition.

Upon choosing the surface of tension as the dividing surface for an isotropic interface ($\zeta = 0$) with constant curvature, the Laplace equation (Eq. (27)) reduces to the familiar form $\Delta P = 2H\gamma$ because then both the bending moment B as well as the torsion moment Θ will vanish as we will show in the following.

B. On the Choice of the Free Energy Function

A correct thermodynamic analysis of a system should include two indispensable introductory steps. First, the boundaries of the system have to be carefully defined, implying that it should be prescribed which parts of the universe belong to the system and which do not. Second, the nature of the system has to be clarified, and it should be clearly stated which of the (independent) thermodynamic variables are to be held constant and which will be allowed to vary. Surprisingly, these rather self-evident rules are often overlooked in the literature.

For an interface made up of components that are *soluble* in the adjacent bulk phases (as is frequently the case), the interface has to be treated as a completely open system, for which the chemical potentials are kept constant at equilibrium. Furthermore, if the pressures on the two sides of the interface are unequal, a small displacement of the interface will always be accompanied by mechanical PV work. In order to reckon with this contribution, the system considered cannot be limited to just the interface itself but, in addition, portions of the bulk phases on both sides of the interface have to be encompassed. Assuming the entire system to be an open one, such that matter can pass through its boundaries, we can, without limiting the generality of our treatment, assume a constant overall volume V. Consequently, the variables to be held constant are T, V, and μ_i, which are the characteristic variables of the grand potential $\Omega = F - \Sigma n_i \mu_i$ of the interface plus those parts of the bulk phases that are considered to belong to the system.

On the other hand, if there is no exchange of matter, whatsoever, between the interfacial system and the adjacent bulk phases, the appropriate free energy to invoke is the Helmholtz free energy F. In such a case, the appropriate system variables are, of course, T, V, and all N_i.

For *partially open* interfacial systems, the correct function of state to invoke is a mixed Ω potential/Helmholtz free energy ($\hat{\Omega}$), where the Ω potential part refers to the k soluble, and the Helmholtz free energy part refers to the $r-k$ insoluble components, that is,

$$\hat{\Omega} = F - \sum_{i=1}^{k} N_i \mu_i \tag{28}$$

Thus, deducting the bulk phase contributions for the interface itself, we have, per surface area unit:

$$\hat{\omega}_s = f_s - \sum_{i=1}^{k} \Gamma_i \mu_i \tag{29}$$

where the summation is restricted to the soluble components only, numbered $i = 1, 2, \ldots, k$. The corresponding full ω_s quantity is, analogously:

$$\omega_s = f_s - \sum_{i=1}^{r} \Gamma_i \mu_i \tag{30}$$

where the sum runs over all the r components of the interface. Quite generally, ω_s can be identified with the work of formation of an interface. In the case of a solid–gas interface, for example, ω_s equals the reversible work per unit area of cleaving the solid in the ambient gas [25], whereas, as we shall soon demonstrate, for a completely fluid interface, ω_s equals the ordinary interfacial tension γ.

C. The Gibbs–Duhem Condition for an Interface

Adhering to the Gibbs scheme, we consider a small portion of a curved interface of the surface areas ΔS, where the principal curvatures, c_1 and c_2 are approximately constant (Fig. 3). At constant temperature, and for an arbitrary dividing surface, the Helmholtz free energy differential of this portion can be written as:

$$d(f_s\Delta S) = \gamma d(\Delta S) + \sum_{i=1}^{k} \mu_i d(\Gamma_i\Delta S) + \Delta S(\zeta d\beta + C_1^G dc_1 + C_2^G dc_2) \quad (31)$$

where it should be observed that the system considered is supposed to be closed with respect to the insoluble components numbered $i=k+1,\ldots,r$. Making use of the ω_s definition provided by Eq. (29), we can rewrite Eq. (31) in the following ways:

$$df_s = (\gamma - \hat{\omega}_s)d\alpha + \sum_{i=1}^{k} \mu_i d\Gamma_i + \zeta d\beta + C_1^G dc_1 + C_2^G dc_2 \quad (32)$$

$$d\hat{\omega}_s = (\gamma - \hat{\omega}_s)d\alpha - \sum_{i=1}^{k} \Gamma_i d\mu_i + \zeta d\beta + C_1^G dc_1 + C_2^G dc_2 \quad (33)$$

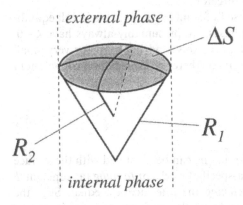

FIG. 3 A small portion ΔS of an interface between an internal phase and an external phase for which the principal curvatures $c_1 = R_1^{-1}$ and $c_2 = R_2^{-1}$ are approximately constant.

where $d\alpha$ stands for the relative surface area increment $d\Delta S/\Delta S$. Furthermore, we can readily switch to the full ω_s potential, resulting in the interfacial Gibbs–Duhem condition:

$$d\omega_s = (\gamma - \omega_s)d\alpha - \sum_{i=1}^{r} \Gamma_i d\mu_i + \zeta d\beta + C_1^G dc_1 + C_2^G dc_2 \qquad (34)$$

where the sum extends over all of the interfacial components, irrespective of whether they are soluble or not. It should be noted, however, that the chemical potentials of the soluble components will be determined primarily by their bulk phase concentrations, whereas the chemical potentials of the insoluble components will be determined, of course, by the conditions prevailing in the interface itself.

In the above expressions, C_1^G and C_2^G stand for the curvature coefficients introduced by Gibbs, which, at constant T, α, and chemical potentials of the soluble components, are defined by $d\omega_s/dc_1$ and $d\omega_s/dc_2$, respectively. In Section II.F, we show that these coefficients actually constitute the principal components of the bending moment tensor $\mathbf{M}_s$ of the interface.

It is worth noting that Eq. (34), which was first derived by Kralchevsky [18] based on some earlier works of Eriksson [25,30] and Rusanov [26], is the most general version of the Gibbs–Duhem condition for an interface obtained so far. Upon studying the text of Gibbs carefully, one finds that he was no doubt fully aware of a somewhat less general version of Eq. (34) where the shearing tension ζ is assumed to be equal to zero. However, not just fluid interfaces but also solid–fluid interfaces and insoluble surface films obey such a less general Gibbs–Duhem condition provided that they fulfil certain minimal rotational symmetry requirements (at least a threefold axis).

Now for *fluid* interfaces, composed of soluble components in full equilibrium with the surrounding bulk phases, we presumably always have $\zeta = 0$. Moreover, not only for large systems of this kind but even for very small systems like the ones primarily considered here (cf. Appendix B), it will hold true that:

$$\left(\frac{d\omega_s}{d\alpha}\right)_{T,\mu_i,c_1,c_2} = 0 \qquad (35)$$

and, hence, it follows from Eq. (34) that ω_s can be identified with the surface tension γ. In other words, the (area-specific) works of *forming* (at constant T and chemical potentials) and *stretching* the interface are equal. Still, the corresponding components of ω_s and γ of the same physical origin can be widely different [31]. In this manner, we obtain a generalized version of the standard Gibbs surface tension equation, valid for an arbitrary dividing surface, with both of the principal curvatures c_1 and c_2 participating:

$$d\gamma = -\sum_{i=1}^{r} \Gamma_i d\mu_i + C_1^G dc_1 + C_2^G dc_2 \tag{36}$$

This equation can, of course, equally well be written in terms of the mean and deviatoric curvatures H and D:

$$d\gamma = -\sum_{i=1}^{r} \Gamma_i d\mu_i + BdH + \Theta dD \tag{37}$$

where the bending moment B stands for the sum of C_1^G and C_2^G and Θ, the torsion moment, for their difference $C_1^G - C_2^G$. Accordingly, by way of definition:

$$B = (\partial\gamma/\partial H)_{T,\mu_i,D} \tag{38}$$

and

$$\Theta = (\partial\gamma/\partial D)_{T,\mu_i,H} \tag{39}$$

Switching to the Gaussian curvature $K = c_1 c_2 = H^2 - D^2$ as the independent variable instead of the deviatoric curvature D, we may write Eq. (37) in the form forwarded by Markin et al. [20]:

$$d\gamma = -\sum_{i} \Gamma_i d\mu_i + C_1 dH + C_2 dK \tag{40}$$

where:

$$C_1 = (\partial\gamma/\partial H)_{T,\mu_i,K} \quad C_2 = (\partial\gamma/\partial K)_{T,\mu_i,H} \tag{41}$$

The coefficients B, Θ, and C_1, C_2 are interrelated through the relationships:

$$B = C_1 + 2C_2 H \tag{42}$$

$$\Theta = -2C_2 D \tag{43}$$

Like B and Θ, both C_1 and C_2 will vanish, of course, when the surface of tension is chosen as the dividing surface.

For a curved interface that is part of an ordinary two-phase system of r components, there will generally be $r + 2$ degrees of freedom, among them the temperature T and two geometrical variables (e.g., H and D). In such a case, one of the differentials $d\mu_1, \ldots, d\mu_r$ in the interfacial Gibbs–Duhem conditions (Eqs. (36), (37), and (40)) will not be an independent differential. However, anyone of the differentials $d\mu_1, \ldots, d\mu_r$ can easily be removed from stage simply by employing the corresponding equimolar dividing surface. To take a specific example, let us consider a small spherical water (Component 1) droplet submerged in an oil (Component 2). Due to the variable Laplace pressure, the water chemical potential will obviously depend on the droplet radius. Thus, a proper exact form of Eq. (37) would be:

$$d\gamma = -\Gamma_{2(l)}d\mu_2 + BdH \tag{44}$$

where $d\mu_2$ and dH can actually be regarded as independent. Alternatively, one can, of course, employ the surface of tension whereby $B = 0$, and $d\mu_1$ and $d\mu_2$ are both considered to be independent.

On the other hand, for surfactant micelles or microemulsion droplets that are formed by aggregation in a one-phase region of a phase diagram, there will be one more degree of freedom, in all $r + 3$: the $r + 1$ thermodynamic state variables plus two geometrical variables, in full agreement with the way we have written the Gibbs–Duhem conditions (Eqs. (36), (37), and (40)) above (where the temperature variable is omitted).

D. The General Laplace Equation According to Boruvka and Neumann

In terms of the mean and Gaussian curvatures H and K, and the new curvature-related parameters C_1 and C_2, the generalized Laplace equation (Eq. (27)) can be rewritten in essentially the same way as the Laplace equation due to Boruvka and Neumann:

$$\Delta P = 2H\gamma + 2D\zeta - C_1\left(2H^2 - K\right) - 2C_2HK - \frac{1}{2}\,\nabla_s^2 C_1 - K\nabla_s^* \cdot (\nabla_s C_2) \tag{45}$$

where, on purely mathematical grounds:

$$K\nabla_s^* \cdot (\nabla_s C_2) = 2H\nabla_s^2 C_2 - \mathbf{b} : \nabla_s\nabla_s C_2 \tag{46}$$

Taking into account that Boruvka and Neumann [10] restricted their treatment to the case of an isotropic surface tension ($\zeta = 0$) and, moreover, that they in effect introduced a special definition of the surface tension of curved interfaces, viz.,

$$\gamma^{BN} = \gamma - C_1 H - C_2 K \tag{47}$$

which, obviously, differs from the common (and more natural) definition of γ as the interfacial Ω potential per unit area, the above Eq. (45) can easily be cast in the same form as the one originally presented by Boruvka and Neumann [loc. cit.]:

$$\Delta P = 2H\gamma^{BN} + 2C_1^{BN}K - \nabla_s^2 C_1^{BN} - K\nabla_s^* \cdot \left(\nabla_s C_2^{BN}\right) \tag{48}$$

where $C_1^{BN} = C_1/2$ and $C_2^{BN} = C_2$.

In his pioneering article from 1956, Buff [7] assumed constant curvature and, furthermore, put the shearing tension $\zeta = 0$. Additionally, following Gibbs, he assumed Θ to be equal to zero even for an arbitrary dividing surface, resulting in the approximate Laplace equation:

$$\Delta P = 2H\gamma - C_1(2H^2 - K) \tag{49}$$

where merely the first and third terms on the right-hand side of Eq. (45) are included. Later, however, Murphy [21] and Melrose [22] derived exact ΔP expressions for the constant curvature case encompassing all of the relevant terms of Eq. (45).

E. The Shape Equation Derived by Ou-Yang and Helfrich

Perhaps the simplest possible model assumption one can do about the curvature coefficients C_1 and C_2 is to put:

$$C_1 = \text{const.}(H - H_0) = 4k_c(H - H_0) \tag{50}$$

where H_0 is the *spontaneous* mean curvature, and:

$$C_2 = \text{const.} = \bar{k}_c \tag{51}$$

The constants k_c and $\bar{k}_c$ introduced in this manner are supposed to depend, but weakly, on the overall thermodynamic state. Contrarily, the spontaneous curvature H_0 may depend strongly on the temperature and the composition. The above extra-thermodynamic relations characterize a "Helfrich interface" for which the interfacial tension γ can be written in integrated form [9]:

$$\gamma = \gamma_0 + 2k_c(H - H_0)^2 + \bar{k}_c K \tag{52}$$

as follows directly from the full version of the Gibbs surface tension equation (Eq. (40)), assuming the chemical potentials to be held constant during the integration, and taking a cylinder of the radius $1/(2H_0)$ to be the initial state. It might be noted, however, that this "Helfrich expression" is so far but formally the same as Helfrich's original expression for the bending free energy f_{curv} of an interface that is *closed* in the thermodynamic sense, whereas presently we are considering a fluid interface of soluble components, which is a completely *open* interfacial system. Thus, the bending constants k_c and $\bar{k}_c$ in Eq. (52) are no true elasticity moduli and, numerically, they are not of the same order as for a closed interface. The reason for this was elucidated by Szleifer et al. [32], who showed that the bending energy of a lipid bilayer was reduced by almost a factor of 100 when the constituent molecules were allowed to exchange between the two sides of the bilayer at the process of bending. A similar effect is obtained, of course, for an open interfacial system in equilibrium, where the constituent molecules exchange with the bulk solution at constant chemical potentials. In such a case, k_c-values on the order of kT have been experimentally recorded as well as theoretically estimated for a number of different systems.

Moreover, in Appendix B, we present some general arguments showing that the Helfrich expression in the form of Eq. (52) can be applied even for aggregates of small size. Hence, in this expression, $-4k_cHH_0$ and $2k_cH^2 + \bar{k}_cK$ stand primarily for the first-order and second-order curvature corrections of the macroscopical interfacial tension, implying that there is no need to separately account for the equilibrium fluctuations of the surface densities that, otherwise, of course, will grow in importance as the aggregate size shrinks.

Upon inserting the above relation (Eq. (52)) in the generalized Laplace equation (Eq. (45)), we obtain the ΔP expression ($\zeta = 0$):

$$\Delta P = 2H\gamma_0 - 4k_c(H - H_0)(H^2 + HH_0 - K) - 2k_c\nabla_s^2 H \tag{53}$$

which, in essence, is the "shape equation" derived in the late 1980s by Ou-Yang and Helfrich [29]. These authors considered a geometrically closed vesicle and employed a carefully elaborated variational approach to minimize the shape free energy. Note that at this level of approximation, the value of the Gaussian bending constant $\bar{k}_c$ does not affect ΔP at all.

For a reversibly formed aggregate of spherical shape, the above ΔP expression yields the simple relation (recall that $\gamma_\infty = \gamma_0 + 2k_cH_0^2$):

$$\Delta P = 2H(\gamma_\infty - 2k_cH_0H) \tag{54}$$

predicting a Laplace pressure that becomes zero for $H = 0$ (which is trivial), and for $H = \gamma_\infty/(2k_cH_0)$ as well. Considering, in particular, large vesicles that tend to form spontaneously only for exceedingly small (flat) bilayer tensions, typically on the order of 10^{-6} N m^{-1} or less [33], the excess pressure inside a vesicle with $H_0 = 0$ of about 10 nm in radius is vanishingly small, at most 100 N m^{-2}, roughly corresponding to the pressure in an air bubble of radius 1 mm submerged in water. Nonetheless, upon taking the entire vesicle size distribution into account, one finds that the *average* excess pressure inside a reversibly formed vesicle must be exactly equal to zero (see Section IV.B).

Let us now briefly go back to the question about employing the generalized Laplace equation (Eq. (45)) as well as the resulting Eq. (53) for a *partially closed* interface as, for example, for the bilayer of a vesicle formed by sonication where the main bilayer lipid (Component 2) does *not* participate in a dynamic equilibrium with the surrounding bulk solution. Referring to our previous discussion about the proper choice of the free energy function, one realizes that the generalized Laplace equation should still be applicable. The same holds true for the general Gibbs equation (Eq. (40)), which in this case can be written as follows by choosing the midplane as dividing surface and by considering local differential changes:

$$d\gamma = -\Gamma_1 d\mu_1 - \Gamma_2^{in} d\mu_2^{in} - \Gamma_2^{out} d\mu_2^{out} + C_1 dH + C_2 dK \tag{55}$$

where the inside and outside chemical potentials μ_2^{in} and μ_2^{out} are defined by the relations:

$$\gamma^{in} = f_s^{in} - \Gamma_1^{in}\mu_1 - \Gamma_2^{in}\mu_2^{in} \tag{56}$$

$$\gamma^{out} = f_s^{out} - \Gamma_1^{out}\mu_1 - \Gamma_2^{out}\mu_2^{out} \tag{57}$$

that correspond with Eq. (30), and where the inside and outside monolayer tensions γ^{in} and γ^{out} are invoked. The chemical potentials μ_2^{in} and μ_2^{out} will depend on the curvature variables H and K, and, in addition, on the state of strain (α).

Suppose we start with a flat bilayer in a certain state of strain, and shape a vesicle without stretching or compressing the bilayer midplane, thus leaving the strain variable α unchanged. In accordance with this scheme, the form of the Gibbs equation (Eq. (40)) can largely be preserved by writing Eq. (55) in the following way:

$$d\gamma = -\Gamma d\mu_1 - \left(\Gamma_2^{in}\frac{\partial\mu_2^{in}}{\partial\alpha} + \Gamma_2^{out}\frac{\partial\mu_2^{out}}{\partial\alpha}\right)d\alpha$$

$$- \left(\Gamma_2^{in}\frac{\partial\mu_2^{in}}{\partial H} + \Gamma_2^{out}\frac{\partial\mu_2^{out}}{\partial H} - C_1\right)dH$$

$$- \left(\Gamma_2^{in}\frac{\partial\mu_2^{in}}{\partial K} + \Gamma_2^{out}\frac{\partial\mu_2^{out}}{\partial\alpha} - C_2\right)dK \tag{58}$$

$$= -\Gamma_1 d\mu_1 - \left(\Gamma_2^{in}\frac{\partial\mu_2^{in}}{\partial\alpha} + \Gamma_2^{out}\frac{\partial\mu_2^{out}}{\partial\alpha}\right)d\alpha + C_1^* dH + C_2^* dK$$

By assuming the Eqs. (50) and (51) to hold for C_1^* and C_2^* as well and by integrating Eq. (58), it is seen that the Helfrich expression Eq. (52)) is applicable also for the partially closed vesicle case under discussion.

As we have already remarked above, however, totally different numerical values are anticipated of the (α-dependent) parameters involved. Consequently, the shape equation (Eq. (53)) should be fulfilled not just for a vesicle that is open in the thermodynamic sense but also for a vesicle made up of a closed bilayer membrane. In particular, for the latter kind of vesicle of spherical shape with $H_0 = 0$, Eq. (54) states that ΔP is determined entirely by the mean curvature H and the tension γ_∞ of the flat bilayer, supposing the latter to be in the same state of (midplane) strain as the curved vesicle bilayer. The excess pressure would hence become zero provided that the flat *tensionless* bilayer actually were the proper reference state.

Finally, we note that by using Eq. (53) and varying the parameters γ_0, H_0, k_c, and ΔP, a number of more complex vesicle shapes have been derived [34], and that theoretical schemes based on the pressure tensor in the interface have been worked out to derive the bending constants k_c and $\bar{k}_c$ [35,36].

F. Different Kinds of Surface Tension and Additional Forms of the Laplace Equation

As discussed above in Section II.A, the local work of deformation per unit surface can be written as:

$$dw_s = \gamma d\alpha + \zeta d\beta + BdH + \Theta dD \tag{59}$$

Except for the term $\zeta d\beta$, this expression was in fact already employed by Gibbs [1]. Gibbs, however, tacitly assumed that $\gamma_1 = \gamma_2$, which implies that $\zeta = 0$, an assumption that presumably holds for a vast majority of interfaces in the liquid state (as opposed to the gel state) but not necessarily for interfaces of a more complicated structure, such as biological membranes or, for the part, rubber membranes and many solid–liquid interfaces. In the above expression, γ denotes the ordinary (thermodynamic) surface tension and is defined as the mean value of the principal components γ_1 and γ_2 of the surface tension tensor $\boldsymbol{\gamma}_s$.

When the dividing surface is chosen to be coincident with the surface of tension, the tensor $\boldsymbol{\gamma}_s$ will generally account for the mechanical tensions, operating in the tangential plane of the interface. However, especially for strongly curved, surfactant-laden interfaces of some thickness with fairly low interfacial tension, it has turned out to be inconvenient to use the surface of tension as the dividing surface, the reason being that the surface of tension is then often located far away from any natural choice dividing surface (cf. Fig. 1). Considering interfaces of *variable* curvature, even the calculation of the location of the surface of tension represents a complicated mathematical problem in its own right. Thus, for a host of interesting interfaces, the surface of tension concept is rather useless. Instead, one has to resort to, for example, the equimolar dividing surface for one of the main components.

At all other dividing surfaces than the surface of tension, the mechanical tensions in the tangential plane at a certain point are given by a surface tension tensor $\boldsymbol{\sigma}_s$ different from $\boldsymbol{\gamma}_s$. This tensor $\boldsymbol{\sigma}_s$ can be written as:

$$\boldsymbol{\sigma}_s = \sigma \mathbf{U}_s + \eta \mathbf{q} \tag{60}$$

where:

$$\sigma = \frac{1}{2}\mathbf{U}_s : \boldsymbol{\sigma}_s = \frac{1}{2}(\sigma_1 + \sigma_2) \quad \eta = \frac{1}{2}\mathbf{q} : \boldsymbol{\sigma}_s = \frac{1}{2}(\sigma_1 - \sigma_2) \tag{61}$$

σ_1 and σ_2 (or, more exactly, σ_1^1 and σ_2^2) being the principal components, assuming, of course, that the tensor $\boldsymbol{\sigma}_s$ is diagonal on the basis of the principal curvatures.

The interrelations between σ and γ, as well as between η and ζ, have been worked out by Gurkov and Kralchevsky [37] who established that in terms of σ and η (instead of γ and ζ), the local work of surface deformation amounts to:

$$dw_s = \left(\sigma + \frac{1}{2}BH + \frac{1}{2}\Theta D\right)d\alpha + \left(\eta + \frac{1}{2}BD + \frac{1}{2}\Theta H\right)d\beta + BdH + \Theta dD \tag{62}$$

A comparison with the previous expression (Eq. (59)) for dw_s then immediately shows that:

$$\gamma = \sigma + \frac{1}{2}BH + \frac{1}{2}\Theta D \tag{63}$$

$$\zeta = \eta + \frac{1}{2}BD + \frac{1}{2}\Theta H \tag{64}$$

with B and Θ being determined by the rather transparent relations:

$$B = \mathbf{U}_s : \mathbf{M}_s = M_1 + M_2 \tag{65}$$

$$\Theta = \mathbf{q} : \mathbf{M}_s = M_1 - M_2 \tag{66}$$

where M_1 and M_2 (or, more exactly, M_1^1 and M_2^2) are the principal components of the bending moment tensor $\mathbf{M}_s$, which is diagonal in the basis of the principal curvatures (i.e., in the principal axes of the curvature tensor $\mathbf{b}$). Note, by the way, that M_1 and M_2 are the same as the curvature-related coefficients C_1^G and C_2^G used by Gibbs (cf. Eq. (36)).

It is worth emphasizing that the pair of equations (Eqs. (63) and (64)) first derived by Gurkov and Kralchevsky in their clarifying paper from 1990 [loc. cit.] are of crucial significance for bridging the gap between the mechanical and thermodynamic treatments of curved interfaces. Moreover, from the above relations (Eqs. (65) and (66)), it is readily seen that B and Θ will vanish when the components M_1 and M_2 of the bending moment tensor become zero (as they do when the surface of tension is chosen as the dividing surface). In passing, we may also note that for a Helfrich interface, we have:

$$\sigma = \gamma_0 - 2k_c H_0(H - H_0) \tag{67}$$

implying that when H equals the spontaneous curvature H_0, σ will be equal to γ_0.

The basic reason for the difference between γ and σ is apparent already from the Eq. (59) for dw_s above, which shows that $dw_s = \gamma d\alpha$ only holds if $d\beta$, dH, and dD are all equal to zero (i.e., in particular if H and D are kept constant). This means that a curved surface should result from the dilation, which implies that different layers of the interface, parallel to the dividing surface, have to be stretched to different degrees, or, differently stated, a dilation is always accompanied by bending, which gives rise to the extra terms in the expression for γ (Eq. (63)), which may also be written as:

$$\gamma = \sigma + \frac{1}{2}\left(M_1 c_1 + M_2 c_2\right) \tag{68}$$

It is obvious that $(M_1 c_1 + M_2 c_2)d\alpha/2$ accounts for the work of bending that accompanies dilation under conditions of constant H and D.

Using Eqs. (63) and (64), the generalized Laplace equation (Eq. (27)) can now be rewritten in terms of σ and η as follows:

$$\Delta P = 2H\sigma + 2D\eta - \nabla_s \nabla_s : \mathbf{M}_s \tag{69}$$

or, because $\sigma = (\sigma_1 + \sigma_2)/2$ and $\eta = (\sigma_1 - \sigma_2)/2$, even more simply as:

$$\Delta P = c_1 \sigma_1 + c_2 \sigma_2 - \nabla_s \nabla_s : \mathbf{M}_s \tag{70}$$

These "mechanical" Laplace conditions evidently imply that for a spherical interface with $\eta = 0$, it will always hold true that:

$$\Delta P = 2H\sigma \tag{71}$$

irrespective of the dividing surface chosen.

Additionally, we note that Eq. (27) can also be put in the form:

$$\Delta P = 2H\gamma + 2D\zeta - \left(M_1 c_1^2 + M_2 c_2^2\right) - \nabla_s \nabla_s : \mathbf{M}_s \tag{72}$$

where:

$$\pi_c = -\left(M_1 c_1^2 + M_2 c_2^2\right) \tag{73}$$

is the *curvature pressure*. Thus, we have:

$$\Delta P = 2H\gamma + 2D\zeta + \pi_c - \nabla_s \nabla_s : \mathbf{M}_s \tag{74}$$

The curvature pressure becomes zero, of course, at the surface of tension where the tensor components M_1 and M_2 vanish. However, as was pointed out earlier, to employ the surface of tension as dividing surface is for the most part inconvenient for strongly curved aggregate surfaces, in particular when ΔP approaches zero.

For the sake of illustration, we may consider the case of a spherical micro-emulsion droplet with curvature-dependent interfacial tension γ for which we get:

$$\Delta P = 2H\gamma + \pi_c = 2H\gamma - BH^2 \qquad (75)$$

or, going back to Eq. (38) and remembering that $H = -1/R$, where R is the radius of the dividing surface:

$$-\Delta P = 2\gamma/R + (\partial\gamma/\partial R)_{T,\mu_i} \qquad (76)$$

showing that ΔP might well become zero for a certain radius provided that the curvature dependence of γ is strong enough, relatively speaking. For that particular value of the droplet radius R, the standard Laplace pressure $2\gamma/R$ and the curvature pressure π_c are of equal magnitude but opposite sign (cf. Ref. 38). Incidentally, for cylindrical geometry, Eq. (76) will have to be replaced by the rather self-evident expression:

$$-\Delta P = \gamma/R + (\partial\gamma/\partial R)_{T,\mu_i} \qquad (77)$$

that may likewise yield $\Delta P = 0$ for a certain cylinder radius provided that γ is curvature-dependent.

G. Generalization to Nondiagonal Surface Stress and Strain Tensors

The tensors $\mathbf{U}_s$ and $\mathbf{q}$ are mutually orthogonal but do not form a complete set of basis tensors. Thus, only such tensors that are diagonal in the basis of the principal curvatures can be expressed as a linear combination of $\mathbf{U}_s$ and $\mathbf{q}$. If we take the analogous case of the Pauli spin matrices, we know that four 2×2 basis matrices are required to express an arbitrary 2×2 matrix. They are the 2×2 unit matrix e and the Pauli spin matrices σ_x, σ_y, and σ_z, of which e and σ_z are diagonal, which means that only diagonal matrices can be expressed as linear combinations of e and σ_z alone. Thus, it was explicitly admitted in the theory of Gurkov and Kralchevsky [37] referred to above that the rate of deformation tensor:

$$d = \frac{1}{2}\dot\alpha\mathbf{U}_s + \frac{1}{2}\dot\beta\,\mathbf{q} \qquad (78)$$

is diagonal in the basis of the principal curvatures as also follows directly from its form. This constitutes a certain limitation of the admissible deformations $\delta\mathbf{R}$. A completely general deformation rate can be represented in the basis of the principal curvatures as:

$$2d = \dot{\alpha} \begin{bmatrix} 1 & 0 \\ 0 & 1 \end{bmatrix} + \dot{\beta} \begin{bmatrix} 1 & 0 \\ 0 & -1 \end{bmatrix} + \dot{\beta}_1 \begin{bmatrix} 0 & 1 \\ 1 & 0 \end{bmatrix} \tag{79}$$

where the three matrices correspond to e (the unit matrix), σ_z, and σ_x. The matrix:

$$q_2 = \begin{bmatrix} 0 & 1 \\ -1 & 0 \end{bmatrix}$$

proportional to σ_y is not included because it is antisymmetrical (the deformation tensor being, by definition, symmetrical) and because it would represent a pure rotation. In tensor form, the above equation can be written in the following way (replacing Eq. (16)):

$$2\mathbf{d} = \dot{\alpha}\mathbf{U_s} + \dot{\beta}\mathbf{q} + \dot{\beta}_1\mathbf{q}_1 \tag{80}$$

where $\mathbf{q}_1$ is obtained from $\mathbf{q}$ by rotation through an angle equal to $45°$ about the surface normal. Then Eq. (15) may be extended by the following relations:

$$\mathbf{q}_1 : \mathbf{q}_1 = 2 \quad \mathbf{U_s} : \mathbf{q}_1 = 0 \quad \mathbf{q} : \mathbf{q}_1 = 0$$
$$\mathbf{q} \cdot \mathbf{q}_1 = \mathbf{q}_2 \quad \mathbf{q} \cdot \mathbf{q}_2 = \mathbf{q}_1 \quad \mathbf{q}_1 \cdot \mathbf{q}_2 = \mathbf{q} \tag{81}$$

and, furthermore, it can be shown that:

$$\mathbf{q} \cdot \mathbf{q}_1 = \mathbf{q} \cdot \mathbf{q} = \mathbf{U_s} \tag{82}$$

In summary, we may say that from a qualitative point of view, Eq. (78) describes deformations of dilation and shear engendered by the bending and twisting of the interface. From this viewpoint, the new term with the tensor $\mathbf{q}_1$ in Eq. (80) describes an additional (rate of) shear, which is independent of (or decoupled from) the changes of the interfacial shape. Such a shear deformation may, for instance, be due to a two-dimensional convective flow of the surface molecules.

Because, in general, for completely arbitrary membranes, there is no reason why σ_s should be diagonal in the basis of the normal curvatures, the definition equation (Eq. (60)) will now be generalized to read:

$$\sigma_s = \sigma\mathbf{U_s} + \eta\mathbf{q} + \eta_1\mathbf{q}_1 \tag{83}$$

where

$$\sigma = \frac{1}{2}\mathbf{U}_s : \sigma_s \quad \eta = \frac{1}{2}\mathbf{q} : \sigma_s \quad \eta_1 = \frac{1}{2}\mathbf{q}_1 : \sigma_s \tag{84}$$

Then:

$$\sigma_s : \mathbf{d} = \sigma\dot{\alpha} + \eta\dot{\beta} + \eta_1\dot{\beta}_1 \tag{85}$$

Furthermore, according to Gurkov and Kralchevsky [loc. cit.], the rate of deformation work per unit surface equals:

$$\frac{\mathrm{d}w_s}{\mathrm{d}t} = \sigma_s : \mathbf{d} + \mathbf{M}_s : \left(\frac{\partial\mathbf{b}}{\partial t} - \mathbf{b}\cdot\mathbf{d}\right) \tag{86}$$

The first term on the right-hand side has already been evaluated above. The second term obviously depends on the nature of $\mathbf{M}_s$. In the general case, the calculus becomes extremely involved. However, in the special case where $\mathbf{M}_s$ is diagonal in the basis of the normal curvatures, which may possibly be a less stringent condition, at least for isotropic interfaces, than the corresponding condition allied to σ_s, the second term in the above equation for $\mathrm{d}w_s/\mathrm{d}t$ will contain no contribution whatever from the term $1/2\beta_1\mathbf{q}_1$ the expression for $\mathbf{d}$, as is easily shown by using the tensor relation:

$$\mathbf{A} : (\mathbf{b}\cdot\mathbf{C}) = (\mathbf{A}\cdot\mathbf{b}) : \mathbf{C} \tag{87}$$

We thus obtain the following slightly modified form of Eq. (62):

$$\begin{aligned}\mathrm{d}w_s = {} &\left(\sigma + \frac{1}{2}BH + \frac{1}{2}\Theta D\right)\mathrm{d}\alpha \\ &+ \left(\eta + \frac{1}{2}BD + \frac{1}{2}\Theta H\right)\mathrm{d}\beta + B\mathrm{d}H + \Theta\mathrm{d}D + \eta_1\mathrm{d}\beta_1\end{aligned} \tag{88}$$

Modifying Eq. (59) in a corresponding manner:

$$\mathrm{d}w_s = \gamma\mathrm{d}\alpha + \zeta\mathrm{d}\beta + \zeta_1\mathrm{d}\beta_1 + B\mathrm{d}H + \Theta\mathrm{d}D \tag{89}$$

we obtain the following relationships:

$$\gamma = \sigma + \frac{1}{2}BH + \frac{1}{2}\Theta D \tag{90}$$

$$\zeta = \eta + \frac{1}{2}BD + \frac{1}{2}\Theta H \tag{91}$$

$$\zeta_1 = \eta_2 \tag{92}$$

where we might note that Eqs. (90) and (91) are the same as the previous Eqs. (63) and (64).

Returning to the minimum conditions, we note again that:

$$\mathbf{d}\mathrm{d}t = \nabla_s \mathbf{s} - \psi \mathbf{b} \tag{93}$$

which means that in addition to the variations in Eqs. (10)–(13), we have the additional variation:

$$\delta\beta_1^{(1)} = \mathbf{q}_1 : \mathbf{d}\mathrm{d}t = \mathbf{q}_1 : \nabla_s \mathbf{s} \tag{94}$$

From Eq. (93), it also follows that:

$$\delta\alpha^{(1)} = \mathbf{U}_s : \mathbf{d}\mathrm{d}t = \mathbf{U}_s : \nabla_s \mathbf{s} - 2H\psi$$

and

$$\delta\beta^{(1)} = \mathbf{q} : \mathbf{d}\mathrm{d}t = \mathbf{q} : \nabla_s \mathbf{s} - 2D\psi$$

Thus, we have recovered Eqs. (10) and (11). In the integrand of the second integral in Eq. (8), we now have an additional term $\zeta_1 \delta\beta_1$, and Eq. (18) has to be replaced by:

$$\boldsymbol{\gamma}_s = \gamma \mathbf{U}_s + \zeta \mathbf{q} + \zeta_1 \mathbf{q}_1 \tag{95}$$

Moreover, Eq. (25) has to be replaced by the two equations:

$$\zeta = \frac{1}{2} \mathbf{q} : \boldsymbol{\gamma}_s \tag{96}$$

$$\zeta_1 = \frac{1}{2} \mathbf{q}_1 : \boldsymbol{\gamma}_s \tag{97}$$

The general Laplace equation (Eq. (27)) remains unaffected, but Eq. (26) will have to be replaced by:

$$\nabla_s \cdot \boldsymbol{\gamma}_s - B\nabla_s H - \Theta\nabla_s D = 0 \tag{98}$$

or

$$\nabla_s \gamma + \nabla_s \cdot (\zeta\mathbf{q} + \zeta_1\mathbf{q}_1) - B\nabla_s H - \Theta\nabla_s D = 0 \tag{99}$$

from which it follows that the Laplace equations (Eqs. (69) and (72)) remain valid. On the other hand, Eq. (70) does not hold any more because $\eta = \mathbf{q}{:}\boldsymbol{\sigma}_s/2$ can no longer be put equal to $(\sigma_1 - \sigma_2)/2$. Instead, the more general expression:

$$\Delta P = \mathbf{b} : \boldsymbol{\sigma}_s - \nabla_s \nabla_s : \mathbf{M}_s \tag{100}$$

has to be employed, which is always valid.

Finally, we note that Eqs. (68) and (73) remain valid, and that Eq. (68) can be written in the following alternative way:

$$\gamma = \sigma + \frac{1}{2}\mathbf{b} : \mathbf{M_s} \tag{101}$$

and Eq. (73) as follows:

$$\pi_c = \mathbf{U_s} : (\mathbf{b} \cdot \mathbf{M_s} \cdot \mathbf{b}) \tag{102}$$

In conclusion, it should be pointed out that both $\mathbf{M_s}$ and σ_s of surfactant-laden interfaces in a fluid state are likely to be diagonal in the basis of the principal curvatures because, locally, there are no preferential directions in the surface except the directions of principal curvature.

H. Comments on the Stability Issue

For an interfacial equilibrium to be stable, it will be necessary, of course, not just that the first variation $\delta^{(1)}\mathcal{F}$ of the relevant free energy function $\mathcal{F}$ equals zero, but, in addition, that the second variation $\delta^{(2)}\mathcal{F}$ is larger than zero. Determining the sign of the latter can occasionally be far from a trivial matter. Moreover, whether or not a certain interfacial organization corresponds to a stable equilibrium situation may crucially depend on what boundary conditions are actually valid. Hence, it is hardly surprising that, so far, we have lacked general guidelines about stability questions.

Let us dwell upon this point by discussing a few examples. First, we shall consider a small *spherical gas bubble* submerged in water (Component 1) that is *supersaturated* with the gaseous component (Component 2). Choosing the surface of tension as the dividing surface, the Laplace condition in its standard form, that is,

$$-\Delta P \equiv P_i - P_e = 2\gamma/R \tag{103}$$

must evidently be fulfilled at equilibrium. Furthermore, assuming the equilibrium to be rapidly established between the gas in the bubble and the gas dissolved in the surrounding water phase, the proper free energy function to invoke (which supposedly has a minimum for the particular radius determined by the Laplace condition) is the change of the Ω potential, or, upon forming the bubble in a fixed overall volume V_{tot}:

$$\mathcal{F} = \Delta\Omega = \Delta PV + \gamma S \tag{104}$$

where V denotes the bubble volume and S is the surface area of the bubble (Fig 4). At equilibrium, this free energy function has to be minimal at constant T, μ_1, and μ_2. Differentiation with respect to R yields the Laplace condition

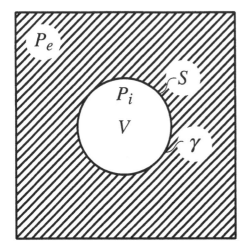

FIG. 4 A spherical gas bubble submerged in water saturated with gas at the pressure of the gas in the bubble. The equilibrium is stable for a thermodynamically closed bubble but unstable for an open bubble.

(Eq. (103)) as anticipated. However, by differentiating a second time and by inserting the Laplace condition, one obtains:

$$\frac{\partial^2 \mathcal{F}}{\partial R^2} = -8\pi\gamma \tag{105}$$

showing that $\mathcal{F}$ has, in fact, a maximum and therefore the equilibrium is *unstable*. Hence, gas bubbles with a smaller radius than what corresponds to the $\mathcal{F}$ maximum will tend to become reduced in size, whereas larger ones will grow, the course of events being governed largely by kinetic factors [39]. This situation evidently parallels Ostwald ripening of small crystallites.

Conversely, assuming instead that the small gas bubble is *closed* with respect to the gaseous component, the proper free energy to employ is rather the following:

$$\mathcal{F} = \Delta\hat{\Omega} = \Delta PV + (\mu_{2i} - \mu_2)N_{2i} + \gamma S \tag{106}$$

where N_{2i} denotes the number of moles of gas enclosed, and the bubble pressure P_i as well as the chemical potential of the gaseous component inside the bubble μ_{2i} will vary with the size of the bubble. Making use of the gas law, we can easily rewrite Eq. (106) in the following manner:

$$\mathcal{F}/4\pi = P_e R^3/3 + \gamma R^2 - (3/4\pi)N_{2i}R_{\text{gas}}T\ln R + \text{const.} \tag{107}$$

Again, putting $\partial \mathcal{F}/\partial R$ equal to zero results in the Laplace equation (Eq. (103)). This time, however, the second derivative becomes (for $R = -2\gamma/\Delta P$):

$$\frac{\partial^2 \mathcal{F}}{\partial R^2} = 4\pi(4\gamma + 3P_e R) \tag{108}$$

and is thus normally positive, which means that the $\mathcal{F}$ minimum for the radius where the Laplace condition is satisfied. Consequently, the equilibrium is stable.

Relatively large bubbles with low excess pressure are likely to behave as if they were closed, whereas the reverse should hold true for very small bubbles. Thus, upon increasing the bubble size, a transition takes place from instability to practically complete stability. Invoking shape deformations will not necessitate any major modifications of the above discussion as the spherical shape is always associated with minimal free energy for a given bubble volume.

A similar scheme is relevant also for an *emulsion droplet*, say of an oil (Component 2) covered with surfactant (Component 3) in water (Component 1). An excess of oil is supposed to be present, which fixes the chemical potential μ_2 of the oil dissolved in the water. As the dividing surface, it is convenient to employ the equimolar dividing surface of the oil, tacitly assuming that we can nevertheless neglect the curvature dependence of the interfacial tension. Then the free energy function to consider is primarily the following:

$$\mathcal{F} = \Delta \hat{\Omega} = \Delta PV + (\mu_{2i} - \mu_2)N_{2i} + \gamma S \tag{109}$$

which is formally the same as the one given by Eq. (106) above, but where μ_{2i} is now the size-dependent chemical potential of the oil inside the droplet. By invoking density fluctuations for a *closed* spherical droplet, containing a fixed mole number N_{2i} of the oil, and by taking into account the Gibbs–Duhem condition:

$$-VdP_i + N_{2i}d\mu_{2i} = 0 \tag{110}$$

from Eq. (109), we can readily derive the ordinary Laplace equation. The corresponding stability condition is found to be:

$$-\frac{\partial P_i}{\partial R} > \frac{2\gamma}{R^2} \tag{111}$$

and is fulfilled for any approximately incompressible liquid for all R values except in the limit $R = 0$. In passing, we may note that the above condition is in fact general enough to encompass the stability condition for an enclosed gas (Eq. (108)).

On the other hand, the chemical potential of the nearly incompressible oil inside the droplet μ_{2i} will, to a good approximation, be larger than μ_2 by the amount $-\Delta P v_2 = (2\gamma/R)v_2$, with v_2 denoting the molar volume of the oil, as follows from integrating Eq. (110). Consequently, by using the obvious relation $Nv_2 = (4\pi/3)R^3$ and differentiating Eq. (109) with respect to R, we find that:

$$\frac{\partial \mathcal{F}}{\partial R} = \Delta P \frac{dV}{dR} + \left(\frac{\mu_{2i} - \mu_2}{v_2} \right) \frac{dV}{dR} + \gamma \frac{dS}{dR} \tag{112}$$

where, however, the first two terms on the right-hand side will cancel. Thus, for an *open* emulsion droplet of practically incompressible oil, it is usually sufficient to invoke the simple free energy function $\mathcal{F} = \gamma S$, and because $\partial \mathcal{F}/ \partial R = 8\pi R\gamma$ is always larger than zero, equilibrium with respect to varying N_{2i} can never be reached for a finite size. Needless to say, however, that the rate of change will be very sluggish when the chemical potential difference becomes extremely small as for large droplets.

An entirely different situation may arise when the interfacial tension γ becomes small enough to exhibit an appreciable *curvature dependence* that, approximately at least, is accounted for by the Helfrich expression, Eq. (52). In the Winsor I case, for instance, where we have small oil droplets dressed by surfactant dispersed in water and where an excess oil phase is present (Fig. 5), the (spherical) droplet free energy function $4\pi R^2\gamma$ passes through a minimum for $R = R_{eq}$. At this particular radius, the Laplace equation written in the form of Eq. (76) yields $\Delta P = 0$, and thus the condition of equal chemical potential in the droplet and the excess phase is satisfied. Around the minimum (where the curvature of the free energy function amounts to $8\pi\gamma_\infty$), equilibrium fluctuations in size and shape occur, which have important entropic implications [40].

The stability analysis of spherical Helfrich interfaces can be carried through in a somewhat broader manner by minimizing the following free energy function of the droplet system:

$$\mathcal{F} = \Delta P V + \gamma S \tag{113}$$

noting that now we have $\gamma = \gamma(R)$. The additional term $\Delta P V$ is relevant for single-phase microemulsions and accounts for the free energy needed to transport the oil (at fixed chemical potential) to a hypothetical pure state from which the droplet then is formed. Differentiation with respect to R yields the Laplace condition in the form noted previously (Eq. (76)):

$$-\left(\frac{2\gamma}{R} + \frac{\partial \gamma}{\partial R} \right) = \Delta P \tag{114}$$

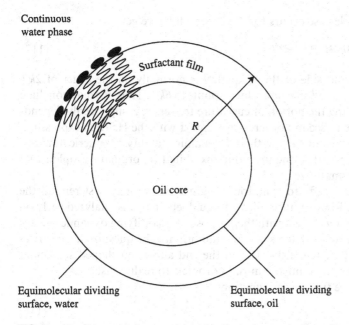

Continuous
water phase

Surfactant film

R

Oil core

Equimolecular dividing
surface, water

Equimolecular dividing
surface, oil

FIG. 5 An oil/water microemulsion droplet composed of an oil core dressed with a surfactant film. When an excess phase is present (Winsor I case), the chemical potential of the oil in the fully equilibrated droplet is the same as in the excess oil phase.

implying that, by means of the derivative term, the pressure inside the droplet is lowered to yield the same chemical potential of the oil inside the droplet as in the outside solution. From the second derivative with respect to the radius R and the Helfrich expression (Eq. (52)), we obtain the following stability criterion:

$$\gamma_\infty + R_{eq}\Delta P > 0 \tag{115}$$

with R_{eq} denoting the equilibrium radius where Eq. (114) is fulfilled. As $\Delta P \geq 0$, the above criterion is readily satisfied for practically all cases of interest.

The stability (minimum) condition with respect to a deformation mode as described by a spherical harmonics Y_{lm} has been derived by Ou-Yang and Helfrich [29]. It has the form:

$$\Delta P = -\left(\frac{2\gamma_\infty}{R_{eq}} + \frac{4k_c H_0}{R_{eq}^2}\right) \leq \frac{2k_c}{R_{eq}^2}\left[\ell(\ell+1) - H_0 R_{eq}\right] \tag{116}$$

The most unstable case occurs for $\ell = 2$, for which we get:

$$-\left(\gamma_\infty R_{eq}^2 + 3k_c H_0 R_{eq}\right) \leq 6k_c \tag{117}$$

where the left-hand side of the inequality is normally on the order of $2kT$ whereas the right-hand side typically amounts to $6kT$. Thus, we may conclude that by introducing the notion of curvature free energy, supposedly depending on curvature to second order in agreement with the Helfrich expression, we can readily account for the thermodynamic stability of spherical microemulsions as opposed to the instability exhibited by ordinary droplets and bubbles of very small size.

Next, turning briefly to cylindrical aggregates, we may first remind the reader of the well-known instability of liquid jets that was analyzed early on by Bohr [41] among others, with the purpose of quantifying dynamic surface tensions of interfacial systems out of equilibrium. The question hence arises about what molecular mechanisms in the end allow rod-shaped surfactant micelles and cylindrical microemulsion droplets to really exist.

Minimizing the free energy function:

$$\mathcal{F} = 2\pi R \gamma L \tag{118}$$

of an *open* Winsor I or Winsor II type of cylindrical droplet system with respect to changes of R at constant cylinder length L, we first recover the Laplace condition in the correct form for the present case, viz.,

$$\frac{\gamma}{R} + \frac{\partial \gamma}{\partial R} = 0 \tag{119}$$

and, furthermore, using the relevant Helfrich expression for cylindrical geometry:

$$\gamma = \gamma_\infty + \frac{2k_c H_0}{R} + \frac{k_c}{2R^2} \tag{120}$$

we find that the equilibrium radius is given by the relation:

$$\gamma_\infty R_{eq}^2 = k_c/2 \tag{121}$$

and that the stability criterion simply requires k_c to be larger than zero. Moreover, mechanical equilibrium in the direction perpendicular to the cylinder axis prevails if:

$$-\Delta P = 2\gamma/R_{eq} \tag{122}$$

which means that γ must in fact be equal to zero as we have assumed from the outset that $\Delta P = 0$. In other words, under full equilibrium conditions, because the work of formation is nil, the cylinder will grow to infinite length.

However, cylindrical aggregates may also form as a result of fluctuations if the associated entropy gain is sufficiently large to compensate for the work of formation expended [42]. In that case, the inside pressure is but slightly higher than the outside pressure; likewise, the chemical potential of the interior component is just a little above its value in the outside solution. Under these circumstances, the Laplace equation and the condition of mechanical equilibrium perpendicular to the cylinder axes will result in the relation:

$$\frac{\gamma}{R} - \frac{\partial \gamma}{\partial R} = 0 \tag{123}$$

which, upon making use of the Helfrich expression, yields the equation:

$$\gamma_\infty + \frac{4k_c H_0}{R} + \frac{3k_c}{2R^2} = 0 \tag{124}$$

from which one can conclude the following, referring, in particular, to Winsor I and Winsor II microemulsion systems, which incorporates excess phases of oil and water, respectively [42]:

(i)   Stable and fully equilibrated droplets of cylindrical shape (for which both γ and ΔP must be zero) will be of infinite length and have a curvature equal to the spontaneous curvature H_0 and an interfacial tension γ_0 equal to zero. Conversely, claiming a priori that γ_0 equals zero will inevitably imply the formation of infinite cylindrical aggregates of a radius equal to $1/2H_0$.

(ii) Cylindrical aggregates of finite length with γ larger than zero, in which the dispersed component has a chemical potential slightly above the corresponding value in the excess phase, may form as a result of equilibrium fluctuations. Their curvature has to be between H_0 and $2/3H_0$, and the value of the γ_0 parameter has to fall within the range:

$$0 \le \gamma_0 \le \frac{2}{3} k_c H_0^2 \tag{125}$$

calling for γ_0 values on the order of 10^{-5} N m^{-1} or less.

By and large, stability with respect to *shape variations* also depends upon having a positive bending constant k_c. According to Ou-Yang and Helfrich [29], a cylindrical aggregate becomes unstable with respect to a sinusoidal variation of the radius as a function of the coordinate z along the axis with the period $2\pi nR$ when:

$$n^2 q^2 - 1 \le \sqrt{2(2H_0 R_{eq} - 1)} \tag{126}$$

where $q = 2\pi R/L$. For a long-enough cylinder for which $R_{eq} = 1/2H_0$, the least stable mode was found to be characterized by $n^2 q^2 = 1$, which means that in such case, the period of the sinusoidal deformation equals the perimeter length $2\pi R$. Nevertheless, it is evident from the above discussion that the notion of curvature elasticity enables us to account not just for the existence of spherical microemulsions and vesicles but likewise for the corresponding cylindrical aggregates.

For an open interface containing surfactant molecules, the bending constants k_c and $\bar{k}_c$ are largely determined by competing entropic factors on the water and oil sides of the interface, respectively, that depend on the curvature of the interface. In the end, it is mostly a matter of what volume is accessible to the head groups and the counterions, in relation to the volume accessible to the hydrocarbon tails.

Finally, we note that the stability with respect to shape deformations of infinite *minimal surfaces* has been proved earlier by the present authors [43]. Interfacial structures of this nature are characterized by the mean curvature H being everywhere equal to zero and a channel-to-channel ratio equal to unity. A special class of infinite periodical surfaces covering a wider range of composition, called *gyroid surfaces*, which are particularly interesting for modelling purposes, have the same mean curvature (equal to or different from zero) at every point in the surface [44]. One might claim, of course, that it would be rather astonishing if the shape stability would depend in a crucial manner on the exact value of H, but so far, a more general proof seems to be lacking.

Extended interfacial structures organized in this gyroid manner (Fig. 6) might be contemplated as models of *bicontinuous* middle phase microemulsions and L$_3$ phases. To have the chemical potentials the same everywhere (and equal to their values in coexisting bulk phases), the pressure has to be same as the external pressure throughout both of the channel systems. Noting that the Gaussian curvature K will vary from point to point, always being less than or equal to zero, and referring once more to the Laplace equation for a Helfrich interface (Eq. (52)), it is seen that this ΔP condition can most readily be met by putting γ_0 equal to zero and H equal to the spontaneous curvature H_0. Additionally, however, there can be no resulting excess in interfacial free energy for a bulk phase of homogeneous pressure. Such a requirement might call for adding a quadratic term in K to the standard Helfrich expression represented by Eq. (52), but this issue is still in need of further clarification [43,45].

Summarizing our discussion of the stability issue, we may first observe that the Laplace equilibrium is generally a stable one for closed interfacial systems (i.e., on a short-enough time scale). The long-term stability, however, is a rather different matter, which will depend upon whether or not there is a free

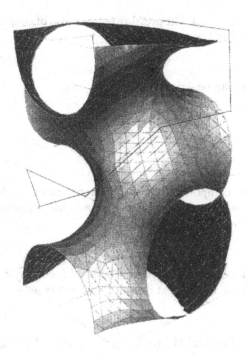

FIG. 6 Translational fundamental domain of the minimum gyroid that divides space into two geometrically equivalent channels. (From Ref. 44.)

energy minimum for an interfacial organization such that the value of the internal pressure P_i complies with the condition of having equal chemical potential throughout the system. Apparently, true long-term stability can only arise when the interfacial tension is low enough to exhibit a noticeable curvature dependence as given by the (approximate) Helfrich expression, Eq. (52). In this regime, a variety of thermodynamically stable surfactant aggregates and lipid membrane structures tend to form.

III. MICROMECHANICAL EXPRESSIONS FOR SURFACE EXCESS QUANTITIES

A. General Relations

All the relevant interfacial quantities can be expressed as integrals over the three-dimensional pressure tensor **P** of the interface regarded as a three-dimensional body. The pressure tensor is simply the negative of the three-dimensional mechanical stress tensor σ (to be distinguished from the two-dimensional mechanical stress tensor σ_s). The pressure tensor must, of course, satisfy the general mechanical equilibrium condition:

$$\nabla \cdot \boldsymbol{P} = 0 \qquad (127)$$

which, for a spherical interface, can be written as:

$$d\left(r^2 P_N\right)/d\left(r^2\right) = P_T \qquad (128)$$

or

$$r^2\left[dP_N/d\left(r^2\right)\right] = P_T - P_N \qquad (129)$$

with P_T and P_N denoting the components of the pressure tensor in the tangential and radial directions, respectively. The corresponding equation for a cylindrical interface has the following form:

$$d(rP_{rr})/dr = P_{\varphi\varphi} \qquad (130)$$

where P_{rr} is the radial component and $P_{\varphi\varphi}$ is the tangential component perpendicular to the cylinder axis. Let us next introduce the excess tensor:

$$\boldsymbol{P}^s = \boldsymbol{P} - P_{\alpha\beta}\mathbf{U} \qquad (131)$$

where $\mathbf{U}$ is the three-dimensional (or bulk) identity tensor, related to $\mathbf{U}_s$ in the following way:

$$\mathbf{U} = \mathbf{U}_s + \boldsymbol{nn} \qquad (132)$$

bearing in mind that $\mathbf{U}_s = \mathbf{a}^1\mathbf{a}_1 + \mathbf{a}^2\mathbf{a}_2$. Furthermore, we introduce the conventional notation:

$$P_{\alpha\beta} = \begin{cases} P_\alpha \text{ if } \lambda < 0 \\ \\ P_\beta \text{ if } \lambda < 0 \end{cases} \qquad (133)$$

where λ is a coordinate along the surface normal that equals zero at the dividing surface, P_α is the homogeneous isotropic pressure on the α-side of the interface, and P_β is the corresponding pressure on the β-side. As elaborated earlier, in particular by Kralchevsky [18], we then have the following micromechanical definitions of the two surface tension tensors:

$$\boldsymbol{\gamma}_s = -\int_{\lambda_1}^{\lambda_2} \boldsymbol{P}^s\chi d\lambda \quad \boldsymbol{\sigma}_s = -\int_{\lambda_1}^{\lambda_2} \boldsymbol{L} \cdot \boldsymbol{P}^s d\lambda \qquad (134)$$

where:

$$\chi = (1 - \lambda H)^2 - \lambda^2 D^2 = 1 - 2\lambda H + \lambda^2 K \qquad (135)$$

is recognized as the local dilation factor for the area of a surface that is parallel with the dividing surface. The tensor $\mathbf{L}$ is defined as:

$$\mathbf{L} = (1 - 2\lambda H)\mathbf{U}_s + \lambda \mathbf{b} \tag{136}$$

and serves the following purpose. In order to calculate the tangential tensions and the bending moments, we need to know the forces acting on a vectorial surface element dS_{orth} of a sectorial strip that is perpendicular to the dividing surface. Then, according to Eliassen [23]:

$$dS_{orth} = \boldsymbol{v} \cdot \mathbf{L}d\lambda dl \tag{137}$$

where $\boldsymbol{v}$ is a vector in the tangent plane of the dividing surface, normal to the abovementioned perpendicular sectorial strip at λ values different from zero. Thus, $\boldsymbol{v} \cdot \mathbf{L}$ is a vector perpendicular to the sectorial strip. Furthermore, because we have the following definitions:

$$\sigma = \frac{1}{2}\mathbf{U}_s : \sigma_s \quad \gamma = \frac{1}{2}\mathbf{U}_s : \boldsymbol{\gamma}_s \tag{138}$$

$$\eta = \frac{1}{2}\mathbf{q} : \sigma_s \quad \zeta = \frac{1}{2}\mathbf{q} : \boldsymbol{\gamma}_s \tag{139}$$

which, due to the orthogonality of $\mathbf{U}_s$ and $\mathbf{q}$, are consistent with:

$$\boldsymbol{\gamma}_s = \gamma\mathbf{U}_s + \zeta\mathbf{q} \tag{140}$$

$$\sigma_s = \sigma\mathbf{U}_s + \eta\mathbf{q} \tag{141}$$

It follows that

$$\gamma = -\frac{1}{2}\int_{\lambda_1}^{\lambda_2} \mathbf{U}_s : \boldsymbol{P}^s\chi d\lambda \tag{142}$$

$$\zeta = -\frac{1}{2}\int_{\lambda_1}^{\lambda_2} \mathbf{q} : \boldsymbol{P}^s\chi d\lambda \tag{143}$$

and

$$\sigma = -\frac{1}{2}\int_{\lambda_1}^{\lambda_2} \mathbf{L} : \boldsymbol{P}^s d\lambda \tag{144}$$

$$\eta = -\frac{1}{2}\int_{\lambda_1}^{\lambda_2} (\mathbf{q} \cdot \mathbf{L}) : \boldsymbol{P}^s d\lambda \tag{145}$$

Furthermore, the bending moment tensor is defined by:

$$\mathbf{M} = \frac{1}{2}\int_{\lambda_1}^{\lambda_2} \mathbf{L} \cdot \boldsymbol{P}^s\chi d\lambda \tag{146}$$

which is a tensor in three dimensions. The corresponding two-dimensional surface tensor equals:

$$\mathbf{M}_s = \frac{1}{2}\left(\mathbf{U}_s\mathbf{U}_s + \mathbf{qq}\right) : \mathbf{M} = \frac{1}{2}\left(B\mathbf{U}_s + \Theta\mathbf{q}\right) \tag{147}$$

as was already anticipated above. Because the bending moment B equals $1/2\mathbf{U}_s{:}\mathbf{M}$ and the torsion moment Θ equals $1/2\mathbf{q}{:}\mathbf{M}$, it follows that:

$$B = \int_{\lambda_1}^{\lambda_2} \mathbf{L} : \boldsymbol{P}^s \lambda \mathrm{d}\lambda \tag{148}$$

$$\Theta = \int_{\lambda_1}^{\lambda_2} (\mathbf{q} \cdot \mathbf{L}) : \boldsymbol{P}^s \lambda \mathrm{d}\lambda \tag{149}$$

It should be mentioned that Eqs. (148) and (149) as well as Eqs. (142) and (143) were first obtained by Ivanov and Kralchevsky [46]. Below, we treat the special cases of spherical and cylindrical interfaces. Most of the formulas quoted have been published earlier (cf. Ref. 37).

B. Spherical Interfaces

As the first example of the application of the above relations, we take the case of a spherical interface. If the radius of the dividing surface is denoted by R, the following relations emerge:

$$\gamma = \frac{1}{R^2}\int_0^\infty \left(P_{\alpha\beta} - P_T\right)r^2\mathrm{d}r \tag{150}$$

$$\sigma = \frac{1}{R}\int_0^\infty \left(P_{\alpha\beta} - P_T\right)r\mathrm{d}r \tag{151}$$

where, now:

$$P_{\alpha\beta} = \begin{cases} P_\alpha \text{ if } r < R \\ P_\beta \text{ if } r > R \end{cases} \tag{152}$$

These relations display γ as being primarily an energetic quantity and σ as a force per unit length, respectively. Furthermore, the bending moment B is given by:

$$B = -\frac{2}{R}\int_0^\infty \left(P_{\alpha\beta} - P_T\right)r(r - R)\mathrm{d}r \tag{153}$$

Consequently:

$$\gamma = \sigma + \frac{1}{2} BH \tag{154}$$

The Laplace equation reads:

$$\Delta P = 2H\gamma - BH^2 = 2H\sigma \tag{155}$$

with $H = -1/R$, where we may note that $-BH^2 = -B/R^2$ equals the curvature pressure π_c. Finally, we have:

$$\mathbf{M} = \mathbf{M}_1 = \mathbf{M}_2 = -\frac{1}{R}\int_0^\infty (P_{\alpha\beta} - P_T)r(r - R)dr = B/2 \tag{156}$$

Whence, from Eq. (154):

$$\gamma = \sigma + Mc \tag{157}$$

where η is, of course, equal to zero for a spherical surface implying that γ and σ are equal at the surface of tension for which $M = 0$. The condition of lateral mechanical equilibrium requires that $\nabla_s\gamma$ and $B\nabla_s H$ both be equal to zero, which is obviously satisfied on a sphere.

The *work of formation* of a spherical aggregate at constant chemical potentials can be written in several different ways:

$$w_f = 4\pi \int_0^\infty (P_\beta - P_T)r^2 dr = 4\pi R^2\gamma - 4\pi \int_0^R (P_\alpha - P_\beta)r^2 dr$$
$$= (4\pi R^2/3)(\gamma + B/R) = (4\pi R^2/3)(\gamma + \pi_c R) \tag{158}$$

or, making use of the P_N component:

$$w_f = (4\pi/3)\int_0^\infty (P_N - P_T)r^2 dr = 2\pi \int_0^R (P_N - P_\beta)r^2 dr \tag{159}$$

The *radius of the surface of tension* $R_{s.o.t.}$ is formally obtained by setting B or π_c equal to zero. Thus, we get:

$$R_{s.o.t.}^3 = \frac{6}{P_\alpha - P_\beta}\int_0^\infty (P_\beta - P_T)r^2 dr$$
$$= \frac{2}{P_\alpha - P_\beta}\int_0^\infty (P_N - P_T)r^2 dr = \frac{3}{P_\alpha - P_\beta}\int_0^\infty (P_N - P_\beta)r^2 dr \tag{160}$$

where we have also used the condition of mechanical equilibrium:

$$\int_0^\infty (P_\beta - P_T)r\,dr = 0 \tag{161}$$

which can be verified by substituting the Laplace equation $\sigma = (P_\alpha - P_\beta)R_{\text{s.o.t.}}/2$ in Eq. (151).

C. Cylindrical Interfaces

Generally, for a cylindrical surface, we have $H = D = -1/2R$ and $\lambda = r - R$, where R is the radius of the dividing surface and:

$$\mathbf{L} = (1 - 2\lambda H)\mathbf{U}_s + \lambda\mathbf{b} \tag{162}$$

For the surface tension γ and the shearing tension ζ, we derive the expressions:

$$\gamma = -\frac{1}{2}\int_{\lambda_1}^{\lambda_2}\left(P_{\varphi\varphi}^s + P_{zz}^s\right)\frac{r}{R}\,dr \tag{163}$$

and

$$\zeta = -\frac{1}{2}\int_{\lambda_1}^{\lambda_2}\left(P_{\varphi\varphi}^s - P_{zz}^s\right)\frac{r}{R}\,dr \tag{164}$$

Furthermore, the principal components of the σ_s tensor are given by:

$$\sigma_{\varphi\varphi} = \int_0^\infty P_{\varphi\varphi}^s\,dr \quad \sigma_{zz} = -\int_0^\infty P_{zz}^s\frac{r}{R}\,dr \tag{165}$$

resulting in:

$$\sigma = -\frac{1}{2}\int_0^\infty\left(P_{\varphi\varphi}^s + \frac{r}{R}P_{zz}^s\right)dr = \frac{1}{2}\left(\sigma_{\varphi\varphi} + \sigma_{zz}\right) \tag{166}$$

and

$$\eta = \frac{1}{2}\left(\sigma_{\varphi\varphi} - \sigma_{zz}\right) = -\frac{1}{2}\int_0^\infty\left(P_{\varphi\varphi}^s - \frac{r}{R}P_{zz}\right)dr \tag{167}$$

Similarly, the principal components of the bending moment tensor are:

$$\mathbf{M}_{\varphi\varphi} = \int_0^\infty P_{\varphi\varphi}^s(r - R)\,dr \quad \mathbf{M}_{zz} = \int_0^\infty P_{zz}^s\frac{r}{R}(r - R)\,dr \tag{168}$$

Thus,

$$B = \mathbf{M}_{\varphi\varphi} + \mathbf{M}_{zz} = \int_0^\infty\left(P_{\varphi\varphi}^s + \frac{r}{R}P_{zz}^s\right)(r - R)\,dr \tag{169}$$

and

$$\Theta = \mathbf{M}_{\varphi\varphi} - \mathbf{M}_{zz} = \int_0^\infty\left(P_{\varphi\varphi}^s - \frac{r}{R}P_{zz}^s\right)(r - R)\,dr \tag{170}$$

From these last two relations, it is seen that:

$$\mathbf{M}_{\varphi\varphi} = \frac{1}{2}(B + \Theta) \quad \mathbf{M}_{zz} = \frac{1}{2}(B - \Theta) \tag{171}$$

Hence, Eqs. (163) and (164) can be rewritten as follows:

$$\gamma = \sigma + \frac{1}{2}(B + \Theta)H = \sigma - M_{\varphi\varphi}/2R \tag{172}$$

$$\zeta = \eta + \frac{1}{2}(B + \Theta)H = \eta - M_{\varphi\varphi}/2R \tag{173}$$

Consequently, for a cylindrical surface:

$$\gamma - \zeta = \sigma - \eta \tag{174}$$

D. Cylindrical Surfactant Aggregates

Simple mechanical equilibrium considerations show that a freely floating cylindrical object must satisfy the condition $\sigma_{\varphi\varphi} = 2\sigma_{zz}$ as will be shown explicitly below. On the other hand, a perfectly fluid interface should be isotropic in the tangent plane of the interface (i.e., $\sigma_{\varphi\varphi} = \sigma_{zz}$). From this, we are forced to conclude that aggregates with a perfectly fluid surface cannot have a cylindrical shape.

A completely different case is presented by solid elastic interfaces (e.g., rubber membranes). Another interesting case is that of a rod-shaped micelle, which is usually considered to have at least approximately a cylindrical shape. Presumably, the entanglement of the hydrocarbon chains endows the interface (which comprises practically all of a micellar aggregate) with sufficient rigidity to support anisotropic tensions in the tangent plane of the dividing surface so that the condition $\sigma_{\varphi\varphi} = 2\sigma_{zz}$ is actually satisfied. In spite of all, such interfaces are usually regarded as fluid because the surfactant molecules are able to easily exchange places. In the following, we treat the idealized case where this crucial condition is supposed to be satisfied exactly. Nonetheless, it might well be that, in reality, elongated surfactant aggregates do not have a perfectly cylindrical shape but rather a sinusoidally varying contour. Mechanical equilibrium with respect to two sections, one at right angles to the cylinder axis and another containing the axis, requires that:

$$\int_0^\infty \left(P_{zz}^s - P_\beta\right) r \, dr = 0 \tag{175}$$

$$\int_0^\infty \left(P_{\varphi\varphi}^s - P_\beta\right) r \, dr = 0 \tag{176}$$

The micromechanical definitions of $\sigma_{\varphi\varphi}$ and σ_{zz} (Eq. (165)) then yield, after simple calculations:

$$\sigma_{\varphi\varphi} = R(P_\alpha - P_\beta) \quad \sigma_{zz} = \frac{1}{2}R(P_\alpha - P_\beta) \tag{177}$$

that is,

$$\sigma_{zz} = \frac{1}{2}\sigma_{\varphi\varphi} \tag{178}$$

Thus, from Eq. (165):

$$\int_0^\infty P^s_{\varphi\varphi}\,dr = 2\int_0^\infty P^s_{zz}\frac{r}{R}\,dr \tag{179}$$

Finally, according to Eq. (164), we have:

$$\zeta = -\frac{1}{2R}\int_0^\infty P^s_{\varphi\varphi}(r - R/2)\,dr \tag{180}$$

An important quantity in the thermodynamics of surfactant micelles is the work of formation at equilibrium (i.e., when the chemical potential of the surfactant monomers in the solution is the same as for the surfactant molecules in the micelles). The work of formation of a cylindrical micelle can be written in two different ways corresponding to two different modes of formation at constant chemical potentials.

In the first mode, the cylindrical part is stretched very slowly in the z-direction so that chemical equilibrium is maintained all the time. Then:

$$w_f(z) = -2\pi\Delta L\int_0^\infty (P^s_{zz} - P_\beta)\,dr \tag{181}$$

From Eq. (175), it follows that:

$$w_f(z) = 0 \tag{182}$$

As already discussed in Section II, and as will be further discussed in Section IV below, this scheme is valid for the formation of *infinite* rod-shaped aggregates.

In the second mode, the cylindrical part is formed by letting a wedge-shaped cylindrical sector (like a piece of cake) grow from $\varphi = 0$ to $\varphi = 2\pi$ in analogy with the growing spherical cone used to calculate the work formation of a spherical droplet. Thus,

$$w_f(\varphi) = -2\pi\Delta L\int_0^\infty \left(P^s_{\varphi\varphi} - P_\beta\right)r\,dr \tag{183}$$

From Eqs. (164), (181), and (183), it is seen that:

$$R\zeta = \frac{1}{4\pi\Delta L}[w_f(\varphi) - w_f(z)] \tag{184}$$

However, the right-hand side of this equation is manifestly independent of the radius R of the dividing surface. Hence, it follows that $R\zeta$ is invariant with respect to R. Furthermore, because, for physical reasons, the work of formation must be the same in the two modes of formation, we must have $w_f(\varphi) = w_f(z) = 0$ and, consequently:

$$\zeta = 0 \tag{185}$$

which implies that:

$$\gamma_1 = \gamma_2 = \gamma \tag{186}$$

Furthermore, from Eqs. (163–165), we deduce that:

$$\gamma = \sigma_{zz} \tag{187}$$

As $\mathbf{M}_s$ is constant on a cylindrical interface, in agreement with Eq. (69), the Laplace equation may be written as:

$$\Delta P = -\sigma_{\varphi\varphi}/R \tag{188}$$

whereas, according to Eq. (74):

$$\Delta P = -\gamma/R - \zeta/R + \pi_c = -\gamma/R + \pi_c \tag{189}$$

where, on the basis of Eq. (73) and the micromechanical equations, it is easy to show that:

$$\pi_c = -\mathbf{M}_{\varphi\varphi}/R^2 = \gamma/R - \sigma_{\varphi\varphi}/R \tag{190}$$

holds, thus demonstrating full consistency.

IV. GENERAL CONDITIONS GOVERNING THE FORMATION OF AGGREGATES OF SURFACTANTS/LIPIDS

In this part, we consider the application of the Laplace condition to various surfactant-based aggregates, such as microemulsions, vesicles, surfactant micelles, and infinite periodical interfacial structures. The purpose is to show that a universal, though somewhat approximate, framework for predicting the size and shape of fully equilibrated aggregates is actually furnished by means of the Ou-Yang–Helfrich equation (Eq. (53)). However, in order to obtain a more complete theoretical account, in addition, one has to invoke the size and shape fluctuations of the aggregates insofar as they constitute small thermodynamic systems.

A. Microemulsions

Making a surface-thermodynamic approach, we can account for the free energy of formation, or, more exactly, the grand potential of formation $\Delta\Omega_{aggr}$ of an arbitrarily shaped microemulsion aggregate at the conditions (temperature and chemical potentials) prevailing in the surrounding solution, by writing [40]:

$$\Delta\Omega_{aggr} = \iint_S \gamma \mathrm{d}S + \sum_j N_j \Delta f_j \tag{191}$$

The first term represents the free energy it takes to form the equilibrated *surfactant film* of the aggregate, and the second term represents the free energy needed to assemble its internal *core* part consisting of j components. For the water/oil (and correspondingly for the oil/water) single-phase microemulsion case, the $N_w \Delta f_w$ term may also be expressed in terms of a pressure difference ΔP_r defined by the relation:

$$V(P_e - P_r) = V\Delta P_r = N_w \Delta f_w \tag{192}$$

where P_r stands for a reference pressure that is equal to the pressure needed to obtain the same chemical potential for pure (incompressible) water, as the chemical potential of the molecularly dispersed water actually present in the solution. V is the volume of the aggregate as determined by means of the particular dividing surface involved, preferably the equimolecular dividing surface for the oil component, and S stands for the surface area of the aggregate.

Incidentally, by assuming spherical geometry, using Eqs. (191) and (192), and by identifying ΔP_r with ΔP, we can easily recover Eq. (113). In such a case, the minimization of Eq. (191) yielding the pertinent Laplace condition will enable us to determine the radius of the fully equilibrated microemulsion droplet as we discuss in more detail the following. In passing, we may note that the free energy of formation of a spherical equilibrium droplet can also be written in the Gibbsian manner as $\gamma_{s.o.t.} A_{s.o.t.}/3$, provided the surface of tension is used as dividing surface, which, however, is rarely appropriate for microemulsion droplets.

Furthermore, referring to the (quasi-statistical–mechanical) multiple equilibrium description of aggregation, for each aggregate characterized by a certain volume and surface area (i.e., by a certain number of molecules making up the core and the surfactant film, respectively), we have in the dilute regime:

$$\Delta\Omega_{\text{aggr}} + kT \ln \phi_{\text{aggr}} = 0 \tag{193}$$

where ϕ_{aggr} is the volume fraction of the particular aggregate in question (cf. Ref. 39.) In other words, because we are actually dealing with a bulk phase, the $\Delta\Omega_{\text{tot}}$ of a certain portion of the solution must not be affected by the formation of additional microemulsion droplets of the kind considered. In fact, supposing the volume V_{tot} of this portion to be conserved, $\Delta\Omega_{\text{tot}}$ should always stay equal to $-P_e V_{\text{tot}}$. Hence, according to Eq. (193), for each kind of aggregate, the free energy of formation $\Delta\Omega_{\text{aggr}}$ is counterbalanced by the (partial) free energy of dilution in the continuous medium, $kT \ln \phi_{\text{aggr}}$. The overall volume fraction of droplets is thus given by the "partition function":

$$\phi_{\text{aggr}}^{\text{tot}} = \sum_{\text{aggregate states}} \exp(-\Delta\Omega_{\text{aggr}})/kT \tag{194}$$

where the sum extends over all stoichiometrically different states.

The scheme outlined here has turned out to make up a convenient approach, first, to treating Winsor I and Winsor II microemulsion systems where an excess phase is present [38], in which case it is sufficient to consider merely the first interfacial term of Eq. (191), and, second, to the corresponding one-phase microemulsions close to the two-phase region (i.e., on the border to what occasionally is referred to as emulsification failure) [47].

Now, by turning to the Laplace equation in Ou-Yang–Helfrich form (Eq. (53)):

$$\Delta P = 2H\gamma_0 - 4k_c(H - H_0)(H^2 + HH_0 - K) - 2k_c \nabla_s^2 H$$

by means of which the relevant free energy minima can be pinpointed, and assuming the spontaneous curvature H_0 of the interfacial film to vary monotonously with the thermodynamic state of the system, we can broadly rationalize the following well-established sequence of microemulsion formation (cf. Fig. 7):

One-phase oil/water droplet microemulsions
Winsor I droplet microemulsions (excess oil phase)
Winsor III bicontinuous microemulsions (excess phases of water and oil, respectively)
Winsor II droplet microemulsions (excess water phase)
One-phase water/oil droplet microemulsions

implying a stepwise change of the interfacial curvature from oil first being the internal phase to water becoming the internal phase. Remembering that

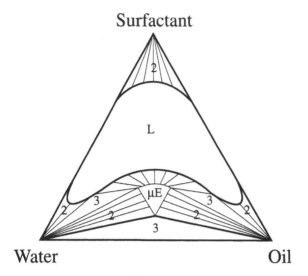

Surfactant

L

μE

Water   Oil

FIG. 7 Idealized phase diagram for microemulsion-forming water/oil/surfactant system under well-balanced conditions. (Redrawn from A. Kabalnov, B. Lindman, U. Olsson, L. Piculell, K. Thuresson, and H Wennerström, Colloid. Polym. Sci. **1996**, *274*, 297.) The Winsor I and Winsor II microemulsions are on the right-hand and left-hand sides, respectively, of the central microemulsion region, whereas the Winsor III microemulsion is located at the downward tip.

$\Delta P_r = \Delta P$ is a constant, the following constant curvature solutions of Eq. (53) are, of course, of particular interest.

(i) *Spherical geometry*, $\Delta P > 0$ and $H_0 \neq 0$, for which case we obtain the solution ($K = H^2$):

$$H^{\mathrm{sph}} = \left(\gamma_\infty + \sqrt{\gamma_\infty^2 - 4k_c H_0 \Delta P}\right)/4k_c H_0 \tag{195}$$

where $\gamma_\infty = \gamma_0 + 2k_c H_0^2$, accounting for the equilibrium size of the droplets in oil/water and water/oil *single-phase* microemulsions. Recall that according to the geometrical sign convention that we have adopted, H^{sph} and H_0 are both negative, for oil/water as well as water/oil droplets. Moreover, it is seen that the droplet radius, as a rule, is distinctly less than $1/H_0$.

(ii) *Spherical geometry*, $\Delta P = 0$ and $H_0 \neq 0$, yielding the simple relation:

$$H^{\mathrm{sph}} = \gamma_\infty/(2k_c H_0) \tag{196}$$

that determines the equilibrium size of the droplets present in *Winsor I and Winsor II microemulsions*. The droplet radius is, for the most part, less than $1/H_0$ but approaches $1/H_0$ when γ_0 becomes much smaller than $2k_cH_0^2$.

Let us next consider in more detail the full $\Delta\Omega_{\text{aggr}}$ function for Winsor II (and correspondingly for Winsor I) droplets obtained on the basis of the Helfrich expression (Eq. (52)) as shown in Fig 8. For every droplet size, the Laplace equation holds in the form:

$$P_i - P_e = 2\gamma/R + (\partial\gamma/\partial R)_{T,\mu i} \tag{197}$$

At the minimum for $R = R_{\text{eq}}$, the pressure inside the droplet P_i becomes exactly equal to the reference pressure P_r, which, for Winsor I and Winsor II cases, is equal to P_e, thus establishing equality between the chemical potentials for the core component in the droplet and in the excess phase. For $R > R_{\text{eq}}$, the internal pressure is higher than in the excess phase and, consequently, the chemical potential is higher, resulting in a tendency for the

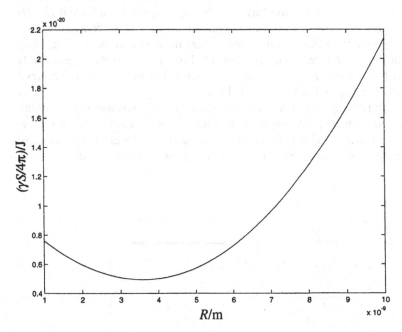

FIG. 8 The excess free energy function $\gamma S/4\pi$ of a spherical Winsor II water/oil microemulsion droplet assuming $\gamma = 4.00 \times 10^{-4} - 2.87 \times 10^{-12}R^{-1} + 1.01 \times 10^{-20}R^{-2}$. This γ function has a minimum for $R = 7.03$ nm, whereas the γS function shown above has its minimum for $R = 3.59$ nm where $\Delta P = 0$.

droplet to dissolve and diminish in size. The reverse obviously holds true for $R < R_{eq}$. In this way, stability is granted. One has to keep in mind that to a very good approximation, these internal pressure variations with droplet size will leave the Helmholtz free energy of the nearly incompressible core unaffected.

(iii) *Cylindrical geometry*, $\Delta P \geq 0$. In this case, it results from the Ou-Yang–Helfrich equation (Eq. (53)) upon putting $K = 0$ that *elongated cylinders* may exist, normally having a curvature a little less than the spontaneous curvature H_0, as discussed earlier in Section II.H [42]. As a matter of fact, we get ($\Delta P = 0$):

$$H^{cyl} = -\sqrt{\gamma_\infty/2k_c} \tag{198}$$

yielding $H^{cyl} = H_0$ as γ_0 tends to zero. The end parts of a cylinder where the curvature changes represent a special problem that might be treated on the basis of Eq. (53) while retaining the last $\nabla_s^2 H$ term (Fig. 9). It is also worth pointing out that very long cylinders may arise when γ_0 becomes small enough and the work of elongation nearly vanishes [43].

(iv) *Infinite interfacial structures*, $\Delta P = 0$, with both oil and water excess phases present. The condition furnished by Eq. (53) will be fulfilled if $H = H_0$ and $\gamma_0 = 0$, irrespective of the value of the Gaussian curvature K, indicating the possibility of forming an infinite constant mean curvature (CMC), triply periodical (*gyroid*) interfacial structure [44]. For comparatively large absolute values of the spontaneous curvature H_0, such a structure closely resembles a network of elongated, interconnected rods (Fig. 10).

A special case is the much discussed infinite periodical *minimal surface* with a mean curvature $H = H_0$ everywhere equal to zero, which separates equal volumes of water and oil. Evidently, interfacial structures of this principal nature can readily account for the water and oil bicontinuity observed for

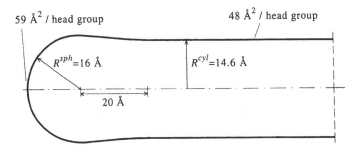

FIG. 9 Spherocylindrical micelle model with slightly swollen end cap where the $\nabla_s^2 H$ term of Eq. (53) contributes (except at the very tip).

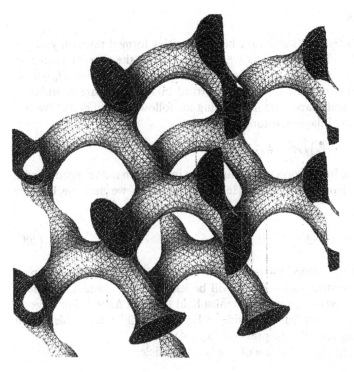

FIG. 10 Constant mean curvature gyroid structure devised by Grosse-Brauckmann (From Ref. 44.) for which the thinner channel system occupies 0.1875 of the total volume. According to Eq. (53), ΔP can attain a small constant value equal to $2H_0\gamma_0$ provided that H adjusts itself to H_0. Note the similarity with a structure composed of interconnected rods. Upon decreasing the channel volume even more, a continuous transition into spherical droplets occurs. (From Ref. 44.) Assuming γ_0 to be different from zero, for the Winsor III case (excess phases of both oil and water present), we will have $H = H_0 = 0$.

Winsor III microemulsions as well as single-phase microemulsions in the middle of the microemulsion channel. Still, the role of the Gaussian curvature bending constant $\bar{k}_c$ in this context remains to be fully clarified. Moreover, it seems likely that higher-order Gaussian curvature terms have to be invoked to account for the stability of these structures (cf. Refs. 43 and 45).

An additional solution of Eq. (53) is, of course, a *planar bilayer* with $H = H_0 = K = 0$. However, interactions among interfaces (that obviously are important for the formation of lamellar liquid crystals) fall evidently beyond the scope of the survey presented in this chapter.

B. Vesicles

For (spherical) lipid or surfactant bilayer vesicles formed reversibly in the course of equilibrium fluctuations, we may assume that the internal water part is free energy-wise, the same as the external water part. Putting $H_0 = 0$, as seems reasonable in view of the symmetrical bilayer structure postulated, assuming spherical shape, and introducing the following extended curvature expansion for the bilayer tension:

$$\gamma = \gamma_0 + \left(2k_c + \overline{k}_c\right)/R^2 + k_4/R^4 \tag{199}$$

which includes a fourth-order curvature term in addition to the second-order term obtained from the original Helfrich expression, we find the Laplace condition:

$$\gamma_0/R - k_4/R^5 = -\Delta P/2 = 0 \tag{200}$$

where now $\gamma_0 = \gamma_\infty$. Hence, the radius R_{eq} of the equilibrium vesicle is given by $\sqrt[4]{k_4/\gamma_\infty}$. For smaller radii, $P_i - P_e$ will be less than zero, and, conversely, $R > R_{eq}$. On the average, however, it must hold true that $\Delta P = 0$. For large R, the second term on the left-hand side of Eq. (200) will be negligible, which means that it agrees with the previous Eq. (54).

The free energy of formation of a spherical vesicle:

$$\Omega_{vesicle} = 4\pi\left(\gamma_\infty R^2 + 2k_c + \overline{k}_c + k_4/R^2\right) \tag{201}$$

will increase in proportion to $\gamma_\infty R^2$ for large radii, per se favoring small, rather than large, vesicles. Nevertheless, when γ_∞ is small enough (less than about 10^{-6} mJ m^{-2}), an appreciable volume fraction of fairly large vesicles may well result due to the entropic advantage associated with the size and shape distributions [48]. In fact, the size distribution of vesicles can peak between 10 and 1000 nm, which obviously means that the vesicle curvature is practically unrelated with the spontaneous curvature $H_0 = 0$ of the bilayer.

As was mentioned in Section II, by varying the parameters H_0, ΔP, γ_0, and k_c, and by utilizing Eq. (53), a variety of nonspherical vesicle and biomembrane shapes were early on derived by Deuling and Helfrich [34]. More recently, Tachev et al. [49] have carried through similar investigations. In particular, nonspherical shapes may arise when γ_0 approaches zero, and the curvature free energy becomes decisive for the resulting shape. The same type of approach will be feasible regardless of whether the bilayer itself is thermodynamically closed or not, the major difference between the two situations rather being the magnitude of the bending constants k_c and $\overline{k}_c$. For thermodynamically open (equilibrated) bilayers, we may expect k_c values of $1kT$, whereas for closed bilayers, k_c values on the order of $100kT$ are anticipated.

Interestingly, in addition to constituting vesicles and lamellar structures, the bilayer building block also appears in the so-called L_3 phase, characterized by a thermally disturbed, triply periodic, minimal surface type of structural organization for which we must have, of course, $\Delta P = 0$.

C. Compact Surfactant Micelles

Surfactant micelles, both of the ordinary and the reverse kind (with a water-rich core), can be treated along lines similar to those already discussed for microemulsions. However, one has to take into account that the aggregates, which we commonly refer to as surfactant micelles, are principally different from microemulsion droplets as they possess the intrinsic nature of an *interfacial complex*. To put it otherwise, a surfactant micelle behaves like a curved interface throughout because it is lacking a well-defined homogeneous core part, which is distinct from its surface. Thus, instead of relying primarily on the Helfrich expression, in the past, detailed theories of surfactant micelles of different geometries have been worked out by invoking molecular models of the main physical factors determining $\Delta\Omega_{\mathrm{mic}}$ of micellar aggregates, and by considering appropriate packing constraints for hydrocarbon chains [50–59].

As an illustrative example, let us consider an ordinary spherical micelle formed by an ionic surfactant for which one can start out from the expression:

$$\Delta\Omega_{\mathrm{mic}}/N_{\mathrm{aggr}} = \varepsilon_{\mathrm{hc}} + \varepsilon_{\mathrm{contact}} + \varepsilon_{\mathrm{el}} + \varepsilon_{\mathrm{conf}} + \varepsilon_{\mathrm{pg}} \qquad (202)$$

where N_{aggr} is the aggregation number, and where (cf. Fig. 11) $\varepsilon_{\mathrm{hc}}$ denotes the free energy gained per surfactant molecule upon aggregating the hydrocarbon chains to form liquid hydrocarbon, whereby hydrocarbon–water contacts are annihilated [51]; $\varepsilon_{\mathrm{contact}}$ is the contact free energy expended to form the hydrocarbon–water interface at the aggregate surface; $\varepsilon_{\mathrm{el}}$ is the (curvature-dependent) electrostatic free energy it takes to charge the interface and assemble the counterions. In the case of monovalent ionic surfactants, it is well known that Poisson–Boltzmann theory furnishes useful and fairly accurate expressions for this contribution for different geometries; $\varepsilon_{\mathrm{conf}}$ is the geometry-dependent conformational free energy that has to be expended when attaching the hydrocarbon chains to the head groups at the aggregate surface [55]; and $\varepsilon_{\mathrm{pg}}$, finally, is the change in head group free energy due to aggregation, other than what is already included in $\varepsilon_{\mathrm{el}}$.

The sum of the last four terms, $\varepsilon_{\mathrm{contact}} + \varepsilon_{\mathrm{el}} + \varepsilon_{\mathrm{conf}} + \varepsilon_{\mathrm{pg}}$ above, is typically on the order of $10\ kT$ per surfactant molecule, whereas the first term $\varepsilon_{\mathrm{hc}}$ is slightly less in magnitude but negative, thus providing the driving force for surfactant aggregation. When combined, these contributions usually give rise

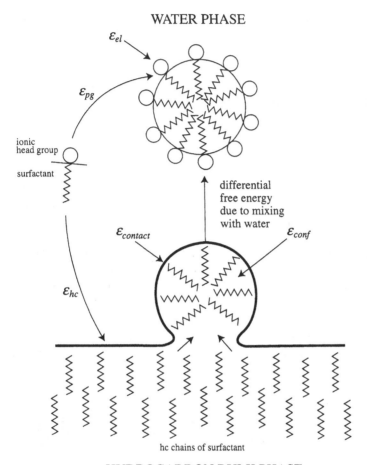

FIG. 11 Drawing that displays the various origins of the excess free energy of an ordinary (spherical, ionic) micelle of aggregation number N_{aggr}. In comparison with a water phase saturated with the corresponding hydrocarbon, the monomer solution is supersaturated with respect to the hydrocarbon moiety, thus providing the driving force for aggregating the hydrocarbon chains and assembling the charged head groups and the counterions. The hydrocarbon chains are drawn in a schematic fashion; in reality, their conformational state is disordered similarly as shown in Fig. 15.

to a shallow free energy minimum, on the order of $0.2kT$ per surfactant molecule at the critical micelle concentration (cmc) that quite satisfactorily can account for the formation of ordinary surfactant micelles (Fig. 12).

Moreover, on the same basis, one can rather readily deal with the other geometries: cylinders, planar bilayers, and curved bilayers, thus enabling theoretical calculations of the relative stabilities of spherical, rod-shaped, and disc-shaped micelles as well as geometrically closed vesicles [60].

However, even for micellar aggregates of this kind, in the spirit of the curvature free energy approach to microemulsions, we may write:

$$\Delta\Omega_{mic} = \iint_S \omega_s dS + \Delta PV \tag{203}$$

thus introducing the work of micelle formation per surface area unit ω_s, that, in principle, incorporates the four last factors listed above ($\varepsilon_{contact}$, ε_{el}, ε_{conf}, and ε_{pg}), whereas the large (shape-independent) negative term ΔPV relates to ε_{hc}—in other words, to the cooperative formation of the hydrocarbon core of the micelle.

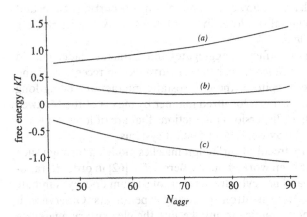

FIG. 12 Free energy per surfactant molecule ε as a function of the aggregation number N_{aggr} of a micelle formed in a SDS/water solution (middle curve). The core contribution (upper curve) gives rise to the ascending branch and is due to the conformational restrictions on the chains imposed by the geometrical constraints. The interfacial contribution ($\varepsilon_{el} + \varepsilon_{contact} + \varepsilon_{pg}$), combined with the hydrophobic term ε_{hc}, generates the descending branch and is accounted for here by the lower curve. The total ε has a minimum for $N_{aggr} = 70$, whereas the overall micellar free energy $N_{aggr}\varepsilon$ has a minimum for $N_{aggr} = 66$, in approximate correspondence with the extended hydrocarbon chain length that equals 16.7 Å.

Furthermore, we may assume that the $\Delta\Omega_{\text{mic}}$ of a surfactant aggregate of a certain size and shape can be calculated by relying on a curvature expansion of Helfrich type in terms of ω_s:

$$\omega_s = \omega_{s0} + 2k_c(H - H_0)^2 + \bar{k}_c K \tag{204}$$

where ω_{s0} refers to a cylindrical aggregate surface of mean curvature equal to H_0. Such a calculation can aim at deriving the equilibrium size upon assuming a certain micellar shape and invoking the proper packing constraints, while making use of the Ou-Yang–Helfrich equation in a form entirely analogous to Eq. (53), that is,

$$\Delta P = 2H\omega_{s0} - 4k_c(H - H_0)\left(H^2 + HH_0 - K\right) - 2k_c \nabla_s^2 H \tag{205}$$

Progressing along such a route, however, comparatively large values of the bending constants are obtained on the order of $100kT$ for k_c as the size changes contemplated will then necessarily imply that the surfactant chemical potential in the micelle varies a great deal.

Alternatively, the effort can be directed toward predicting the shape of the predominant, fully equilibrated micellar aggregate of a certain core volume by means of solving Eq. (205) in a similar manner as has been done for bilayer membranes [49]. In such a case, however, shape changes occurring at constant surfactant chemical potential would be involved, implying k_c values ranging between about 1 and $10kT$.

Patchwise treatment of surfactant aggregates and surfactant-laden curved interfaces, as implicitly assumed here, has been advanced in recent years and is, in principle, feasible in spite of the electrostatic interactions being long range as they can be handled in a localized manner by means of the Maxwell tensor [61], whereas other (dispersion) interactions that are of long range per se for the most part are fairly weak in surfactant systems.

A rather successful treatment of surfactant micelles modeled through such a Helfrich approach has been worked out by Bergström [62] in order to rationalize the nonspherical and noncylindrical forms of sodium dodecyl sulphate (SDS) micelles inferred from neutron scattering experiments. Otherwise, it has mostly been common practice to invoke just the elementary aggregate shapes: spheres, cylinders, and flat bilayers (with dimensions corresponding to the extended hydrocarbon chain length), more recently supplemented with various infinite triply periodical interfaces. Yet, it is becoming increasingly evident that, in reality, we are confronted with a wider range of geometries, at least for aggregates of the ordinary kind with cores made up of hydrocarbon chains. The geometrical layout of Bergström's general micelle model is shown in Fig. 13.

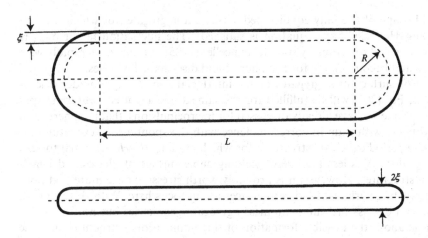

FIG. 13 A general tablet-shaped micelle modeled as being composed of a central rectangular bilayer part of thickness ξ, width $2R$, and length L, and semicircular bilayer disc parts of radius R at the two ends. The entire bilayer part is surrounded by half-cylindrical rims of radius ξ. Assuming $L = 0$, a disc-shaped micelle is obtained, whereas putting $R = 0$ yields a spherocylindrical micelle with nonswollen ends. In the limit $R,L = 0$, an ordinary spherical micelle is generated. An advantage of this model is that it automatically encompasses a whole range of micellar shapes, which is a desirable feature when analyzing SANS data. Yet, it is evident that the geometrical outline eventually will have to be modified (swollen ends and rims) in order to comply with the shape equation [Eq. (205)].

Among the various options as to aggregate size and shape that are open to a certain surfactant system, those aggregates that have the lowest free energy $\Delta\Omega_{mic}$ of the respective equilibrium aggregate, and, additionally, the shallowest minimum allowing a multitude of aggregates, but slightly differing in size and shape, will predominate as a result of equilibrium fluctuations. Generally, we can write for the volume fraction of a micellar species:

$$\phi_{mic} = S_{fluct}\exp\left[-\iint_S \omega_s dS/kT\right]\exp[-\Delta PV/kT] \qquad (206)$$

where the size fluctuations give rise to the preexponential factor* and the integral in the exponent extends over the dividing surface defining the size

---

*For compact surfactant micelles, merely the size fluctuations contribute to S_{fluct} as the shape fluctuations leave the aggregate stoichiometry unaffected.

and shape of the fully equilibrated surfactant aggregate (for which the Ou-Yang–Helfrich shape equation is satisfied). The fluctuation factor S_{fluct} is rather small for ordinary spherical micelles (≈ 25), but increases rapidly with size and becomes large for rod-shaped and disc-shaped micelles.

For surfactant aggregates of the interfacial complex type under discussion, Eq. (205) will be fulfilled for the same sequence of elementary shapes as the one we have already discussed for microemulsions, that is, spheres and cylinders with radii in correspondence with the spontaneous curvature, infinite periodical CMC structures for which ω_{s0} and $H-H_0$ are equal to zero but where K is less than zero, yielding cubic surfactant phases and lamellar structures. However, it is probably worth stressing once more that noninteracting surfactant-laden interfaces are in focus here. With this in mind, we can discuss just interactionless aggregate geometries but not the whole issue about the possible formation of three-dimensional structures of these aggregates.

D. Reverse Surfactant Micelles

The reverse micellar aggregates with a water-rich core likewise belong to the scheme presented here, albeit after a few minor changes. Thus, the large negative term ε_{hc}, causing the aggregation of ordinary micelles, has to be replaced by ε_{core}, which quantifies the free energy gained per surfactant molecule upon assembling the core mixture of head groups and water from the continuous apolar solvent phase. Moreover, one has to invoke a free energy contribution ε_{mix} associated with the demixing of surfactant hydrocarbon chains and the surrounding apolar solvent, due to the aggregation. In this context, it is advantageous to choose the equimolecular dividing surface for the apolar solvent as the dividing surface (Fig. 14). Eq. (20) still applies to $\Delta P < 0$ and V is equal to the volume of aggregated surfactant and the associated water molecules.

Interestingly, in addition to spherical reverse micelles, even very long, *threadlike micelles* with a water-rich core have been documented [63], indicating that a comparatively rigid zone of tightly packed hydrocarbon chains is present next to the polar head groups, granting mechanical stability as discussed previously in Section II.D.

E. The Critical Packing Parameter

Rather than to focus on the mostly entropic factors that are decisive for the spontaneous curvature H_0, it has been customary in recent years to simply

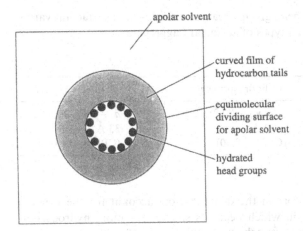

FIG. 14 Aggregation into the reverse type of compact surfactant micelles is due to the circumstance that (hydrated) polar head groups have lower contact free energy among themselves than when they are dispersed in an apolar solvent. Note that the size of a reverse aggregate is related to the spontaneous curvature H_0 similarly as for ordinary micelles. Obviously, the aggregate size is in this case only remotely related to the length of the hydrocarbon chain of the aggregating surfactant. The drawing is schematic; in reality, the outer part of the hydrocarbon film is more or less mixed with the apolar solvent, whereas the inner part is subject to high tangential pressures due to restriction of the hydrocarbon chain conformations.

refer to the dimensionless critical packing parameter (CPP) of a surfactant defined as:

$$\frac{\text{Hydrocarbon chain volume}}{\text{head group area} \times \text{hydrocarbon chain length}} = v_{\text{chain}}/(al_{\text{max}}) \qquad (207)$$

to obtain a rough guide about possible aggregate geometries by consulting in essence the following table [64]:

CCP	<1/3	1/3–1/2	1/2–1	>1
Preferred geometry	Spherical micelles	Cylindrical micelles	Bilayers, discs, lamellar structures	Reversed micelles

Regrettably, there are no valid statistical–thermodynamic arguments in support of this rather deceptive scheme. All that can safely be concluded is, in fact, that the expected head group area a of a single-chain surfactant varies as follows for the standard types of surfactant aggregates:

Spherical micelle	Cylindrical micelle	Planar bilayer
$a = v_{chain}/l_{max}$ ≈ 60 Å^2 (CPP $= 0.33$)	$a = v_{chain}/(0.85 l_{max})$ ≈ 50 Å^2 (CPP $= 0.40$)	$a = v_{chain}/(0.63 l_{max})$ 33 Å^2 (CPP $= 0.60$)

where the numerical factors in the denominators account for the average hydrocarbon chain length, which becomes shorter the more hydrocarbon chains there are that have to stretch to reach the center of an aggregate. Note, however, that the CPP values do fall within the anticipated intervals.

Owing to the rather subtle variations of the various local free energies upon changing the surface area per head group, the curvature, and the solution parameters, it is generally a delicate matter to predict what detailed shape is going to predominate in the end. Moreover, the occurrence of long rod-shaped micellar aggregates (and large vesicles) as the favored kinds of aggregates is mostly due to the size fluctuations becoming thermodynamically important as an additional source of entropy. Thus, neither the CPP parameter nor the notion of a spontaneous curvature constitutes a sufficient basis for accurately predicting aggregate geometries.

For instance, to account in a CPP manner for the well-known succession of aggregate shapes sphere, cylinder, bilayer, which frequently is observed upon adding salt to a micellar solution of an ionic surfactant, one has to invoke the notion of an "effective" head group area a_{eff} of the aggregated surfactant that diminishes as the salt concentration is raised. Hence, due to the variation of a_{eff}, the CPP parameter is supposed to cross the borderlines between the realms of existence of spherical, cylindrical, and planar geometries, respectively. In reality, however, the salt effect in question can be traced back to the circumstance that the electrostatic free energy ε_{el} is lowered more upon adding salt for a less curved than for a strongly curved charged surface [60].

F. Micromechanical/Surface-Thermodynamic Approach to Treating Surfactant Micelles

Several years ago, the present authors elaborated combined micromechanical and surface-thermodynamic treatments of spherical, rod-shaped, and disc-shaped surfactant micelles [57]. Choosing spherical geometry (Fig. 15) to

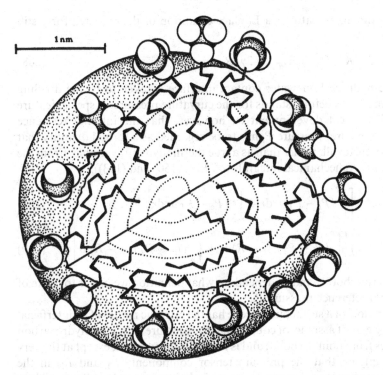

FIG. 15 Model of a spherical SDS micelle devised by Israelachvili (J. Israelachvili, *Intermolecular and Surface Forces*, 2nd Ed.; Academic Press: London, 1991.) displaying a fairly well-defined hydrocarbon–water interface. The hydrocarbon chains packed inside the core to normal density are subject to geometrical constraints, which are taken into account in the (mean field) calculations of their configurational state due to Gruen and de Lacey [55] and others. The sodium counterions have been omitted. Besides, in the standard Poisson–Boltzmann treatment of the curved electrostatic double layer, the sulphate head groups are represented by a smeared out, evenly distributed surface charge located at, or somewhat outside, the hydrocarbon–water contact interface. Furthermore, the counterions are, as a rule, treated as point charges. The packing of the hydrocarbon chains and the accumulation of counterions give rise to a surface pressure that counteracts the interfacial tension of the hydrocarbon–water interface (≈ 50 mJ m^{-2}).

exemplify the main features, a Laplace condition of the expected form still holds, viz.,

$$-\Delta P = 2\gamma/R + (\partial\gamma/\partial R)_{T,\mu_i} = 2\gamma/R - \pi_c \tag{208}$$

where the hydrocarbon–water contact interface is taken to be the dividing surface, and π_c, as before, stands for the curvature pressure. A special feature of this theoretical development is, however, that the internal reference (hydrocarbon core) pressure P_i has to be regarded as a formal quantity that can be chosen at will, rather than as a true thermodynamic variable. From the following micromechanical integrals (cf. Eqs. (150) and (153)):

$$\gamma = \left(1/R^2\right)\left[\int_0^R (P_i - P_T)r^2 dr + \int_R^\infty (P_e - P_T)r^2 dr\right] \tag{209}$$

$$\pi_c = \left(2/R^3\right)\left[\int_0^R (P_i - P_T)r(r - R)dr + \int_R^\infty (P_e - P_T)r(r - R)dr\right] \tag{210}$$

it appears that both γ and π_c depend on what definition we prefer to make of the internal reference pressure P_i.

As the whole of a surfactant micelle has the intrinsic nature of an interfacial complex, we must assume, of course, that the pressure within the hydrocarbon core varies from point to point, and is generally anisotropic, except at the very center, implying that the pressure tensor components P_T and P_N in the tangential and normal directions, respectively, are, as a rule, different. Thus, there is no natural definition of P_i. Nonetheless, it turns out to be convenient to put P_i equal to the average tangential pressure within the micelle core, that is,

$$P_i = \left(3/R^3\right)\int_0^R P_T r^2 dr \tag{211}$$

Accordingly, the core contribution to γ will vanish, yielding simply:

$$\gamma = \gamma_e = \int_R^\infty (P_e - P_T)r^2 dr \tag{212}$$

On the other hand, in terms of γ and π_c, the work of forming a spherical equilibrium micelle at fixed chemical potentials is found to be given by the expression (Eq. (158)):

$$\Omega_{mic} = 4\pi R^2(\gamma + \pi_c R)/3 = 4\pi\int_0^\infty (P_e - P_T)r^2 dr \ (R = R_{eq}) \tag{213}$$

that is necessarily invariant with respect to both the choice of dividing surface as well the internal reference pressure.

Generally, the "micelle tension" $\gamma + \pi_c R$ can be obtained from the free energy per surfactant molecule in the micellar state ε by differentiating with respect to the head group area a:

$$\partial \varepsilon / \partial a = \gamma + \pi_c R \tag{214}$$

while keeping the environmental variables fixed. Here, it is worth noting that the constant terms ε_{hc} an ε_{pg} will not contribute to $\gamma + \pi_c R$. Furthermore, it is seen from Eq. (213) that for the micelle of equilibrium size, we will have:

$$\varepsilon / a = (\gamma + \pi_c R)/3 \quad (R = R_{eq}) \tag{215}$$

where the various physical mechanisms involved as a rule give rise to different contributions to the left-hand and right-hand sides, respectively.

Taking a (spherical) SDS micelle of equilibrium size in the c.m.c. range as a typical example, theoretical estimates [57] show the following:

	hc	contact	el	conf	pg	total
Contribution to ε/a [mJ m^{-2}]	−77.2	50.0	30.9	6.6	−9.1	1.2
Contribution to $(\gamma + \pi_c R)$ [mN m^{-1}]	0	50.0	−26.0	−20.5	0	3.5

corresponding to an overall work of micelle formation of about $12kT$. Here, $\gamma = \gamma_e$ covers just the contributions related with the hydrocarbon–water contact and the curvature-dependent electrostatics, whereas a major part of the curvature pressure term $\pi_c R$ stems from the curvature-dependent conformational free energy of the aggregated hydrocarbon chains packed inside the spherical dividing surface. Conversely, this latter free energy source makes just a minor contribution to ε/a, which for an ionic surfactant is dominated by the contributions due to the ε_{hc}, $\varepsilon_{contact}$, and ε_{el} terms. As is seen from the table, the various components of ε/a ("surface free energy") and $\gamma + \pi_c R$ ("micelle tension"), respectively, may be widely different. Yet, it must necessarily hold true for a spherical micelle of equilibrium size that the total work of formation is one and the same, irrespective of whether it is computed as $4\pi R^2 \varepsilon/a$ or $4\pi R^2 (\gamma + \pi_c R)/3$ for $R = R_{eq}$.

The volume fraction of spherical micelles ϕ_{mic} can be written as follows (cf. Eq. (206)):

$$\phi_{mic} = S_{fluct}\exp\left[-N_{aggr}\varepsilon/kT\right] = S_{fluct}\exp\left[-4\pi R^2(\gamma + \pi_c R)/3kT\right] \tag{216}$$

where the exponentials should be evaluated for the equilibrium size with radius R_{eq}, and S_{fluct} is the fluctuation factor (≈ 25) that accounts for the entropic advantage inherent with having a size distribution, and where ΔP is identified as $\varepsilon_{hc}/v_{chain}$. Accordingly, from experimental assessment of the average micelle size R_{eq}, the width of the size distribution (yielding S_{fluct}), and the intensity of the size distribution peak (ϕ_{mic}), one can, in fact, derive an experimental value of $\gamma + \pi_c R$ for $R = R_{eq}$ to be compared with a theoretically derived value.

An additional result of general interest emanating from the micromechanical approach is embodied in the Laplace condition in the form:

$$P_N(R) = P_e = 2\gamma_e/R - \pi_{c,e} \tag{217}$$

where $P_N(R)$ can be shown to correspond to the *average hydrostatic pressure* inside the micellar core. Hence, $P_N(R)$ is independent of P_i, being largely determined by the Laplace term $2\gamma_e/R$ on the right-hand side of Eq. (217). Typically, for a C_{12} straight-chain surfactant, $2\gamma_e/R$ amounts to about 400 atm, a figure that is broadly compatible with data on depressed gas solubilities in micellar solutions in comparison with the corresponding hydrocarbon liquids [65].

V. CONCLUSION

As we have shown in this chapter, the development of surface thermodynamics during the past half century to explicitly encompass strongly curved interfaces has actually provided us with a proper framework for treating surfactant-based aggregates theoretically in a universal manner. Contrary to what one might believe at first, the small size of these aggregates does not cause any major obstacle when it comes to accounting for the local interactions and entropy factors responsible for their excess free energies. However, in order to devise a complete theoretical scheme, it is also necessary to invoke the size/shape fluctuations as these imply additional entropic advantages favoring aggregate formation.

A particularly fruitful, although admittedly approximate, approach to curved interfaces was initiated by Ou-Yang and Helfrich in the late 1980s for bilayer vesicles, and was later on shown to result from combining the generalized Laplace equation due to Neumann and Boruvka with the Helfrich quadratic curvature free energy expression. This expression introduces the parameters γ_0, H_0, k_c, and $\bar{k}_c$ (i.e., the cylindrical interfacial tension γ_0 at the spontaneous curvature H_0, and the bending constants k_c and $\bar{k}_c$). In principle, one can, of course, calculate these parameters through modeling of the various molecular factors involved. Yet, for many purposes, it has turned out to be sufficient not to go beyond the level of the Helfrich parameters in order to rationalize experimental findings. On this basis, one is in fact able to account for the occurrence of most of the common types of surfactant-based aggregates: microemulsion droplets, vesicles, as well as surfactant micelles of various shapes. We might add, however, that a number of recent observations indicate that the original Helfrich expression has to be complemented with higher-order curvature terms to enable fully satisfactory treatments, in particular of the interfacial structures that arise when the spontaneous curvature approaches zero.

ACKNOWLEDGMENTS

The authors are most grateful to Peter A. Kralchevsky and Magnus Bergström for scrutinizing a preliminary version of the manuscript, and for several helpful suggestions and comments. We also wish to thank Per Claesson for putting the necessary facilities to conclude this chapter at our disposal, and to Maud Norberg and Kathy Arvidsson for technical assistance.

APPENDIX A. ON THE GENERAL VALIDITY OF EQ. (8)

Let us write Eq. (33) in terms of the curvature variables H and D instead of the principal curvatures c_1 and c_2. Thus, we have:

$$d\hat{\omega}_s = (\gamma - \hat{\omega}_s)d\alpha - \sum_{i=1}^{k} \Gamma_i d\mu_i + \zeta d\beta + B dH + \Theta dD \tag{A1}$$

where:

$$\mathcal{F} \equiv \hat{\Omega}_s = \iint_S \hat{\omega}_s \, dS \tag{A2}$$

and the summation refers to soluble components only. Now, all μ_i $(i = 1,\ldots,k)$ of the soluble components should be kept constant and, furthermore:

$$\delta\mathcal{F}_s = \delta\iint_S \hat{\omega}_s \, dS = \iint_S (\delta\hat{\omega}_s + \hat{\omega}_s\delta) = dS \tag{A3}$$

where, by definition, $\delta dS = \delta\alpha dS$. Hence:

$$\delta\mathcal{F}_s = \iint_S (\delta\hat{\omega}_s + \hat{\omega}_s\delta\alpha)dS \tag{A4}$$

or

$$\delta\mathcal{F}_s = \iint_S (\gamma\delta\alpha + \zeta\delta\beta + B\delta H + \Theta\delta D)dS \tag{A5}$$

Adding the ΔP contribution from the bulk phases, we obtain Eq. (8), which accordingly holds for any interface that is equilibrated with respect to its soluble components.

APPENDIX B. THE GRAND POTENTIAL OF OPEN SYSTEMS OF SMALL SIZE

Let us consider an ensemble of $\mathcal{N}$ water/oil microemulsion droplets, all of a given volume and surface area S in a μ_w, μ_{surf}, T environment. The number of surfactant molecules s in the surfactant film of a droplet is supposed to fluctuate in the course of time (i.e., the surface density of surfactant in the film fluctuates about its equilibrium value). The grand partition function Ξ_{film} of the total surfactant film extending over the entire interfacial area $\mathcal{N} S = S_{tot}$ is related to the grand potential Ω_{film} of the film in accordance with the general relation:

$$\Omega_{film} = -kT \ln \Xi_{film}(T, \mu_{surf}, \mathcal{N}) \tag{B1}$$

On the other hand, supposing the droplets constitute independent, open subsystems, we can write:

$$\Xi_{film} = \xi_f^{\mathcal{N}}(T, \mu_{surf}) \tag{B2}$$

with ξ_f denoting the single droplet film partition function defined by:

$$\xi_f = \sum_{s=0}^{max} q(s) \lambda^s \tag{B3}$$

where:

$$q(s) = \sum_j e^{-\varepsilon_j(s)/kT} \tag{B4}$$

is the partition function of a droplet film containing s surfactant molecules, and λ, as usual, stands for the absolute activity. Consequently, from Eqs. (B1) and (B2), we have:

$$\varepsilon_f \equiv \Omega_{film}/\mathcal{N} = -kT \ln \xi_f(T, \mu_{surf}) \tag{B5}$$

and, in particular, for a spherical droplet:

$$\gamma = -\left(\frac{kT}{4\pi R^2}\right) \ln \xi_f(T, \mu_{surf}) = \Omega_{film}/S_{tot} \tag{B6}$$

Now, the entire film system is a macroscopical one and, hence, Ω_{film} can be calculated solely from the maximum term of the partition function Ξ_{film}, whereas the droplet film partition function ξ_f will include additional contributions from the terms about the maximum term. Nonetheless, exactly the same values of ε_f/S or γ are obtained, irrespective of whether we employ the Ξ_{film} of the large (macroscopic) system or ξ_f of the small droplet system, the ultimate reason being that Ξ_{film} is actually an extensive property, proportional to $\mathcal{N}$ and S_{tot} at constant T and chemical potentials. Conversely, we are justified in quantifying ε_f (or $4\pi R^2 \gamma$) without invoking any density or concentration fluctuations, simply as if the system, in spite of its small size, were actually a large *macroscopical* system. On the other hand, the *size* fluctuations (core volume and surfactant film area) necessarily imply entropy of the mixing type of contributions for the overall droplet system that we must explicitly account for in one way or the other.

To further illuminate this rather subtle but important point, for the sake of simplicity, we may envisage an ideal Langmuir adsorption system where the adsorbent is made of single crystal faces (in all $\mathcal{N}$ of them) of the same kind, each with M adsorption sites exposed to an adsorbing gas. On the average, N^* adsorbate molecules will be adsorbed on every crystal face. Let us assume a Gaussian distribution about the maximum term t_{N^*} of the grand partition function:

$$t_N = t_{N^*} e^{-\left(N - N^*\right)^2 / 2\sigma_N^2} \tag{B7}$$

implying that we can write ξ_{face} in the form:

$$\xi_{face} = t_{N^*} \sqrt{2\pi} \sigma_N \tag{B8}$$

where σ_N^2 is the variance. However, for this Langmuir case where adsorbate–adsorbate interactions are absent, the grand partition function can also be written as a sum over N in the following way:

$$\xi_{face} = \sum_{N=0}^{M} \frac{M!(q\lambda)^N}{N!(M-N)!} \tag{B9}$$

where q is the molecular partition function of the adsorbate. Thus the largest term is:

$$t_{N^*} = \frac{M!(q\lambda)^{N^*}}{N^*!(M-N^*)!} = \frac{1}{\sqrt{2\pi}} \sqrt{\frac{M}{N^*(M-N^*)}} \frac{M^M (q\lambda)^{N^*}}{(N^*)^{N^*}(M-N^*)^{M-N^*}} \tag{B10}$$

where we have inserted a more exact version of the Stirling approximation than the one commonly used in the thermodynamic limit, without the $(2\pi N)^{1/2}$ factor. But for a binomial distribution that approximates the Gaussian distribution of Eq. (B7) above, it holds true that:

$$\sigma_N = \sqrt{Mp(1-p)} \tag{B11}$$

where p in this case equals the average fraction of sites occupied $\Theta = N^*/M$. Upon combining Eqs. (B8), (B10), and (B11), it is seen that even for a small Langmuir system, the grand partition function is given by an expression of the familiar "macroscopical" form:

$$\xi_{\text{face}} = \frac{M^M (q\lambda)^{N^*}}{(N^*)^{N^*} (M - N^*)^{M-N^*}} = \Omega_{\text{ads}}^{1/\mathcal{N}} \tag{B12}$$

where the subscript "ads" refers to the overall adsorption system, implying that the effect of the "small size" broadening about the maximum term is cancelled exactly by the finite size corrections, which have to be made of the "large system" Stirling estimates of the factorials involved. Evidently, this means that the grand partition function of a small Langmuir adsorption system can be handled as if the system were actually large, by means of its *thermodynamic* maximum term.

For the Langmuir case under discussion, the site partition function ξ_{site} (as distinct from ξ_{face} that equals ξ_{site}^M) is given exactly by:

$$\xi_{\text{site}} = 1 + q\lambda \tag{B13}$$

where a curvature dependence is not normally presumed. However, widening the scope of the model by considering solid adsorbent surfaces of spherical shape, a curvature-dependent surface free energy $\omega_s(R)$ will result upon supposing the energy levels determining the molecular partition function q to be functions of R. Then the depth of the adsorption potential and its shape around the minimum as functions of curvature will become decisive.

Similarly, in the case of the microemulsion droplet system, a curvature dependence of γ would arise from the partition function, with $q(s)$ being curvature-dependent. Most, if not all, effects of this nature are, in principle, covered by means of the bending constants in the Helfrich expression k_c and $\bar{k}_c$ and the spontaneous curvature H_0.

REFERENCES

1. Gibbs, J.W. The Collected Works of J. Willard Gibbs,. Longmans: New York, 1928; Vol. 1, 210.
2. de Laplace, P.S. *Traité de Mécanique Celeste. Supplements en Livre X*; 1805–1806. Chelsea: New York, 1966. English translation by Bowditch, 1839.
3. Verschaffelt, J.E. Acad. R. Belg. Bull. Classe Sci. 1936, *22*, 373, 390, 402.
4. Guggenheim, E.A. Trans. Faraday Soc 1940, *36*, 398.
4a. Guggenheim, E.A. *Thermodynamics*. North-Holland Publ. Co., Amsterdam, 1957.
5. Tolman, R.C. J. Chem. Phys. 1948, *16*, 758.
5a. Tolman, R.C. J. Chem. Phys. 1949, *17*, 333.

6. Hill, T.L. J. Phys. Chem. 1952, *56*, 526.
7. Buff, F.P. J. Chem. Phys. 1956, *25*, 146.
7a. Buff, F.P. *Handbuch der Physik,* Band X, p. 281 Springer: Berlin, 1960.
8. Ono, S.; Kondo, S. *Handbuch der Physik, Band X*; Springer: Berlin, 1960.
9. Helfrich, W. Z. Naturforsch., C 1973, *28*, 693.
10. Boruvka, L.; Neumann, A.W. J. Chem. Phys. 1977, *66*, 5464.
11. Evans, E.A.; Skalak, R. CRC Crit. Rev. Bioeng. 1979, *3*, 181.
12. Kirchhoff, G. Crells J. 1850, *40*, 51.
13. Love, A.E.H. Phil. Trans. R. Soc. Lond. A, 1888, *179*, 491.
14. Ljunggren, S.; Eriksson, J.C.; Kralchevsky, P.A. J. Colloid Interface Sci. 1997, *191*, 424.
15. Kralchevsky, P.A.; Eriksson, J.C.; Ljunggren, S. Adv. Colloid Interface Sci. 1994, *48*, 19.
16. Podstrigach, Y.S.; Povstenko, Y.Z. *Introduction to the Mechanics of Surface Phenomena in Deformable Solids (in Russian)*; Nauka Dumka: Kiev, 1985. see also Ref. 15.
17. Rusanov, A.I.; Shchekin, A.K. Colloids Surf., A Physicochem. Eng. Asp. 2001, *192*, 357.
18. Kralchevsky, P.A. J. Colloid Interface Sci. 1990, *137*, 217.
19. Kralchevsky, P.A.; Nagayama, K. *Particles at Fluid Interfaces and Membranes: Attachment of Colloid Particles and Proteins to Interfaces and Formation of Two-Dimensional Arrays*; Elsevier: Amsterdam, 2001; Chapter 4.
20. Markin, V.S.; Koslov, M.M.; Leikin, S.L. J. Chem. Soc. Faraday Trans. 2. 1988, *84*, 1149.
21. Murphy, C.L. Ph.D. Thesis, University of Minnesota, 1966; University Microfilms, Ann Arbor, MI, 1984.
22. Melrose, J.C. Ind. Eng. Chem. 1968, *60*, 53.
23. Eliassen, J.D. Ph.D. Thesis, University of Minnesota, 1963; University Microfilms, Ann Arbor, MI, 1983.
24. Shuttleworth, R. Proc. Phys. Soc. Lond. A 1950, *63*, 444.
25. Eriksson, J.C. Surf. Sci. 1969, *14*, 221.
26. Rusanov, A.I. *Phasengleichgewichte und Grenzflächenerscheinungen*; Akademie-Verlag: Berlin, 1978; 312 pp.
27. Neogi, P.; Kim, M.; Friberg, S.E. J. Phys. Chem. 1987, *91*, 605.
28. C.E. Weatherburn. *Differential Geometry of Three Dimensions*; Cambridge, 1930.
29. Ou-Yang, Z.; Helfrich, W. Phys. Rev., A 1989, *39*, 5280.
30. Eriksson, J.C. J. Colloid Interface Sci. 1971, *37*, 659.
31. Eriksson, J.C.; Ljunggren, S. Colloids Surf., A Physicochem. Eng. Asp. 1989, *38*, 179.
32. Szleifer, I.; Kramer, D.; Ben-Shaul, A.; Gelbart, W.M.; Safran, S.A. J. Chem. Phys. 1990, *92*, 6800.
33. Bergström, M.; Eriksson, J.C. Langmuir 1996, *12*, 624.
34. Deuling, H.J.; Helfrich, W. J. Phys. (Paris) 1976, *37*, 1335.
35. Barneveld, P.A.; Scheutjens, J.M.H.M.; Lyklema, J. Langmuir 1933, *8*, 3122.
36. Lekkerkerker, H.N.W. Physica, A 1990, *167*, 384.
37. Gurkov, T.D.; Kralchevsky, P.A. Colloids Surf. 1990, *47*, 45.

38. Eriksson, J.C.; Ljunggren, S. Prog. Colloid Polym. Sci. 1990, *81*, 41.
39. Ljunggren, S.; Eriksson, J.C. Colloids Surf., A Physicochem. Eng. Asp. 1997, *129–130*, 151.
40. Eriksson, J.C.; Ljunggren, S.; Kegel, W.K.; Lekkerkerker, H.N.W. Colloids Surf., A Physicochem. Eng. Asp. 2001, *183–185*, 347.
41. Bohr, N. Philos. Trans., A 1909, *209*, 281.
42. Eriksson, J.C.; Ljunggren, S. J. Colloid Interface Sci. 1991, *145*, 224.
43. Ljunggren, S.; Eriksson, J.C. Langmuir 1992, *8*, 1300.
44. Grosse-Brauckmann, K. J. Colloid Interface Sci. 1997, *187*, 418.
45. Anderson, D.; Wennerström, H. J. Phys. Chem. 1989, *93*, 4243.
46. Ivanov, I.B.; Kralchevsky, P.A. In *Thin Liquid Films*; Ivanov, I.B., Ed.; Marcel Dekker: New York, 1988; 91 pp.
47. Eriksson, J.C.; Ljunggren, S. Langmuir 1995, *11*, 1145.
48. Bergström, M.; Eriksson, J.C. Langmuir 1988, *14*, 288.
49. Tachev, K.D.; Angarska, J.K.; Danov, K.D.; Kralchevsky, P.A. Colloids Surf., B Biointerfaces 2000, *19*, 61; see also Ref. 19.
50. Hill, T.L. *Thermodynamics of Small Systems*; Benjamin: New York, 1963–1964; Vols. I and II.
51. Tanford, C. J. Phys. Chem. 1974, *78*, 2469.
52. Ruckenstein, E.; Nagarajan, R. J. Phys. Chem. 1975, *79*, 2622.
52a. Nagarajan, R.; Ruckenstein, E. Langmuir 1991, *7*, 2934.
53. Israelachvili, J.N.; Mitchell, J.J.; Ninham, B.N. J. Chem. Soc. Faraday Trans. 2. 1976, *72*, 1525.
54. Gunnarsson, G.; Jönsson, B.; Wennerström, H. J. Phys. Chem. 1980, *84*, 3114.
55. Gruen, D.W.R.; de Lacey, E.H.B. In *Surfactants in Solution*; Mittal, K.L. Lindman, B., Eds.; Plenum Press: New York, 1984; Vol. I, 279 pp.
55a. Gruen, D.W.R. J. Phys. Chem. 1985, *89*, 153.
56. Puvvada, S.; Blankschtein, D. J. Chem. Phys. 1990, *92*, 3710.
56a. Shiloach, A.; Blankschtein, D. Langmuir 1998, *14*, 1618.
57. Eriksson, J.C.; Ljunggren, S.; Henriksson, U. J. Chem. Soc. Faraday Trans., 2, 1985, *81*, 833.
57a. Eriksson, J.C.; Ljunggren, S. ibid., 1985, *81*, 1209.
57b. Ljunggren, S.; Eriksson, J.C. ibid., 1986, *82*, 913.
58. Fennel Evans, D.F.; Wennerström, H. *The Colloidal Domain*, 2nd Ed.; Wiley-VCH: New York, 1999; Chapter 4.
59. Rusanov, A.I. *Micellization in Surfactant Solutions, Chemistry Reviews, Part 1*; Harwood Academic Publ.: Amsterdam, 1997; Vol. 22.
60. Eriksson, J.C.; Ljunggren, S. Langmuir 1990, *6*, 895.
61. Ljunggren, S.; Eriksson, J.C. J. Chem. Soc. Faraday Trans. 2. 1988, *84*, 329.
62. Bergström, M. J. Chem. Phys. 2000, *113*, 5559.
63. Schurtenberger, P.; Cavaco, C. Langmuir 1994, *10*, 100.
64. Jönsson, B.; Lindman, B.; Holmberg, K.; Kronberg, B. *Surfactants and Polymers in Aqueous Solution*; Wiley: New York, 1998; 82 pp.
65. Bolden, P.L.; Hoskins, J.C.; King, A.D. Jr. J. Colloid Interface Sci. 1983, *91*, 454.

Index

Printed in the United States
by Baker & Taylor Publisher Services